P9-ELT-035

HISTORICAL ELECTRONICS MUSEUM. INC

MANUAL FOR
INTEGRATED CIRCUIT USERS

MANUAL FOR INTEGRATED CIRCUIT USERS

John D. Lenk

Consulting Technical Writer

Reston Publishing Company, Inc.

A Prentice-Hall Company

Reston, Virginia

Reston Publishing Company, Inc., A Prentice-Hall Company
P.O. Box 547
Reston, Virginia 22090

10 9 8 7 6 5 4 3 2 1

ISBN: **0-87909-482-6**

Library of Congress Catalog Card Number: 72-96757
Printed in the United States of America.

Dedicated to

Irene, Karen, Mark and **Lambie**

CONTENTS

PREFACE

As the title implies, this handbook is written for integrated circuit (IC) *users*, rather than for designers of integrated circuits. The book is, therefore, written on the basis of using existing, commercial integrated circuits to solve design and application problems. Typical users include design specialists who want to integrate electronic units and systems, or technicians who must service equipment containing ICs. Other groups that can make good use of this approach to integrated circuits are the experimenters and hobbyists.

There are two very common, although not necessarily accurate, concepts concerning ICs. First, it is assumed that the basic approach to IC design involves stating design parameters and requirements for a particular circuit to an IC manufacturer, and then having them fabricate an IC to meet these exact requirements. While this approach is satisfactory for some highly specialized circuits, and particularly where cost is no factor (such as the aerospace industry where the government is footing the bill), the approach may generally be wasteful and often unnecessary. On the other hand, it is often assumed that existing commercial ICs are limited in application; that such ICs are designed with only one or two uses in view.

There is a middle ground. Except for certain very special circuits, there are a number of commercial ICs that can be adapted to meet most circuit requirements. Also, new IC modules are being developed by various manufacturers. Likewise, although most off-the-shelf ICs are manufactured with certain specific uses in mind, these ICs are certainly not limited to only those uses.

Thus, the approach found in this handbook serves a two-fold purpose: (1) to acquaint the readers with ICs, in general, so that the users can select commercial units to meet their particular circuit requirements, and (2) to show readers the many other uses for existing ICs, not found on the manufacturer's data sheet.

This handbook assumes that the reader is already familiar with basic electronics, including solid-state, but has little or no knowledge of ICs. For this reason, Chapter 1 provides an introduction to ICs. Such topics as the how and why of IC fabrication techniques, a comparison of IC to discrete component circuits, a description of the basic physical types, and circuit types, found in commercial ICs, as well as some basic design considerations are covered.

With basics out of the way, Chapter 2 discusses practical considerations for ICs. No matter what IC is used, it must be mounted (both in breadboard form for design, and in final production form), leads must be soldered and unsoldered, power must be applied, and heat sinks may be required. These subjects are described in detail.

As the reader will soon find out, there are two basic IC types: linear and digital. Linear ICs are discussed in Chapters 3 through 5. Digital ICs are covered in Chapter 6. Considerable emphasis is placed on interpreting manufacturer's datasheets. For example, in Chapter 3, a typical IC datasheet is analyzed, characteristic by characteristic.

Often, experimenters must work with ICs on which complete data is not available. Under these circumstances, it is necessary to test the IC under simulated operating conditions. For this reason, detailed test data is included in Chapter 7.

With any IC, it is possible to apply certain approximations or rules-of-thumb for the selection of external component values. These rules can then be stated in basic equations, requiring only simple arithmetic for the solutions. This book starts with rules-of-thumb for the selection of external components on a trial-value basis, assuming a specified goal and a given set of conditions. The handbook concentrates on simple, practical approaches to IC use, not on IC analysis. Theory is kept to a minimum.

The values of external components used with ICs depend upon IC characteristics, available power sources, the desired performance (voltage amplification, stability, etc.) and existing circuit conditions (input/output impedances, input signal amplitude, etc.) The IC characteristics are to be found in the manufacturer's data. The overall circuit characteristics can then be determined, based on a reasonable expectation of the IC characteristics. Often, the final circuit is a result of many tradeoffs between desired performance and available characteristics. This handbook discusses the problem of tradeoffs from a simplified, practical standpoint.

Since the book does not require advanced math or theoretical study, it is ideal for the experimenter. On the other hand, the book is suited to schools where the basic teaching approach is circuit analysis, and a great desire exists for practical design.

The author has received much help from many organizations and individuals prominent in the field of integrated circuits. He wishes to thank them all.

The author also wishes to express his appreciation to Mr. Joseph A. Labok of Los Angeles Valley College for his help and encouragement.

John D. Lenk

1. INTRODUCTION TO INTEGRATED CIRCUITS

Typically, an *integrated circuit*, or *IC*, consists of transistors, resistors, and diodes etched into a semiconductor material. The material is usually silicon, and is finally sold or used in the form of a "chip." Since all of the components are fabricated on the same chip, construction of an IC is called "monolithic." All of the devices are interconnected (by techniques similar to those used in printed circuit boards) to perform a definite function or operation. Thus, the IC concept is one of a complete (or nearly complete) circuit, rather than a group of related semiconductor devices.

To make the IC package an operable unit, it must be connected to a power source, an input, and an output. In most cases, the output must also be connected to external components such as capacitors and coils, since it is not practical to combine these relatively large parts on the very tiny semiconductor chip.

1-1. PACKAGING INTEGRATED CIRCUITS

In theory, an integrated circuit semiconductor chip could be connected directly to the power source, input, etc. However, this is not practical because of the very small size of the chip. IC chips are almost always microminiature. Instead of direct connection, the chip is mounted in a suitable container and connected to the external circuit through leads on the container.

There are three basic IC packages: the *transistor package*, the *flat-pack*, and the *dual-in-line package*. Some typical examples of these are shown in Fig. 1-1.

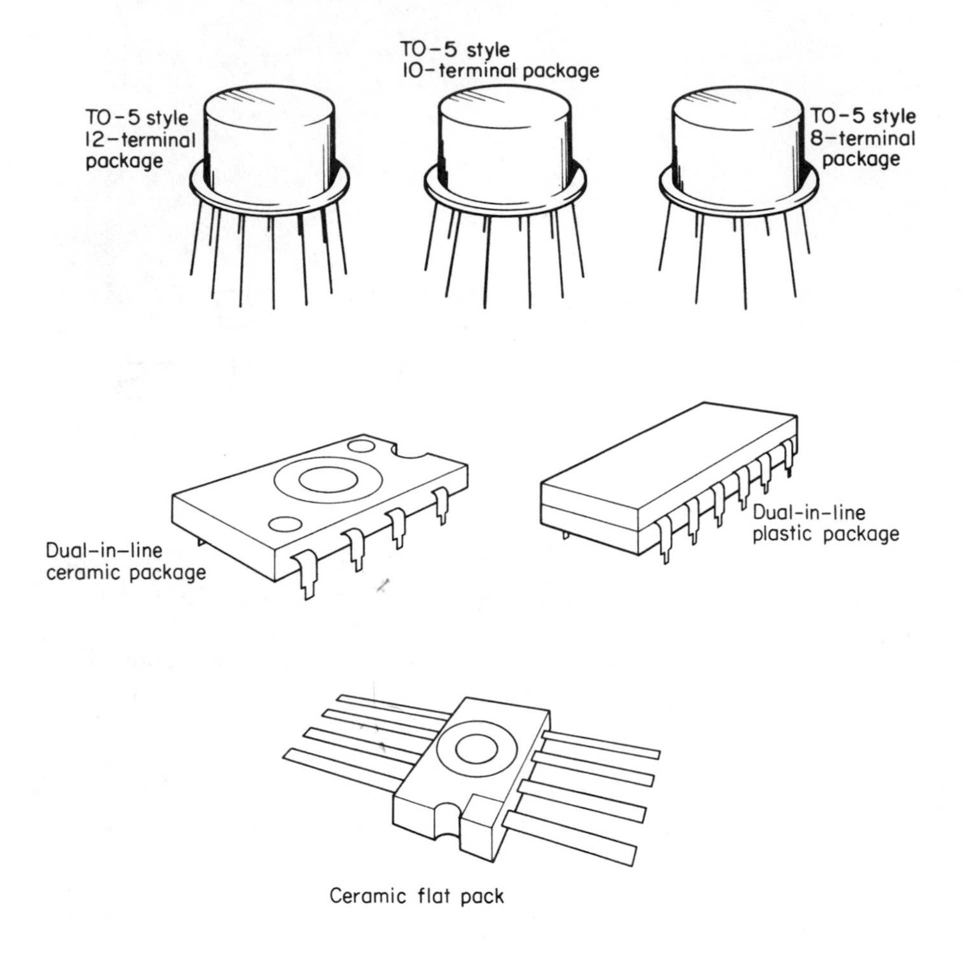

FIG. 1-1. Typical IC packages.

In the transistor package, the chip is mounted inside a transistor case such as a TO-5 case. Instead of the usual three leads found on a transistor case (emitter, collector, and base), there are 8, 10, 12, or more leads to accommodate the various power source and input/output connections required in a complete circuit.

In the flat-pack, the chip is encapsulated in a rectangular case with terminal leads extending through the sides and ends.

In the dual-in-line package (DIP), the chip is encapsulated in a rectangular case longer than the flat-pack. In general, the DIP has replaced the flat-pack for most applications.

Although there has been some attempt at standardization for IC terminal connections, the various manufacturers still use their own systems. It is therefore necessary to consult the data sheet for the particular IC when making connections from an external circuit.

1-2. INTERNAL CONSTRUCTION OF INTEGRATED CIRCUITS

There are many methods for fabrication of the semiconductor chip of an IC, and new methods are being developed constantly. Because of the many methods and because we are primarily interested in using existing IC units we will not discuss all of the methods. Instead, we will describe one popular technique that is similar to the method used for fabrication of *silicon planar transistors*.

As shown in Fig. 1-2, the starting material for a planar transistor is a uniform single crystal of *N*-type silicon. (*P*-type silicon could also have been used for the starting material.)

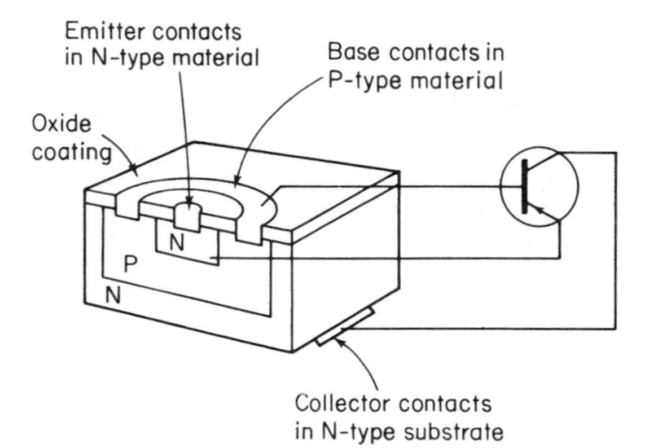

FIG. 1-2. Basic planar transistor.

The collector of the planar transistor is formed by the *wafer* (or *substrate*) of *N*-type silicon, which has been *passivated* (coated) with an oxide layer. A circular trench is etched out of the oxide. The trench is filled with a *P*-type crystal by a diffusion process requiring precise time and temperature control. The *P*-type material forms the transistor base element. Another disc-shaped area is etched at the center and filled (by diffusion) with an *N*-type crystal, which forms the transistor emitter. The result is an *NPN* transistor. Metalized contacts are attached to the three elements. The oxide layer prevents shorts between the metalized contacts and protects the emitter-base and collector-base from contamination.

A single transistor is shown in Fig. 1-2. The same basic process is used to fabricate many electrically-isolated transistors on a single silicon substrate. The first step is to diffuse two (or more) regions of similar crystal-line material into a substrate of dissimilar material, as shown in Fig. 1-3. Here, two *N*-

Single Uniform Crystal
(P-type substrate)

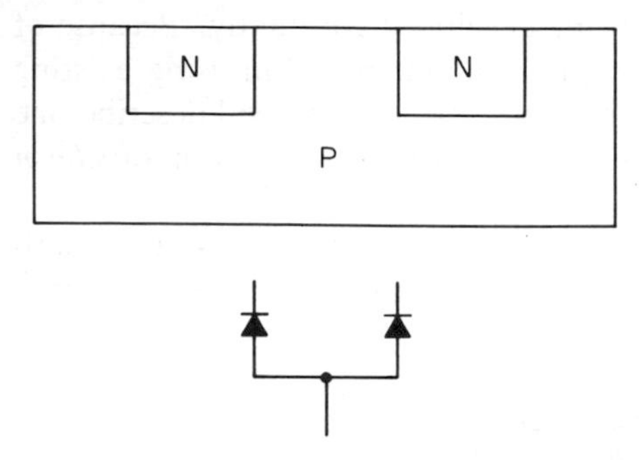

FIG. 1-3. Diffusion of N-type areas into P-type substrate to produce two diodes with common anodes and isolated cathodes.

type regions are diffused into a *P*-type substrate. Without further processing, this would result in two diodes with common anodes, but isolated cathodes.

Transistors are formed by diffusion of additional *N*-type and *P*-type regions, as shown in Fig. 1-4. The silicon wafer is then coated with an insulating oxide layer, and the oxide is opened (etched) selectively to permit metalized contacts and interconnections between elements (and between transistors), as required. With the contact arrangement shown in Fig. 1-5, two separate and electrically-isolated *NPN* transistors are formed in a *P*-type substrate.

When resistors are required in the integrated circuit, the *N*-type emitter diffusion is omitted, and two contacts are made to a *P*-type region (formed concurrently with the transistor base diffusion), as shown in Fig. 1-6. Here, an *NPN* transistor with a resistance connected to the emitter is integrated in a *P*-type substrate.

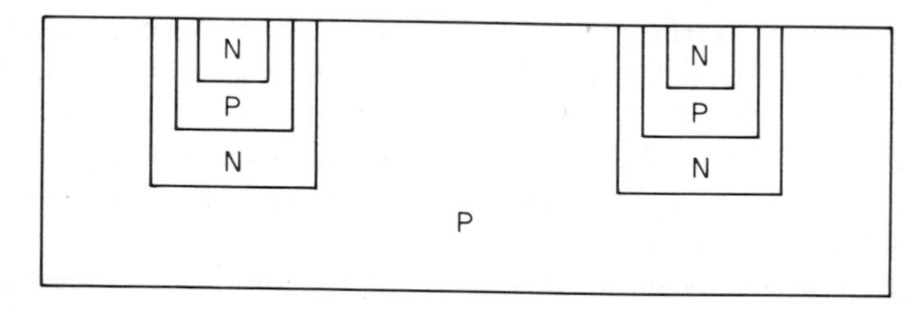

FIG. 1-4. Diffusion of P-type and N-type materials into a P-type substrate to form two transistors.

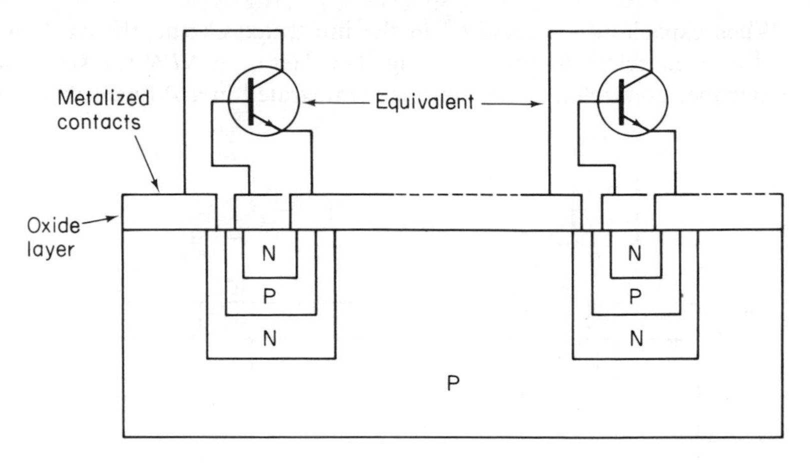

FIG. 1-5. Addition of metalized contacts to transistor elements formed in P-type semiconductor chip.

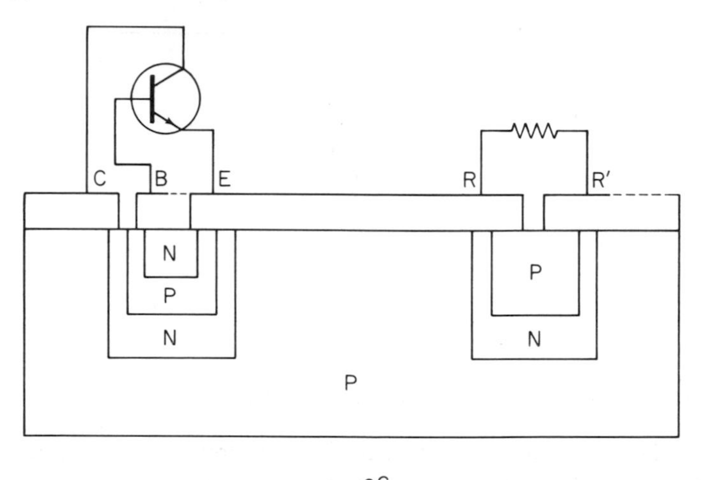

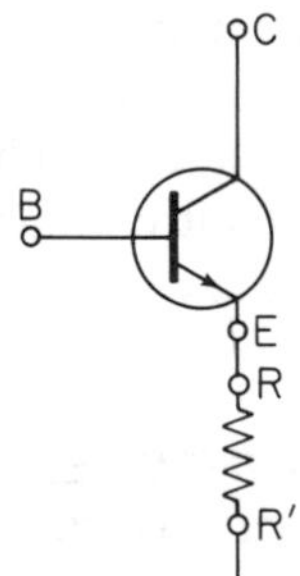

FIG. 1-6. Connection of contacts to P-type region to form integrated transistor and resistor.

When capacitors are required in the integrated circuit, the oxide itself is used as a dielectric, as shown in Fig. 1-7. Here, an *NPN* transistor with a capacitance connected to the emitter is integrated in a *P*-type substrate.

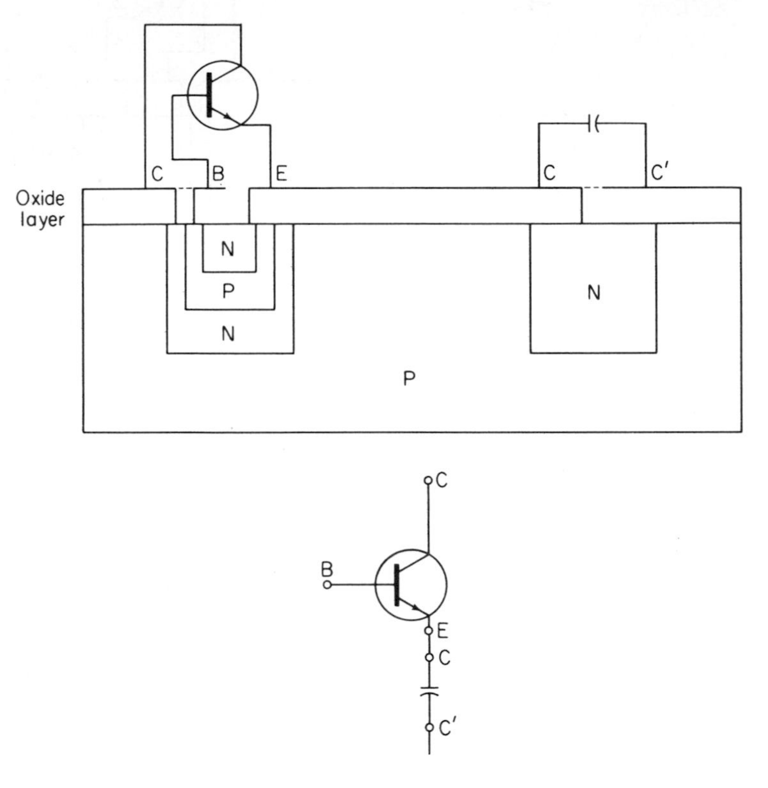

FIG. 1-7. Use of oxide as a dielectric to form integrated capacitor.

Figure 1-8 shows a very simple integrated circuit with a combination of the three types of elements on a single chip. It is not uncommon to have several dozen components on a single chip. Figure 1-9 shows the physical arrangement of a typical IC semiconductor chip. The IC shown in Fig. 1-9 is a complete voltage regulator circuit containing approximately two dozen transistors, 18 resistors, and ten diodes.

1-3. DIFFERENCES BETWEEN DISCRETE AND INTEGRATED CIRCUITS

Although the basic circuits used in ICs are similar to those of discrete transistors, there are certain differences. For example, inductances (coils) are never found as part of an IC. It is impossible to form a useful inductance on a material that contains transistors and resistors. Likewise,

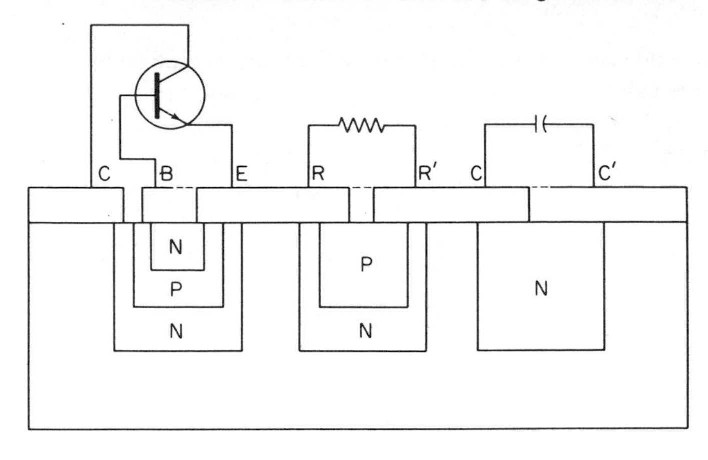

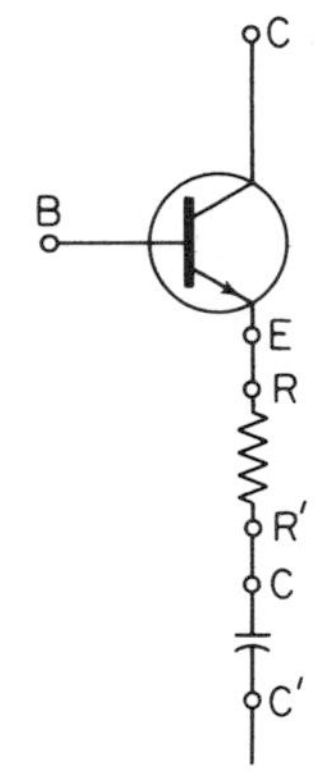

FIG. 1-8. Simple IC with three basic elements on a single chip.

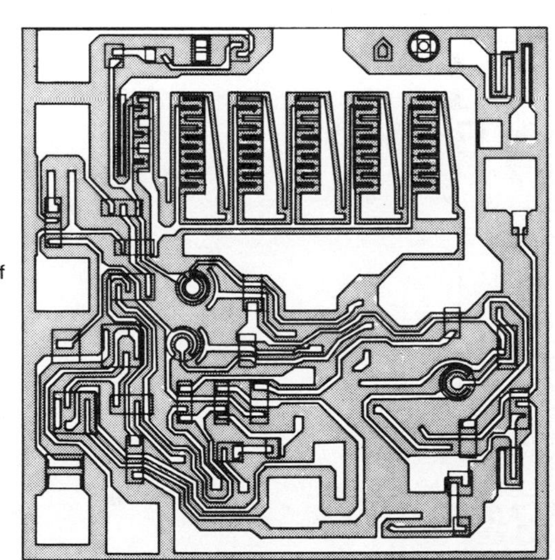

FIG. 1-9. Physical arrangement of typical IC semiconductor chip.

large value capacitors (about 100 pF) are not formed as part of an IC. When a large value capacitor, or an inductance of any type is a necessary part of a circuit, these components are part of the external circuit.

Integrated circuits often use direct-coupled circuits to eliminate capacitors. Figure 1-10 shows how a transistor, Q_3, is used to eliminate the need for a

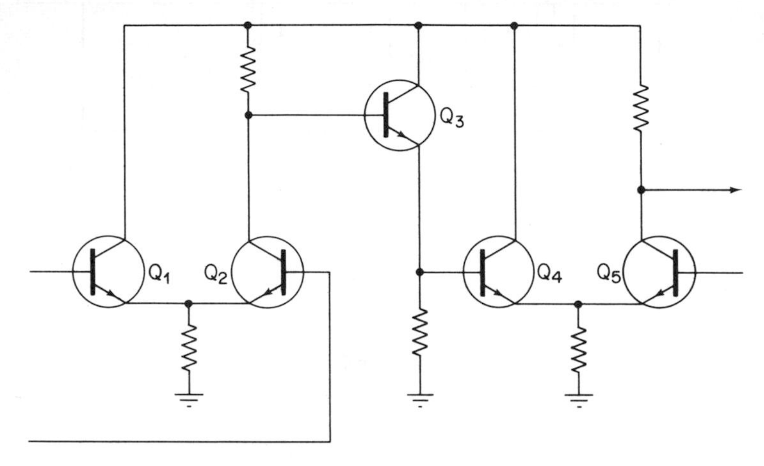

FIG. 1-10. Transistor Q_3 substitutes for capacitor in typical IC.

capacitor between Q_1–Q_2 and Q_4–Q_5. By eliminating the capacitor, the frequency range of the circuit is also extended.

Transistors are often used in place of resistors in IC packages. Usually, such a transistor is a *field effect transistor* (FET) since the basic FET acts somewhat like a resistor. Figure 1-11 shows how an FET can be substituted

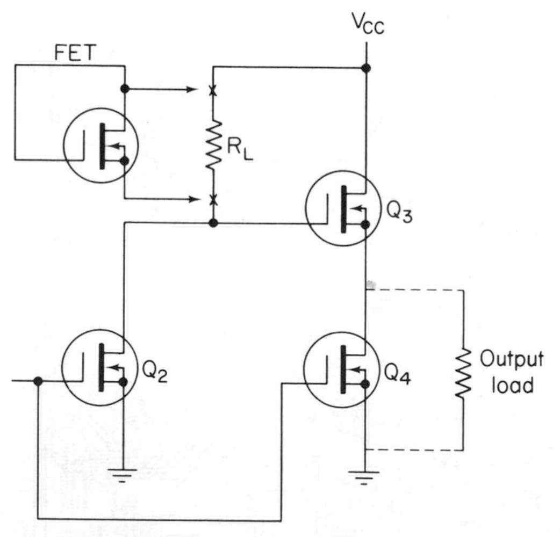

FIG. 1-11. How a FET can be substituted for a resistor in an IC.

for a resistor in an IC. In this circuit, the FET gate is returned to one side of the supply. With such an arrangement, the FET takes up less space than a corresponding resistor and provides a much higher power dissipation capability. Integrated transistors can also be connected to form diodes.

Although IC transistors are essentially the same as discrete transistors (except for some added capacitance produced across the substrate and transistor junctions), integrated resistors are significantly different from discrete versions. Discrete resistors are normally made in standard forms, and different values are obtained by variations in the resistivity of the material. In integrated circuits, the resistivity of the material cannot be varied. Thus, the value of the resistor depends primarily on its physical shape. An IC resistance value, R, is determined by the product of its diffusion-determined sheet or chip resistance R_S, and the ratio of its length L to its width W (that is, $R = R_S \times L/W$). As a result, small-value resistors are short and squat, whereas large-value resistors are long and narrow.

The value of an integrated capacitor C is equal to the product of its area A, and the ratio of the dielectric constant E to the thickness D of the oxide layer (that is, $C = A \times E/D$). Because D is kept constant, capacitor values vary directly with area.

As a point of reference, a 1000-Ω IC resistor occupies about twice as much area as an IC transistor, whereas a 10-pF capacitor occupies three times the area of a transistor.

1-4. SUPPLY VOLTAGES FOR INTEGRATED CIRCUITS

Typical integrated circuit supply voltages are generally less than 35 volts. However, future development of IC fabrication and packaging methods may increase this figure. The limiting factor in maximum IC supply voltages is usually the collector-to-emitter breakdown voltage of the transistors.

Typically, supply voltages are applied to integrated circuits in the same polarity required for *NPN* transistor stages. That is, the collector voltage is positive, with the emitter at ground, zero voltage, or at some negative value. If the polarity is reversed, the normally reverse-biased collector-to-substrate insolation junction will conduct very heavily and cause a portion of the metalized contact pattern to be destroyed. For this reason, a protective diode is often used in the dc supply line.

Any length of ribbon or wiring on a printed-circuit board has inductance that can develop significant voltages in response to high-frequency or fast-rise currents. Such voltages are added to the dc supply voltage of an integrated circuit mounted on the board. In addition, the supply-lead metalization of the integrated circuit may develop high-frequency signals as a result of stray internal feedback. Adequate compensation for both effects can be achieved by bypassing the supply leads of high-gain units with small external capacitors.

The practical considerations for connecting power supply leads to integrated circuits are discussed fully in Chapter 2.

1-5. TEMPERATURE CONSIDERATIONS FOR INTEGRATED CIRCUITS

The basic temperature considerations for integrated circuits are essentially the same as those for discrete component circuits. The heat dissipated by the circuit components must be transferred to the outside of the package without the temperature at any point on the circuit chip becoming excessive.

In an integrated circuit, the heat is dissipated on top of the silicon chip. The heat sources, therefore, are highly localized, with the exact distribution determined by the circuit layout. Because the silicon chip mounted on a metal header is a good heat conductor, the heat rapidly diffuses throughout the chip, and the entire chip may be considered to be essentially at the same temperature. It is more meaningful to examine the dissipation capability of the overall chip than to determine the limits of each of the various regions on the chip. The dissipation capability of a monolithic silicon circuit chip is determined primarily by the encapsulating material, the chip mount, the terminating leads, and the volume and area of the integrated circuit package.

Generally, heat transfer in an integrated circuit is conducted through the silicon chip and through the case. The effects of internal free convection and radiation, and lead conduction are small and may be neglected. The value of thermal resistance from pellet to case is dependent upon the pellet dimension, the package configuration, and the location of the selected case reference point.

The effect of temperature on an IC transistor is different from that on an IC resistor. The effect for resistors is generally detrimental, while the effect for transistors is generally beneficial.

Integrated transistors on the same circuit chip have a number of advantages over discrete units as a result of their proximity. In fabrication, adjacent transistors receive almost identical processing and thus are closely matched in characteristics. Because of the close spacing, minimum temperature differences occur between components, and this close match is maintained over a wide operating range. In addition, integrated circuits can contain many more transistors per given area than discrete components. (Typically, six IC transistors occupy the same area as one discrete transistor.)

Integrated resistors have a relatively large value variation with temperature. This temperature dependence makes it difficult to achieve close tolerances on absolute values of resistors. However, the ratios of IC resistors can be closely controlled during the fabrication process. As a result, it is desirable that IC design be made dependent on ratios rather than on absolute values of resistors.

The practical considerations for power dissipation and thermal design problems of integrated circuits are discussed in Chapter 2.

1-6. BASIC INTEGRATED CIRCUIT TYPES

There are two basic types of integrated circuits: digital and linear.

1-6.1 Digital integrated circuits

Digital ICs are the integrated circuit equivalents of basic transistor logic circuits. Like their discrete component counterparts, digital ICs are used in computers, digital telemetry, etc., and form such circuits as gates, counters, choppers, multivibrators, shift registers, etc. A digital IC is a complete functioning logic network, usually requiring nothing more than an input, output, and power source. However, as discussed in later chapters, the interconnection of digital IC packages requires some analysis in design.

Digital circuits are generally repetitive and concerned with only two levels of voltage or current. Thus they do not require accurate control of transitional-region characteristics (transconductance linearity for example). Therefore, digital circuits are standardized into a few basic designs and are produced in large quantities as low-cost off-the-shelf devices.

1-6.2 Linear Integrated circuits

Linear ICs are the integrated circuit equivalents of basic transistor circuits. Examples are amplifiers, oscillators, mixers, frequency multipliers, modulators, limiters, detectors, and so on. Although linear ICs are complete functioning circuits, they often require additional external components (in addition to a power supply) for satisfactory operation. Typical examples of such external components are: a resistor to convert a linear amplifier into an operational amplifier, a resistor-capacitor combination to provide frequency compensation, and a coil-capacitor combination to form a filter (for band pass or band rejection).

Linear ICs are of two basic types: They can be versatile general-purpose devices that may be adapted to provide many different types of circuit functions in a variety of applications. Alternatively, they can be special-purpose devices that simultaneously provide multiple circuit functions in specialized high-volume applications, such as home-entertainment units.

Special-purpose linear ICs are usually designed to replace several stages of discrete-component circuits. Typical examples include IF strips in AM or FM radio receivers, sound circuits (IF amplifier-limiters, discriminator and audio voltage amplifiers) in intercarrier television receivers, remote amplifiers for remote-control television receivers, and similar types of specialized multi-stage circuits typical of the home-instrument consumer industry. With few exceptions, special-purpose ICs can provide multiple circuit functions at performance levels equal to or greater than those of their discrete-component counterparts.

The high gain required of the amplifier sections in special-purpose ICs can be provided by cascades of *balanced differential amplifiers*, which are the basic building blocks for most linear integrated circuits. (The balanced differential amplifier is discussed further in later sections of this chapter and in chapters throughout this book.) Of course, the differential amplifier must be augmented by other circuits, such as voltage regulators or reference-voltage supplies, FM detectors, Darlington pairs, phase splitters, and buffer stages, to provide the multiple-circuit functions required in the specific application.

1-6.3 Component arrays in integrated circuits

In addition to the basic digital and linear ICs, solid-state component arrays are available in IC form. Such arrays may consist of groups of unconnected active devices, of diode quads, of transistor Darlington pairs, or of individual circuit stages. The components of an array, which are fabricated simultaneously in the same way on a silicon chip, have nearly identical characteristics. The characteristics of the various components track each other with temperature variations because of the proximity of the components, and the good thermal conductivity of silicon.

The IC arrays are especially suited for applications in which closely matched device or circuit characteristics are required and in which a number of active devices must be interconnected with external components, such as tuned circuits, large-value or variable resistors, and large bypass or filter capacitors. For example, diode arrays are particularly useful in the design of bridge rectifiers, balanced mixers or modulators, gating circuits, and other configurations that require identical diodes.

Transistor arrays make available closely matched devices that may be used in a variety of circuit applications (for example, push-pull amplifiers, differential amplifiers, multivibrators, and dual-channel circuits). The individual transistors in the array may also be used in circuit stages that are located in different signal channels, or in cascade or cascode circuits. Arrays of individual circuit stages are very useful in equipment that has two or more identical channels, such as stereo amplifiers, or they may be interconnected by the use of external coupling elements to form cascade circuits.

1.7 THE BASIC LINEAR IC

The most common form of linear IC is the *operational amplifier*. Such amplifiers are high-gain, direct-coupled circuits where the *gain and frequency* response are controlled by external feedback networks. It is possible to convert almost any amplifier circuit into an operational amplifier (or op-amp) by the addition of one external feedback resistor connected between output and input. This technique is discussed fully in Chapter 3 and throughout the book.

By use of external feedback networks, the op-amp can be used to produce a broad range of intricate transfer functions and thus may be adapted for use in many widely differing applications. Although the op-amp was originally designed to perform various mathematical functions (differentiation, integration, analog comparisons, and summation), it may also be used for many other applications. For example, the same op-amp, by modification of the feedback network, may be used to provide the broad, flat frequency-gain response of video amplifiers or the peaked responses of various types of shaping amplifiers. This capability makes the op-amp the most versatile configuration used for linear ICs.

1-7.1 Balanced differential operational amplifier

The most common type of linear IC op-amp uses a *balanced differential amplifier* circuit. Figure 1-12 shows a typical circuit. The complete circuit shown is contained in a ceramic flat pack (less than one-half inch square), which has 14 terminal leads. Note that *no internal capacitors* are used, only resistors and transistors.

The basic purpose of such a circuit is to produce an output signal that is linearly proportional to the *difference between two signals applied to the input.* The circuit shown will provide an overall open-loop gain of approximately 2500.

In a theoretical differential amplifier, no output is produced when the signals at the inputs are identical. Thus, differential amplifiers are particularly useful for op-amps because signals common to both inputs (known as *common mode signals*) are eliminated or greatly reduced. A common mode signal (noise, power line pickup, stray voltages, for example) drives both inputs in-phase with equal amplitude ac voltages, and the amplifier behaves

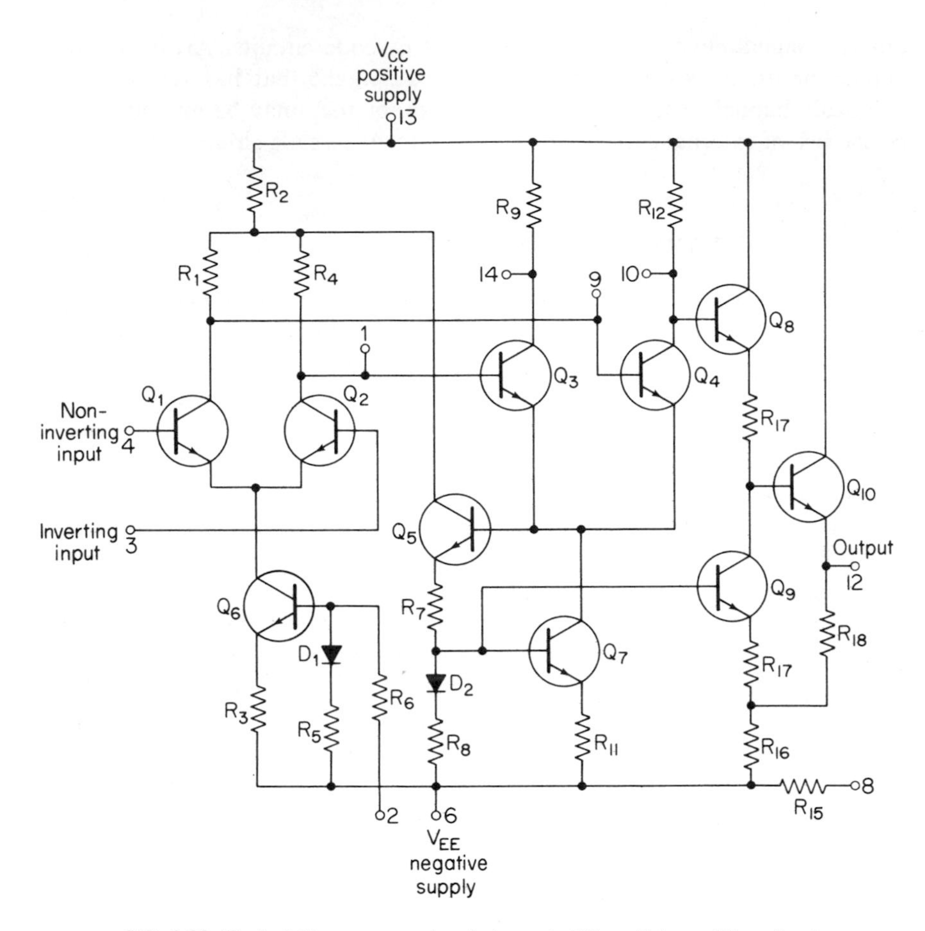

FIG. 1-12. Typical IC op-amp using balanced differential amplifier circuit.

as though no signal were present. The ability of a differential amplifier to prevent conversion of a common mode signal into an output is expressed by its *common mode rejection ratio* (CMR or CMRR), as discussed in Chapter 3 and throughout this book.

1-7.2 Operational amplifier circuit description

The following is a brief description of the circuit shown in Fig. 1-12. As illustrated, the operational amplifier consists basically of two differential amplifiers and a single-ended output circuit in cascade (the output of one stage feeding the input of another stage). The pair of cascaded differential amplifiers provide most of the gain.

The inputs to the op-amp are applied to the bases of the pair of emitter-coupled transistors Q_1 and Q_2 in the first differential amplifier. The inverting input is applied to the base of transistor Q_2, and the noninverting input is applied to the base of Q_1. With a signal applied only to the inverting input and with the noninverting input grounded, the output is an amplified and inverted version of the input. For example, if the input is a positive pulse, the output is a negative pulse. If the noninverting input is used with the inverting input grounded, the output is an amplified version of the input (without inversion).

Transistors Q_1 and Q_2 develop driving signals for the second differential amplifier. Transistor Q_6, a constant-current-sink device, is included in the first differential stage to provide bias stabilization for Q_1 and Q_2. If there is an increase in supply voltage, which would normally increase the current through Q_1 and Q_2, the reverse-bias on Q_6 increases. This reduces the Q_1–Q_2, current and offsets the effect of the initial supply voltage increase. Diode D_1 provides thermal compensation for the first differential stage. If there is an increase in operating temperature of the IC, which would normally increase the current through Q_1 and Q_2, there is a corresponding increase in current flow through D_1, since the diode is fabricated on the same silicon chip as the transistors. The increase in D_1 current also increases the reverse-bias of Q_6, thus offsetting the initial change in temperature.

The emitter-coupled transistors, Q_3 and Q_4, in the second differential amplifier are driven push-pull by the outputs from the first differential amplifier. Bias stabilization for the second differential amplifier is provided by current-sink transistor Q_7. Compensating diode D_2 provides the thermal stabilization for the second differential amplifier, and also for the current-sink transistor Q_9 in the output stage.

Transistor Q_5 develops the negative feedback to reduce common-mode error signals that are developed when the same input is applied to both input terminals of the operational amplifier. Transistor Q_5 samples the signal that is developed at the emitters of transistors Q_3 and Q_4. Because the second differential stage is driven push-pull, the signal at this point will be zero when the first differential stage and the base-emitter circuits of the second stage are matched, and there is no common-mode input.

A portion of any common-mode, or error, signal that appears at the emitters of transistors Q_3 and Q_4 is developed by transistor Q_5 across resistor R_2 (the common collector resistor for transistors Q_1, Q_2, and Q_5) in the proper phase to reduce the error. The emitter circuit of transistor Q_5 also reflects a portion of the same error signal into current-sink transistor Q_7 in the second differential stage so that the initial error signal is further reduced.

Transistor Q_5 also develops feedback signals to compensate for common-mode effects produced by variations in the supply voltage. For example, a decrease in the positive supply voltage results in a decrease of the voltage at the emitters of Q_3 and Q_4. This negative-going change in voltage is reflected by the emitter circuit of Q_5 to the bases of current-sink transistors Q_7 and Q_9. Less current then flows through these transistors. The decrease in collector current of Q_7 results in a reduction of the current through Q_3 and Q_4, and the collector voltages of these transistors tend to increase. This tendency partially cancels the decrease that occurs with the reduction of the positive supply voltage. The partially cancelled decrease in the collector voltage of Q_4 is coupled directly to the base of Q_8 and is transmitted by the emitter of Q_8 to the base of Q_{10}. At this point, the decrease in voltage is further cancelled by the increase in the collector voltage of current-sink transistor Q_9.

In a similar manner, transistor Q_5 develops the compensating feedback to cancel the effects of an increase in the positive supply (or of variations in the negative supply voltage). Because of the feedback stabilization provided by transistor Q_5, the IC op-amp circuit of Fig. 1-12 provides high common-mode rejection, and has excellent open-loop stability and a low sensitivity to power-supply variations. All of these characteristics are critical to IC op-amps, as is discussed in Chapter 3.

In addition to their function in the cancellation of supply voltage variations, transistors Q_8, Q_9, and Q_{10} are used in an emitter-follower type of single-ended output circuit. The output of the second differential amplifier is directly coupled to the base of Q_8, and the emitter circuit of transistor Q_8 supplies the base-drive input for output transistor Q_{10}. A small amount of signal gain in the output circuit is made possible by the connection from the emitter of output transistor Q_{10} to the emitter circuit of transistor Q_9 (at the junction of R_{16} and R_{17}). If this connection were omitted, transistor Q_9 could be considered as merely a constant-current sink for drive transistor Q_8. Because of the connection, however, the output circuit can provide a signal gain of 1.5 from the collector of differential amplifier transistor Q_4 to the output. Although this small amount of gain may seem insignificant, it does increase the output-swing capabilities of the operational amplifiers.

The output from the op-amp circuit is taken from the emitter of the output transistor Q_{10} so that the dc level of the output signal is substantially lower than that of the differential amplifier output at the collector of transistor Q_4. In this way, the output circuit shifts the dc level at the output so that it is effectively the same as that at the input when no signal is applied. This problem of dc level shifting is discussed further in Section 1–7.3.

Resistor R_{15} in series with terminal 8 increases the ac short-circuit load capability of the operational amplifier when terminal 8 is shorted to terminal 12 so that R_{15} is connected between the output and the negative supply.

1-7.3 DC level shifting problems in IC op-amps

In any cascade direct-coupled amplifier, either discrete-component or IC, the dc level rises through successive stages towards the supply voltage. In linear ICs, the dc voltage builds up through the *NPN* stages in the positive direction and must be shifted negatively if large output signal swings are to be obtained. For example, if the supply voltage is 10 V and the output is at 9 V under no-signal conditions, the maximum output voltage swing is limited to less than 1 V.

In multistage high-gain ICs, such as op-amps and special-purpose multifunction circuits that use external feedback, it is especially important to provide for compensation of the dc level shift. Such amplifiers must have equal (and preferably zero) input and output dc levels so that the dc coupling of the feedback connection does not shift any bias point. For example, if the input is at 0 V, but the output is at 3 V, this 3 V is reflected back through an external feedback resistor and changes the input to a 3-V level.

The use of an output stage, such as shown in Fig. 1-12, is a commonly used technique to prevent a shift in dc level between the output and input of an IC. Transistor Q_8 operates as an input buffer, and transistor Q_9 is essentially a current sink for Q_8. The shift in dc level is accomplished by the voltage drop across resistor R_{14} produced by the collector current of transistor Q_9. The emitter of the output transistor Q_{10} is connected to the emitter of Q_9. Feedback through R_{18} results in a decrease in the voltage drop across R_{14} for negative-going output swings, and an increase in this voltage drop for positive-going output swings.

If properly designed, the circuit shown in Fig. 1-12 can provide substantial voltage gain, high input impedance, low output impedance, and an output swing *nearly equal* to the supply voltages, in addition to the desired shift in dc level. Moreover, feedback may be coupled from the output to the input to compensate for dc common-mode effects that result from variations in the supply voltages.

1-7.4 Summary of advantages for differential IC op-amps

The balanced differential amplifier is considered the optimum configuration for general-purpose linear ICs by most manufacturers. This circuit configuration is generally preferred over other possible types (a feedback pair for example) for the following reasons:

Advantage can be taken from the exceptional balance between the differential inputs that results from the inherent match in base-to-emitter voltage and short-circuit current gain of the two (differential-pair) transistors that are processed in exactly the same way and are located very close to each other on the same very small silicon chip.

The differential amplifier circuit uses a minimum number of capacitors. Generally no internal capacitors are used in IC differential amplifiers.

The use of large resistors can usually be avoided, and the gain of the differential amplifier circuit is a function of resistance ratios, rather than of actual resistance values.

The differential amplifier is much more versatile than other possible circuit configurations and can be readily adapted for use in a variety of component applications. For example, in the circuit of Fig. 1-12, there are many connections to internal circuit components, in addition to the inputs, output, and power supplies. This is typical of differential amplifier ICs. The additional connections (such as at the collectors of Q_1, Q_2, Q_3, Q_4 and Q_9) provide a variety of input-output points for the *phase-lead* or *phase-lag compensation* techniques described in Chapter 3.

1-8. THE BASIC DIGITAL IC

Because of the great variety of digital ICs, no single circuit can be considered "typical," as is the case for linear ICs. For that reason, we will consider a number of digital IC configurations. Some of these forms first appeared as discrete component circuits; however, most of the forms are the result of packaging logic elements as ICs.

The following section summarizes the basic digital IC types in common use. The author makes no attempt to promote one form over another, but simply summarizes the facts (capabilities and limitations) about each form. Thus, the user can make an informed and intelligent comparison of the logic forms and select those that are best suited for his particular needs. In addition to the summary in this chapter, individual digital IC types are discussed in detail throughout Chapter 6.

1-8.1 Digital logic terms

It is assumed that the reader is familiar with the basics of digital logic at a level found in the author's HANDBOOK OF LOGIC CIRCUITS.* Not all manufacturers of IC logic elements use the same terms to describe characteristics. Likewise, the same terms are used to describe different characteristics. To avoid confusion, the following definitions will apply to digital IC logic described in this book.

Fan-In. The term *fan-in* is defined as the number of independent inputs on an element. For example, a three-input NAND gate has a fan-in of 3. If an IC has more than one independent element, the term fan-in applies to each independent element. For example, if an IC contains six two-input NAND gates, each gate has a fan-in of 2, but the complete IC package does not have a fan-in of 12, as might be supposed.

* John D. Lenk, *HANDBOOK OF LOGIC CIRCUITS*, Reston Publishing Company, Reston, Virginia, 1972.

Load. Each input to a logic element represents a *load* to the circuit trying to drive it. If there are several logic elements (say, several gates) in a given IC package, the load is increased by one unit for each element. Load therefore means the number of elements being driven from a given input. For example, assume that an IC package has three inputs, each connected to a separate gate, and that the three gate outputs are applied to a common inverter within the IC package. The load at any given input is 2 (one for the gate and one for the common inverter).

Fan-Out. *Fan-out* is defined as the number of logic elements, or loads, that can be driven directly from an output without any further amplification or circuit modification. For example, the output of a NAND gate with a fan-out of three can be connected directly to three elements with a load of one for each element, or to one circuit with a load of three. The term *drive* is often used in place of fan-out.

Delay. Any solid-state element (diode, transistor, and so on) will offer some delay to signals or pulses. That is, the output pulse will occur some time after the input pulse. Thus, every logic element has delay. Likewise, an IC that contains several logic elements in series has an even greater delay. This delay is known as *propagation delay time*, *delay time*, or simply *delay*.

There are many systems for indicating the delay. Some IC manufacturers spell out a specific time (usually in nanoseconds). Other manufacturers specify the delay for a circuit as the number of elements between a given input and a given output.

The problems presented by delay become obvious when it is realized that all logic circuits operate on the basis of coincidence. For example, an AND gate will produce a true output only when both inputs are true simultaneously. If input *A* is true, but switches back to false before input *B* arrives (due to some delay of the *B* input), the inputs do not occur simultaneously, and the AND gate output is false. The problem is compounded when delay occurs in sequential networks (multivibrators, counters, etc.), which depend upon timing as well as coincidence.

Many of the failures that occur in IC digital design are the result of undesired delay. For this reason, the procedures for measuring delay in IC packages, as well as individual digital elements, are stressed in Chapters 6 and 7.

1-8.2 Direct-coupled transistor logic (DCTL)

The DCTL gate was among the first discrete units to be integrated. A simple DCTL parallel gate is shown in Fig. 1-13a. A more practical DCTL parallel gate is shown in Fig. 1-13b. Note that this circuit is sometimes incorrectly referred to as a resistor-transistor logic network (RTL), since only resistors and transistors are involved. However, DCTL is a better name.

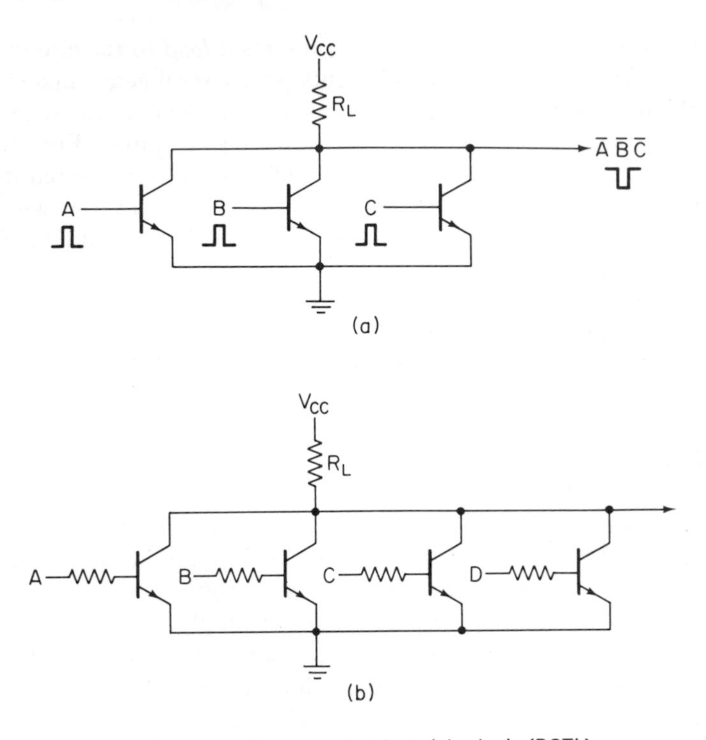

FIG. 1-13. Direct-coupled transistor logic (DCTL).

In this type of gate circuit, there is no separate transistor to provide the inversion and amplification function. Instead, these functions are distributed among the four input transistors (one transistor and one resistor for each input). The circuit of Fig. 1-13b is a NOR gate in that if any input is high, the output will be low. This circuit was used in earlier day of ICs and is still found in IC form.

1-8.3 Resistor-transistor logic (RTL)

Figure 1-14 shows two bonafide RTL circuits. Note that, unlike the DCTL circuits, the base-emitter junctions of all transistors are biased. This keeps current drain at a minimum. (One major problem with DCTL is excessive current drain.) The circuit of Fig. 1-14a is a four-input NOR gate. The switching function is provided by the four input resistors, while the inversion is accomplished by the transistor.

The circuit of Fig. 1-14b is a buffer. Each of the three outputs is an amplified version of the input, without inversion. This occurs because the input transistor inverts the input and thus applies this to all three output transistors which, in turn, invert the pulse back to its original form.

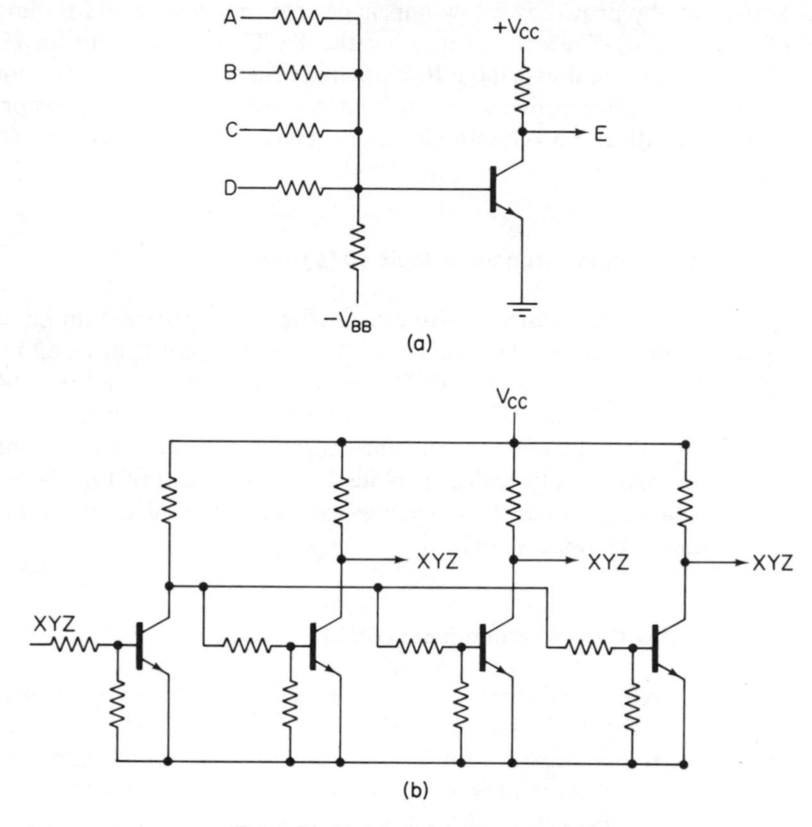

FIG. 1-14. Resistor-transistor logic (RTL).

1-8.4 Resistor-capacitor transistor logic (RCTL)

As shown in Fig. 1-15, the addition of a "speed-up" capacitor across the base resistor of an RTL circuit improves the high frequency charac-

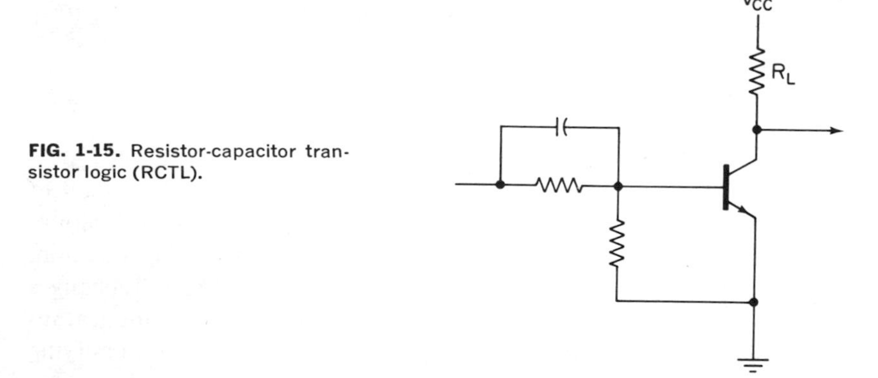

FIG. 1-15. Resistor-capacitor transistor logic (RCTL).

teristics of a gate by providing a low impedance for the leading and trailing edges of input pulse. The disadvantage of the RCTL circuit, from an IC standpoint, is that it requires a large IC chip area. Also, the RCTL does not track the temperature-sensitive saturation characteristics of the transistor, and it is noise-sensitive. Low-amplitude noise signals will easily pass through the capacitor.

1-8.5 Diode-transistor logic (DTL)

The *diode-transistor logic circuit* (Fig. 1-16) provides an input diode arrangement for an AND function with an output transistor as an inverter. The total circuit is then a NAND gate for positive logic and a NOR gate for negative logic. The DTL circuits have moderately high speeds with good noise immunity. Low-amplitude noise signals will not overcome the diode resistance (forward voltage drop). Note that in the circuit of Fig. 1-16a, the speed-up capacitor is used. This increases the speed of the gate, while the diodes act to block the noise signals.

1-8.6 Current mode logic (CML)

Current mode logic, also referred to as *current steering logic* (CSL), is shown in Fig. 1-17. This type of circuit uses one transistor for every input and an additional transistor for biasing. Because the circuit operates in the current mode (some current is always being drawn, transistors are never full-on or full-off), current flows either through the bank of input transistors or through the biasing transistor. Thus, both the inverted and non-inverted outputs are always present.

The presence of both outputs from a single circuit is advantageous in certain situations, as is discussed in later chapters. Also, because the circuit does not operate in the saturated mode (full-on or full-off), the circuit is much faster. The circuit of Fig. 1-17 is both OR and NOR. If any input goes positive, the OR output also goes positive, and the NOR output goes negative.

1-8.7 Transistor-transistor logic (TTL or T^2L)

The TTL circuits shown in Fig. 1-18 use transistors at the inputs for the AND function. Otherwise, TTL is similar to DTL. In the circuit of Fig. 1-18a, a bank of four transistors provides for the switching function, and another transistor provides for the inverting and amplifying function. This is the basic discrete component form of the TTL. In ICs, TTL changes slightly (to take advantage of the integration) in that the bank of input transistors is replaced with a multiemitter transistor (Fig. 1-18b). The amplifying

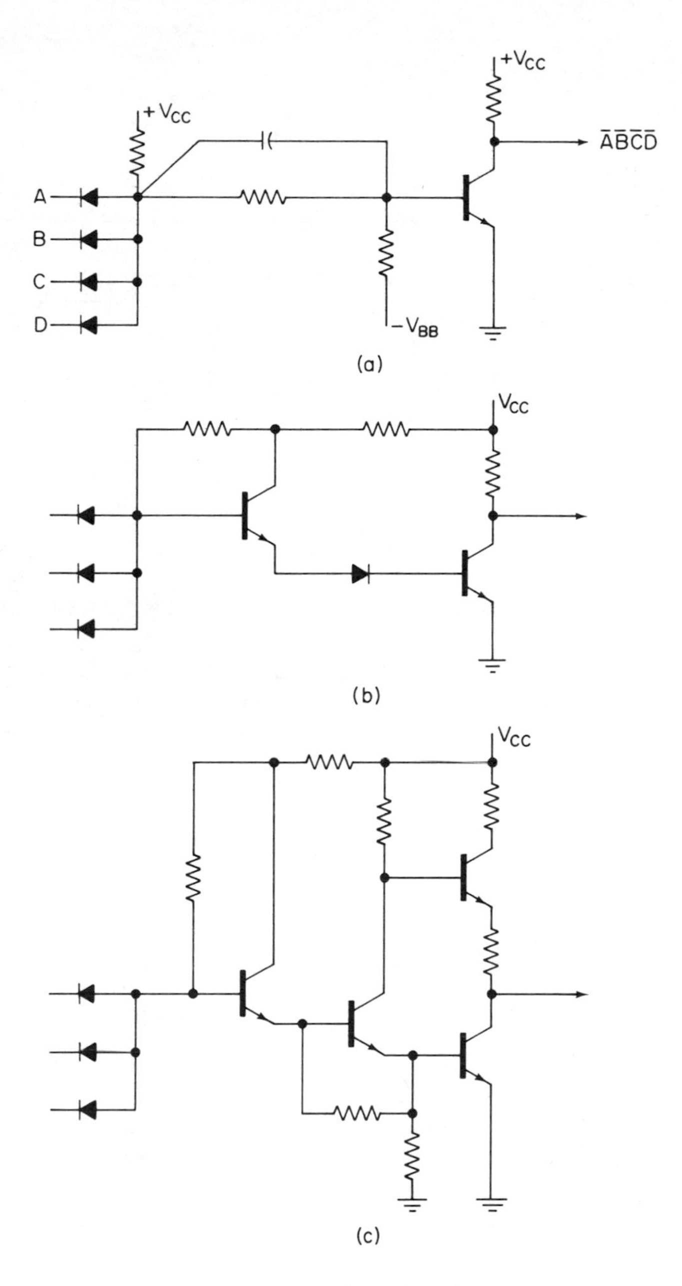

FIG. 1-16. Diode-transistor logic (DTL).

transistor is also replaced by a more elaborate scheme using transistors in a push-pull configuration. In this form, the output can supply a large amount of current both when it is high and low. Thus, the circuit has a high fan-out.

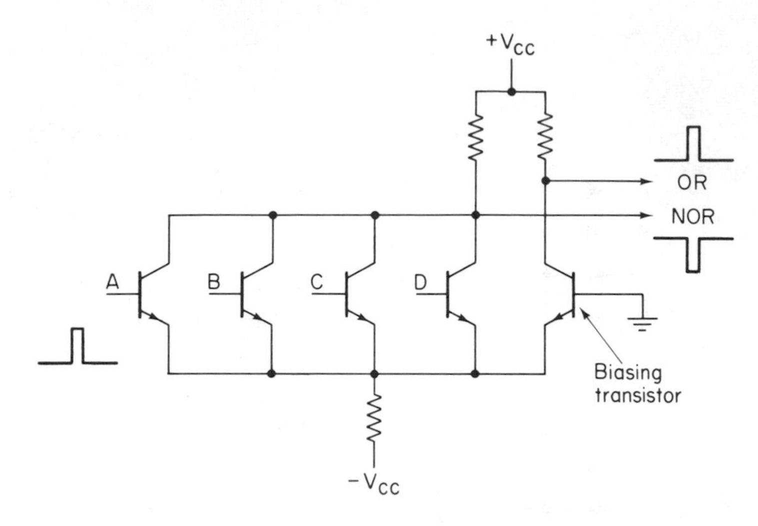

FIG. 1-17. Current mode logic (CML).

The TTL circuit is faster than DTL circuits and switches at approximately the same logic levels. TTL has a wide frequency range and is therefore more sensitive to noise than DTL. TTL and DTL are logically and electronically compatible. Thus, the use of DTL in all cases, except when very high speed TTL is required, is a good logic and electrical combination, recommended by many IC logic designers.

1-8.8 High-threshold logic (HTL)

High-threshold logic is intended for applications in noisy environments where slow speeds can be tolerated. A typical noise margin is about 5 V. The basic HTL gate circuit is similar to DTL circuits, except that a Zener diode replaces the coupling diode.

The basic HTL gate is shown in Fig. 1-19a. Note that this is similar to the DTL shown in Fig. 1-19b. The basic difference is in diode D_1, resistor values, and collector supply voltage (V_{CC}).

In DTL, D_1 is a base-emitter diode operated in its forward direction and having a drop of approximately 0.75 V. The input threshold level of DTL is a net of the two forward diode drops (the input diode offsets a diode drop in the other direction), or about 1.5 V.

In HTL, diode D_1 is a base-emitter junction that is operated in its reverse direction (commonly called Zener operation). Conduction occurs when the junction has approximately 6.7 V across it. Thus, the threshold voltage for HTL is one forward diode drop, plus one reverse diode drop, or about 7.5 V. This is at least 6 V higher than a DTL. Since 90 percent of the electrical noise

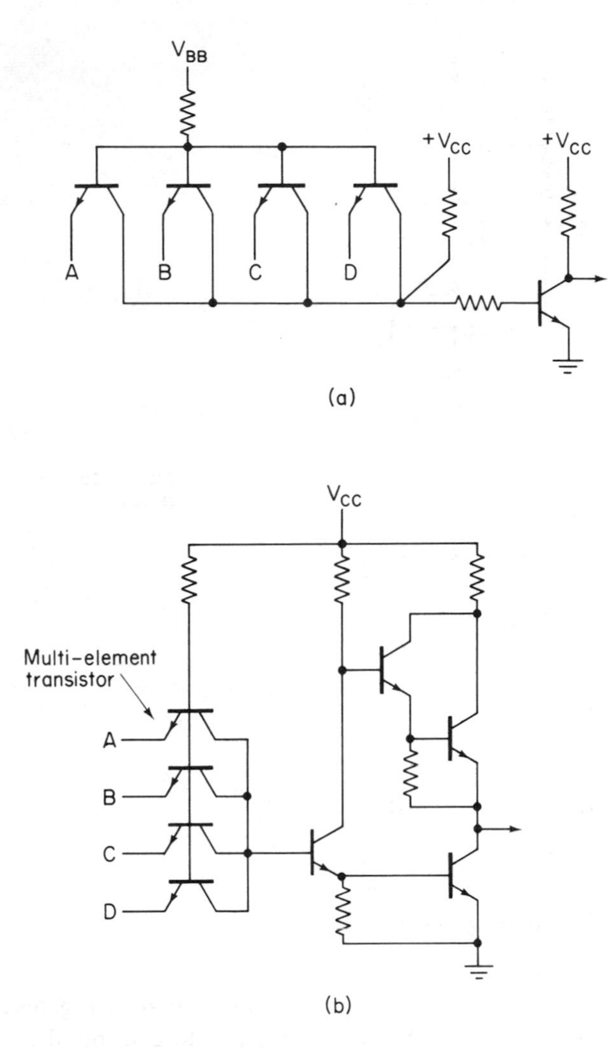

FIG. 1-18. Transistor-transistor logic (TTL).

found in a typical logic system is less than 5 V, the HTL threshold prevents most noise from entering the circuits.

1-8.9 Emitter-coupled logic (ECL)

Emitter-coupled logic, shown in Fig. 1-20, operates at very high speeds. Logic systems can be operated at frequencies of 300 MHz and higher using ECL. Another advantage of ECL is that both a true and complementary output is produced. Thus, both OR and NOR functions are available at the

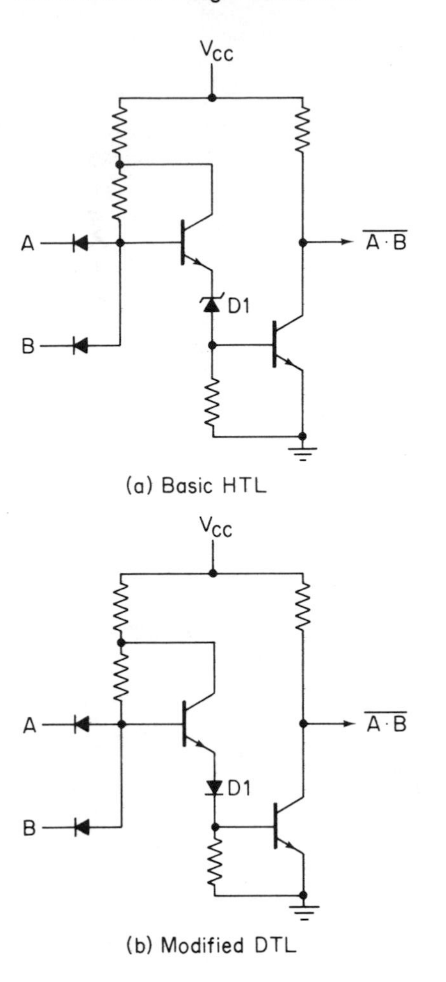

FIG. 1-19. High threshold logic (HTL).

output. Note that when the NOR functions of two ECL gates are connected in parallel, the outputs are ANDed, thus extending the number of inputs. For example, as shown in Fig. 1-20, when two two-input NOR gates are ANDed, the results are the same as a four-input NOR gate (or a four-input NAND gate in negative logic). When the OR functions of two ECL gates are connected in parallel, the outputs are ANDed, resulting in an OR/AND function.

1-8.10 Comparison of basic logic forms

Figure 1-21 shows a comparison of the five logic forms in common use. Typical operating voltages, power dissipations, propagation delay, noise immunity (threshold level), output voltage logic swing (voltage range between typical "1" and "0" outputs) are given for each of the five

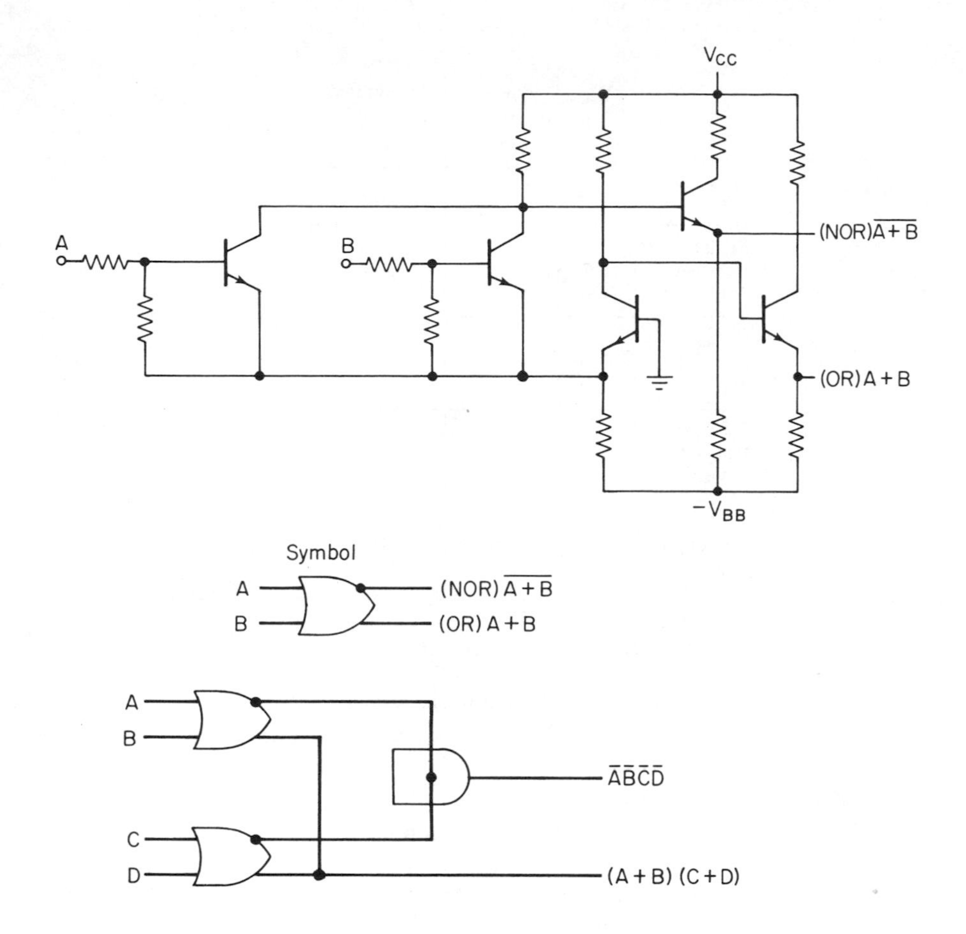

FIG. 1-20. Emitter-coupled logic (ECL)

digital IC types. Keep in mind that all of the values are for "typical" IC packages and are included here as a "starting point" for logic circuit design, not as hard and fast design parameters.

1-8.11 Field effect transistors in digital ICs

Field effect transistors (FETs) are being used to simplify logic circuits. The most common use of FETs in digital ICs is the complementary inverter shown in Fig. 1-22. Type C (enhancement mode only) MOSFETs (metal-oxide-semiconductor FET) are used in this circuit. The logic levels for the inverter are $+V$ for a 1 and ground for a 0.

With a true input ($+V$), the *P*-channel stage has zero-gate voltage and is essentially cut off. The *P*-channel conducts little *drain current* (generally

	V_{CC} (volts)	Power dissipation (mw)	Propagation delay (ns)	Noise Immunity (volts)	Logic swing (volts)
Emitter coupled logic	8.0	25	2	1.3	2
Resistor transistor logic	3.6	12	25	0.5	1
Diode transistor logic	5.0	8	30	1.2	4.5
Transistor transistor logic	5.0	15	10	1.2	3.5
High threshold logic	15	30	85	6	13

FIG. 1-21. Comparison of basic logic forms.

identified as I_{DSS}, and typically a few picoamperes for a *C*-type FET). The *N*-channel element is forward-biased and its *drain voltage* (with only a few picoamperes of I_{DSS} allowed to flow) is near ground or false (0). The load capacitance, C_L, represents the output load and any stray capacitance.

With a false input (ground), the *N*-channel element is cut off and permits only I_{DSS} to flow. The *P*-channel element is forward-biased, and its V_{DS} (drain-source voltage) is low. Thus, the drain terminal of the *P*-channel is near +V, and C_L is charged to +V.

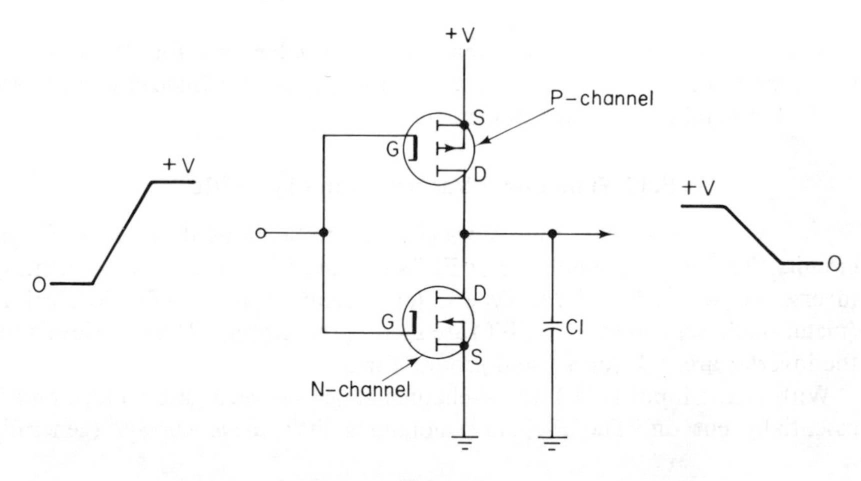

FIG. 1-22. Basic FET complementary inverter logic circuit.

The power dissipation is extremely low since both stable states, true and false (or 1 and 0) are conducting only leakage current. Power is dissipated only during switching, an ideal situation for logic circuits. In addition to the low power dissipation, another advantage of FETs for digital ICs is that no coupling elements are required. The input to a MOSFET is an insulated gate, which in effect resembles a capacitor.

Without the need for a coupling capacitor function, it is relatively simple to fabricate MOSFET logic element in IC form. OR, NOR, AND and NAND gates with either positive or negative logic can be implemented with MOSFETs in digital IC packages. Thus, almost any logic circuit combination can be produced. FET logic can also be used over a wide range of power supply voltages.

A flip-flop, a two-input NOR gate, and a two-input NAND gate are shown in Figs. 1-23, 1-24, and 1-25, respectively.

In the NOR gate circuit of Fig. 1-24, the gates of Q_1 and Q_3 are tied together to form input 1. Elements Q_1 and Q_3 act together as an inverter circuit in that they form a push-pull combination. Element Q_2 acts as a series resistance, which is either extremely high or extremely low, in the inverter formed by Q_1 and Q_3. Likewise, Q_1 acts as a series resistance in the inverter formed by Q_3 and Q_4.

Input 2 is the control input for the second inverter. The output of the gate is at +10 V only when elements Q_1 and Q_2 are switched on. This occurs only if both inputs 1 and 2 are at ground. Thus, the output is a logic 1 only when both inputs are at a logic 0.

Since a NOR gate using positive logic becomes a NAND gate when using negative logic, the NOR gate can be converted to a NAND gate by interchanging *P*- and *N*-channel elements and flipping the circuit upside down. This is done with the two-input NAND gate of Fig. 1-25.

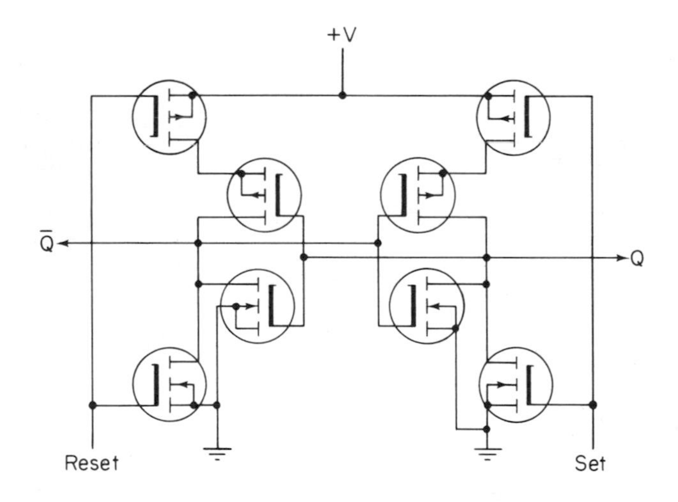

FIG. 1-23. Complementary FET RS flip-flop (Motorola).

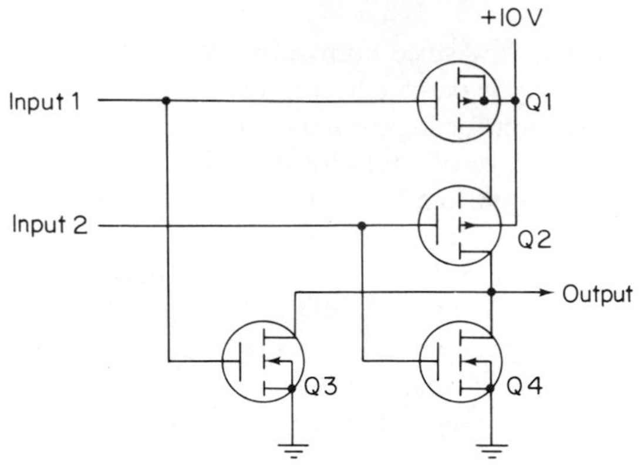

FIG. 1-24. Two-input NOR gate using MOSFETs (Motorola).

0 = 0 V		1 = +10 V
Input 1	Input 2	Output
0	0	1
0	1	0
1	0	0
1	1	0

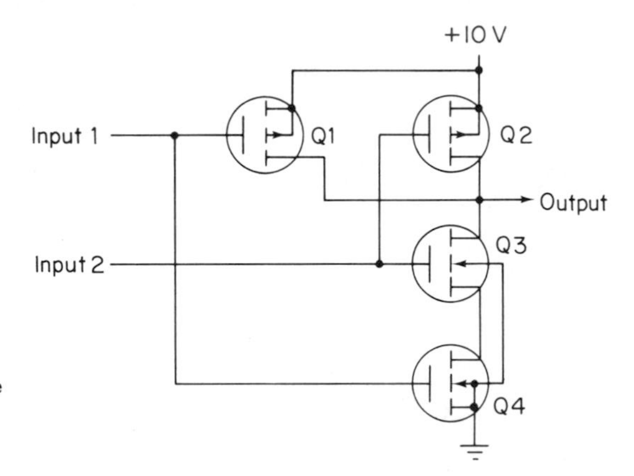

FIG. 1-25. Two-input NAND gate using MOSFETs (Motorola).

0 = 0 V		1 = +10 V
Input 1	Input 2	Output
0	0	1
0	1	1
1	0	1
1	1	0

2. PRACTICAL CONSIDERATIONS FOR INTEGRATED CIRCUITS

At first glance it may appear that there are few practical considerations for ICs, both digital and linear. If you are to believe the data found in some engineering literature, an IC requires only that power be applied for the package to be ready for immediate use. While this is essentially true, there are a number of points to be considered in selecting and using ICs.

For example, each of the three basic package types has certain advantages and disadvantages. These should be considered when selecting ICs to meet specific design requirements. Once the package type has been chosen, there are various alternate methods of mounting the IC to be considered.

For the technician who must work with ICs, removal and replacement of the packages (unsoldering leads, bending leads to fit existing mounting patterns, etc.) can present problems. For the experimenter or design engineer who must work with ICs in the breadboard stage, there are problems of oscillation, circuit "ringing," and noise. As is the case with any solid-state circuit, there are always temperature and power-supply limitations to be considered.

All of these problems have some effect on both digital and linear ICs. In some cases, the problems have greater effect on one type of IC (either digital or linear). In other cases, the problems are common to both types. Such common problems are discussed in this chapter as a starting point for the reader who is interested in the overall practical side of ICs. Considerations that apply to a specific IC type or circuit are discussed in related chapters.

2-1. SELECTING THE IC PACKAGE

As discussed in Chapter 1, there are three basic types of IC packages, TO-5 (metal can) style, flat-pack, and dual-in-line (DIP). There is a further division in that the dual-in-line packages are available in both ceramic and plastic.

In some cases, the designer has no choice of package style since the particular IC he wants is available in only one style. For example, where the IC operates at high power, and dissipates considerable heat, the metal can is required, because it permits the use of heat-sinks and/or possible mounting directly on a metal chassis. In other cases, the choice of package style is wide open. The following notes are included to help the user make this choice.

If cost is a factor, or if a large volume of ICs is required, the dual-in-line is generally the best choice. DIPs are ideally suited for mounting on printed circuit (PC) boards, since there is more spacing between the leads (typically 0.1 inch) than with other package types. During production, DIPs can be inserted (manually or automatically) into mounting holes on PC boards, and soldered by various mass-production techniques.

The choice of ceramic versus plastic is again a matter of cost versus reliability. In general, a ceramic IC will provide a better hermetic seal (to protect the silicon chip), but at a higher cost than plastic. There are exceptions to this rule of course. For example, a ceramic IC with poor plating on the leads could fail faster than a plastic IC with good plating.

The real weak spot in any IC package is at the point where the *leads enter the case or body*. Usually, there is a glass or plastic seal at these points. The seals can be broken, thus exposing the chip and unplated metal inside the package if the leads are bent or twisted during production or repair. Moisture and other undesired elements can then enter the IC package. While this may not cause immediate failure, it will almost certainly cause ultimate failure. If nothing else, the exposed bare metal will corrode, and affect the IC performance.

One method used to offset the effects of possible lead-seal damage during production is to bake the entire PC board after all of the ICs have been mounted and soldered. This technique applies only to plastic ICs. Typically, the bake cycle is about one hour at 125 °C (or at lower temperatures for longer periods of time, depending on the heat characteristics of the IC and of other components mounted on the same board). The effect of the prolonged bake is to remove any moisture and to improve the seal. Of course, the technique will be of no value to an IC with a seriously damaged lead-seal.

If reliability is the major factor, the ceramic flat-pack is generally the best choice. Flat-packs have an excellent history of reliability. Likewise, flat-packs are smaller and lighter than DIPs, all other factors being equal. Ceramic flat-packs are usually the choice for any airborne application, except where high power is involved. (This requires a metal can.)

As in the case of any solid-state device, it is always good practice to check the manufacturer's test data regarding failure rates of package types in special environments. For example, if the IC must be operated in the presence of thermal shock (extreme high temperatures followed by extreme low temperatures), check the failure rates (due to thermal shock) of different package types. Often, the manufacturer can provide considerable data regarding failure rates, reaction to environments, etc., that will prove very helpful in selecting IC package types.

2-2. MOUNTING AND CONNECTING INTEGRATED CIRCUITS

Once the package type has been selected, the IC must be mounted and connected to other components. The selection of a particular method for mounting and connection of ICs depends on the type of IC package, the equipment available for mounting and interconnection, the connection method used (soldering, welding, crimping, etc.) the size, shape, and weight of overall equipment package, the degree of reliability and ease of replacement, and on the never-to-be-forgotten cost factor. The following sections summarize mounting and connection methods for the three basic package types.

2-2.1 Breadboard mounting and connection

During the breadboard stage of design, any of the IC packages can be mounted in commercially available sockets. This will eliminate soldering and unsoldering the leads during design and test. Such sockets are generally made of Teflon or some similar material, and are usually designed for mounting on a PC board. However, some IC sockets are designed for metal chassis mounting. In other cases, the IC can be soldered to the socket that is in the form of a plug-in PC card. The card is then plugged into or out of the circuit during design and testing. Sockets manufactured by Azimuth Electronics, Barnes Development Co., Jefferson Products, Inc., and Sealectro Corp. are typical of the temporary mounts required during design. In some cases, the same mounts can be used on a permanent basis if required.

2-2.2 Ceramic flat-pack mounting and connection

Figure 2-1 shows five methods for making solder connections to flat packs. Note that there is a notch in one end of the package. This is a reference point to identify the lead numbering. Usually, but not always, this reference notch is nearest to Lead No. 1. Always consult the manufacturer's

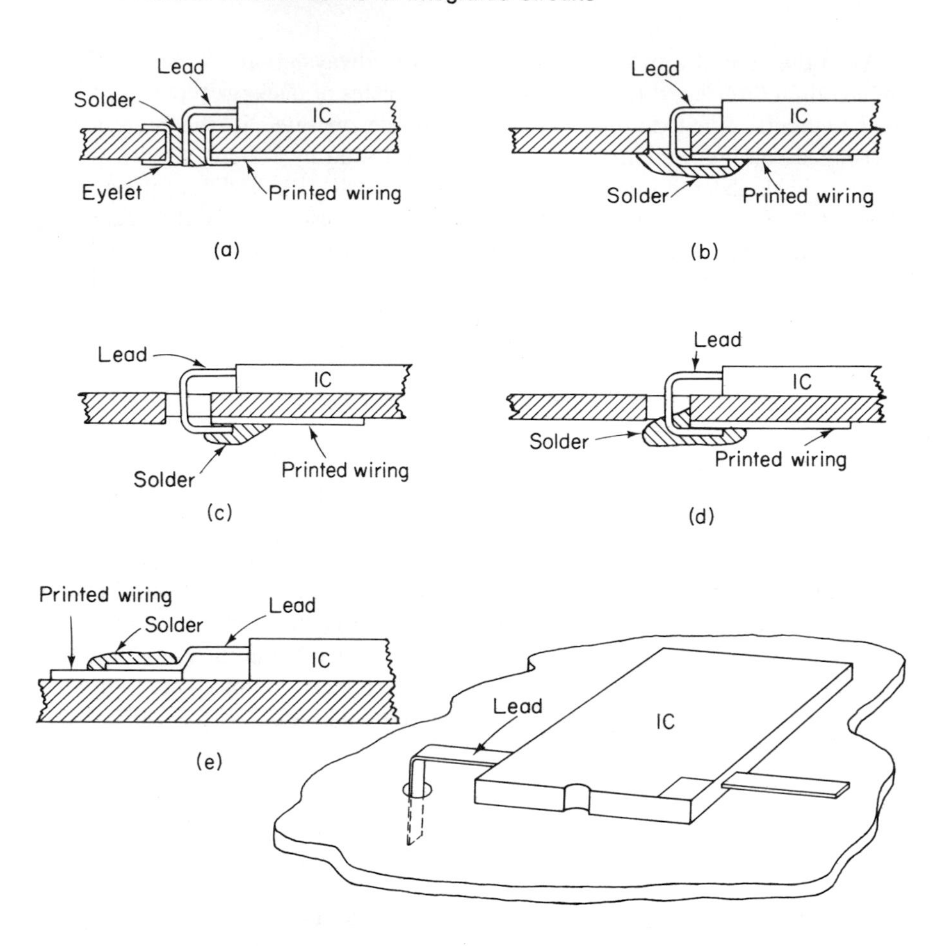

FIG. 2-1. Typical soldering techniques for flat-pack ICs.

data regarding IC lead numbering. Although there has been some attempt at standardization, most manufacturers prefer their own lead identification system.

In the *straight-through method* (Fig. 2-1a), the leads are bent downward at a 90° angle and are inserted in the circuit-board holes. When assembled during manufacture, all leads are connected simultaneously by dip soldering or wave soldering. During repair, the leads are soldered one at a time. A disadvantage of the straight-through method is that the IC package must be held firmly in position during the soldering operation.

The *clinched-lead, full-pad soldering method* (Fig. 2-1b) requires an additional operation (clinching the lead), but has the advantage that the IC does not have to be held in position during soldering.

The *clinched-lead, offset-pad* (Fig. 2-1c) and the *clinched-lead, half-pad* (Fig. 2-1d) methods are variations of the clinched-lead full-pad method. Replacement is easier with the half-pad and offset-pad methods since the hole is not filled with solder.

In the *surface-connection method* (Fig. 2-1e), the connections are made on the package side of the board. This method is often used when ICs must be mounted on both sides of the board. No holes are required in the board. The *reflow solder technique* is often used when surface-connection ICs are assembled during production. With reflow soldering, both the surface contacts and the leads are tinned and covered with solder. Then the leads are set on the surface contacts, and heat is applied to all contacts (or all contacts on one side). The heat causes the solder to reflow and make a good connection between leads and surface contacts.

The mounting patterns shown in Fig. 2-1 use *in-line lead and pad arrangements*. Although such arrangements simplify lead forming, they result in very close spacing between leads (typically 32 mils) and require the use of high-precision production techniques in both board manufacture and assembly of ICs on the board, particularly when the leads must be inserted through holes in the PC board. Another disadvantage of the in-line arrangement is the limited space available for routing circuit conductors between adjacent solder pads.

Some of these disadvantages can be overcome by the use of *staggered lead arrangements*, shown in Fig. 2-2c. In these staggered arrangements, the lead holes and terminal pads for adjacent leads on the same edge of a flat package are offset by some convenient distance from the in-line axis. Although a staggered lead arrangement requires somewhat more PC board area per IC than the in-line arrangement, staggered leads provide several advantages: (1) tolerances are far less critical; (2) larger terminal pads can be used; (3) even with larger pads, more space is available for routing circuit conductors between adjacent terminal connections; and (4) larger lead holes can be used to simplify lead insertion.

In a staggered lead arrangement, a good compromise between loss of available PC board area and gain in the number of conductors that can be routed between adjacent circuit terminal pads can be achieved by use of an offset between adjacent pads.

The staggered lead arrangement provides great flexibility in circuit wiring configuration. For example, if offset lead holes 30 mils in diameter and solder pads 80 mils in diameter are used, at least one 8-mil wide printed conductor can be routed between adjacent solder pads, using the dimensions of Fig. 2-2c. If 60-mil diameter solder pads are used, up to two 8-mil wide printed conductors can be routed between adjacent pads.

For some applications, it may be necessary or desirable to use *welded connections* rather than soldered connections. Figure 2-3 shows three methods

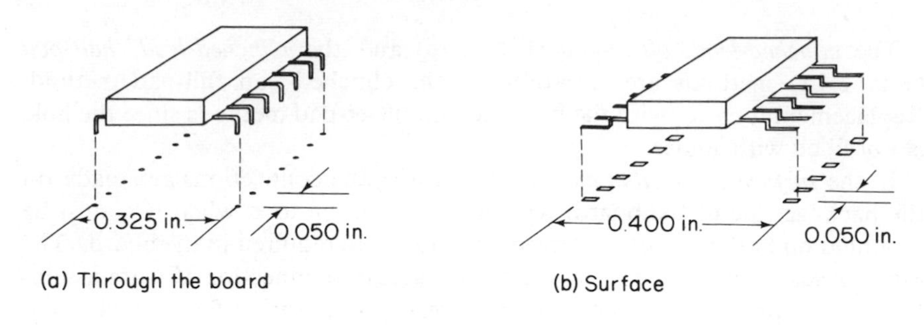

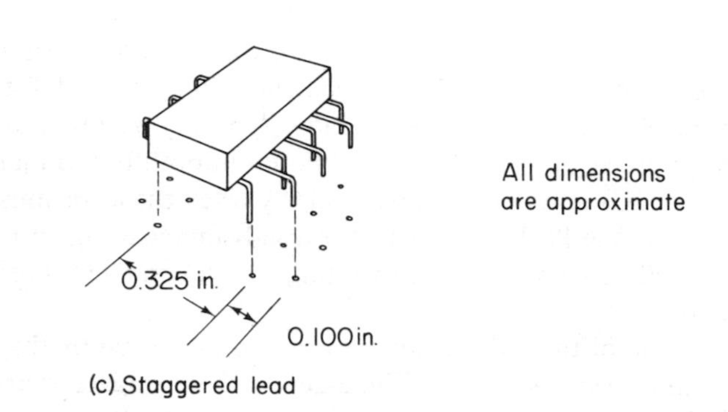

FIG. 2-2. Typical mounting patterns for flat-pack ICs.

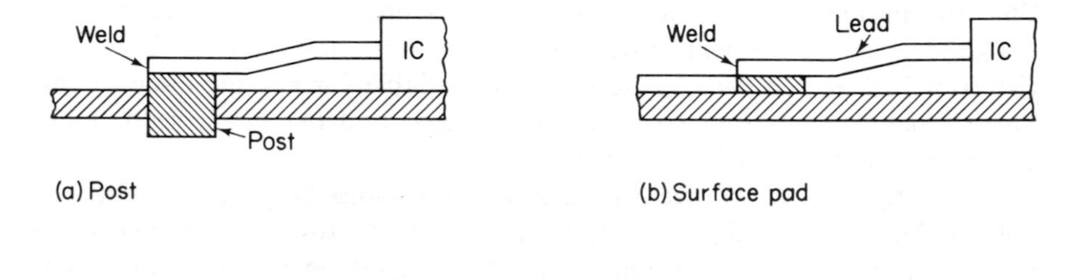

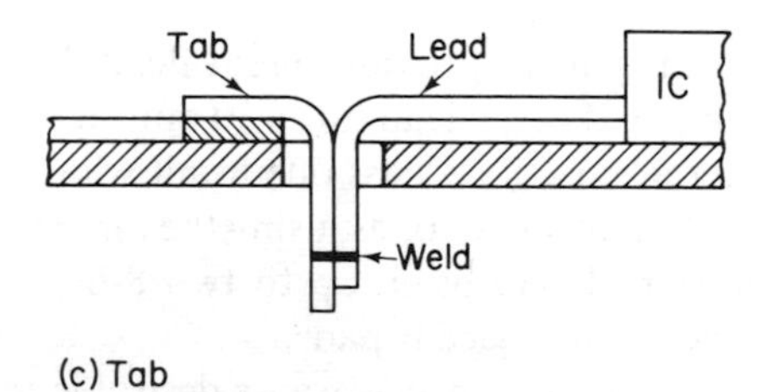

FIG. 2-3. Typical welding techniques for flat-pack ICs.

that may be used for making such welded connections. In general, the mechanical space considerations described for solder connections apply equally well to welded connections.

Although for some applications welding may provide more reliable connections than soldering, the former has a disadvantage in that with conventional welding equipment only one connection can be made at a time.

The "tab" method of welding shown in Fig. 2-3c is good from a maintenance standpoint because the IC can be easily removed if replacement becomes necessary. The tab method uses "cross-wire" resistance welding in which the weld is made at the ends of the IC lead and the terminal tab. The IC can thus be removed simply by clipping the leads just above the weld point. However, the replacement process is not as simple. The IC lead must be soldered to the tab.

2-2.3 TO-5 style package mounting and connection

The most direct method for mounting TO-5 style packages is shown in Fig. 2-4a. Here, the leads are simply inserted in the proper plated-through holes in the PC board, and connection is completed by dip- or wave-soldering.

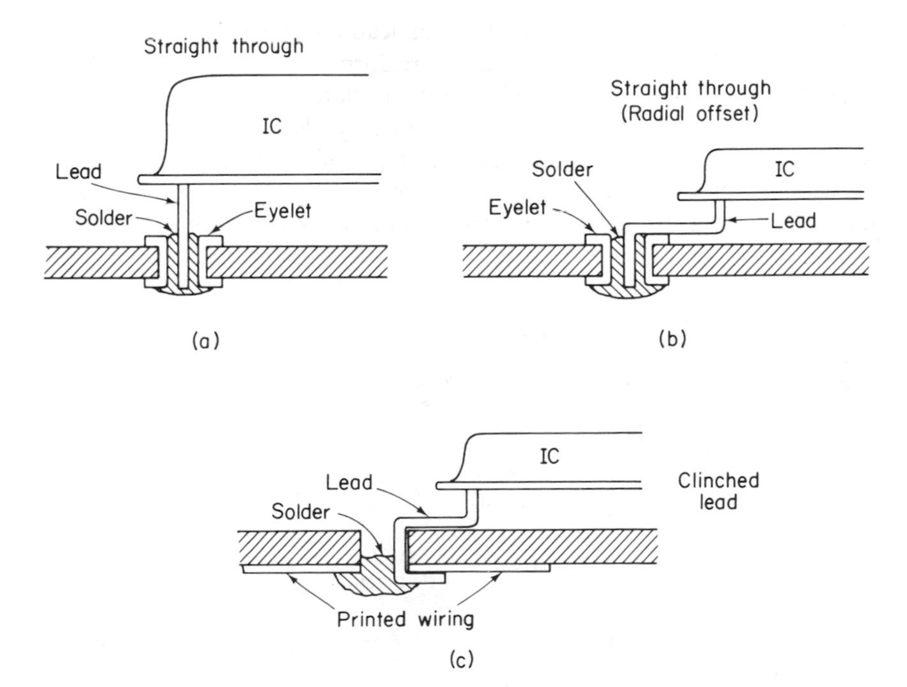

FIG. 2-4. Typical soldering techniques for TO-5 style ICs.

Although the method of Fig. 2-4a requires minimum handling of the IC (trimming of terminal leads to approximate lengths may be necessary), it does require extremely precise drilling and "through-plating" of the lead holes and preparation of the solder pads. The method also has disadvantages; namely, that automatic insertion of the device leads in such limited space can present problems, and that the device must be held in position during the soldering operations.

The methods shown in Fig. 2-4a and 2-4b make it possible to achieve effective "wicking" of the solder around the lead. Figure 2-4c shows a method in which the holes are not plated through and the leads are clinched before the soldering operation is performed. Because the electrical connection depends upon the solder being on the pad and not in the hole, the lead-hole diameter can be made larger and, therefore, this method permits easier lead insertion (possibly with an automatic tool during production). Also, the clinching of the leads helps to hold the IC in position during soldering.

Figure 2-5 shows the lead hole arrangement for straight-through mounting of the ten-lead TO-5 style package. This arrangement provides approximately 11 mils clearance between adjacent pads, the smallest clearance possible without danger of shorting. This separation is insufficient to accommodate a printed conductor of conventional size (8 mils wide with 8 mils spacing).

Figure 2-5 also shows a radially-offset lead-circle pattern for a 12-lead TO-5 style package in which the leads are formed to increase the effective lead-circle diameter (530 mils). This configuration permits the use of 80-mil diameter pads with sufficient spacing between pads to accommodate three 8-mil wide printed conductors. This radially-offset pattern also permits clinched-lead mounting and represents a good compromise from the standpoint of mounting arrangement, reliability, ease of maintenance, and cost.

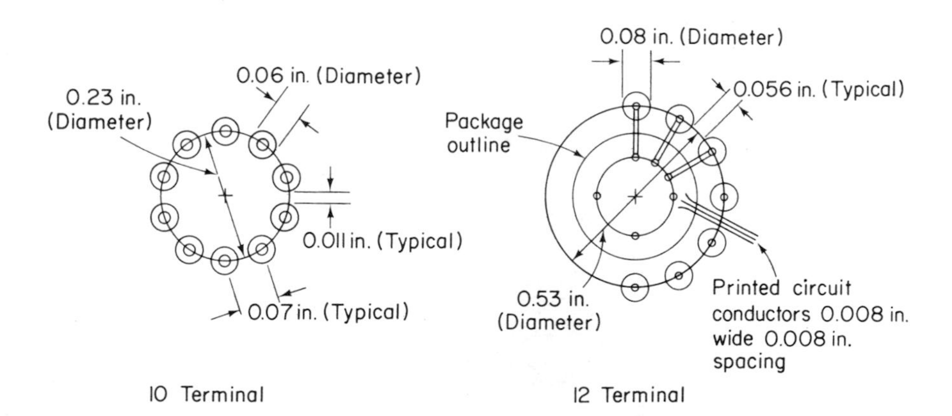

FIG. 2-5. Typical mounting patterns for TO-5 style ICs.

2-2.4 Dual-in-line package mounting and connection

Figure 2-6 shows the mounting arrangement used for integrated circuits in dual-in-line packages. Because the package configurations are very similar, the mounting arrangement and terminal-sorting techniques used for these circuits are much the same as those used in the in-line method (Fig. 2-2a) for the flat-pack ICs. The DIP IC is longer however, and the soldering operation is made simpler because of the greater spacing provided between lead terminals.

The terminals of the DIP may be soldered to a PC board by use of any of the through-the-board techniques (shown in Figs. 2-1a through 2-1d) used for the in-line method of mounting ceramic flat-packs. The DIP terminal leads are larger than those of the flat-pack. The larger-size terminals are more rigid and more easily inserted in the mounting holes (either in the PC board or in IC sockets).

Another significant feature of the DIP is the sharp step increase in width of the terminals near the package end. This step forms a shoulder upon which the package rests when mounted on the board. Thus, the package is not mounted flush against the board. As a result, it is possible to run printed circuit wiring directly under the package. Also, convection cooling of the package is increased, and the circuit can be more easily removed if replacement is required.

2-3. WORKING WITH INTEGRATED CIRCUITS

It is assumed that the reader is familiar with common hand tools used in electronics, such as diagonal wire cutters, long needle-nose pliers, soldering tools (both pencil-type and soldering guns), insulated probes, nut

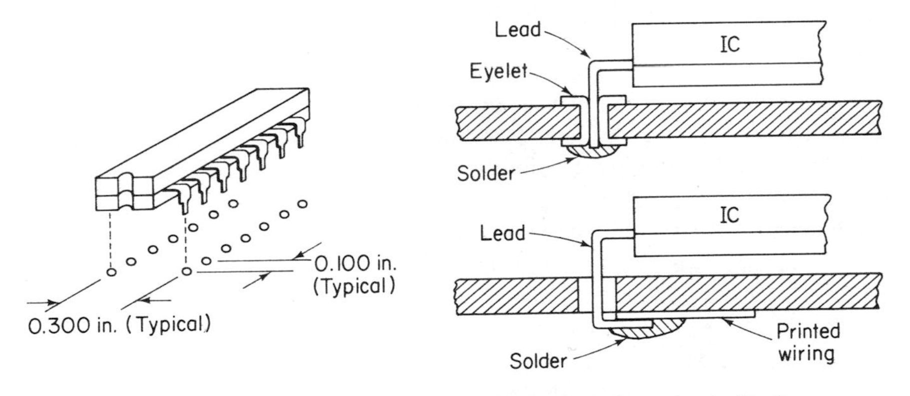

FIG. 2-6. Typical mounting pattern and soldering techniques for dual-in-line ICs.

drivers, and wrenches. These same tools are used in IC work. However, certain additional tools and techniques are also required (or will make life much easier if used).

The main problem in working with ICs, besides the small size, is *heat*. Typically, ICs are made of silicon, which cannot withstand high temperatures for any length of time. A temperature of 200 °C is about tops for any IC. The leads cannot be subjected to continuous and excessive heat when soldering and unsoldering leads.

Practically all ICs are miniature. This demands the use of small soldering tools, pliers with fine points, and very delicate handling in general. The problems of the fine wire leads and the seals where the leads enter the package are always present when working with ICs.

The following paragraphs summarize the major problems found in working with ICs during broadboard design, production, and repair.

2-3.1 Lead bending

In any method of mounting ICs that involves bending or forming of the leads, it is extremely important that the leads be supported and clamped *between* the bend and the seal, and that bending be done with extreme care to avoid damage to lead plating. Long-nose pliers can be used to hold the lead, as shown in Fig. 2-7. In no case should the radius of the bend be less than the diameter of the lead. In the case of rectangular leads, the radius of the bend should be greater than the lead thickness. It is also extremely important that the ends of the bent leads be perfectly straight and parallel to assure proper insertion through the holes in the PC board.

2-3.2 Working with printed circuit boards

Integrated circuits are often mounted on printed circuit or etched circuit boards. Likewise, ICs may be mixed with other components (resistors, transistors, coils, etc.) on the same PC board. In turn, the board may be bolted to a main chassis or plugged into a connector on the chassis. In either case, the component leads must pass through eyelets (sometimes known as *soldering* cups, or simply *holes*) in the board. The eyelets make contact with the printed or etched wiring. The component leads must be soldered to the eyelets as shown in Fig. 2-8.

When properly executed, removal and replacement of components on a PC board is not impossible, although it may appear that way in the case of ICs. The following techniques are recommended to replace components on etched or printed boards. The techniques will be satisfactory for most practical applications. Of course, you must modify the procedure as necessary to fit the particular equipment.

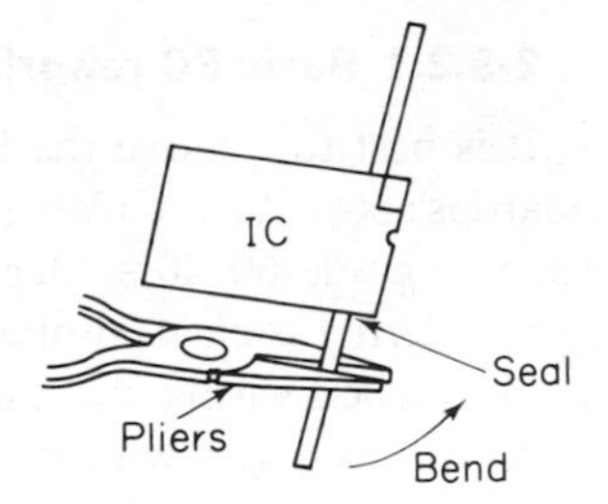

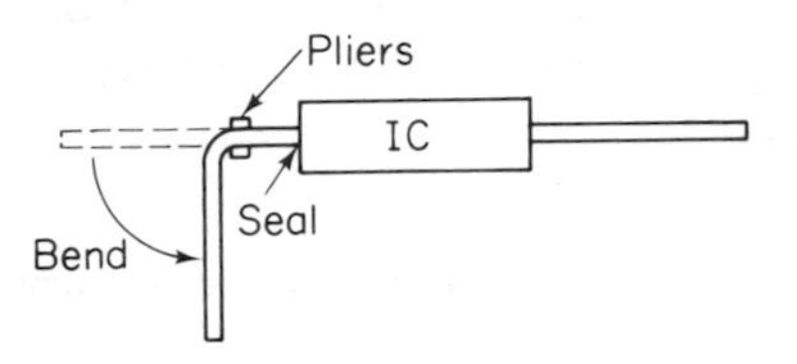

FIG. 2-7. Lead bending techniques for ICs.

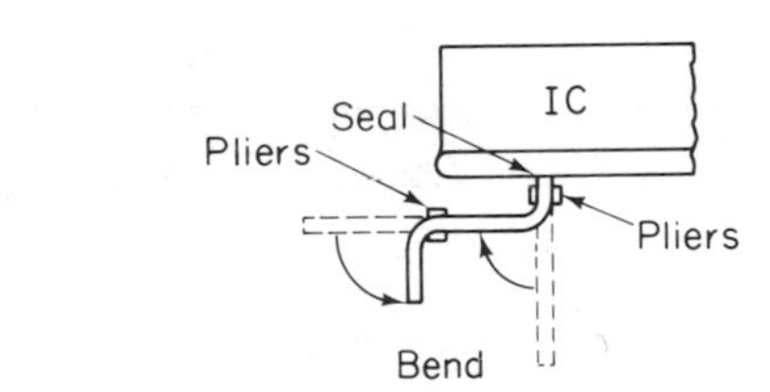

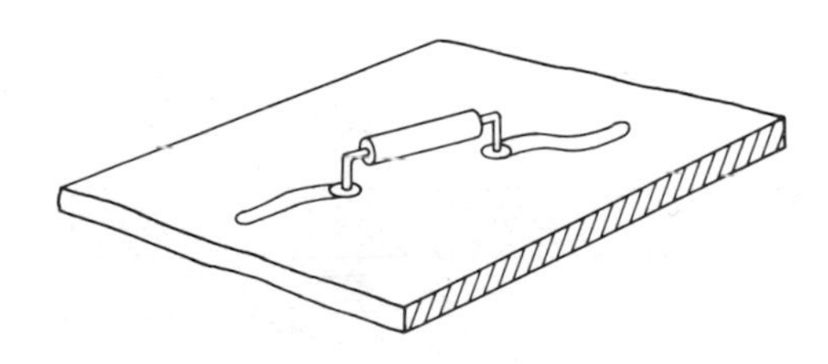

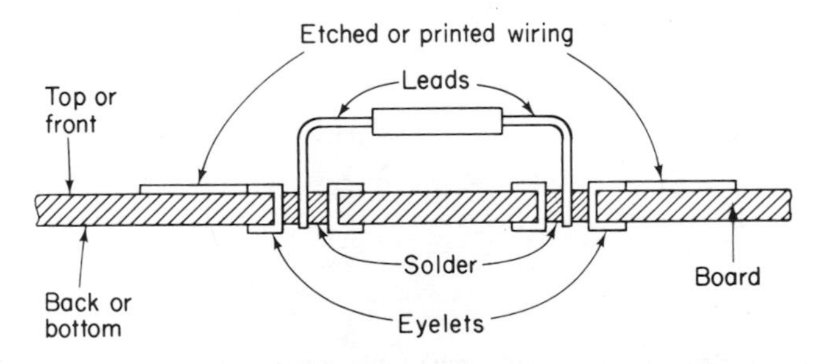

FIG. 2-8. Component mounting on printed circuit board.

2-3.2.1 Basic PC rework procedure

It is best to remove the board from the equipment, unless the back of the board is accessible. Unless specified otherwise in the service literature, use electronic grade 60/40 solder and a 15-W pencil soldering tool. A higher-wattage soldering tool, if applied for too long a period, can ruin the bond between the etched wiring and the insulating base material by charring the glass epoxy laminate (or whatever is used for the base). However, a 40- to 50-W soldering tool can be used if the touch-and-wipe technique is followed (described later in this chapter). The author prefers a chisel tip for the soldering tool, about $\frac{1}{16}$-inch wide. However, this is a matter of personal preference.

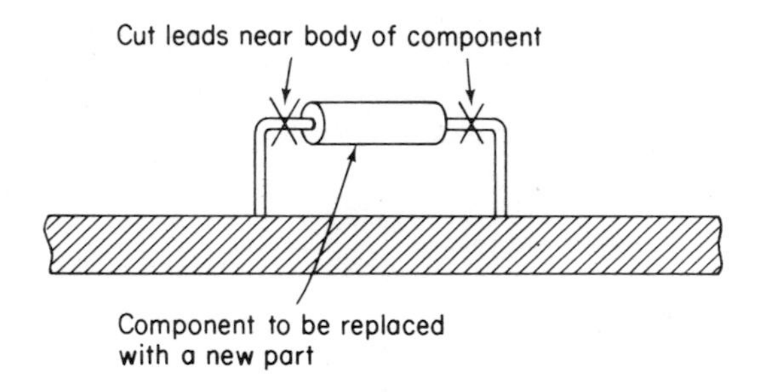

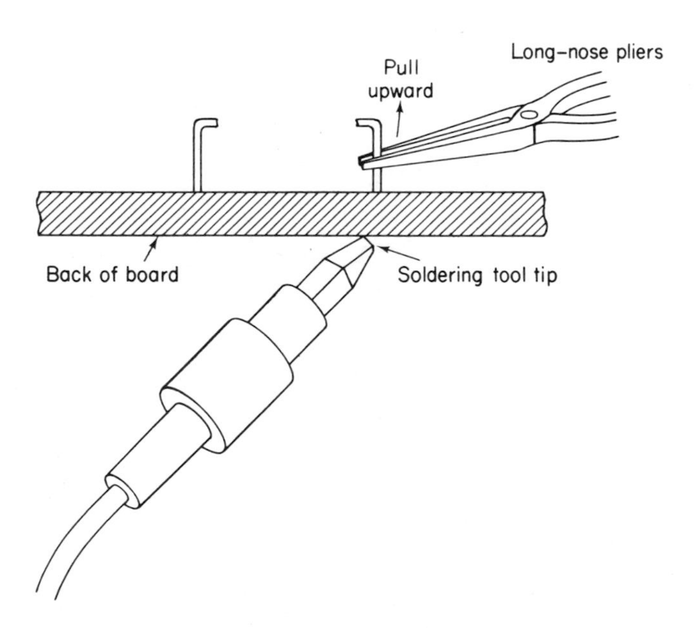

FIG. 2-9. Basic printed circuit rework procedure (when component is to be replaced).

If the component is to be removed and replaced with a new part, cut the leads near the body of the component as shown in Fig. 2-9. This will free the leads for individual unsoldering. Grip the lead with long-nose pliers; apply the tip of the soldering tool to the connection at the back of the board; and then pull gently to remove the lead.

If the component is to be removed for test and possibly reinstalled, do not cut the leads. Instead, grip the lead from the front with long-nose pliers and apply the soldering tool to the connection at the back of the board. Lift the lead straight out as shown in Fig. 2-10.

Note that the pliers will provide a heat sink for the component being removed. This will keep the component from heating if it is necessary to apply the soldering tool for a long time (which should be avoided). If an IC or transistor with a metal case is to be unsoldered, an alligator clip attached to the case (shown in Fig. 2-11) will provide a temporary heat sink. Soldering heat sinks are also available commercially.

When the lead comes out of the board, the lead should leave a clean hole. If it does not, the hole should be cleaned by reheating the solder and placing a sharp object such as a toothpick or enameled wire into the hole to clean out the old solder. Some technicians prefer to blow out the hole with compressed air. This is *never* recommended when the board is in or near the equipment, since the solder spray could short other circuits on other parts in the equipment.

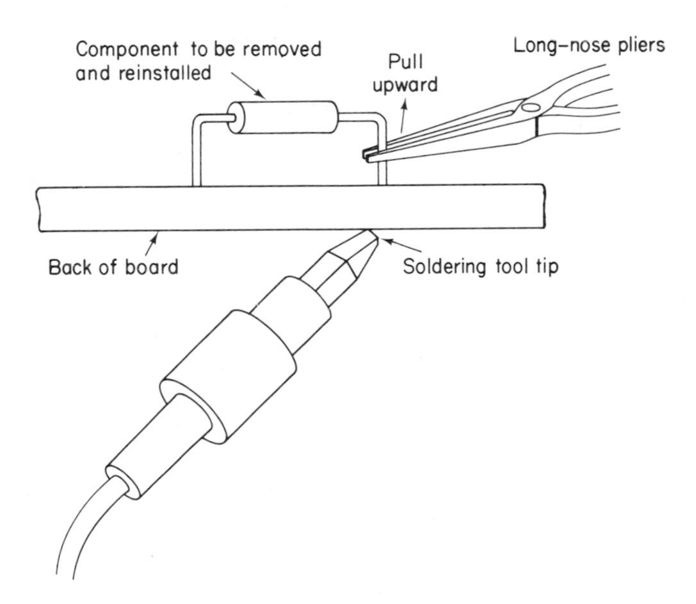

FIG. 2-10. Basic printed circuit rework procedure (when component is to be removed and reinstalled).

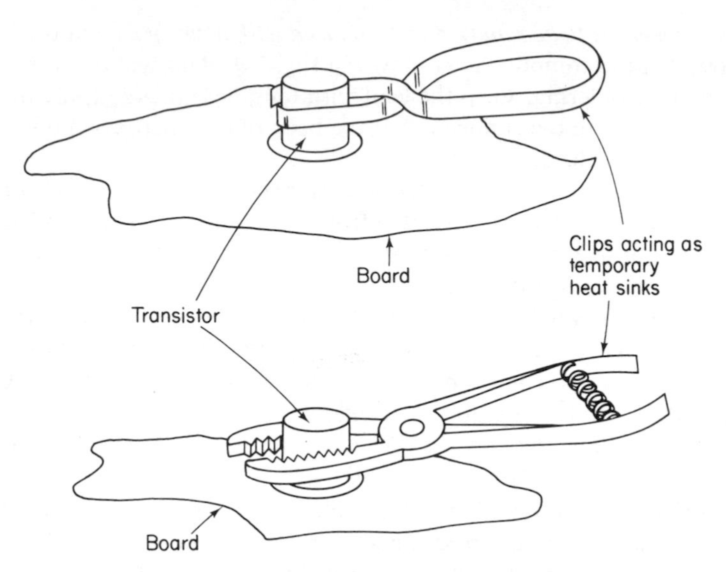

FIG. 2-11. Using clips as transistor heat sinks during soldering and desoldering.

2-3.2.2 Removal and replacement of ICs on PC boards

It may appear that an IC cannot be removed from a PC board or socket without destroying or seriously damaging the IC. In any package style, all leads must be unsoldered before the IC can be removed. This brings up obvious problems. You cannot use the standard desoldering procedure discussed in foregoing sections (pulling each lead with long-nose pliers while applying heat to the connection). That is, such a procedure is not possible in some cases (due to arrangement of the leads) and is highly impractical in most cases.

There are two practical solutions to the problem. First, you can use a *desoldering tool tip* that will contact all of the lead connections simultaneously. Such tips are shown in Fig. 2-12. There are commercial versions of IC desoldering tips, or you can make your own by cutting or grinding a conventional tip to fit the particular need.

As an alternative procedure for removing ICs, you can use the *solder gobbler*, which was developed originally as a desoldering tool for ICs. There are many versions of solder gobblers (also known as *vacuum desolderers*). The tool shown in Fig. 2-13 is hand-operated and is used mostly for repair work. There are vacuum-pump-operated models available, used mostly for assembly-line work.

As shown in Fig. 2-13, the solder gobbler tool consists of a soldering tool, collector tip, and bulb. In use, the bulb is squeezed, the collector tip is placed

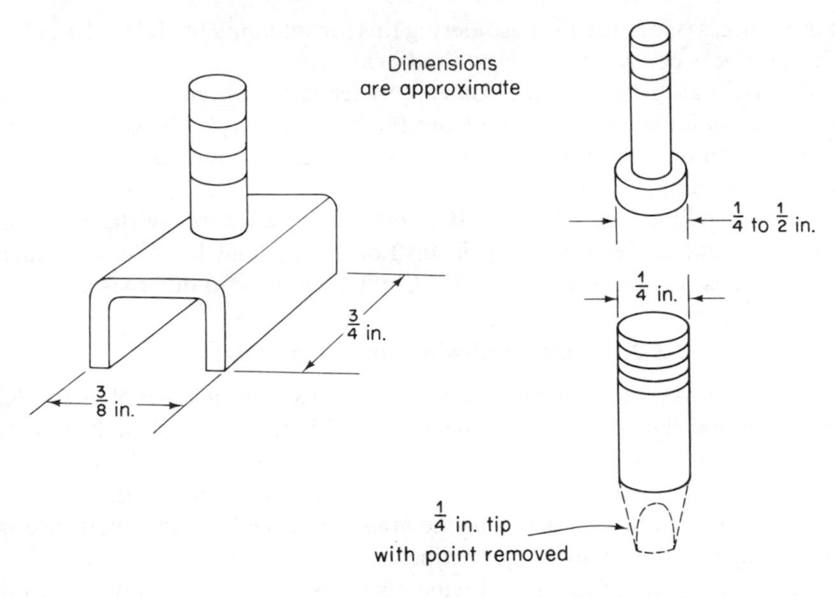

FIG. 2-12. Typical desoldering tool tips for IC removal.

on the solder area, and when the solder is molten, the bulb is released, drawing the solder into the collector tip. Then the solder can be forced from the tip by squeezing the bulb again. The tool can also be used for soldering.

The main advantage of the special desoldering tip is speed. The main disadvantage is the excess heat that might damage the IC semiconductor material. When using a special tip that covers all contacts simultaneously, be ready to remove the IC immediately (i.e., just as soon as the solder is molten).

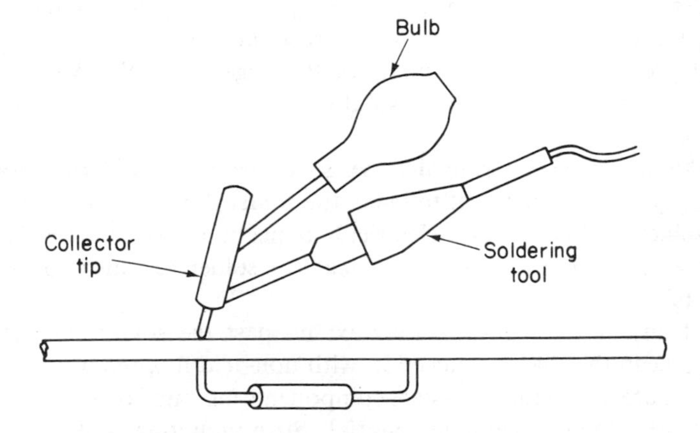

FIG. 2-13. Typical hand-operated desoldering tool (solder gobbler).

It is not necessary to use the desoldering tips for soldering the IC back in place. This produces considerable unrequired heat.

The main advantage of the solder gobbler is the absence of excess heat. Also, the solder will be removed from the hole (usually). Of course, it takes more time to remove an IC with a solder gobbler than it does to remove one with a desoldering tip.

When installing a replacement IC, you must exactly follow the mounting pattern (including lead bending, if any) of the original IC. We say *exactly* since there is rarely enough space in IC equipment to do otherwise.

2-3.2.3 Touch-and-wipe soldering of ICs

As an alternate to desoldering tips or solder gobblers for ICs, the *touch-and-wipe* method can be used. With this method, a hair-bristle soldering brush, which has its bristles shortened to about $\frac{1}{4}$ inch in length, is used. A conventional soldering tool of 50-W or less is employed.

The tip of the tool is touched to the area where the IC lead comes through the board, with the brush ready to wipe the melted solder away. The solder is removed by a series of touch-and-wipe operations. The tool is touched to the board only long enough to melt the solder and is removed as the brush is wiped across the joint. After a few touch and wipe operations, the IC lead should be free of the solder pad or eyelet on the board.

2-3.2.4 Miscellaneous soldering techniques

ICs are often used with discrete components. The following notes apply to such components.

In the soldering of large metal terminals (switch terminals, potentiometer terminals, etc.) found in solid-state equipment, ordinary 60/40 electronic-type solder is satisfactory. However, a larger soldering tool is required for such terminals. The 15-W soldering tool recommended for ICs, and printed or etched circuits is usually too small for the large terminals. A 40- to 50-W soldering tool will do most jobs in solid-state electronics. A soldering gun is also useful.

When soldering large terminals, use good electronic soldering practices; that is, apply only enough heat to make the solder flow freely and apply only enough solder to form a good electrical connection. Too much solder may impair operation of the circuit. An excess of solder can also cover a cold solder joint.

Clip off any excess wire that may extend past the solder connection. If necessary, clean the solder connection with non-acid flux-remover solvent.

Many ICs are used with solid-state components mounted on strips. Usually, the strips are made of a ceramic material. Strip mounting is used instead of (or in addition to) printed circuit mounting, expecially on equipment whose parts are not of the plug-in type.

Typical strip mounting is shown in Fig. 2-14. Often, the notches in these strips are lined with a silver alloy. This should be verified by reference to the service literature. Application of excessive heat or repeated use of ordinary 60/40 tin/lead solder can break the silver-to-ceramic bond. Occasional use of ordinary solder is permissible, but for general repair work, solder containing about 3 percent silver should be used.

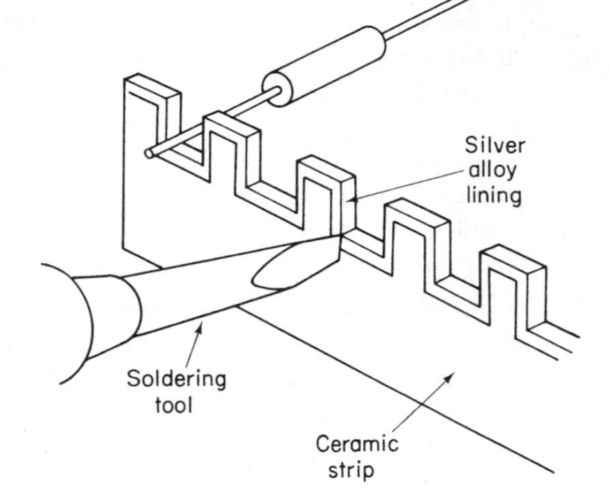

FIG. 2-14. Strip soldering techniques.

When removing or installing a part mounted on ceramic terminal strips:

1. Use a 50-W soldering tool. The tip should be tinned with silver-bearing solder.
2. Apply heat by touching one corner of the soldering tool tip to the base of the notch as shown in Fig. 2-14. Do not force the tip of the tool into the terminal notch. This may chip or break the ceramic strip.
3. Use the minimum amount of heat required to make the solder flow freely.
4. Apply only enough solder to form a good bond. Do not attempt to fill the notch with solder.

If it is necessary to hold a bare wire in place while soldering (a bare wire is sometimes used as a bus when several ICs are mounted on the same board), a handy tool for this purpose can be made by cutting a notch into one end of a wooden tool as shown in Fig. 2-15. There are commercial versions of such

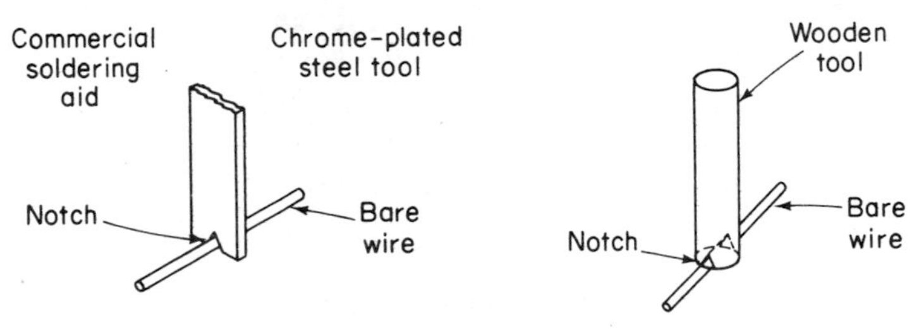

FIG. 2-15. Typical wire-holding tools.

wire-holding tools. Usually, the commercial soldering aids are made of chrome-plated steel. (The solder will not stick to the chrome plating.)

2-3.3 Handling MOSFET ICs

Special precautions are usually necessary when handling MOSFET ICs (refer to Sec. 1-8.11). In circuit, a MOSFET IC is just as rugged as any other IC. Out of circuit, the MOSFET is subject to damage from static charges when handled. Often the MOSFET is shipped with its leads all shorted together to prevent damage in shipping and handling (there will be no static discharge between leads with the shorting arrangement).

To test a MOSFET out of circuit, the leads from the test device should be connected before the short is removed from the MOSFET leads.

When soldering or unsoldering a MOSFET IC, you must keep the soldering tool tip at ground potential (no static charge). Connect a clip lead from the barrel of the soldering tool to a power-line ground.

> *WARNING:* Make certain that the chassis (or whatever the soldering tool is clipped to) is at power-line ground. In some obsolete or defective equipment, the chassis or panel is above ground (typically by one-half the line voltage).

Remove power from the circuit before inserting or removing a MOSFET IC (or a plug-in module containing a MOSFET). The voltage transients developed when terminals are separated may damage the MOSFET IC. This same precaution is taken when using ICs with conventional transistors. However, the chances of damage are greater with MOSFET ICs, due to their delicate (out-of-circuit) nature.

2-4. LAYOUT OF INTEGRATED CIRCUITS

Because the layout of ICs is such a broad field, no attempt is made here to cover the entire subject. Instead, we shall summarize the problems that apply to all types of ICs.

From a practical standpoint, each type of IC has its own problems. For example, most linear ICs have high gain and are thus subject to oscillation if undesired feedback is not controlled by good circuit layout. However, because of the differential input used by most linear ICs (refer to Chapter 3), the pick up of signal noise is usually not a major problem. On the other hand, digital ICs rarely oscillate due to low gain, but are subject to noise signals. Proper circuit layout can minimize the generation and pick up of such noise. The following paragraphs describe those circuit layout problems that IC users must face at one time or another.

It is assumed that the readers are already familiar with good design practices applicable to all electronic equipment. It should be noted that most ICs are mounted on printed-circuit cards or boards (there are exceptions, of course). All of the problems that apply to discrete components on PC boards also apply to ICs mounted in the same way.

2-4.1 Layout of Digital ICs

All logic circuits are subject to noise. Any circuit, discrete or IC, will produce erroneous results if the noise level is high enough. Therefore, it is recommended that noise and grounding problems be considered from the very beginning of layout design.

Wherever dc distribution lines run an appreciable distance from the supply to a logic chassis (or a PC board), both lines (positive and negative) should be bypassed to ground with a capacitor, at the point at which the wires enter the chassis.

The values for power-line bypass capacitors should be on the order of 1 to 10 μF. If the logic circuits operate at higher speeds (above about 10 MHz), add a 0.01 μF capacitor in parallel with each 1–10-microfarad capacitor. Keep in mind that even though the system may operate at low speeds, there are harmonics generated at higher speeds. The high-frequency signals may produce noise of the power lines and interconnecting wiring. A 0.01-μF capacitor should be able to bypass any harmonics present in most logic systems.

If the digital ICs are particularly sensitive to noise, as is the case with the TTL logic form, extra bypass capacitors (in addition to those at the power and ground entry points) can be used effectively. The additional power- and ground-line capacitors can be mounted at any convenient point on the board or chassis, provided that there is no more than a 7-inch space between any IC and a capacitor (as measured along the power or ground line). Use at least one additional capacitor for each 12 IC packages and possibly as many as one capacitor for each six ICs.

The dc lines and ground return lines should have large enough cross sections to minimize noise pickup and dc voltage drop. Unless otherwise recommended by the IC manufacturer, use AWG No. 20 or larger wire for all digital IC power and ground lines.

In general, all leads should be kept as short as possible, both to reduce noise pickup and to minimize the propagation time down the wire. Typically, present-day logic circuits operate at speeds high enough so that the propagation time down a long wire or cable can be comparable to the delay time through a logic element. This propagation time should be kept in mind during the layout design.

Do not exceed 10 inches of line for each nanosecond of fall time for the fastest logic pulses involved. For example, if the clock pulses (usually the fastest in the system) have a fall time of 3 nanoseconds, no logic signal line (either printed-circuit or conventional wire) should exceed 30 inches. (Pulse fall time is discussed in Chapter 7.)

The problem of noise can be minimized if ground planes are used. (That is, if the circuit board has solid metal sides.) Such ground planes surround the active elements on the board with a noise shield. Any logic system that operates at speeds above approximately 30 MHz should have ground planes. If it is not practical to use boards with built-in ground planes, run a wire around the outside edge of the board. Connect both ends of the wire to a common or "equipment" ground.

Do not run any logic signal line near a clock line for more than about 7 inches because of the possibility of cross-talk in either direction.

If a logic line must be run a long distance, design the circuits so that the long line feeds a single gate (or other logic element) rather than several gates. External loads to be driven must be kept within the current and voltage limits specified on the IC data sheet.

Some digital IC manufacturers specify that a resistor (typically 1 kΩ) be connected between the gate input and the power supply (or ground, depending upon the type of logic), where long lines are involved. Always check the IC data sheet for such notes.

The general rules for digital IC layout are given in Fig. 2-16.

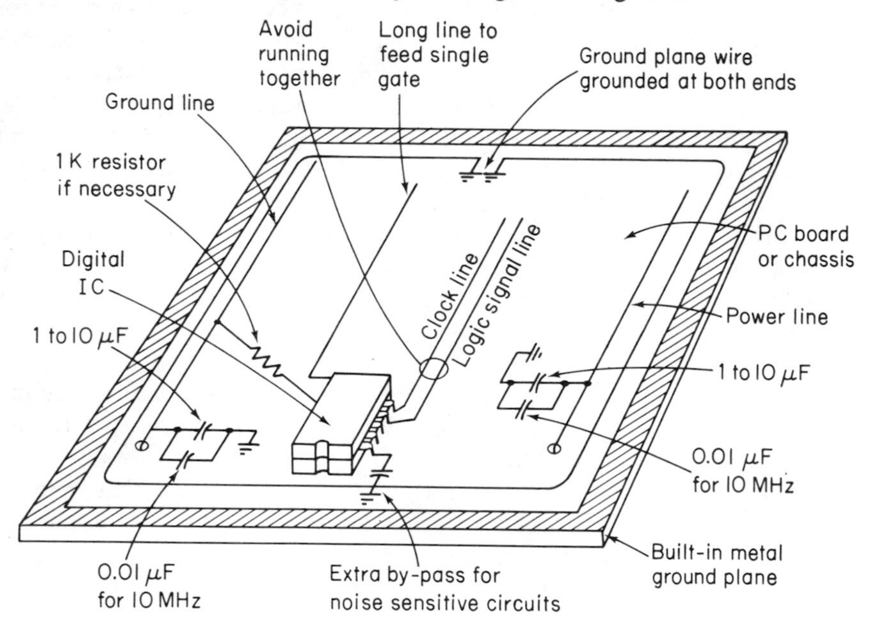

FIG. 2-16. General rules for digital IC layout.

2-4.2 Layout of linear ICs

The main problem with layout of linear ICs is undesired oscillation due to feedback. Virtually all of today's IC operational amplifiers are capable of producing high gain at high frequencies. Since the ICs are physically small, the input and output terminals are close, creating the ideal conditions for undesired feedback. To make matters worse, most linear ICs are capable of passing frequencies higher than those specified on the data sheet.

For example, an operational amplifier to be used in the audio range (say up to 20 kHz with a power gain of 20 dB) could possibly pass a 10-MHz signal with some slight gain. This higher-frequency signal could be a harmonic of signals in the normal operating range and, with sufficient gain, could feed back to the input and produce undesired oscillation.

In laying out any linear IC, particularly in the breadboard or experimental stage, always consider the circuit as being radio-frequency (RF), even though the IC is not supposed to be capable of RF operation, and the circuit is not normally used with RF.

As is discussed in Sec. 2-7 of this chapter, all linear IC power-supply terminals should be bypassed to ground. This provides a path for any RF. The layout should be such that the capacitors are as near to the IC terminals as possible. Do not mount the capacitors at the power-supply end of the line.

Note that the lead between the IC terminal and the capacitors has some inductance. This inductance can combine with the capacitor to form a resonant circuit. If the circuit resonates at some frequency (including fundamental, harmonic, or sub-harmonic), the signal could be passed by the IC to produce oscillation.

Keep IC input and output leads as short as practical. Use shielded leads wherever practical. Use one common tie point near the IC for all grounds. Resonant circuits can also be formed by poor grounding or by ground loops in general.

As a general rule, ICs mounted on PC boards (particularly with ground planes) tend to oscillate less than when conventional wiring is used. For that reason, an IC may oscillate in the breadboard stage, but not when mounted in final layout form.

Once all of the leads have been connected to an IC and power is applied, monitor all IC terminals for oscillation with an oscilloscope before signals are applied. Except in a few rare cases, there should be no evidence of ac or RF signals at any IC terminal under no-signal conditions. Of course, there can be power-line hum, noise, etc., that is not the fault of the IC or the layout. (IC testing procedures are discussed in Chapter 7.)

The general rules for linear IC layout are given in Fig. 2-17.

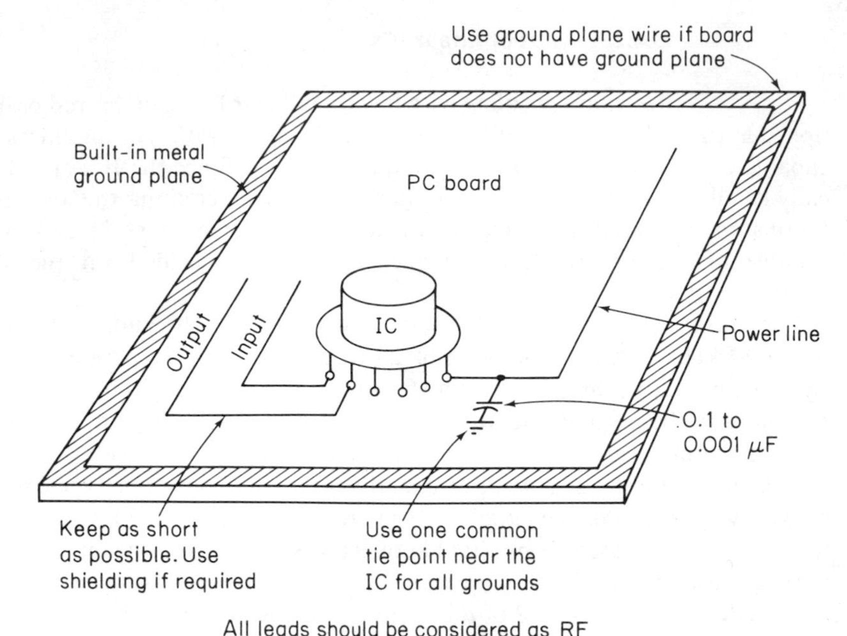

FIG. 2-17. General rules for linear IC layout.

2-5. POWER DISSIPATION PROBLEMS IN INTEGRATED CIRCUITS

As discussed in Sec. 1-5, the basic rules for ICs regarding power dissipation and thermal considerations are essentially the same as those for discrete transistor circuits.

The maximum allowable power dissipation (usually specified as P_d or P_D, or "maximum device dissipation" on IC data sheets) is a function of the maximum storage temperature T_S, the maximum ambient temperature T_A, and the thermal resistance from the semiconductor chip (or pellet) to case θ_{PC}. The basic relationship is:

$$P_D = \frac{T_S - T_A}{\theta_{PC}}$$

All IC datasheets do not necessarily list all of these parameters. It is quite common to list only the maximum power dissipation for a given ambient temperature and then show an "derating" factor in terms of maximum power decrease for a given increase in temperature.

For example, a typical IC might show a maximum power dissipation of 110 mW at 25 °C, with a derating factor of 1 mW/°C. If such an IC is operated at 100 °C, the maximum power dissipation is: 100 − 25, or 75 °C increase; 110 − 75 = 35 mW.

In the absence of specific datasheet information, the following typical temperature characteristics can be applied to the basic IC package types:

No IC should have a temperature in excess of 200 °C.

Ceramic flat pack:
thermal resistance = 140 °C per watt
maximum storage temperature = 175 °C
maximum ambient temperature = 125 °C

TO-5 style package:
thermal resistance = 140 °C per watt
maximum storage temperature = 200 °C
maximum ambient temperature = 125 °C

Dual-in-line (ceramic):
thermal resistance = 70 °C per watt
maximum storage temperature = 175 °C
maximum ambient temperature 125 °C

Dual-in-line (plastic):
thermal resistance 150 °C per watt
maximum storage temperature = 85 °C
maximum ambient temperature = 75 °C

2-5.1 Working with power ICs

Until recently, most ICs were low power (typically less than 1 watt). This is still essentially true today. However, there is an increasing use of power ICs. For example, there are commercial ICs that offer a complete, packaged audio amplifier circuit capable of 100 watts output. Power ICs usually use the TO-5 style package, or some similar enclosure. In any event, the package is metal and is used with some type of heat sink (either an external heat sink or the metal chassis).

The datasheets for power ICs usually list sufficient information to select the proper heat sink. Also, the datasheets or other literature often make recommendations as to mounting for the power IC. Always follow the IC manufacturer's recommendations. In the absence of such data and to make the reader more familiar with the terms used, the following sections summarize considerations for power ICs.

2-5.2 Maximum power dissipation

From a user's standpoint, an IC is a complete, pre-designed, functioning circuit that cannot be altered in regard to power dissipation.

That is, if the power-supply voltages, input signals, output loads, and ambient temperature are at their recommended levels, the power dissipation will be well within the capabilities of the IC. With the possible exception of the data required to select or design heat sinks, the user need only follow the datasheet recommendations. Of course, if the IC must be operated at a temperature higher than the rated ambient, the power dissipation must be derated as previously described (Sec. 2-5).

2-5.3 Thermal resistance

ICs designed for power applications usually have some form of thermal resistance specified to indicate the power dissipation capability, rather than a simple "maximum device dissipation." Thermal resistance can be defined as the increase in temperature of the semiconductor material (transistor junctions), with regard to some reference, divided by the power dissipated, or °C/W.

Power IC datasheets often specify thermal resistance at a given temperature. For each increase in temperature from this specified value, there will be a change in the temperature-dependent characteristics of the IC. Since there is a change in temperature with changes in power dissipation, the semiconductor chip (or transistor junction) temperature also changes, resulting in a characteristic change. Thus, IC characteristics can change with ambient temperature changes and with changes produced by variation in power dissipation. As is discussed in Sec. 2-5.4, most ICs incorporate circuits to offset the effects of temperature.

In power ICs, thermal resistance is normally measured from the semiconductor chip (or pellet) to the case. This results in the term θ_{PC}. On those ICs whose cases are bolted directly to their mounting surfaces with an integral threaded bolt or stud, the terms θ_{MB} (thermal resistance to mounting base) or θ_{MF} (thermal resistance to mounting flange) are used. These terms take into consideration only the thermal paths from junction to case (or mount).

2-5.4 Thermal runaway

The main problem in operating any semiconductor device (transistor or IC) at or near its maximum power dissipation limits is a condition known as *thermal runaway*. When current passes through a transistor junction, heat will be generated. If all of this heat is not dissipated by the case (an impossibility), the junction temperature will increase. This, in turn, causes more current to flow through the junction, even though the voltage and other circuit values, remain the same. In turn, this causes the junction temperature to increase even further, with a corresponding increase in current flow. If the heat is not dissipated by some means, the transistor will burn out and be destroyed.

Most ICs are designed to prevent thermal runaway. The usual arrangement is to place a diode in the reverse bias circuit for one or more transistors in the IC. The diode is fabricated on the same semiconductor chip as the transistors and thus has the same temperature characteristics. The circuit is arranged so that the reverse bias is increased (by a decrease in diode resistance) when there is an increase in temperature. When temperature increases because of an increase in transistor current (or vice versa), the diode resistance changes and increases the reverse bias. This offsets the initial change in current due to temperature changes.

Many different temperature compensation circuits have been developed by IC manufacturers. However, the IC user need not be concerned as long as the datasheet limits are observed, since the circuit is already designed into the IC.

2-5.5 Operating ICs without heat sinks

If an IC is not mounted on a heat sink (as is typical for about 90 percent of the ICs in use today), the thermal resistance from case-to-ambient is so large in relation to that from junction-to-case (or mount) that the total thermal resistance from junction-to-ambient air is primarily the result of the case-to-ambient term.

The thermal resistance for a TO-5 style case is 140 °C. Assuming that the ambient temperature is 25 °C and that the maximum temperature for any IC is 200 °C, the maximum permissible temperature rise is 175 °C (200 − 25 = 175). Thus, the maximum theoretical power dissipation for any TO-5 style IC is:

$$\text{Maximum power dissipation} = \frac{\text{maximum permissible temperature increase}}{\text{thermal resistance}}$$

or

$$= \frac{175}{140} = 1.25\ \text{W}$$

In practice, this would be an absolute maximum figure. Rarely, if ever, are ICs operated above 1 W without heat sinks.

2-5.6 Operating ICs with heat sinks

After about 1 W (or less) it becomes impractical to increase the size of the case to make the case-to-ambient thermal resistance term comparable to the junction-to-case term. For this reason, most power ICs are designed for use with an external heat sink. Sometimes the chassis or mounting area serves as the heat sink. In other cases, a heat sink is attached to the case. Either way, the primary purpose of the heat sink is to increase the effective heat-dissipation area of the case and provide a low heat-resistance path from case to ambient.

To properly design (or select) a heat sink for a given application, the thermal resistance of both the IC and heat sink must be known. For this reason, power IC datasheets usually specify the junction-to-ambient thermal resistance, which must be combined with the heat-sink thermal resistance to find the total power dissipation capability. Note that some power IC datasheets specify a "maximum case temperature," rather than thermal resistance. As is discussed in Sec. 2-5.7, maximum case temperature can be combined with heat-sink thermal resistance to find maximum power dissipation.

Heat-sink ratings. Commercial fin-type heat sinks can be used with TO-5 style ICs. Such heat sinks are especially useful when the ICs are mounted in Teflon sockets, which provide no thermal conduction to the chassis or printed-circuit board. Commercial heat sinks are rated by the manufacturer in terms of thermal resistance, usually in terms of °C/W. When heat sinks involve the use of washers, the °C/W factor usually includes the thermal resistance between the IC case and sink. Without a washer, only the sink-to-ambient thermal resistance is given (usually as θ_{SA}). In either case, the thermal resistance factor represents temperature increase (in °C) divided by wattage dissipated. For example, if the heat sink temperature rises from 25 to 100 °C (a 75 °C increase) when 25 W are dissipated, the thermal resistance is $75/25 = 3$. This can be listed on the datasheet as an θ_{SA} of 3, or simply as 3 °C/W.

All other factors being equal, the heat sink with the lowest thermal resistance (°C/W) is best. That is, a heat sink with 1 °C/W is better than a 3 °C/W heat sink. Of course, the heat sink must fit the IC case and the space around the IC. Except for these factors, selection of a suitable heat sink should be no particular problem.

Calculating heat sink capabilities. The thermal resistance of a heat sink can be calculated if the following factors are known: material, mounting provisions, exact dimensions, shape, thickness, surface finish, and color. Even if all of these factors are known, the thermal resistance calculations are still only approximate.

As a very approximate rule-of-thumb:

$$\text{Heat-sink thermal resistance (in °C/W)} = \sqrt{\frac{1500}{\text{area}}}$$

where the area (total area exposed to the air) is in square inches, material is $\frac{1}{8}$-inch thick aluminum, and the shape is that of a flat disc.

From a practical design standpoint, it is better to accept the manufacturer's specification for a heat sink. The heat-sink thermal resistance actually consists of two series elements: the thermal resistance from the case to the heat sink that results from conduction (case-to-sink) and the thermal resistance from the heat sink to the ambient air, caused by convection and radiation.

Practical heat sink considerations. To operate an IC at its full power capabilities, there should be no temperature difference between the case and ambient air. This occurs only when the thermal resistance of the heat sink is zero, and the only thermal resistance is that between the junction and case. It is not possible to manufacture a heat sink with zero resistance. However, the greater the ratio of junction-to-case/case-to-ambient, the more closely the maximum power limit can be approached.

When ICs are to be mounted on heat sinks, some form of electrical insulation is usually required between the case and heat sink. As is discussed in Sec. 2-7, many IC cases are not at electrical ground; instead, they can be connected to some other point (above or below ground) in the internal circuit.

Because good electrical insulators usually are also good thermal insulators, it is difficult to provide electrical insulation without introducing some thermal resistance between case and heat sink. The best materials for this application are mica, beryllium oxide (Beryllia), and anodized aluminum, with typical °C/W ratings of 0.4, 0.25, and 0.35, respectively.

The use of a zinc-oxide-filled silicon compound (such as Dow Corning #340 or Wakefield #120) between the washer and chassis, together with a moderate amount of pressure from the top of the IC helps to decrease thermal resistance.

If the IC is mounted within a fin-type heat sink, an insulated cap (such as Beryllia) should be used between the case and heat sink. Figure 2-18 shows both methods of mounting ICs and heat sinks.

When a washer is added between the IC case and heat sink, a certain amount of capacitance is introduced. In general, this capacitance will have no effect on operation of ICs unless the frequency is about 100 MHz. Rarely, if ever, do power ICs ever operate above the audio range. Thus, few problems should be encountered.

2-5.7 Calculating power dissipation for ICs with heat sinks

The maximum dissipation capability of an IC used with a heat sink is dependent upon three factors: (1) the sum of the series thermal resistances from the transistor junctions (semiconductor chip) to ambient air, (2) the maximum junction temperature, and (3) the ambient temperature.

The following are some examples of how power dissipation can be calculated.

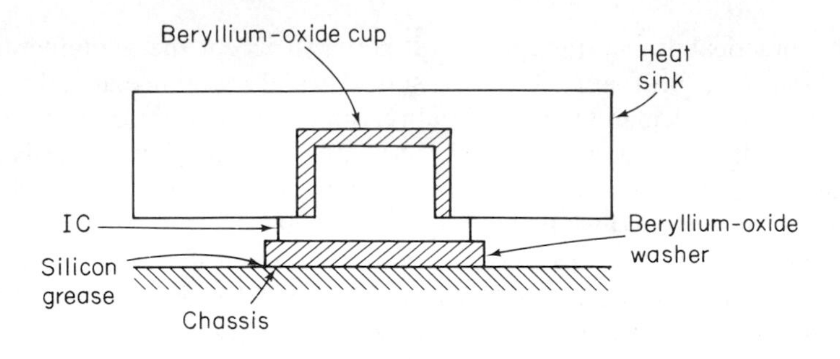

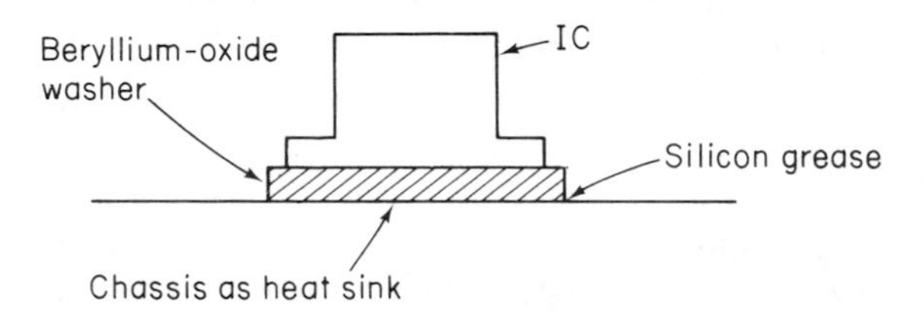

FIG. 2-18. Typical mounting arrangements for IC heat sinks.

Assume that it is desired to find the maximum power dissipation of an IC/heat-sink combination. The following conditions are specified: a maximum junction temperature of 200 °C (typical for any silicon semiconductor), a junction-to-case thermal resistance of 2 °C/W for the IC, and a heat sink with a thermal resistance of 3 °C/W and an ambient temperature of 25 °C.

Note that the thermal resistance of the heat sink includes any thermal resistance produced by the washer between the IC case and heat sink. (If this factor is not known, add a rule-of-thumb factor of 0.5 °C/W thermal resistance for any washer between case and heat sink.

First, find the total junction-to-ambient thermal resistance:

$$\text{junction-to-case (2)} + \text{sink-to-ambient (3)} = \text{junction-to-ambient (5)}$$

Next, find the maximum permitted power dissipation:

$$\frac{\text{maximum junction temperature (200)} - \text{ambient (25)}}{\text{junction-to-ambient (5)}}$$

$$= \frac{200 - 25}{5} = 35 \text{ W (maximum)}$$

If "maximum case temperature" is specified instead of maximum junction temperature, the calculations for maximum power dissipation are as follows: Assume that a maximum case temperature of 130 °C is specified instead of a 200 °C maximum junction temperature.

First, subtract the ambient temperature from the maximum permitted case temperature:

$$130° - 25° = 105\ °C$$

Then, divide the case temperature by the heat-sink thermal resistance:

$$105/3 = 35 \text{ W maximum power}$$

2-6. EFFECTS OF TEMPERATURE EXTREMES ON INTEGRATED CIRCUITS

The effects of temperature extremes (either high or low) will vary with the type of circuit involved, case style, and fabrication techniques of the manufacturer. Thus, no fixed rules can be made. However, the following general rules can be applied to most ICs.

In some instances, the IC will fail to operate at temperature extremes, but will return to normal when the operating temperature is returned to the "normal" range.

In other cases, the IC will fail to operate properly once it has been subjected to a temperature extreme. In effect, the IC is destroyed once it is operated or stored at an extreme temperature.

The effects of high-temperature extremes are generally worse than those of low temperatures. This is primarily because high temperatures can cause thermal runaway (see Sec. 2-5.4).

In general, high temperatures cause the IC characteristics to change. An increased operating temperature also produces increased leakage currents, increased sensitivity to noise, increased unbalance in balanced circuits, increased "switching spikes" or transient voltages for transistors in digital ICs, and an increase in the ever present possibility of burn out.

If the power-supply voltages, input signals, output loads, and ambient temperatures specified on the datasheet are observed, there should be no danger of temporary failure (or total destruction) for any IC. However, as a final check, multiply the rated thermal resistance by the maximum device dissipation and then add the actual ambient temperature. If the result is less than 200, the IC should be safe.

For example, the typical thermal resistance of a dual-in-line plastic IC is 150 (refer to Sec. 2-5). Assume that the IC has a maximum device dissipation of 600 mW, and that the actual ambient temperature is 50 °C. Under these conditions:

$$150 \times 0.6 = 90; 90 + 50 = 140.$$

Since 140 (which would be the junction temperature of the transistors within the IC) is less than 200, the IC should be safe. In practice, the ambient temperature could probably go up to about 100 or 110 °C without permanent damage, but this would be approaching the danger point.

When an IC is operated at its low temperature extreme, the IC is likely to "underperform." Usually, the IC will not be destroyed by extremely low temperature (except where the low temperature is prolonged). However, the IC will simply not perform as specified. For example, at low temperatures gain and power output will be different for operational amplifiers and other linear ICs; operating speed will be reduced for digital ICs; and the drive or output load capabilities of digital ICs will be reduced.

In the absence of any specific information from the manufacturer, the following rule can be applied to derating an IC when operated at low temperatures: Derate the characteristic by 1 percent for each degree below the rated low operating temperature limit. Make certain to use the low operating temperature limit, not the low storage temperature limit.

For example, if the rated operating low temperature limit is −5 °C and the IC operates at −25 °C, derate all performance characteristics by 20 percent (25 − 5 = 20). In the case of digital ICs, allow for a 20 percent reduction in drive or output load capability as well as circuit speed.

In no event should the IC be operated below the rated storage temperature. As a general rule, the low storage temperature limit is 10 to 20 °C below the operating limit. Always consult the datasheet for operating and storage temperature limits.

2-7. POWER SUPPLIES FOR INTEGRATED CIRCUITS

As a general rule, a linear IC requires connection to both a positive and negative power supply. This is because most linear ICs use one or more differential amplifiers in their circuits. Digital ICs, on the other hand, rarely require two power supplies. Generally, digital ICs require only one *positive* power supply (except for ECL).

When two power supplies are required for a linear IC, the supplies are usually equal or symmetrical (such as +9 V and −9 V, +20 V and −20 V). This is the case with the IC of Fig. 2-19, which normally operates with +18 V and −18 V. A few linear ICs use unsymmetrical power supplies (+9 V and −4.5 V), and there are linear ICs that require only a single supply. However, such cases are the exception. And, in some cases, it is possible to operate a linear IC that normally requires two supplies from a single supply by means of special circuits (external to the IC). Such circuits are discussed in Sec. 2-7.6.

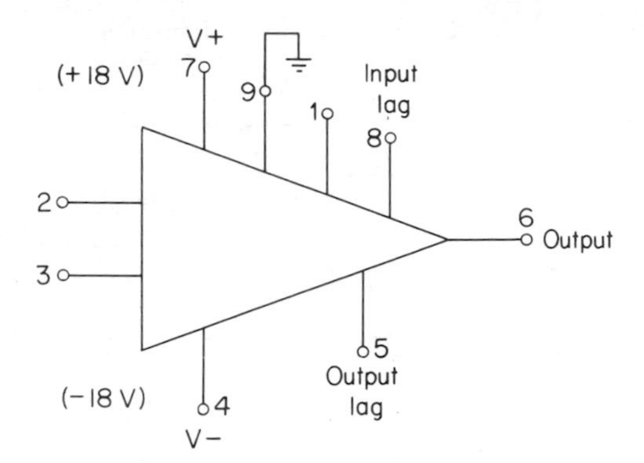

FIG. 2-19. Typical linear IC op-amp operating with symmetrical 18-V power supplies.

2-7.1 Labeling of IC power supplies

Unlike most discrete transistor circuits in which it is usual to label one power supply lead positive and the other negative without specifying which (if either) is common to ground, it is necessary that all IC power-supply voltages be referenced to a common or ground (which may or may not be physical or equipment ground).

As in the case of discrete transistors, manufacturers do not agree on power-supply labeling for ICs. For example, the circuit of Fig. 2-19 uses V+ to indicate the positive voltage and V− to indicate the negative voltage. Another manufacturer might use the symbols V_{EE} and V_{CC} to represent negative and positive, respectively. As a result, the IC datasheet must be studied carefully *before applying* any power source.

2-7.2 Typical IC power-supply connections

Figure 2-20 shows typical power-supply connections for both linear and digital ICs. The digital IC shown requires only one 5.2-V supply, with the positive connected to V_{CC} and the negative connected to V_{EE}. Keep in mind that either V_{CC} or V_{EE} could be at physical (equipment) ground. For example, if V_{CC} is at the physical ground, V_{EE} will still be negative, but will be 5.2 V below ground. Likewise, if V_{EE} is at physical ground, V_{CC} is positive and is 5.2 V above ground.

The protective diodes shown are recommended for any power-supply circuit in which the leads could be accidently reversed. The diodes permit current flow only in the appropriate direction.

The linear IC of Fig. 2-20 requires two power sources (of 18 V each) with the positive lead of one and the negative lead of the other tied to ground or common.

The two capacitors shown in Fig. 2-20 provide for decoupling of the power supply (signal bypass). Usually, disc ceramic capacitors are used.

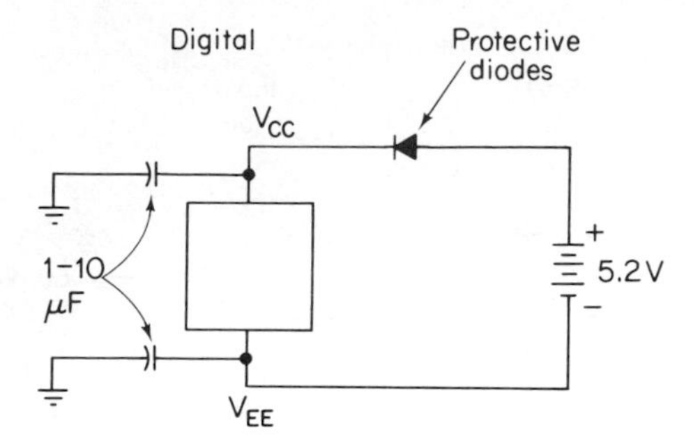

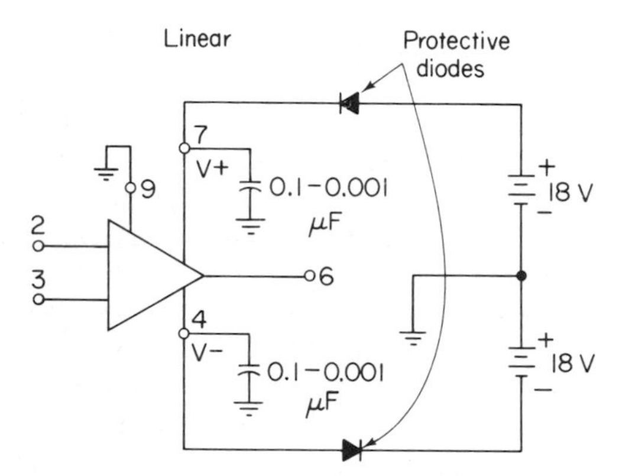

FIG. 2-20. Typical power supply connections for digital and linear ICs.

The capacitors should always be connected as close to the IC terminals as is practical, not at the power-supply terminals. A guideline for linear IC power-supply decoupling capacitors is to use values between 0.1 and 0.001 μF.

In addition to the capacitors shown in Fig. 2-20, the ICs (both linear and digital) may require additional capacitors on the power lines. (Refer to Sec. 2-4.1.)

2-7.3 Grounding metal IC cases

The metal case of the linear IC shown in Fig. 2-20 is connected to terminal 9 *and to no other point in the internal circuit.* Thus, terminal 9 can and should be connected to equipment ground, as well as to the common or ground of the two power supplies.

The metal case of some ICs, both digital and linear, may be connected to a point in the internal circuit. If so, the case will be at the same voltage as the point of contact. For example, the case might be connected to pin 4 of the IC shown in Figs. 2-19 and 2-20. If so, the case would be below ground (or "hot") by 18 V. If the case were mounted directly on a metal chassis that was at ground, the IC and power supply would be damaged. Of course, not all ICs have metal cases; likewise, not all metal cases are connected to the internal circuits. However, this point must be considered *before* using a particular IC.

2-7.4 Calculating current required for linear ICs

The datasheets for linear ICs usually specify a nominal operating voltage (and possibly a maximum operating voltage), as well as a "total device dissipation." These figures can be used to calculate the current required for a particular IC. Use simple dc Ohm's law and divide the power by the voltage to find the current. However, certain points must be considered.

First, use the actual voltage applied to the IC. The actual voltage should be equal to the nominal operating voltage, but in no event higher than the maximum voltage.

Second, use the total device dissipation. The datasheet may also list other power dissipations, such as "device dissipation," which is defined as the dc power dissipated by the IC itself (with output at zero and no load). The other dissipation figures will always be smaller than the total power dissipation.

2-7.5 Calculating power required for digital ICs

Digital ICs usually operate with pulses. Thus, current is maximum in either of two states (0 or 1, high or low, etc.), but not in both states. Most digital IC datasheets list the current drain for the maximum (or "worst-case") condition. As an example, I_{PDL} indicates the current drain when the logic signals (pulses) are low. If only I_{PDL} is listed for a digital IC, it can be reasonably assumed that the "low state" produces maximum current drain. I_{PDH} indicates the "high-state" current drain and, with no other listing on the datasheet, should be the maximum current drain. If both I_{PDL} and I_{PDH} are listed, it is obvious that the higher of the two indicates the maximum current drain state.

When the current drains for both states are listed, some manufacturers recommend that the current drains be averaged to calculate power. For example, if the I_{PDL} is 10 mA and the I_{PDH} is 50 mA, add the two currents together and divide by 2 for an average current of 30 mA ($10 + 50 = 60$; $60/2 = 30$).

The current requirements for digital ICs are also affected by the operating speed of the logic circuits and the type of loads into which the IC must operate.

A digital IC will require more current as the operating speed is increased. Generally, the datasheet will list a "nominal operating speed" and a "maximum operating speed," together with the current drain at the "nominal speed" figure. Of course, the IC should never be operated beyond the maximum limit. When operating between the nominal and maximum speeds, the additional current can be approximated by adding 0.5 to 1 mA for each 1 MHz of speed increase. For example, assume that the nominal speed is 15 MHz with a maximum speed of 25 MHz, the nominal or average current drain is 30 mA, and the IC is operated at the maximum speed. Then: $25 - 15 = 10$ MHz; $10 \times 0.5 = 5$ mA; $10 \times 1 = 10$ mA; $30 + 5 = 35$; $30 + 10 = 40$, and the IC will require between 35 and 40 mA.

Capacitive loads generally cause a digital IC to draw more current than pure resistive loads. However, it is not practical to judge the direct effect of a capacitive load on an IC, so no general rules are given. The problem of loads and drives for digital ICs are discussed more thoroughly in Chapter 6.

2-7.6 Power-supply tolerances for ICs

Typically, digital ICs power supplies must be kept within ± 5 to ± 10 percent, whereas linear ICs will generally operate satisfactorily with ± 20 percent power sources. These tolerances apply to actual operating voltage, not to maximum voltage limits. The currents (or power consumed) will vary proportionately.

Power-supply ripple and regulation are both important. Generally, solid-state power supplies with filtering and full feedback regulation are recommended. As in the case of discrete transistors, ripple (and any other power-supply noise) must be kept to a minimum for noise-sensitive circuits (such as TTL digital ICs and high-gain linear ICs). Ideally, ripple (and all other noise) should be 1 percent, or less.

The effects of operating ICs beyond voltage tolerance are essentially the same as those experienced when the IC is operated at temperature extremes. That is, a high power-supply voltage will cause the IC to "overperform," whereas low voltages will result in "underperformance." A low voltage will usually not result in damage to the IC, as is the case with operating an IC beyond the maximum rated voltage.

2-7.7 Operating ICs from a single power supply

A linear IC operational amplifier is generally designed to operate from symmetrical positive and negative power-supply voltages. This results in a high common-mode rejection capability, as well as good low-frequency operation (typically a few Hz down to dc). If the loss of very low frequency operation can be tolerated, it is possible to operate linear IC op-amps from a single power supply, even though designed for dual supplies. Except for the low frequency loss, the other IC operating characteristics should be unaffected.

The following notes describe a technique that can be used with most IC op-amps to permit operation from a single power supply, with a minimum of design compromise. The same maximum device ratings that appear on the datasheet are applicable to the IC when operating from a single polarity power supply and must be observed for normal operation. Likewise, all of the considerations discussed thus far in this chapter apply to single supply operation.

For example, power-supply decoupling capacitors are still required. The importance of decoupling capacitors, whether with single-supply or dual-supply operation, cannot be overemphasized. Today's IC op-amps are high-gain, high-frequency devices. Stray signals coupled back through the power supply can create instability problems. The decoupling capacitors should be placed as close as possible physically to the device to minimize the effects of the inductance of the power-supply leads. Circuit interconnections should be laid out in such a manner that the lead lengths are short enough to minimize pickup.

The technique described here is generally referred to as the "split Zener" method. The main concern in setting up for single-supply operation is to maintain the relative voltage levels. With an IC designed for dual supply operation, there are three reference levels: +V, 0, and −V. For example, if the datasheet calls for plus and minus 10 V supplies, the three reference levels are: +10 V, 0 V, and −10 V.

For single-supply operation, these same reference levels can be maintained by using ++V, +V, and ground (that is +20 V, +10 V and 0 V), where ++V represents a voltage level double that of +V. This is illustrated in Fig. 2-21 where the IC is connected in the split Zener mode. Note that there is no change in the *relative* voltage levels even though the various IC terminals are at different voltage levels (with reference to ground). Terminal 4 (normally connected to the −10 V supply) is at ground. Pin 3 (normally at ground or common) is set at one-half the total Zener voltage (+10 V). Pin 6 (normally connected to the +10 V supply) is set at the full Zener voltage (+20 V).

With single supply, the differential input terminals (1 and 2), which are normally at ground in a dual-supply system, must also be raised up one-half the Zener voltage (+10 V). Under these circumstances, the output terminal (5) will also be at one-half the Zener voltage, plus or minus an offset voltage error due to input offset voltage, input offset current, and impedance unbalance. (Refer to Chapter 3 for information on these op-amp characteristics.)

To minimize offset errors due to unequal voltage drops caused by the input bias currents across unequal resistances, it is recommended that the value of the input offset resistance R_4 be equal to the parallel combination of R_2 and R_3. This is in keeping with the standard op-amp practice as discussed in Chapter 3.

As with any op-amp, the deviation between absolute Zener level will also contribute to an error in the output voltage level. Typically, this is on the order of 50 to 100 μV per volt of deviation of Zener level. Except in rare cases, this deviation should be of little concern.

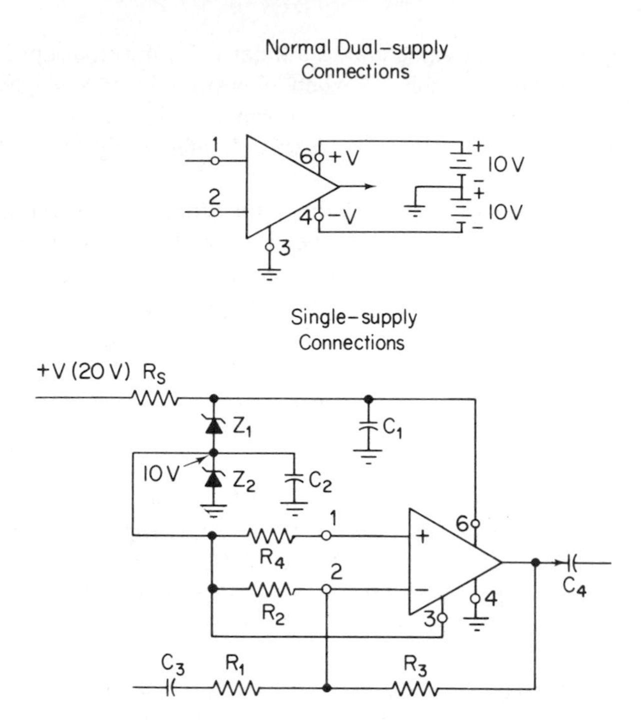

FIG. 2-21. Connections for single power supply operation (with ground reference).

Note that the IC of Fig. 2-21 has a ground reference terminal (terminal 3). Not all IC op-amps have such terminals. Some ICs have only +V and −V terminals or leads even though the two levels are referenced to a common ground. That is, there is no physical ground terminal or lead on the IC, only +V and −V terminals. Figure 2-22 shows the split Zener connections for single supply operation with such ICs. Here, the input terminals (A and B) are set at one-half the total Zener supply voltage; the −V terminal is set at ground; and the +V terminal is at the full Zener voltage (+20 V).

Figures 2-21 and 2-22 both show connection to positive power supplies. Negative power supplies can also be used. With a negative supply, the +V terminal is connected to ground, the −V terminal is connected to full Zener supply (−20 V), with the input terminals and IC ground terminal (if any) connected to one-half the Zener supply. Of course, the polarity of the Zener diodes must be reversed.

Figures 2-21 and 2-22 both show a series resistance R_S for the Zener diodes. This is standard practice for Zener operation. The approximate or trial value for R_S is found by:

$$\frac{(\text{maximum supply voltage} - \text{total Zener voltage})^2}{\text{safe power dissipation of Zeners}}$$

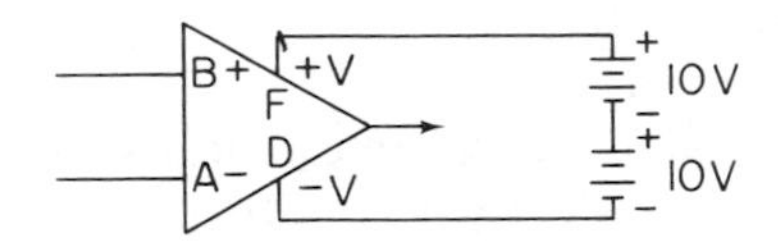

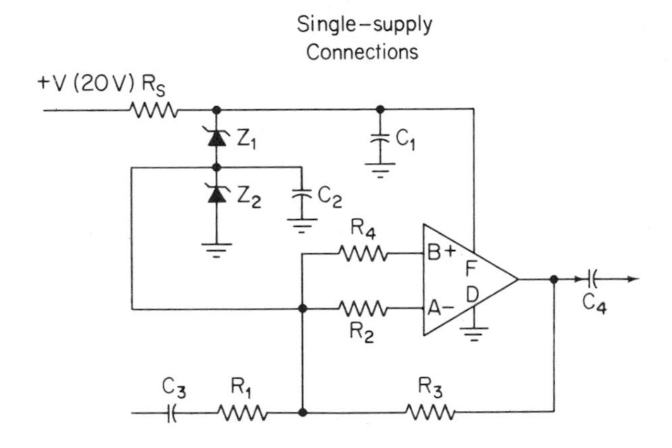

FIG. 2-22. Connections for single supply operation (without ground reference).

For example, assume that the total Zener voltage is 20 V (10 V for each Zener), that the supply voltage may go as high as 24 V, and that 2-W Zeners are used. Under these conditions:

$$\frac{(24-20)^2}{2} = \frac{(4)^2}{2} = \frac{16}{2} = 8\ \Omega \text{ for } R_S$$

2-7.7.1 Effects on circuit operation using a single supply

From a user's standpoint, operation of an IC with a single supply is essentially the same as with the conventional dual power supply. The following notes describe the basic differences in operational characteristics of the IC with both types of power supplies. It is recommended that those readers not already familiar with basic op-amp theory study Chapter 3.

The normal IC frequency compensation techniques are the same for both types of supplies. The high-frequency limits are essentially the same. However, the low-frequency limit of an IC with a single supply is set by the values of capacitors C_3 and C_4. These capacitors are not required for dual supply operation. Capacitors C_3 and C_4 are required for single-supply operation since both the input and output of the IC are at a voltage level equal to one-half the total Zener voltage (or 10 V using our example). Thus, the IC op-amp cannot be used as a dc amplifier with the single-supply system. In a dual-supply system, the inputs and output are at 0 V.

The closed-loop gain is the same for both types of supplies and is determined by the ratio of R_3/R_1.

The values of decoupling capacitors C_1 and C_2 are essentially the same for both types of supplies. However, it may be necessary to use slightly larger values with the single-supply system, since the impedance of the Zeners is probably different than that of the power supply (without Zeners).

The value of R_2 should be between 50 and 100 kΩ for a typical IC op-amp. Values of R_2 much higher or lower than these limits can result in decreased gain or in an abnormal frequency response. From a practical standpoint, choose trial values using the guidelines and then run gain and frequency response tests.

The value of R_4, the input offset resistance, is chosen to minimize offset error due to impedance unbalance. As an approximate trial value, the resistance of R_4 should be equal to the parallel combination of R_2 and R_3. That is, R_4 is approximately equal to $R_2R_3/R_2 + R_3$.

3. THE BASIC IC OPERATIONAL AMPLIFIER

The designation *operational amplifier* or *op-amp* was originally used for a series of high-performance dc amplifiers that formed a basic part of analog computers. These amplifiers were used to perform mathematical operations applicable to analog computation (summation, scaling, subtraction, integration, etc.). Today, the availability of inexpensive IC amplifiers has made the packaged operational amplifier useful as a replacement for any low-frequency amplifier.

In this chapter, we shall discuss the basic IC operational amplifier. (The many applications for op-amps are discussed in Chapter 4.) Here, we shall concentrate on how to interpret IC op-amp datasheets, design considerations for frequency response and gain, and a specific design example for a typical IC op-amp.

Most of the basic design information for a particular IC can be obtained from the datasheet. Likewise, a typical IC datasheet may describe a few specific applications for the IC. However, IC datasheets generally have two weak points. First, they do not show how the listed parameters relate to design problems. Second, they do not describe the great variety of applications for which a basic IC operational amplifier can be used.

In any event, it is always necessary to interpret IC datasheets. Each manufacturer has its own system of datasheets. It is impractical to discuss all datasheets here. Instead, we shall discuss the typical information found on IC datasheets and see how this information affects simplified design.

3-1. TYPICAL IC OPERATIONAL AMPLIFIER

IC operational amplifiers generally use several *differential* stages in cascade to provide common mode rejection and high gain. Thus, they generally require both positive and negative power supplies. Since a differential amplifier has two inputs, it provides phase inversion for degenerative feedback and can be connected to provide in-phase or out-of-phase amplification.

A conventional op-amp requires that the output be fed back to the input through a resistance or impedance. The output is fed back to the negative or inverting input so as to produce degenerative feedback (to provide the desired gain and frequency response). As in any amplifier, the signal shifts in phase as it passes from input to output. This phase shift is dependent upon frequency. When the phase shift approaches 180°, it adds to (or cancels out) the 180° feedback phase shift. Thus, the feedback is in phase with the input (or nearly so) and will cause the amplifier to oscillate. This condition of phase shift with increased frequency limits the bandwidth of an op-amp. The condition can be compensated by the addition of a phase shift network (usually an *RC* circuit, but sometimes a single capacitor).

3-1.1 Packages, power sources, and thermal design problems for IC op-amps

Linear IC op-amps are available in the three basic package types (flat-pack, TO-5, and dual in-line) described in Chapter 1. All of the problems concerning mounting, power dissipation, power supplies, etc., covered in Chapter 2 are applicable to IC op-amps.

3-1.2 Typical IC op-amp circuit

The circuit diagram and the equivalent circuit (or symbol), as they appear on the datasheet of a typical IC op-amp, are shown in Fig. 3-1. This Motorola circuit is a three-stage amplifier with the first stage a differential-in, differential-out amplifier designed for high gain, high common mode rejection, and input over-voltage protection. (The input diodes prevent damage to the circuit should the input terminals be accidently connected to the power-supply leads, or to some other undesired voltage source.)

The second stage is a differential-in, single-ended-out amplifier with low gain and high common mode rejection. Common-mode feedback is used from the second stage back to the first stage to further aid in the control of a common-mode input signal. Using these two differential amplifier stages and a common-mode feedback, you can obtain a typical common mode rejection of 110 dB. The third stage is a single-ended amplifier that provides high-gain

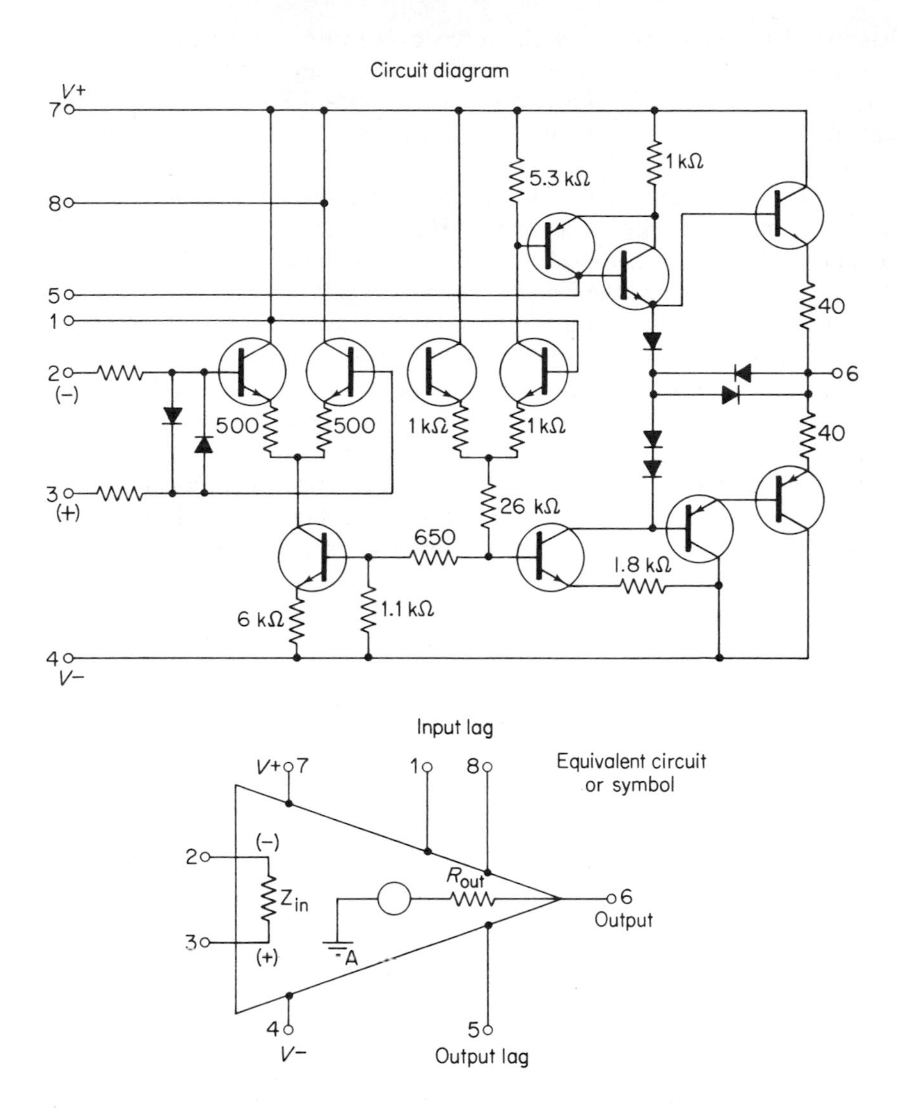

FIG. 3-1. Typical IC op-amp circuit and symbol.

voltage translation to a ground reference, output current drive capabilities, and output short-circuit protection.

3-2. DESIGN CONSIDERATIONS FOR FREQUENCY RESPONSE AND GAIN

Most of the design problems encountered using IC operational amplifiers are the result of tradeoffs between gain and frequency response (or bandwidth). The open-loop (without feedback) gain and frequency response are characteristics of the basic IC package, but they can be modified with external *phase compensation* networks. The closed-loop (with feedback) gain and frequency response are primarily dependent upon *external feedback* components.

The two basic operational amplifier configurations, inverting feedback and non-inverting feedback, are shown in Figs. 3-2 and 3-3, respectively. The equations shown on Figs. 3-2 and 3-3 are classic guidelines. The equations do not take into account the fact that open-loop gain is not infinitely high and the output or load impedance is not infinitely low. Thus, the equations contain built-in inaccuracies and *must be used as guides only*.

When *loop gain* is large, the inaccuracies in the equations will decrease. Loop gain is defined as the ratio of open-loop gain to closed-loop gain, as shown in Fig. 3-4.

The relationships in Fig. 3-4 are based on a theoretical operational amplifier. That is, the open-loop gain rolls off at 6 dB per octave, or 20 dB per decade. (The term 6 dB/octave means that the gain drops by 6 dB each time frequency

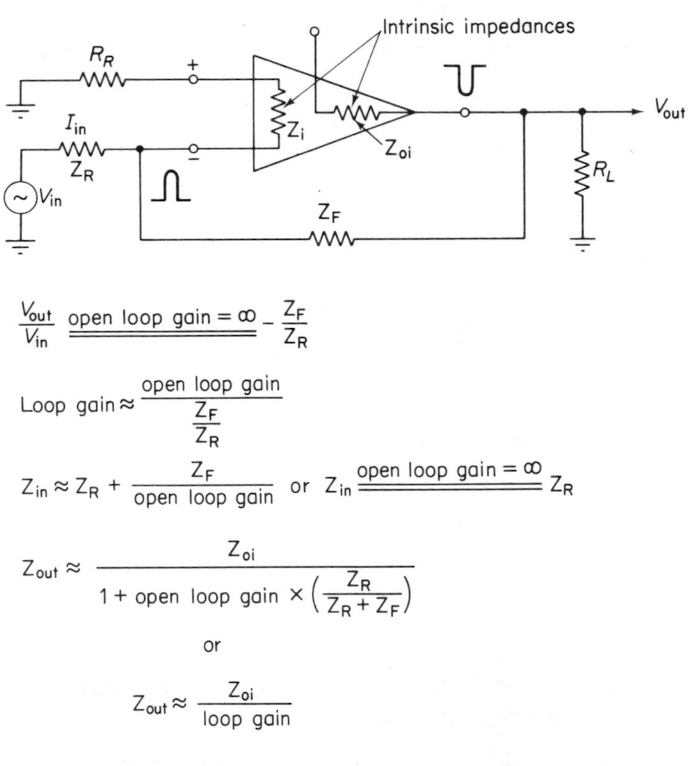

Intrinsic impedances
R_R
+
−
I_{in}
Z_R
Z_i
Z_{oi}
V_{in}
Z_F
R_L
V_{out}
$\frac{V_{out}}{V_{in}} \overset{\text{open loop gain} = \infty}{=\!=\!=} -\frac{Z_F}{Z_R}$
$\text{Loop gain} \approx \frac{\text{open loop gain}}{\frac{Z_F}{Z_R}}$
$Z_{in} \approx Z_R + \frac{Z_F}{\text{open loop gain}}$ or $Z_{in} \overset{\text{open loop gain} = \infty}{=\!=\!=} Z_R$
$Z_{out} \approx \frac{Z_{oi}}{1 + \text{open loop gain} \times \left(\frac{Z_R}{Z_R + Z_F}\right)}$
or
$Z_{out} \approx \frac{Z_{oi}}{\text{loop gain}}$

FIG. 3-2. Inverting feedback op-amp.

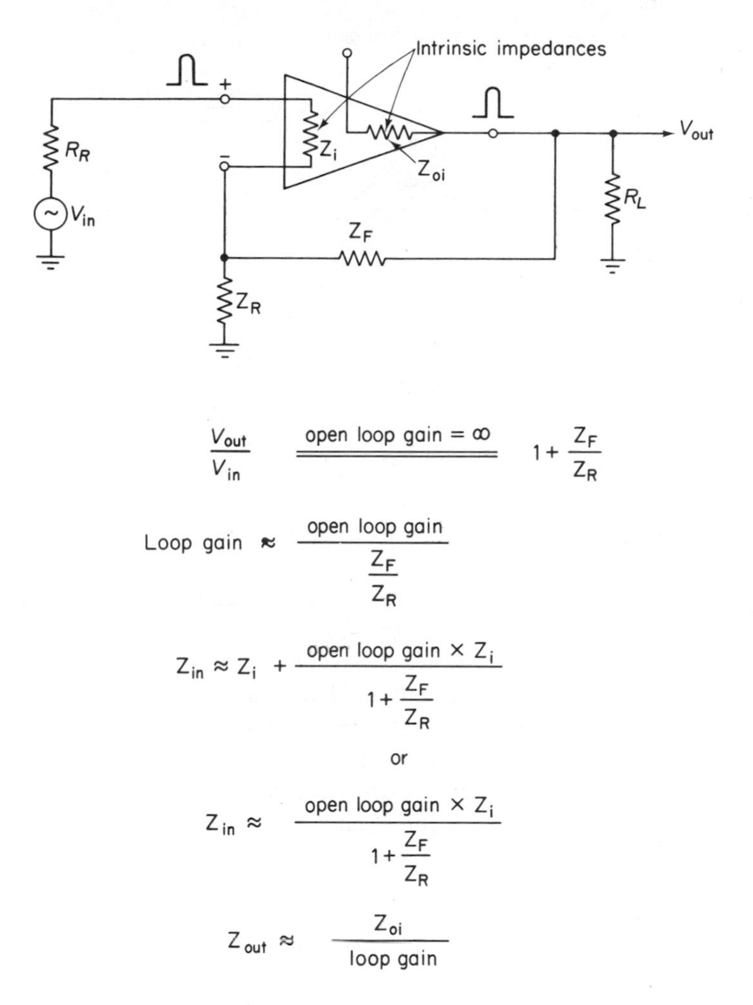

Intrinsic impedances
+
−
Z_i
Z_{oi}
V_{out}
R_R
V_{in}
Z_F
R_L
Z_R
$\frac{V_{out}}{V_{in}}$ open loop gain = ∞ $1+\frac{Z_F}{Z_R}$
Loop gain ≈ $\frac{\text{open loop gain}}{\frac{Z_F}{Z_R}}$
$Z_{in} \approx Z_i + \frac{\text{open loop gain} \times Z_i}{1+\frac{Z_F}{Z_R}}$
or
$Z_{in} \approx \frac{\text{open loop gain} \times Z_i}{1+\frac{Z_F}{Z_R}}$
$Z_{out} \approx \frac{Z_{oi}}{\text{loop gain}}$

FIG. 3-3. Non-inverting feedback op-amp.

is doubled. This is the same as a 20-dB drop each time the frequency is increased by a factor of 10.)

If the open-loop gain of an amplifier is as shown in Fig. 3-4, any stable closed-loop gain could be produced by the proper selection of feedback components, provided the closed-loop gain is less than the open-loop gain. The only concern would be a tradeoff between gain and frequency response. For example, if a gain of 40 dB (10^2) is desired, a feedback resistance 10^2 times larger than the input resistance is used. The gain is then flat to 10^4 Hz, and rolls off at 6 dB/octave to unity gain at 10^6 Hz. If 60-dB (10^3) gain is required, the feedback resistance is raised to 10^3 times the input resistance. This reduces

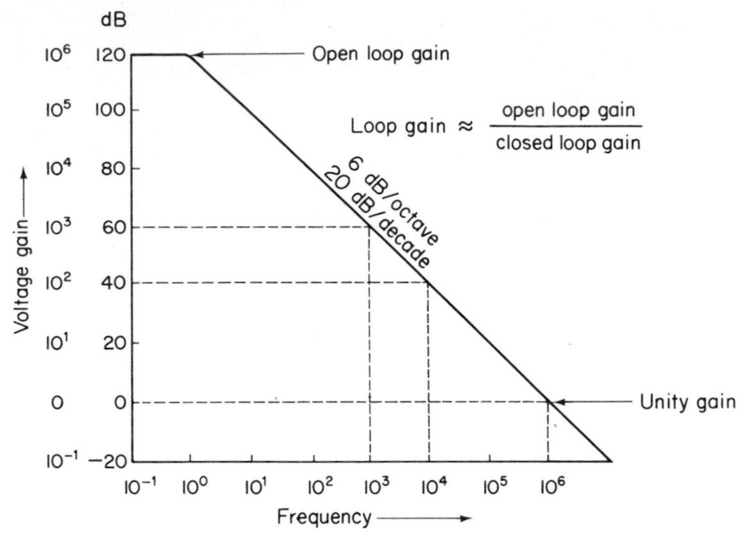

FIG. 3-4. Frequency response curve of theoretical op-amp.

the frequency response. Gain is flat to 10^3 Hz (instead of 10^4 Hz), followed by roll off of 6 dB/octave down to unity gain.

The open-loop frequency response curve of a *practical* IC op-amp more closely resembles that shown in Fig. 3-5. Here, gain is flat at 60 dB to about 200 kHz, then rolls off at 6 dB/octave to 2 MHz. As frequency increases, rolloff continues at 12 dB/octave (40 dB/decade) to 20 MHz (where gain is about unity or 0), then rolls off at 18 dB/octave (60 dB/decade).

Some IC datasheets provide a curve similar to that shown in Fig. 3-5. If the data is not available, it is possible to test the IC package under laboratory

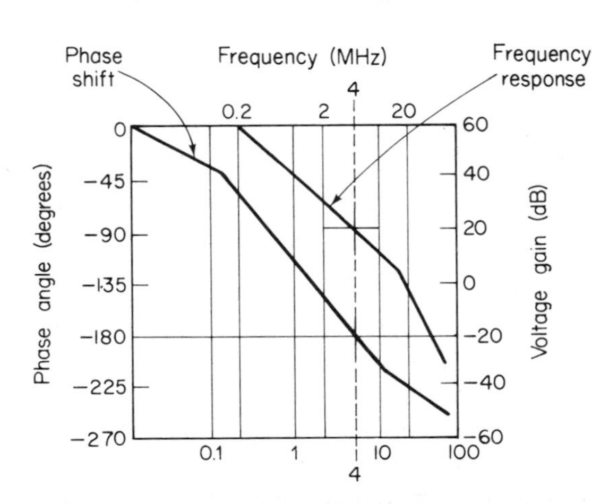

FIG. 3-5. Frequency response and phase shift characteristics.

conditions and draw an actual response curve (frequency response and phase shift). The necessary procedures are described in Chapter 7.

The sharp rolloff at high frequencies, in itself is not a problem in op-amp use (unless the IC must be operated at a frequency very near the high end). However, note that the *phase response* (phase shift between input and output) changes with frequency. The phase response of Fig. 3-5 shows that a negative feedback (at low frequency) can become positive and cause the amplifier to be unstable at high frequencies (possibly resulting in oscillation). In Fig. 3-5, a 180° phase shift occurs at approximately 4 MHz. This is the frequency at which open-loop gain is about +20 dB.

As a guide, when a selected closed-loop gain is equal to or less than the open-loop gain at the 180° phase-shift point, the circuit will be unstable. For example, if a closed-loop gain of 20 dB or less is selected, a circuit with curves of Fig. 3-5 will be unstable. (Note the point where the −180° phase angle intersects the phase shift line. Then draw a vertical line up to cross the open loop gain line.)

The closed-loop gain must be more than the open-loop gain at the frequency where the 180° phase shift occurs, but less than the maximum open-loop gain. Using Fig. 3-5 as an example, the closed-loop gain would have to be greater than 20 dB, but less than 60 dB.

3-2.1 IC phase compensation methods

IC design problems created by excessive phase shift can be solved by the use of compensating techniques that alter response so that excessive phase shifts no longer occur. There are three basic methods of phase compensation.

The closed-loop gain can be altered by means of capacitors and/or inductances in the feedback circuit. These elements change the pure resistance to an impedance that changes with frequency, thus providing a different amount of feedback at different frequencies and a shift in phase of the feedback signal. This offsets the undesired open-loop phase shift. Phase shift compensation by closed-loop methods are generally not recommended since these methods create impedance problems at both the high- and low-frequency limits of operation. Closed-loop compensation will be discussed only where the op-amp is to be used in a *bandpass* function, as described in Chapter 4.

The open-loop input impedance can be altered by means of a resistor and capacitor, as shown in Fig. 3-6. The input impedance of the series C and R decreases as frequency increases, thus altering open-loop gain. As shown in Fig. 3-6, this arrangement causes the rolloff to start at a lower frequency, but produces a stable rolloff similar to that of the "ideal" curve of Fig. 3-4. With the circuit properly compensated, a desired closed-loop gain can be produced by selection of external resistors in the normal manner.

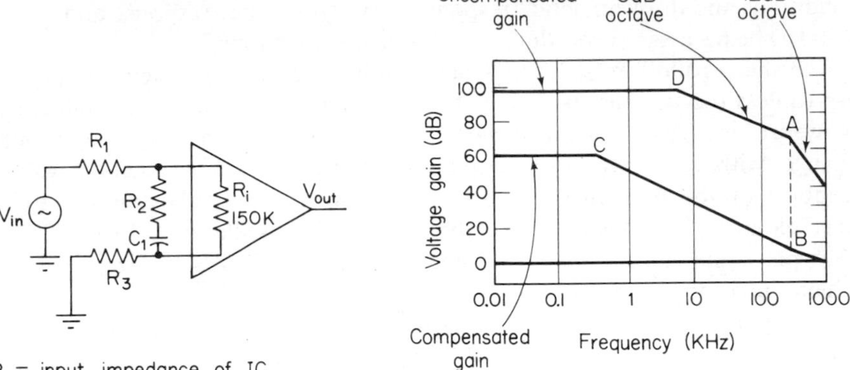

R_i = input impedance of IC

Freq. B = freq. A

$$R_1 = R_3$$

$$R_1 + R_3 = \left(\frac{\text{uncompensated gain (dB)}}{\text{compensated gain (dB)}} - 1\right) R_i$$

$$R_2 = \frac{R_1 + R_3}{\left(\frac{\text{freq. D}}{\text{freq. C}} - 1\right)\left(1 + \frac{R_1 + R_3}{R_i}\right)}$$

$$C_1 = \frac{1}{6.28 \times \text{freq. D} \times R_2}$$

$$\text{Compensated gain} = \frac{\text{uncompensated gain} \times R_i}{R_i + R_1 + R_3}$$

$$\text{Frequency D} = \frac{1}{6.28 \times R_2 \times C_1}$$

FIG. 3-6. Frequency response compensation by modification of open-loop input impedance of op-amp.

The open-loop gain can be altered by one of several phase-compensation methods, as shown in Figs. 3-7, 3-8, *and* 3-9.

In the method, in Fig. 3-7 (known as *phase-lead compensation*), the open-loop gain is changed by an external capacitor (usually connected between collectors in one of the high-gain stages).

In Fig. 3-8 (generally referred to as *RC rolloff*, *straight rolloff*, or *phase-lag compensation*) the open-loop gain is altered by means of an appropriate external *RC* network connected across a circuit component. In this method, the rolloff starts at the corner frequency produced by the *RC* network.

In Fig. 3-9 (known as *Miller-effect rolloff* or *Miller-effect phase-lag compensation*), the open-loop gain is altered by an *RC* network connected between the input and output of an inverting gain stage in the IC op-amp.

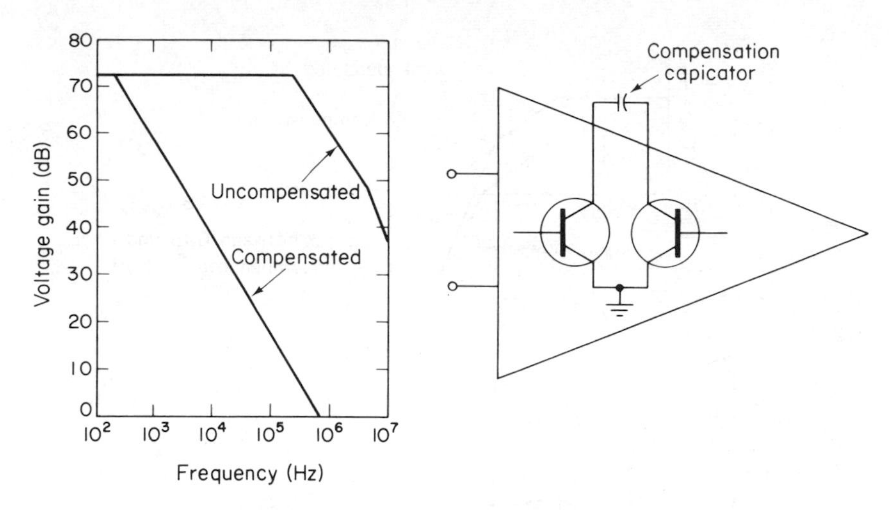

FIG. 3-7. Frequency response compensation with external capacitor (phase-lead compensation).

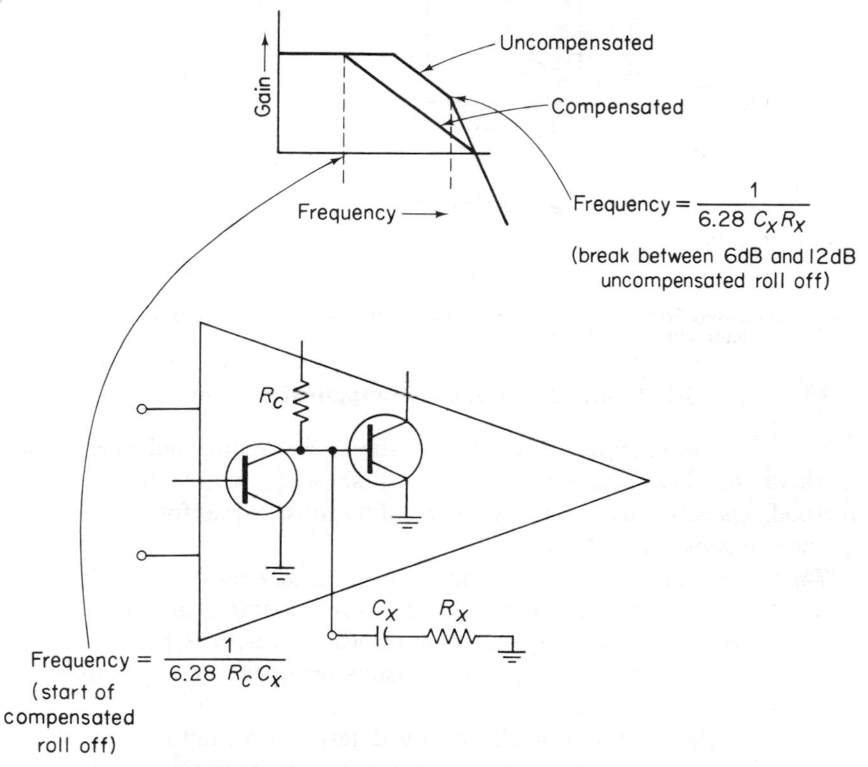

FIG. 3-8. Frequency response compensation with external capacitor and resistor (RC roll-off or phase-lag compensation).

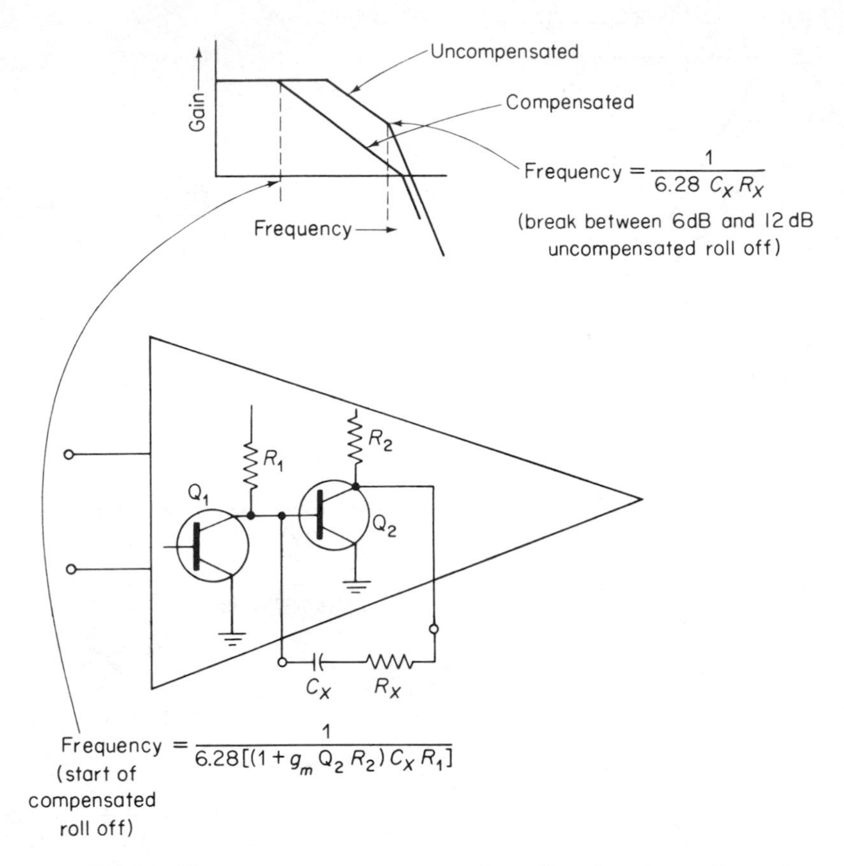

FIG. 3-9. Frequency response compensation with external capacitor and resistor (Miller-effect roll off.)

3-2.2 Selecting a phase compensation scheme

A comprehensive IC datasheet will recommend one or more methods for phase compensation and will show the relative merits of each method. Usually, this is done by means of response curves for various values of the compensating network.

The recommended phase compensation methods and values should be used in all cases. Proper phase compensation of an IC op-amp is at best a difficult, trial-and-error job. By using the IC datasheet values, it is possible to take advantage of the manufacturers' test results on production quantities of a given IC.

If the datasheet is not available or if the datasheet does not show the desired information, it is still possible to design a phase compensating network using the rule-of-thumb equations.

The first step in phase compensation (when not following the datasheet) is to connect the IC to an appropriate power source, as discussed in Chapter 2. Then test the IC for open-loop frequency response and phase shift as described in Chapter 7.

Draw a response curve similar to that in Fig. 3-5. On the basis of actual open-loop response and the equations of Figs. 3-6 through 3-9, select trial values for the phase compensating network. Then repeat the frequency response and phase shift tests. If the response is not that desired, change the values as necessary.

A careful inspection of the equations in Figs. 3-6 through 3-9 will show that it is necessary to know certain internal characteristics of the IC before an accurate prediction of the compensated frequency response characteristics can be found. For example, in Figs. 3-8 and 3-9 the values for R and C of the compensation network are based on the uncompensated open-loop frequency at which gain changes from a 6-dB/octave drop to a 12-dB/octave drop. This can be found by test of the uncompensated IC. However, to predict the frequency at which the compensated response will start to roll off or the gain after compensation requires a knowledge of internal-stage transconductance (or gain) and stage load. This information is usually not available and cannot be found by simple test.

An exception to this is the modification of the open-loop input impedance, as shown in Fig. 3-6. The only IC characteristic required here is the input impedance (or resistance). This is almost always available on the datasheet. If not, it can be found by a simple test described in Chapter 7.

3-2.3 Design example for modification of open-loop input impedance of an IC

The connections, frequency plots, and equations for modification of open-loop input impedance are shown in Fig. 3-6.

The first step is to note the frequency at which the uncompensated rolloff changes from 6 to 12 dB (point A). The compensated rolloff should be zero (unity gain, point B) at the same frequency.

Draw a line up to the left from point B that increases at 6 dB/octave. For example, if point B is at 350 kHz, the line should intersect 35 kHz as it crosses the 20-dB gain point. (In a practical circuit, the rolloff would start at a slightly lower frequency, about 28–30 kHz at 20-dB gain, since the rolloff point is rounded rather than a sharp corner.)

Any combination of compensated gain and rolloff starting frequency (point C) can be selected along the line. For example, if the roll off starts at 10 kHz, the gain is about 30 dB, and vice versa.

Assume that the circuit of Fig. 3-6 is used to produce a compensated gain of 60 dB, with rolloff starting at 280 Hz and dropping to zero (unity gain at

350 kHz). The IC to be used has an uncompensated gain of about 94 dB, with a rolloff pattern similar to that of Fig. 3-6. The typical input impedance is 150 kΩ. Uncompensated gain and input impedance can be found by reference to the datasheet or by actual test.

Using the compensated gain equation of Fig. 3-6, the relationship is:

$$60\text{ dB} = \frac{(94\text{ dB})(150{,}000)}{(150{,}000 + R_1 + R_3)}$$

Therefore,

$$R_1 + R_3 = \left(\frac{94}{60} - 1\right) 150{,}000$$

$$= 0.57 \times 150{,}000$$

$$R_1 + R_3 = 85{,}500$$

If $R_1 = R_3$, $R_1 = R_3 = 42{,}750$

The nearest standard value is 43 kΩ.

Using the equation of Fig. 3-6, the value of R_2 is:

$$R_2 = \frac{85{,}500}{\left(\frac{5000}{280} - 1\right)\left(1 + \frac{85{,}500}{150{,}000}\right)}$$

$$= \frac{85{,}500}{16.85 \times 1.57}$$

$$= 3000\ \Omega \text{ nearest standard value}$$

The value of C_1 is:

$$C_1 = \frac{1}{(3000)(6.28)(5000)}$$

$$= 0.01\ \mu\text{F nearest standard value}$$

If the circuit of Fig. 3-6 shows any instability in the open-loop or closed-loop condition, try increasing the values of R_1 and R_3 (to reduce gain); then select new values for R_2 and C_1.

3-3. INTERPRETING IC DATASHEETS

Most of the basic design information for a particular IC op-amp can be obtained from the datasheet. There are some exceptions to this rule.

For certain applications it may be necessary to test the IC under simulated operating conditions. However, it is always necessary to interpret datasheets. Each manufacturer has their own system of datasheets. It is impractical to discuss all datasheet formats here. Instead, we will discuss typical information found on datasheets and see how this information affects the IC op-amp user.

3-3.1 Open-loop voltage gain

The open-loop voltage gain (A_{VOL} or A_{OL}) is defined as the ratio of a change in output voltage to a change in input voltage at the input terminals. Open-loop gain is always measured without feedback and usually without compensation.

Open-loop gain is frequency dependent (gain decreases with increased frequency). Open-loop gain is also temperature dependent (gain decreases with increased temperature, as shown in Fig. 3-10).

As previously discussed, open-loop gain can be modified by several compensation methods. A typical IC datasheet will show the results of such compensation, usually by means of graphs such as the one shown in Fig. 3-11.

After compensation is applied, the IC can be connected in the closed-loop configuration. The voltage gain under closed-loop conditions is dependent upon external components (the ratio of feedback resistance to input resistance). Thus, closed-loop gain is usually not listed as such on IC datasheets. However, the datasheet may show some typical gain curves with various ratios of feedback (Fig. 3-12). If available, such curves can be used directly to select values of feedback components.

When a capacitor is used to compensate or modify voltage gain, the *slew rate* (or *slewing rate*) of an IC can be affected. The slew rate of an op-amp is the maximum rate of change of the output voltage (V_0 or E_0), with respect to

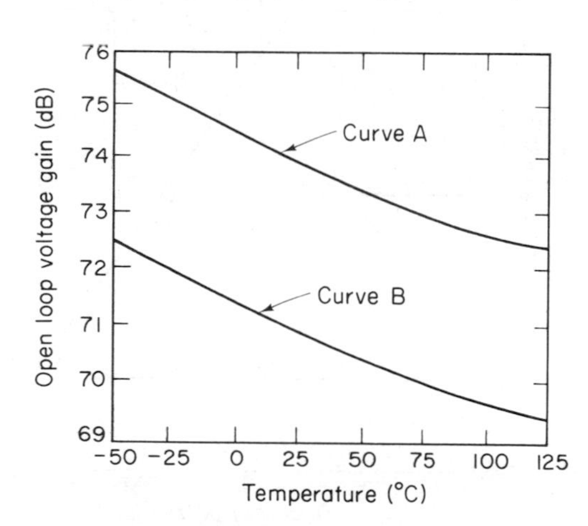

FIG. 3-10. Open-loop gain of an IC as a function of temperature.

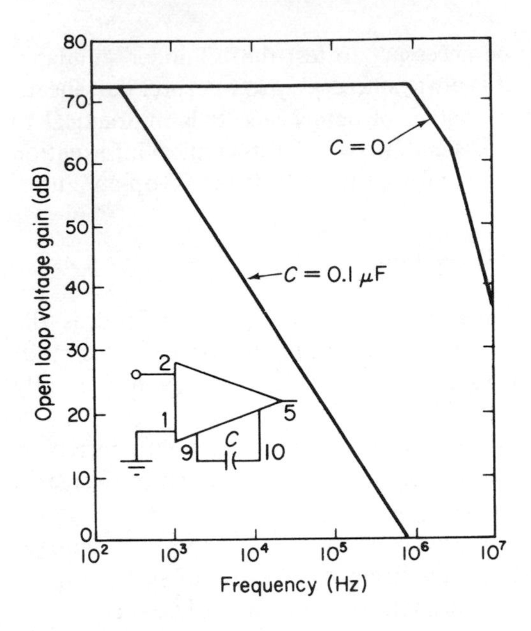

FIG. 3-11. Open-loop voltage gain of IC with and without compensation.

time, that the IC is capable of producing while maintaining its linear characteristics.

Slew rate is expressed in terms of:

$$\frac{\text{difference in output voltage}}{\text{difference in time}} \quad \text{or} \quad \frac{\Delta V_0}{\Delta t}$$

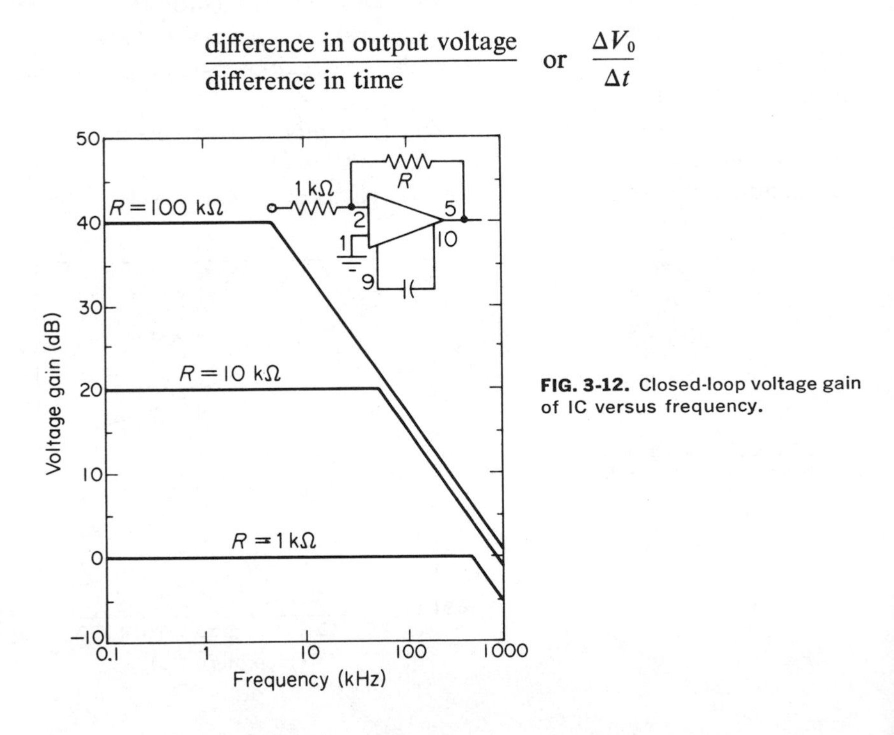

FIG. 3-12. Closed-loop voltage gain of IC versus frequency.

Usually, slew rate is listed in terms of volts per microsecond. For example, if the output voltage from an IC is capable of changing 100 V in 1 microsecond, then the slew rate is 100. If, after compensation, the IC output changes a maximum of 50 V in 1 microsecond, the new slew rate is 50.

Slew rate of an IC is a direct function of the compensation capacity. At higher frequencies, the current required to charge and discharge the capacitor can limit available current to succeeding stages or loads and thus result in lower slew rates. (This is one reason why IC datasheets usually recommend the compensation of early stages in the amplifier where signal levels are still small, and little current is required.)

The major effect of slew rate on design is in output power of the IC amplifier. All other factors being equal, a lower slew rate results in lower power output.

Slew rate decreases as compensation capacitance increases. This is shown by the IC datasheet curve of Fig. 3-13. Therefore, where high frequencies are involved, the lowest value of compensation capacitor should be used.

Figure 3-14 shows the minimum compensation capacitor value that can be used with different closed-loop gain levels. The curves of Fig. 3-14 are typical of those found on IC datasheets in which the slew rate is of particular importance.

3-3.2 Output impedance

Output impedance (Z_{out}) is defined as the impedance seen by a load at the output of the IC amplifier (See Fig. 3-15). Excessive output im-

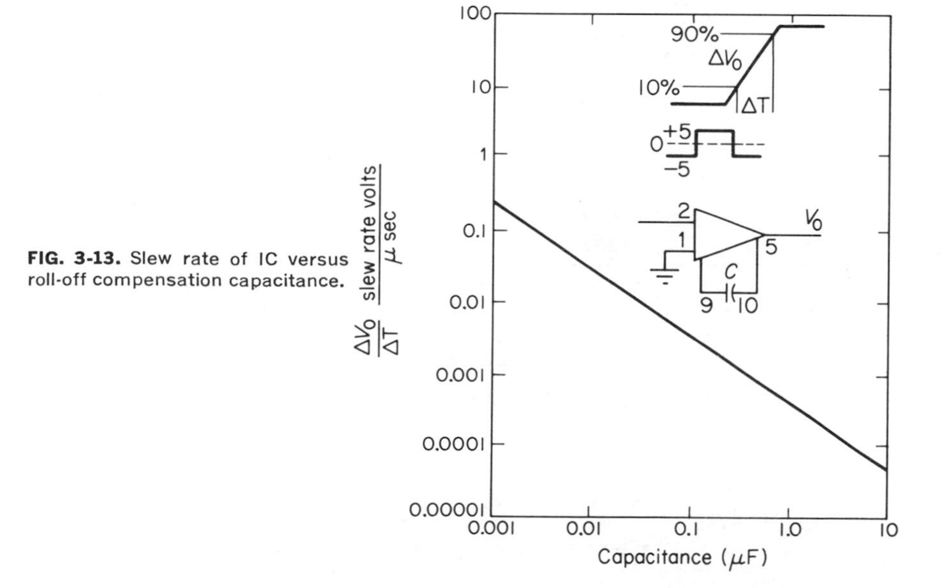

FIG. 3-13. Slew rate of IC versus roll-off compensation capacitance.

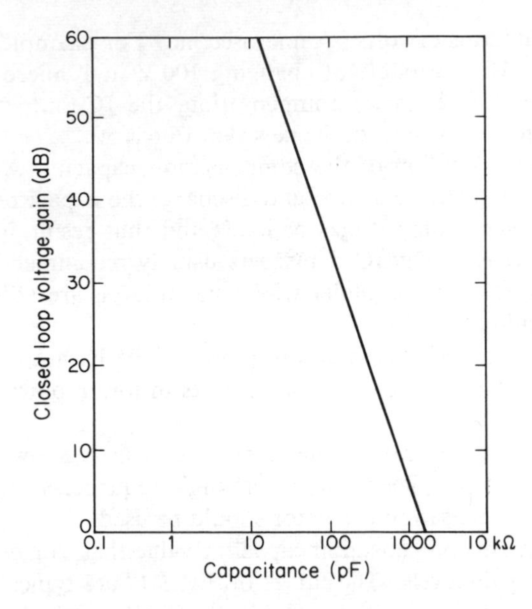

FIG. 3-14. Closed-loop voltage gain of IC versus minimum roll-off capacitance.

pedance can reduce the gain since, in conjunction with the load and feedback resistors, output impedance forms an attenuator network. In general, output impedance of ICs used as op-amps is less than 200 Ω. Generally, input resistances are at least 1000 Ω, with feedback resistance several times higher than 1000 Ω. Thus, the output impedance of a typical IC will have little effect on gain.

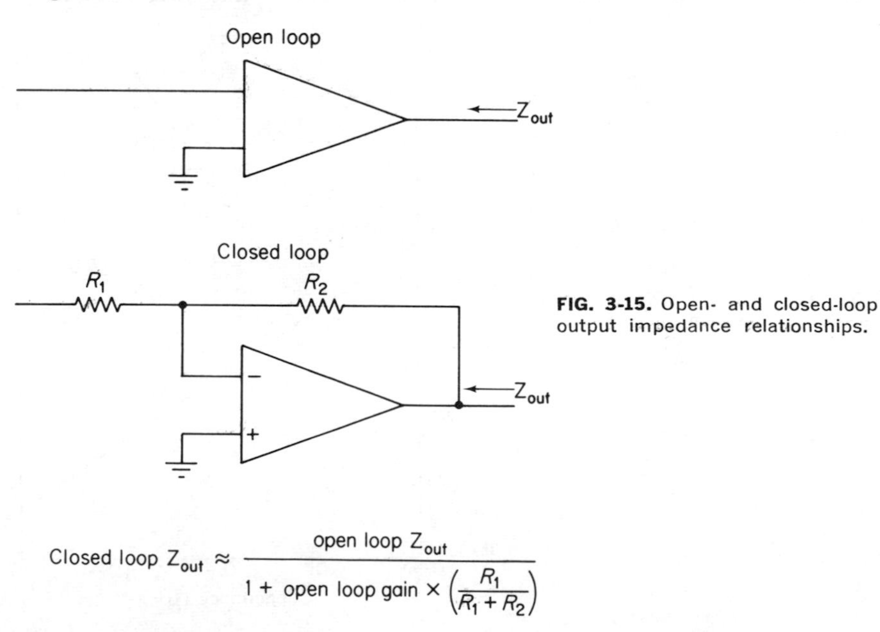

FIG. 3-15. Open- and closed-loop output impedance relationships.

If the IC is serving primarily as a voltage amplifier (as is usually the case) the effect of output impedance will be at a minimum. Output impedance has a more significant effect in design of power devices that must supply large amounts of load current.

Closed-loop output impedance is found by using the equation in Fig. 3-15. Thus, it will be seen that output impedance will increase as frequency increases, since open-loop gain decreases.

3-3.3 Input impedance

Input impedance (Z_{in}) is defined as the impedance seen by a source looking into one input of the IC amplifier with the other input grounded (see Fig. 3-16). The primary effect of input impedance on design is to reduce amplifier loop gain. Input impedance will change with temperature and frequency. Generally, input impedance is listed on the datasheet at 25 °C and 1 kHz.

If the input impedance is quite different from the impedance of the device driving the IC, there will be a loss of input signal due to the mismatch. How-

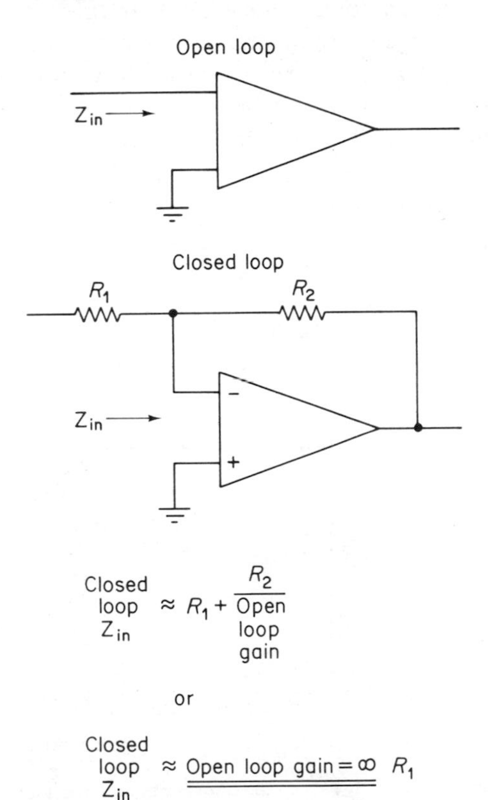

FIG. 3-16. Open- and closed-loop input impedance relationships.

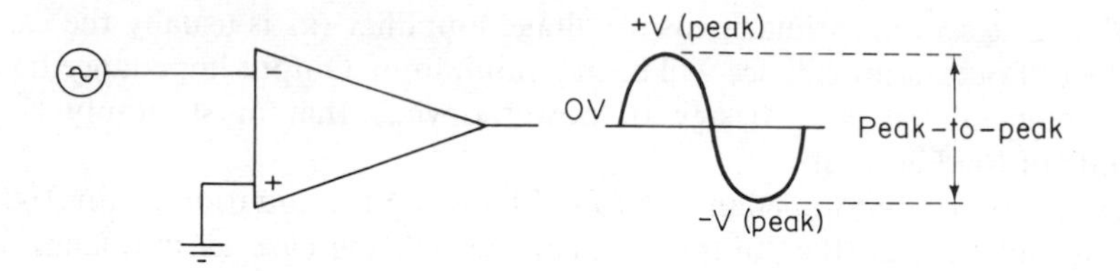

FIG. 3-17. Output voltage swing relationships.

ever, in practical terms, it is not possible to alter the IC input impedance. Thus, if impedance match is critical, either the IC or driving source must be changed to effect a match.

3-3.4 Output voltage swing

Output voltage swing (V_0 or P-P V_0) is defined as the peak output voltage swing (referred to zero) that can be obtained without clipping (see Fig. 3-17). A symmetrical voltage swing is dependent upon frequency, load current, output impedance, and slew rate. Generally, an increase in frequency, load current, and output impedance will decrease the possible output voltage swing. An increase in slew rate, however, will increase possible output voltage swing capabilities. Since slew rate is related to compensation techniques (a high-value compensation capacitor reduces slew rate), output voltage swing is also related to compensation (a high-value compensation capacitor reduces output voltage swing at a given frequency).

3-3.5 Bandwidth and frequency range

Bandwidth for an IC is usually expressed in terms of open-loop operation. The common term is BW_{OL} at -3dB (see Fig. 3-18). For example,

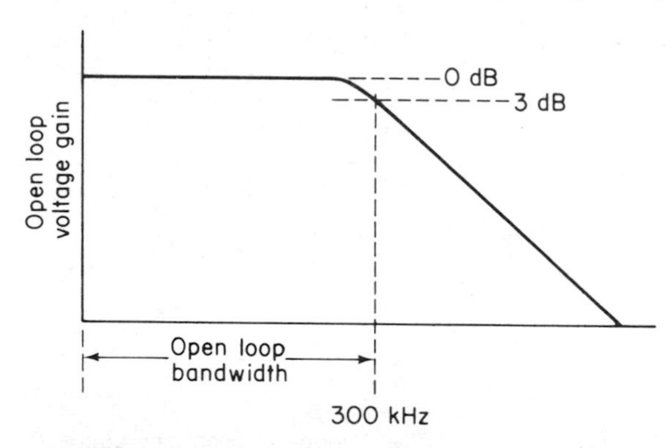

FIG. 3-18. Bandwidth and open-loop gain relationships.

a BW_{OL} of 300 kHz indicates that the open-loop gain of the IC will drop to a value 3dB below the flat- or low-frequency level at a frequency of 300 kHz.

The frequency range of an IC is often listed as "useful frequency range" (such as dc to 15 MHz). Useful frequency range for an IC is similar to the f_T (total frequency) term used with discrete transistors. Generally, the high-frequency limit specified for an IC is the frequency at which gain drops to unity.

3-3.6 Input common mode voltage swing

Input common-mode voltage swing (V_{ICM}) is defined as the maximum peak input voltage that can be applied to either input terminal of the IC without causing abnormal operation or damage (see Fig. 3-19). Some IC datasheets list a similar term: *common mode input voltage range* (V_{CMR}). Usually, V_{ICM} is listed in terms of peak voltage, with positive and negative peaks being equal. V_{CMR} is often listed for positive and negative voltages of different value (such as +1V and −3V).

In practical use, either of these parameters limit the differential signal amplitude that can be applied to the IC input. So long as the input signal does not exceed the V_{ICM} or V_{CMR} values (in either the positive or negative direction), there should be no problem.

Note that some IC datasheets list "single-ended" input voltage signal limits where the differential input is not to be used.

3-3.7 Common mode rejection ratio

IC manufacturers do not agree on the exact definition of *common mode rejection.* One manufacturer defines common mode rejection (CMR or CM_{rej}) as the ratio of differential gain (usually large) to common mode gain (usually a fraction). Another manufacturer defines CMR as the change in output voltage to the change in the input common-mode voltage producing it, divided by the open-loop gain. For example, as shown in

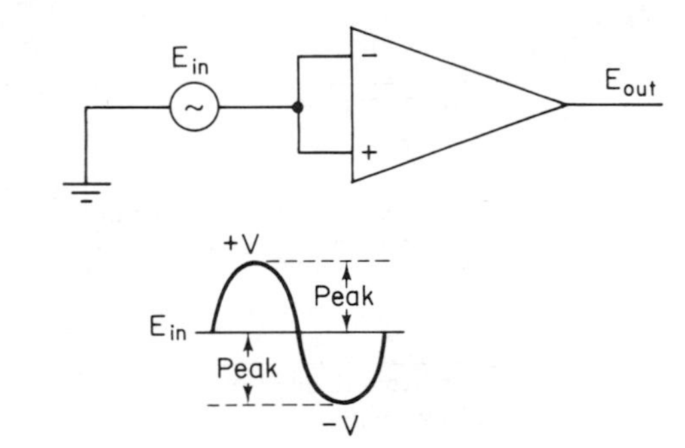

FIG. 3-19. Input common-mode voltage swing relationships.

Fig. 3-20, assume that the common mode input (applied to both input terminals simultaneously) is 1 V. The resultant output is 1 mV, and the open-loop gain is 100. The CMR is then:

$$\frac{\left(\frac{0.001}{1}\right)}{100} = 100{,}000 \qquad (100\text{ dB})$$

Another method for calculating the CMR is to divide the output signal by the open-loop gain to find an "equivalent differential input signal." Then the common mode input signal is divided by this equivalent differential input signal. Using the same figures as in the previous CMR calculation:

$$1\text{ mV}/100 = 0.00001 \text{ equivalent differential input signal}$$

$$1\text{ V}/0.00001 = 100{,}000\ (100\text{ dB})$$

No matter what basis is used for calculation, the CMR is an indication of the degree of circuit balance of the differential stages of the amplifier, since a common mode input signal applied to the input terminals should be amplified identically on both sides of the IC (in theory). A large output for a given "common-mode" input is an indication of large unbalance or poor common-mode rejection.

As a guide, the CMR should be *at least* 20 dB greater than the open-loop gain, and preferably much higher.

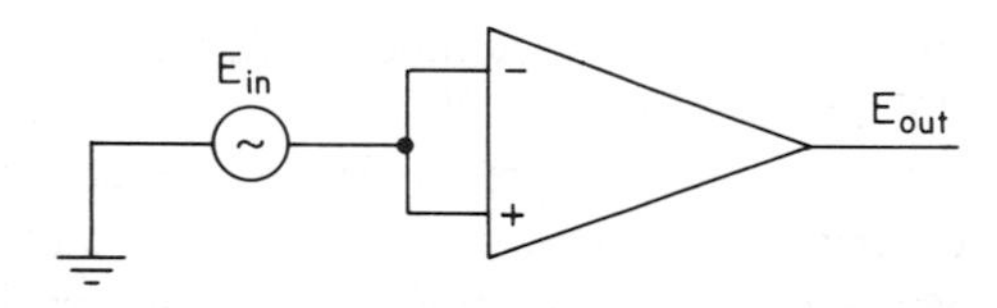

$$\text{Common mode rejection} = \frac{\left(\frac{E_{out}}{E_{in}}\right)}{\text{Open loop gain}}$$

or

$$\frac{E_{out}}{\text{Open loop gain}} = \text{Equivalent differential input signal}$$

$$\text{Common mode rejection} = \frac{E_{in}}{\text{Equivalent differential input signal}}$$

FIG. 3-20. Common mode rejection ratio relationships.

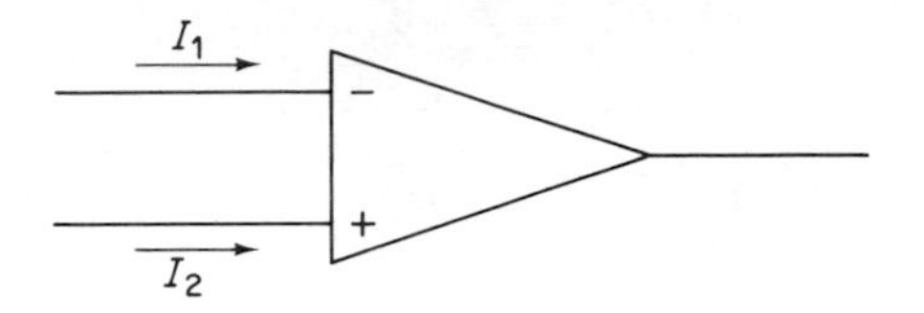

FIG. 3-21. Input bias current.

$$\text{Input bias current} = \frac{I_1 + I_2}{2}$$

Note that the CMR decreases as frequency increases in all IC op-amps. The CMR can also be temperature-sensitive, but this is usually not of major importance in practical use.

3-3.8 Input bias current

Input bias current (I_i or I_b) is defined as the average value of the two input bias currents of the IC differential input stage (see Fig. 3-21). Input bias current is a function of the large signal current gain of the input stage.

In use, the only real significance of input bias current is that the resultant voltage drop across the input resistors can restrict the input common-mode voltage range at higher impedance levels. The input bias current produces a voltage drop across the input resistors. This voltage drop must be overcome by the input signal.

Input bias current decreases as temperature increases.

3-3.9 Input offset current

Input offset current (I_{io}) is defined as the difference in input bias current into the input terminals of an IC (see Fig. 3-22). Input offset current is an indication of the degree of matching of the input differential stage.

When high impedances are used in design, input offset current can be of greater importance than input offset voltage. If the input bias current is different for each input, the voltage drops across the input resistors (or input impedance) will not be equal. If the resistance is large, there will be a large unbalance in input voltages. This condition can be minimized by means of a

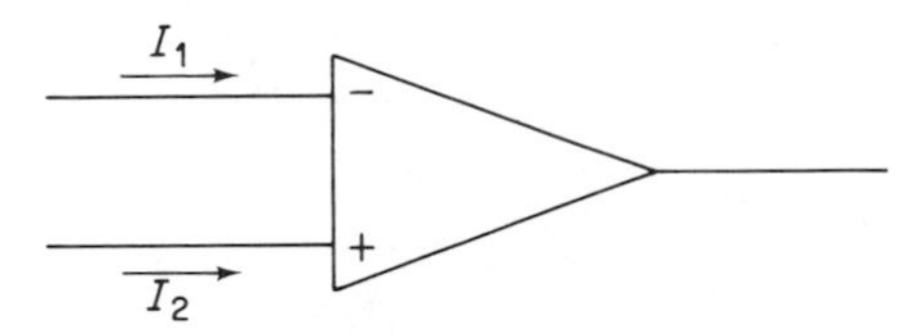

FIG. 3-22. Input offset current.

$$\text{Input offset current} = I_1 - I_2 \text{ or } I_2 - I_1$$

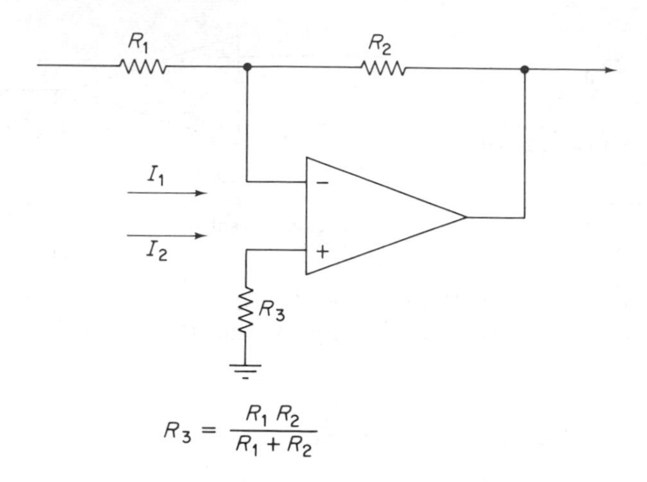

FIG. 3-23. Minimizing input offset current (and input offset voltage).

resistance connected between the non-inverting (or +) input and ground, as shown in Fig. 3-23. The value of this resistor (R_3) should equal the parallel equivalent of the input and feedback resistors (R_1 and R_2), as shown by the equation.

In practical design, the trial value for R_3 is based on the equation of Fig. 3-23. The value of R_3 is then adjusted for minimum voltage difference at both terminals (under normal operating conditions), but with no signal.

3-3.10 Input offset voltage

Input offset voltage (V_{io}) is defined as the voltage that must be applied at the input terminals to obtain zero output voltage (see Fig. 3-24). Input offset voltage indicates the matching tolerance in the differential amplifier stages. A perfectly matched amplifier requires zero input voltage to produce zero output voltage. Typically, input offset voltage is on the order of 1 mV for an IC op-amp.

The effect of input offset voltage on design is that the input signal must overcome the offset voltage before an output will be produced. For example, if an IC has a 1-mV input offset voltage and a 1-mV signal is applied, there is

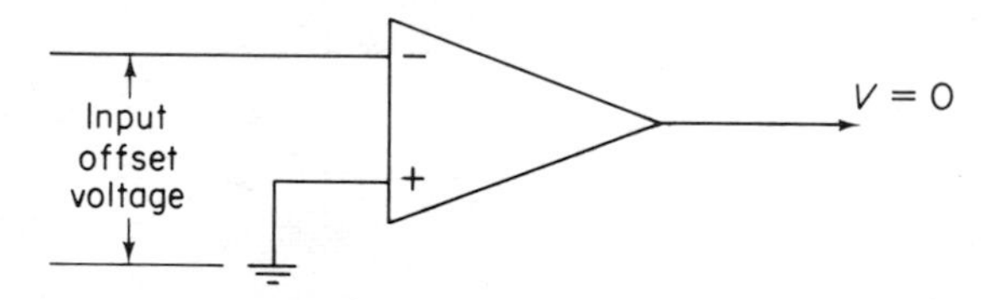

FIG. 3-24. Input offset voltage.

no output. If the signal is increased to 2 mV, the IC will produce only the peaks.

Input offset voltage is increased by amplifier gain. In the closed-loop condition, the effect of input offset voltage is increased by the ratio of feedback resistance to input resistance plus unity (or one). For example, if the ratio is 100 to 1, the effect of input offset voltage is increased by 101.

Some IC op-amps include provisions to neutralize any offset. Typically, an external voltage is applied through a potentiometer to a terminal on the IC. The voltage is adjusted until the offset, at the input and output, is zero. However, since this is a special circuit, no general rules can be included concerning voltage, potentiometer values, or external connections. The datasheet must be consulted.

For an IC without offset compensation, the effects of input offset voltage can be minimized by minimizing input offset current, as described in Sec. 3-3.9 and Fig. 3-23.

3-3.11 Power supply sensitivity

Power supply sensitivity ($S+$ and $S-$) is defined as the ratio of change in input offset voltage to the change in supply voltage producing it, with the remaining supply held constant (see Fig. 3-25). Some IC datasheets list a similar parameter: *input offset voltage sensitivity*. In either case, the parameter is expressed in terms of mV/V or μV/V, representing the change (in mV or μV) of input offset voltage to a change (in volts) of one power supply. Usually, there is a separate sensitivity parameter for each power

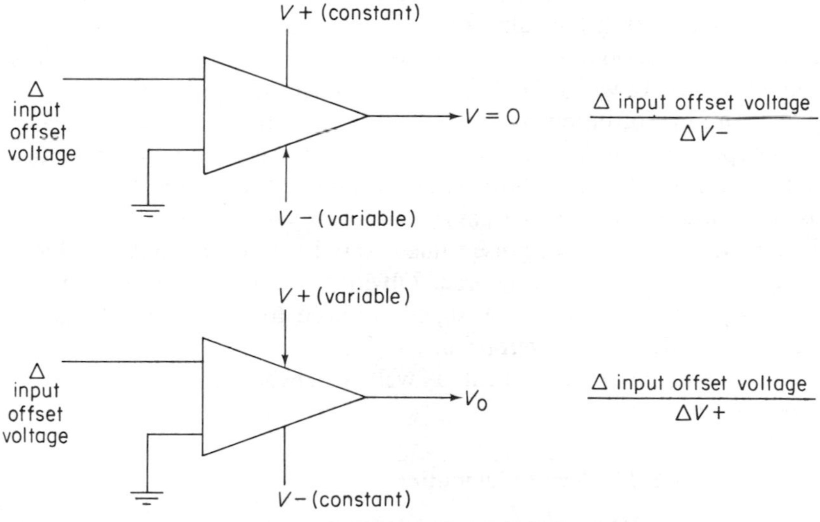

FIG. 3-25. Power supply sensitivity.

supply, with the opposite power supply assumed to be held constant. For example, a typical sensitivity listing is 0.1 mV/V for a positive supply. This implies that with the negative supply held constant, the input offset voltage will change by 0.1 mV for each 1 V change in positive supply voltage.

The effects of power-supply sensitivity (or input offset voltage sensitivity) is obvious. If an IC has considerable sensitivity to power-supply variations, overall performance is affected by each supply-voltage change. The power-supply regulation must be increased to provide correct operation with minimum input signal levels.

3-3.12 Average temperature coefficient of input offset voltage

This parameter (TCV_{io}) is dependent upon the temperature coefficients of various components within the IC. Temperature changes affect stage gain, match of differential amplifiers, etc., and thus change input offset voltage.

From a practical user's standpoint, TCV_{io} need be considered only if the parameter is large, and the IC must be operated under extreme temperatures. For example, if input-offset voltage doubles with an increase to a temperature that is likely to be encountered during normal operation, the higher input offset voltage should be considered the "normal" value for design.

3-3.13 Input noise voltage

There are many systems for measuring noise voltage in an IC op-amp, and equally as many methods used to list the value on datasheets. Some datasheets omit the value entirely. In general, noise is measured with the IC in the open-loop condition, with or without compensation, and with the input shorted or with a fixed-resistance load at the input terminals.

The input and/or output voltage is measured with a sensitive voltmeter or oscilloscope. Input noise is on the order of a few microvolts; output noise is usually less than 100 mV. Output noise is almost always greater than input noise (because of the amplifier gain).

Except in cases where the noise value is very high or the input signal is very low, amplifier noise can be ignored. Obviously, a 10-μV noise at the input will mask a 10-μV signal. If the signal is raised to 1 mV with the same IC op-amp, the noise will be unnoticed.

Noise is temperature-dependent, as well as dependent upon the method of compensation used.

3-3.14 Power dissipation

An IC op-amp datasheet usually lists two power dissipation ratings. One value is the *total device dissipation* which includes any load current. The other value is *device dissipation* (P_D or P_T), which is defined as the dc power dissipated by the IC itself (with output at zero and no load).

The device dissipation must be subtracted from the total dissipation to calculate the load dissipation.

For example, if an IC can dissipate a total of 300 mW (at a given temperature, supply voltage, and with or without a heat sink) and the IC itself dissipates 100 mW, the load cannot exceed 200 mW (300 − 100 = 200 mW).

3-4. TYPICAL IC OPERATIONAL AMPLIFIER CIRCUIT DESIGN

Figure 3-26 is the working schematic of a closed-loop IC op-amp, complete with external circuit parts. The design considerations discussed in all previous sections of this chapter are applicable to the circuit of Fig. 3-26. The following paragraphs provide a specific design example for the circuit.

3-4.1 IC characteristics

The IC shown in Fig. 3-26 has the following characteristics listed in its datasheet:

Supply voltage: +10 and −10 V maximum, +6 and −6 V nominal
Total device dissipation: 600 mW, derating 5 mW/°C
Temperature range: −55 to +125 °C (operating)
Input offset voltage: 1.0 mV
Input offset current: 0.5 μA

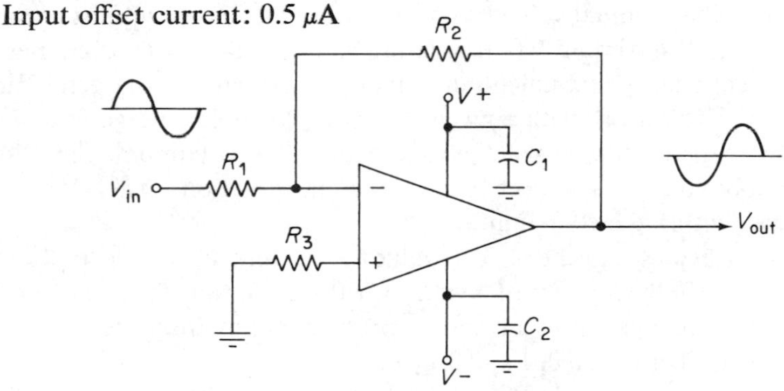

$$\text{Voltage gain} = \frac{V_{out}}{V_{in}} = \frac{R_2}{R_1} \qquad C_1 = C_2 = 0.1\text{–}0.001\ \mu\text{F}$$

$$R_3 = \frac{R_1 R_2}{R_1 + R_2}$$

FIG. 3-26. Basic IC operational amplifier.

Input bias current: 5.0 μA
Input offset voltage sensitivity: 0.25 mV/V
Device dissipation: 100 mW
Open-loop voltage gain: 60 dB
Open-loop bandwidth: 300 kHz
Common-mode rejection: 100 dB
Maximum output voltage swing: 7 V (peak-to-peak)
Input impedance: 15 kΩ
Output impedance: 200 Ω
Useful frequency range: dc to 15 MHz
Maximum input signal: +1, −4V

3-4.2 Design example

Assume that the circuit of Fig. 3-26 is to provide a voltage gain of 100 (40 dB), the input signal is 10 mV (RMS), the input source impedance is not specified, the output load impedance is 100 Ω, the ambient temperature is 25 °C, the frequency range is dc to 200 kHz, and the power supply is subject to 10 percent variation.

Note that the frequency and phase compensation components and values are omitted from Fig. 3-26. As discussed in Sec. 3-2, the compensation method and values found on the IC datasheet must be used. Even these datasheet values should be confirmed by frequency response tests of the final circuit. If no compensation values are available, use the guideline values found in Sec. 3-2 and test the results over the frequency range of interest.

Supply voltage. The positive and negative supply voltages should both be 6 V since this is the nominal value listed. Most IC datasheets will list certain characteristics as "maximum" (temperature range, total dissipation, maximum supply voltage, maximum input signal), and then list the remaining characteristics as "typical" with a given "nominal" supply voltage.

In no event can the supply voltage exceed the 10-V maximum. Since the available supply voltage is subject to a 10 percent variation, or 6.6-V maximum, it is well within the 10-V limit.

Decoupling or bypass capacitors. The values of C_1 and C_2 should be found on the datasheet. In the absence of a value, use 0.1 μF for any frequency up to 10 MHz. If this value produces a response problem at any frequency (high or low), try any value between 0.001 and 0.1 μF.

Closed-loop resistances. The value of R_2 should be 100 times the value of R_1 in order to obtain the desired gain of 100. The value of R_1 should be selected so that the voltage drop across R_1 (with the nominal input bias current) is comparable to the input signal (never larger than the input signal). A 100-Ω value for R_1 will produce a 0.5-mV drop with the nominal 5 μA bias current. Such a 0.5-mV drop is less than 10 percent of the 10-mV input signal. Thus, the fixed drop across R_1 should have no appreciable effect on the input signal. With a 100-Ω value for R_1, the value of R_2 must be 10 kΩ. (100 × 100 gain = 10,000).

Offset minimizing resistance. The value of R_3 can be found using the equation of Fig. 3-26 once the values of R_1 and R_2 have been established. Note that the value of R_3 works out to about 99 Ω, using the Fig. 3-26 equation:

$$\frac{R_1 R_2}{R_1 + R_2} = \frac{100 \times 10{,}000}{100 + 10{,}000} = 99\ \Omega$$

Thus, a simple trial value for R_3 is always *slightly less* than the R_1 value. The final value of R_3 should be such that the no-signal voltages at each input are equal.

Comparison of circuit characteristics. Once the values of the external circuit components have been selected, the characteristics of the IC and the closed-loop circuit should be checked against the requirements of the design example. The following is a summary of the comparison.

Gain. The closed-loop gain should always be less than the open-loop gain. The required closed-loop gain is 100 (40 dB) at a frequency up to 200 kHz. That is, the closed-loop circuit should have a flat frequency response of from 40 dB to at least 200 kHz. The open-loop gain is 60 dB, dropping 3 dB down at 300 kHz. Thus, the closed-loop gain is well within tolerance.

Input voltage. The peak input voltage must not exceed the rated maximum input signal. In this case, the rated maximum is +1 and −4 V, whereas the input signal is 10 mV (RMS) or approximately 14 mV (peak). This is well below the +1-V maximum limit.

When the rated maximum input signal is an uneven positive and negative value, always use the lowest value for total swing of the input signal. In this case, the input swings from +14 mV to −14 mV. An input signal that started from zero could swing as much as +1 and −1 V without damaging the IC. An input signal that started from −2 V could swing as much as ±2 V (from zero to −4 V).

Output voltage. The peak-to-peak output voltage must not exceed the rated maximum output voltage swing (with the required input signal and selected amount of gain).

In this case, the rated maximum output is 7 V peak-to-peak, whereas the actual output is approximately 2800 mV (10 mV RMS input = 28 mV peak-to-peak times gain of 100 = 2800 mV peak-to-peak output). Thus, the anticipated output should be well within the rated maximum.

With a gain of 100, the *input* could go as high as 70 mV peak-to-peak (25 mV RMS).

Output impedance. Ideally, the closed-loop output impedance should be as low as possible and always less than the load impedance. The closed-loop output impedance can be found using the equation of Sec. 3-3.2. In this case, the output impedance would be:

$$\frac{200}{1 + 1000 \times \dfrac{100}{100 + 10{,}000}}$$

or 20 Ω approximately

Output power. The output power of an IC op-amp is usually computed on the basis of RMS output voltage and output load.

In this case, the output voltage is 1 V RMS (10 mV × 100 gain = 1000 mV or 1 V). The load resistance or impedance is 100 Ω as stated in the design assumptions. Thus, the output power is:

$$\frac{E^2}{R}\ \frac{1^2}{100} = 0.01\ \text{W} = 10\ \text{mW}$$

Since the IC dissipation is 100 mW and the total device dissipation is 600 mW, a 10 mW output is well within tolerance. If the case temperature is maintained at 25 °C, it is possible to have a 500-mW power output.

4. IC OP-AMP APPLICATIONS

In this chapter, we shall discuss many applications for the basic IC op-amp. For the readers' convenience, the same format is used for each application (where practical).

First, a working schematic is presented for the circuit, together with a brief description of its function. Where practical, the working schematic also includes the operational characteristics of the circuit (in equation form), as well as rule-of-thumb relationships of circuit values (also in equation form).

Next, design considerations, such as desired performance, use with external circuits, amplification, operating frequency, etc., are covered. This is followed by reference to the equations (on the working schematic) and procedures for determining external component values that will produce the desired results.

Finally, a specific design problem is stated and a design example is given. The value of each external circuit component is calculated in step-by-step procedures, using the rules of thumb established in the design considerations and/or the working schematic equations.

Where applicable, reference is made to Chapter 7 in which procedures for testing the completed circuit are given.

The reader will note that the power-supply and phase/frequency compensation connections are omitted from the schematics in this section. In all of the applications, it is assumed that the IC is mounted and connected to a power source as described in Chapter 2. Likewise, it is assumed that a suitable phase/frequency compensation scheme has been selected for the IC, as described in Chapter 3. The primary concern in this chapter is to provide universal applications for the great number of IC op-amps on the market.

Unless otherwise stated, all of the design considerations for the basic IC op-amp described in Chapters 2 and 3 apply to each application covered here.

4-1. OP-AMP SUMMING AMPLIFIER (ADDER)

Figure 4-1 is the working schematic of an IC op-amp used as a summing amplifier (or analog adder). Summation of a number of voltages can be accomplished using this circuit. (Voltage summation was one of the original uses for operational amplifiers in computer work.) The output of the circuit is the sum of the various input voltages (a total of four in this case) multiplied by any circuit gain. Generally, gain is set so that the output will be at some given voltage value when all inputs are at their maximum value. In other cases, the values are selected for unity gain.

4-1.1 Design example

Assume that the circuit of Fig. 4-1 is to be used as a summing amplifier to sum four voltage inputs. Each of the voltage inputs varies from 2 to 50 mV (RMS). The output must be at least 1 V (RMS) with full input on all four channels, but must not exceed 2 V (RMS) at any time.

To simplify design, make resistors R_1 through R_4 the same value. Note that the input bias current will then be divided equally and produce the same voltage drop across each resistor.

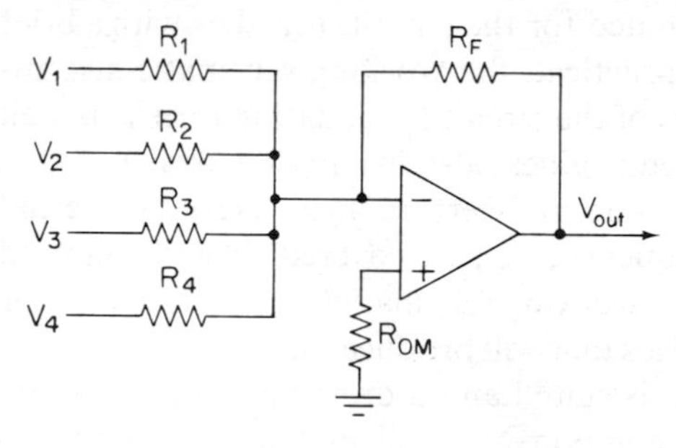

$$V_{out} = \left[\frac{R_F}{R_1} V_1 + \frac{R_F}{R_2} V_2 + \frac{R_F}{R_3} V_3 + \frac{R_F}{R_4} V_4\right]$$

$$E_{nom} \approx < 0.1 \times \text{minimum signal}$$

$$R_1 = \frac{\text{selected } E_{nom}}{\text{nominal input bias current thru } R_1}$$

When $R_1 = R_2 = R_3 = R_4 = R_F$

Then $V_{out} = \left[V_1 + V_2 + V_3 + V_4\right]$

$$R_{OM} = \frac{1}{\frac{1}{R_1} + \frac{1}{R_2} + \frac{1}{R_3} + \frac{1}{R_4} + \frac{1}{R_F}}$$

FIG. 4-1. Summing amplifier using IC op-amp

The values of R_1 through R_4 should be selected so that the voltage drop (E_{NOM}) (with nominal input bias current) is comparable (preferably less than 10 percent) with the minimum input signal. Assume a 5μ-A input bias current for the IC. With four equal resistances, this results in 1.25 μA through each resistor. A 100-Ω value for R_1 through R_4 will produce a nominal 125-μV (E_{NOM}) drop across each resistor, which is less than 10 percent of the 2-mV (2000-μV) minimum input signal.

The total (or maximum possible) signal voltage at the IC input is 200 mV (4 × 50 mV). Thus, the value of R_F should be between five and ten times that of R_1 through R_4 to get a minimum 1 V and a maximum 2 V (1 V/200 mV = 5; 2 V/200 mV = 10).

With a 100-Ω value for R_1 through R_4, the value of R_F could be 500 to 1000 Ω. Assume that the 1000-Ω value is selected. Under these conditions, the circuit output is ten times the sum of the four input voltages, or 80 mV (all four inputs minimum) to 2 V (all four inputs maximum).

The value of the offset minimizing resistance R_{OM} is found using the equation of Fig. 4-1 once the values of R_1 through R_4 and R_F have been established.

$$R_{OM} = \frac{1}{\frac{1}{100} + \frac{1}{100} + \frac{1}{100} + \frac{1}{100} + \frac{1}{1000}} \approx 25\ \Omega$$

4-2. OP-AMP SCALING AMPLIFIER (WEIGHTED ADDER)

Figure 4-2 is the working schematic of an IC op-amp used as a scaling amplifier or weighted adder. A scaling amplifier is essentially the same as a summing amplifier, except that in the former the inputs to be summed are "weighted" or compensated to produce a given output range, or a given relationship between inputs.

For example, assume that there are two inputs to be summed and that one input has a nominal voltage range five times that of the other. Now assume that it is desired that the output voltage range be the same for both inputs. This can be done by making the input resistance for the low voltage input one-fifth the value of the high-input resistance. Since the feedback resistance is the same for both inputs, and gain (or output) is determined by the ratio of feedback-to-input resistance, the low input is multiplied five times as much as the high input. Thus, the output range is the same for both inputs.

In other cases, a scaling amplifier is used to make two equal inputs produce two outputs of different voltage ranges. Likewise, the scaling amplifier can be weighted so that two unequal inputs are made more (or less) unequal by a given *scale factor*. For example, if the inputs are normally 7 to 1, they can be made 3 to 1, 8 to 1, or any practical value within the limits of the IC op-amp.

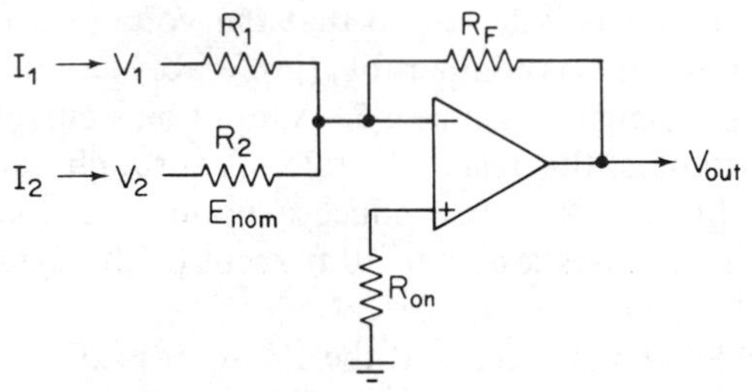

$$V_{out} = \left[\frac{R_F}{R_1}V_1 + \frac{R_F}{R_2}V_2\right]$$

$$R_1 E_{nom} = <0.1 \times \text{minimum signal across } R_1$$

$$R_2 E_{nom} = <0.1 \times \text{minimum signal across } R_2$$

$$R_1 = R_2 \times \text{scale factor}$$

$$I_1 + I_2 = \text{input bias of IC}$$

$$R_{om} = \frac{1}{\dfrac{1}{R_1} + \dfrac{1}{R_2} + \dfrac{1}{R_F}}$$

$$R_F = R_1 \times \text{desired gain of } V_1$$

$$R_1 = \frac{0.1 \times \text{minimum } V_1 \text{ input}}{I_1}$$

$$R_2 = \frac{0.1 \times \text{minimum } V_2 \text{ input}}{I_2}$$

FIG. 4-2. Scaling amplifier (weighted adder) using IC op-amp.

4-2.1 Design example

Assume that the circuit of Fig. 4-2 is to be used as a scaling amplifier for two inputs. One input has a nominal voltage range of 2 to 50 mV. The other input has a higher voltage range, 4 to 100 mV. The output range is to be the same for both inputs. That is, a full swing (2 to 50 mV) of the low input must produce the same output swing as a full swing of the high input (4 to 100 mV). In no case can the output exceed 2 V.

Obviously, R_1 (low input) and R_2 (high input) can not be the same value. The value of resistor R_2 must be twice that of R_1. The input bias current will then divide unequally between the two resistors. The current through R_1 is twice that of R_2. The fixed, no-signal voltage drop across each resistor is the same for both R_1 and R_2.

The values of R_1 and R_2 are selected so that the voltage drop (with nominal input bias current) is comparable (preferably 10 percent or less) to the minimum input signal. Assume a 6-μA input bias current.

Since the current through R_1 is twice that of R_2, the R_1 current is $\frac{2}{3}$ of the total, or 4 μA/($\frac{2}{3} \times 6 = 4$), whereas the R_2 current is 2($\frac{1}{3} \times 6 = 2$).

Since the lowest input voltage is 2 mV (2000 μV), design starts with R_1. Ten percent of 2000 μA is 200 μV. Using this 200-μV value and the 4-μA input bias, you can calculate the resistance of R_1 to be 50 Ω (200/4 = 50). With R_1 at 50 Ω, R_2 must be 100 Ω.

The total (or maximum possible) signal voltage at the IC input is 150 mV (50 mV + 100 mV). However, since R_2 is twice the value of R_1 (with R_F fixed), the gain for the low input is twice that of the high input. Thus, the maximum 50-mV input has the same effect as a 100-mV input. The effective maximum input at the IC is 200 mV (100-mV effective low input + 100-mV actual high input), with one-half the output voltage being supplied by each input.

The value of R_F should be such that the maximum output is 1 V (one-half of the required 2 V) with a maximum signal at either input. With an input at R_2 of 100 mV and an R_2 resistance of 100 Ω, the value of R_F is 1000 Ω (1 V/200 mV = 10; 10 × 100 = 1000). This value of R_F checks with the R_1 value to provide a gain of 20 for the low input (1000/50 = 20). A gain of 20 also results in an output of 1 V for the low input. Thus, the maximum output with both inputs at maximum is 2 V.

The value of the offset minimizing resistance R_{OM} is found using the equation of Fig. 4-2, once the values of R_1, R_2 and R_F have been found.

$$R_{OM} = \frac{1}{\frac{1}{50} + \frac{1}{100} + \frac{1}{1000}} \approx 32\ \Omega$$

4-3. OP-AMP DIFFERENCE AMPLIFIER (SUBTRACTOR)

Figure 4-3 is the working schematic of an IC op-amp used as a difference amplifier and/or subtractor. One signal voltage is subtracted from another through simultaneous applications of signals to both inputs.

If the values of all resistors are the same, the output is equal to the voltage at the positive (non-inverting) input, less the voltage at the negative (inverting) input. The output also represents the difference between the two input voltages. Thus, the circuit can be used as a subtractor or difference amplifier.

Generally, all resistors (R_1 through R_4) are made the same value when the circuit is to be used as a subtractor. Using all four resistors of the same value provides the greatest accuracy.

When some gain is required, the circuit is usually considered a difference amplifier (rather than a subtractor). The gain is directly proportional to the

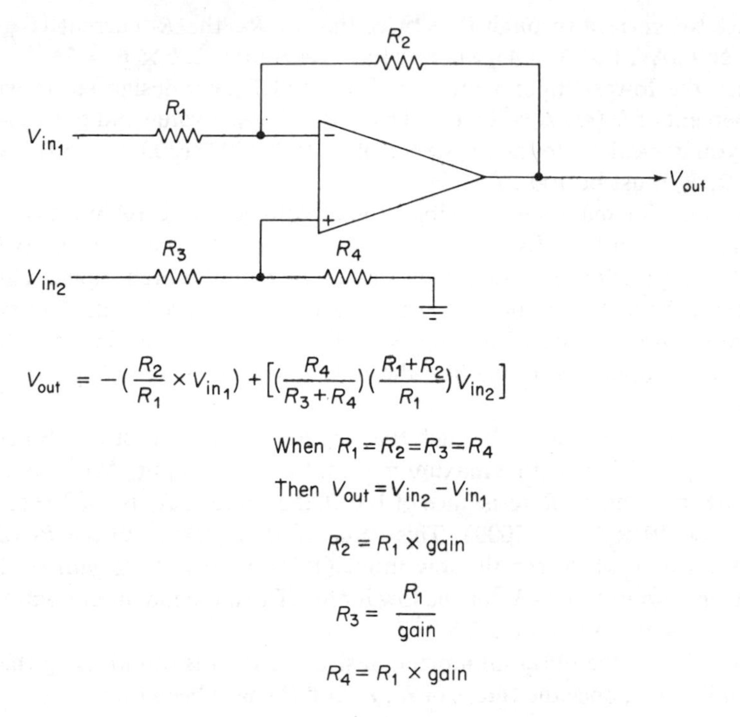

FIG. 4-3. Difference amplifier (subtractor) using IC op-amp.

ratio of R_2/R_1. Under these conditions, the output is approximately equal to the algebraic sum (or difference) of the gains for the two input voltages, as shown by the equation of Fig. 4-3.

4-3.1 Design example

Subtractor. Assume that the circuit of Fig. 4-3 is to be used as a subtractor. Each of the voltage inputs varies from 2 to 50 mV. The voltage at the V_1 input is to be subtracted from the V_2 input. The output is not to exceed approximately 50 mV.

No gain is required. Thus, all resistor values are the same. Design is based on the value of R_1.

The value of R_1 is selected so that the voltage drop (with nominal input bias current) is comparable (preferably 10 percent or less) with the minimum input signal. Assume a 2-μA input bias current.

The lowest input voltage across R_1 is 2 mV (2000 μV). Ten percent of 2000 μV is 200 μV. Using this 200-μV value and the 2-μA input bias, you can calculate the resistance of R_1 to be 100 Ω (200/2 = 100). With R_1 at 100 Ω, R_2, R_3 and R_4 must be 100 Ω.

Difference amplifier. Assume that the circuit of Fig. 4-3 is to be used as a difference amplifier. Again, each of the voltage inputs varies from 2 to 50 mV. The output voltage is to be approximately the value of the V_2 input voltage less the V_1 input voltage, multiplied by a factor of 10. Assume the same 2-μA input bias current as in the case of the straight subtractor circuit.

To provide a gain of 10, the value of R_2 must be ten times the value of R_1. Assume an arbitrary 100-Ω value for R_1 (to provide a voltage drop equal to 10 percent of the minimum input signal, with a 2-μA input bias current). With R_1 at 100 Ω, R_2 must be 1000 Ω to provide a gain of 10.

With a 100-Ω value for R_1 and a gain of 10, the values of R_3 and R_4 are 10 and 1000 Ω, respectively (using the equations of Fig. 4-3).

Now assume that the V_2 input is 50 mV and that the V_1 input is 30 mV. The output should be the difference (20 mV) times the gain (10), or 200 mV. However, using the equation of Fig. 4-3 and the values of R_1 through R_4 established in this example, note that the approximate output is slightly higher. In practical applications, the two input voltages and the output voltages are measured. Then, if necessary, the value of R_3 is trimmed slightly to produce the precise difference voltage output. The same results can be obtained if the value of R_4 is trimmed, but it is usually more practical to work with R_3.

4-4. OP-AMP UNITY GAIN AMPLIFIER (VOLTAGE FOLLOWER OR SOURCE FOLLOWER)

Figure 4-4 is the working schematic of an IC op-amp used as a *unity gain amplifier* (also known as a *voltage follower* or a *source follower*). There is no feedback or input resistance in the circuit. Instead, the output is

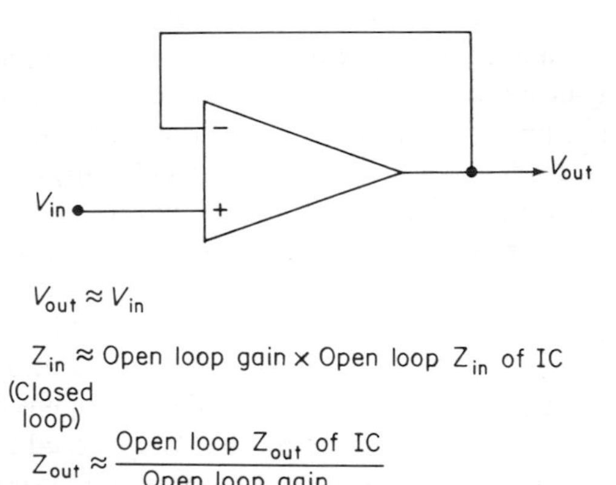

−
+
V_{in}
V_{out}
$V_{out} \approx V_{in}$
$Z_{in} \approx$ Open loop gain × Open loop Z_{in} of IC
(Closed loop)
$Z_{out} \approx \frac{\text{Open loop } Z_{out} \text{ of IC}}{\text{Open loop gain}}$
(Closed loop)

FIG. 4-4. Unity gain amplifier using IC op-amp.

fed back directly to the inverting input. Signal input is applied directly to the non-inverting input. With this arrangement, the output voltage equals the input voltage (or it may be slightly less than the input voltage). However, the input impedance is very high, with the output impedance very low (as shown by the equations of Fig. 4-4).

4-4.1 Design example

Assume that the circuit of Fig. 4-4 is to provide unity gain with high input impedance and low output impedance. Also assume that the IC as an open-loop gain of 1000 (60 dB), an output impedance of 200 Ω, and an input impedance of 15 kΩ. With these characteristics, the closed-loop input impedance is:

$$Z_{in} \approx 1000 \times 15{,}000 = 15\ \mathrm{M}\Omega$$

The closed-loop output impedance is:

$$Z_{out} \approx 200/1000 = 0.2\ \Omega$$

4-5. OP-AMP HIGH INPUT IMPEDANCE AMPLIFIER

Figure 4-5 is the working schematic of an IC op-amp used as a high input impedance amplifier. The high input impedance and low output impedance features of the unity gain amplifier (Sec. 4-4) are combined with modest gain, as shown by the equations of Fig. 4-5.

Note that the circuit of Fig. 4-5 is similar to that of the basic op-amp (Chapter 3), except that in the former there is no input offset compensating resistance (in series with the non-inverting input). This results in a tradeoff of higher input impedance, with some increase in output offset voltage. In the basic op-amp, an offset compensating resistance is used to nullify the input offset voltage of the IC. This (theoretically) results in no offset at the output. The output of the basic op-amp is at zero volts in spite of the tremendous gain. In the unity gain amplifier, there is no offset compensating resistance, but since there is no gain, the output is at the same offset as the input. Typically, input offset is on the order of 1 mV. This figure should not be critical for the output of a typical unity gain amplifier application.

In the circuit of Fig. 4-5, the offset compensation resistance is omitted. The output is therefore offset by an amount equal to the input offset voltage of the IC, multiplied by the closed-loop gain. However, since the circuit of Fig. 4-5 is to be used for modest gains, modest output offset results.

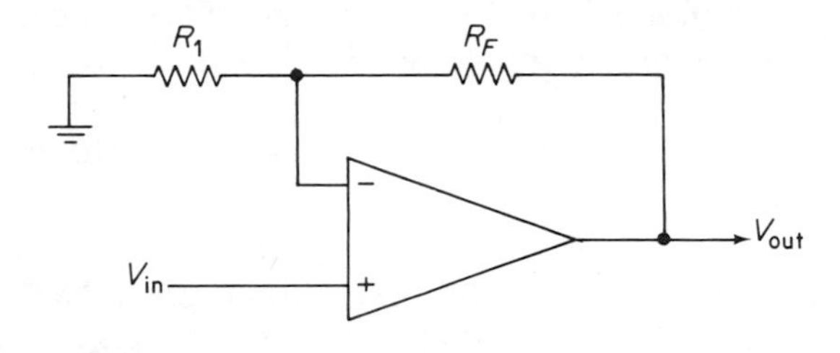

FIG. 4-5. High input impedance amplifier using IC op-amp

$$V_{out} = \frac{R_1 + R_F}{R_1} \times V_{in}$$

$$Z_{out} \text{ (Closed loop)} \approx \frac{\text{Open loop } Z_{out} \text{ of IC}}{1 + \text{Open loop gain} \left(\dfrac{R_1}{R_1 + R_F}\right)}$$

$$R_F \approx (\text{gain} - 1) \times R_1$$

$$Z_{in} \text{ (Closed loop)} \approx \text{Open loop } Z_{in} \text{ of IC} \times \text{Closed loop gain}$$

4-5.1 Design example

Assume that the circuit of Fig. 4-5 is to provide a gain of 10, with high input impedance and low output impedance. Also assume that the IC has an open-loop gain of 1000 (60 dB), an output impedance of 200 Ω, an input impedance of 15 kΩ, and an input offset voltage of 1 mV.

The value of R_1 is chosen on the basis of input bias current and voltage drop, as described previously in this chapter. Assume an arbitrary value of 100 Ω for R_1.

With 100 Ω for R_1 and a gain of 10, the value of R_F is:

$$R_F \approx (10 - 1) \times 100 = 900\ \Omega$$

With a gain of 10, the closed-loop input impedance is:

$$Z_{in} \approx 15{,}000 \times 10 = 150\ \text{k}\Omega$$

The closed loop impedance is:

$$Z_{out} \approx \frac{200}{1 + 1000 \times \left(\dfrac{100}{100 + 900}\right)} \approx 2\ \Omega$$

With a gain of 10 and an input offset voltage of 1 mV (uncompensated), the output offset voltage is:

$$1 \text{ mV} \times 10 = 10 \text{ mV}$$

4-6. OP-AMP UNITY GAIN AMPLIFIER WITH FAST RESPONSE (GOOD SLEW RATE)

One of the problems of a unity gain amplifier is that the slew rate is very poor. That is, the amplifier response time is very slow, and the power bandwidth is decreased, as discussed in Sec. 3-3.1. The reason for poor slew rate with unity gain is that most IC datasheets recommend a large-value compensating capacitor for unity gain. For example, a typical datasheet recommendation for compensating capacitance is 0.01 μF (with a gain of 100). All other factors being equal (but with unity gain), the same datasheet recommends a compensating capacitance of 1 μF.

There are several methods used to provide fast response time (high slew rate) and a good power bandwidth with unity gain. Two such methods are described here.

4-6.1 Design examples

Using IC datasheet phase compensation. The circuit of Fig. 4-6 shows a method of connecting an IC op-amp for unity gain, but with high slew rate (fast response time and good power bandwidth). With this circuit, the phase compensation recommended on the IC datasheet is used, but with modification. Instead of using the unity gain compensation (usually a high-value compensating capacitor), use the datasheet phase compensation recommended for a gain of 100 (generally a much lower value capacitor); then select values of R_1 and R_3 to provide unity gain. That is, R_1 and R_3 must be the same value. As shown by the equations, the values of R_1 and R_3 must be approximately 100 times the value of R_2 (when the datasheet phase compensation for a gain of 100 is chosen). Thus, R_1 and R_3 must be fairly high values for practical design.

Assume that the circuit of Fig. 4-6 is to provide unity gain, but with a slew rate approximately equal to that which results when a gain of 100 is used. Also assume that the input/output signal is 5 V and that the IC has an input bias current of 5 μA.

Use the compensation scheme (values and connections recommended on the datasheet) for a gain of 100.

Next, select values of R_1, R_2, and R_3. The value of R_1 (and, consequently, that of R_3) is selected so that the voltage drop with nominal input bias current

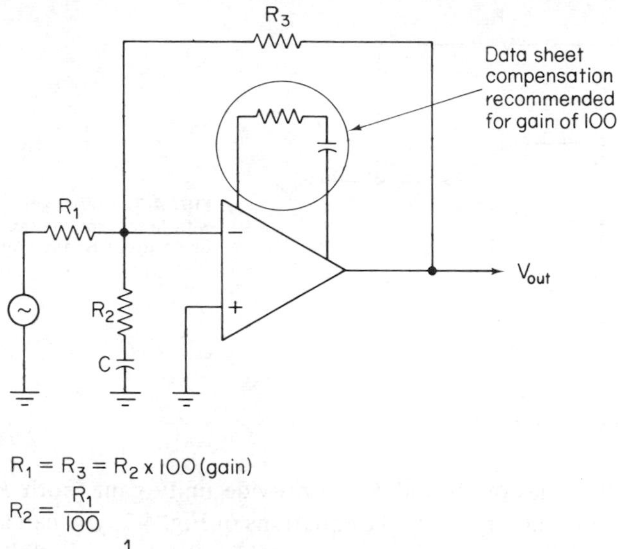

FIG. 4-6. Unity gain IC op-amp with fast response (good slew rate) using IC datasheet phase compensation

is comparable (preferably 10 percent or less) to the input signal. Ten percent of the 5-V input is 0.5 V. Using the 0.5-V value and the 5-μA input bias, you can calculate the resistance of R_1 to be 100 kΩ (0.5 V/5 μA = 100,000). With R_1 at 100 kΩ, R_3 must also be 100 kΩ, and R_2 must be 1 kΩ (100,000/100 = 1 kΩ).

The value of C_1 is found using the equations of Fig. 4-6 once the value of R_2 and the open-loop rolloff point are established. Assume that the datasheet shows a 5-kHz rolloff point (where flat response changes to a 6-dB/octave rolloff; point D of Fig. 3-6).

$$C = \frac{1}{6.28 \times 5000 \times 1000} \approx 0.03\ \mu\text{F}$$

Using input phase compensation. The circuit of Fig. 4-7 shows a method of connecting an IC op-amp for unity gain, but with high slew rate, using input compensation. With this circuit, the phase compensation recommended on the IC datasheet is not used. Instead, the input phase compensation system in Sec. 3-2.3 is used.

The first step is to compensate the IC by modifying the open-loop input impedance, as described in Sec. 3-2.3.

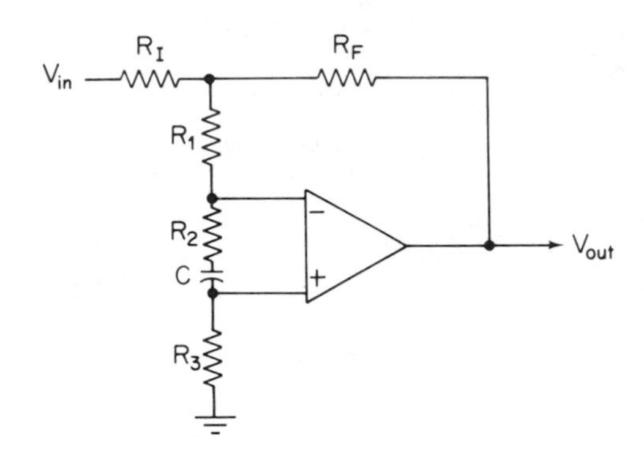

R_1, R_2, R_3, C = see text
$R_I = R_F = 0.25 \times R_1$

FIG. 4-7. Unity gain IC op-amp with fast response (good slew rate) using input phase compensation.

Next, select values of R_I and R_F to provide unity gain (both R_I and R_F must be the same value). Using the equations of Fig. 4-7, you can assume the values of R_I and R_F to be approximately one-fourth the value of R_1 (and R_3). Assuming that the values of R_1 and R_3 are 43 kΩ (using the example of Sec. 3-2.3), 10 kΩ is a good trial value for R_I and R_F.

4-7. OP-AMP NARROW BANDPASS AMPLIFIER

Figure 4-8 is the working schematic of an IC op-amp used as a narrow bandpass amplifier (or tuned peaking amplifier).

Circuit gain is determined by the ratio of R_1 and R_F in the usual manner. However, the frequency at which maximum gain occurs (or the narrow band peak) is the resonant frequency of the L_1C_1 circuit. Capacitor C_1 and inductance L_1 form a parallel-resonant circuit that rejects the resonant frequency. Therefore, there is minimum feedback (and maximum gain) at the resonant frequency.

4-7.1 Design example

Assume that the circuit of Fig. 4-8 is to provide 20-dB gain at a peak frequency of 100 kHz.

The value of R_1 is chosen on the basis of input bias current and voltage drop, as described previously in this chapter. Assume an arbitrary value of 3.3 kΩ for R_1. The value of R_{OM} is then the same, or slightly less.

With a value of 3.3 kΩ for R_1, the value of R_F is 33 kΩ (or R_1 times 10) for a 20-dB gain.

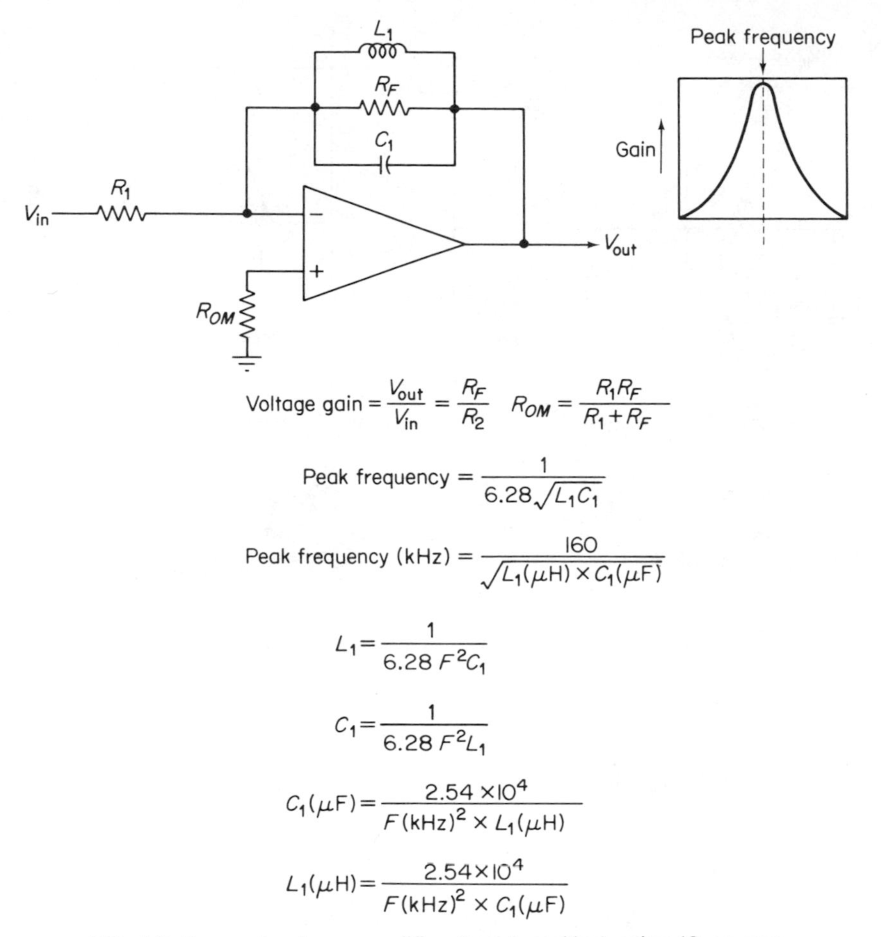

L_1
R_F
C_1
R_1
V_{in}
R_{OM}
V_{out}
Peak frequency
Gain
Voltage gain $= \frac{V_{out}}{V_{in}} = \frac{R_F}{R_2}$ $R_{OM} = \frac{R_1 R_F}{R_1 + R_F}$
Peak frequency $= \frac{1}{6.28\sqrt{L_1 C_1}}$
Peak frequency (kHz) $= \frac{160}{\sqrt{L_1(\mu H) \times C_1(\mu F)}}$
$L_1 = \frac{1}{6.28\, F^2 C_1}$
$C_1 = \frac{1}{6.28\, F^2 L_1}$
$C_1(\mu F) = \frac{2.54 \times 10^4}{F(kHz)^2 \times L_1(\mu H)}$
$L_1(\mu H) = \frac{2.54 \times 10^4}{F(kHz)^2 \times C_1(\mu F)}$

FIG. 4-8. Narrow bandpass amplifier (tuned peaking) using IC op amp.

Any combination of L_1 and C_1 can be used, provided that the resonant frequency is 100 kHz. For frequencies below 1 MHz, the value of C_1 should be between 0.001 and 0.01 μF. Assume an arbitrary 0.0015 μF for C_1. Using the equations of Fig. 4-8, you can find the value of L_1 to be approximately 1700 μH.

4-8. OP-AMP WIDE BANDPASS AMPLIFIER

Figure 4-9 is the working schematic of an IC op-amp used as a wide bandpass amplifier.

Maximum circuit gain is determined by the ratio of R_R and R_F. That is, the gain of the passband or flat portion of the response curve is set by R_F/R_R.

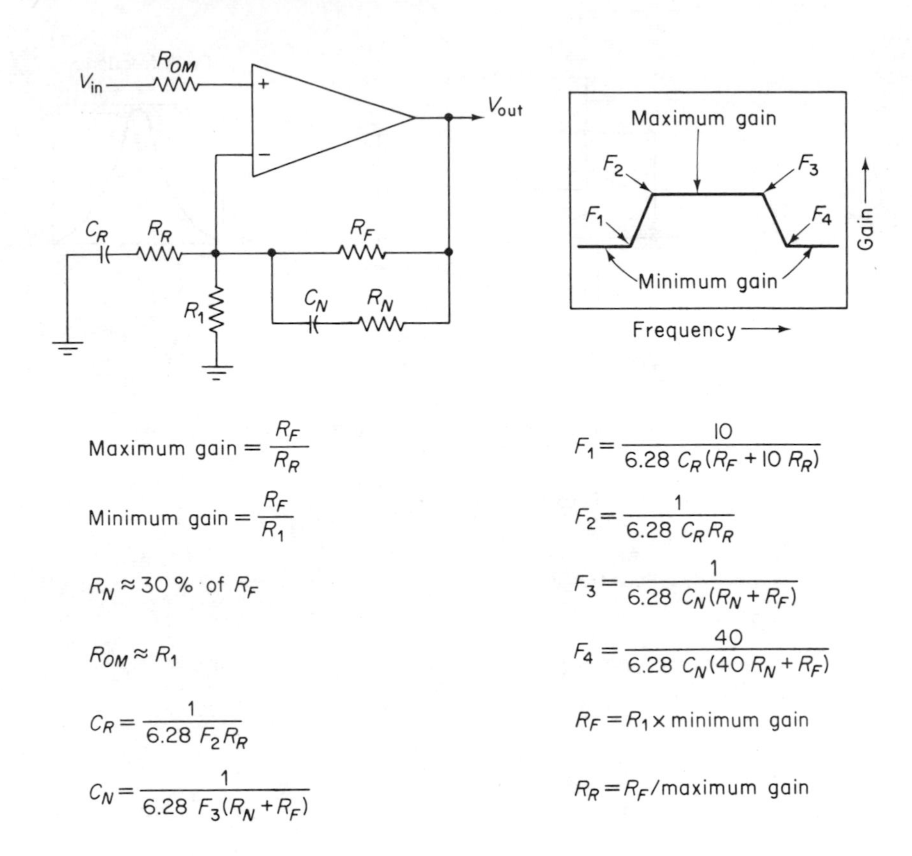

$$\text{Maximum gain} = \frac{R_F}{R_R}$$

$$\text{Minimum gain} = \frac{R_F}{R_1}$$

$$R_N \approx 30\% \text{ of } R_F$$

$$R_{OM} \approx R_1$$

$$C_R = \frac{1}{6.28\, F_2 R_R}$$

$$C_N = \frac{1}{6.28\, F_3 (R_N + R_F)}$$

$$F_1 = \frac{10}{6.28\, C_R (R_F + 10\, R_R)}$$

$$F_2 = \frac{1}{6.28\, C_R R_R}$$

$$F_3 = \frac{1}{6.28\, C_N (R_N + R_F)}$$

$$F_4 = \frac{40}{6.28\, C_N (40\, R_N + R_F)}$$

$$R_F = R_1 \times \text{minimum gain}$$

$$R_R = R_F/\text{maximum gain}$$

FIG. 4-9. Wide bandpass amplifier using IC op-amp.

Minimum circuit gain is determined by the ratio of R_1 and R_F in the usual manner.

The frequencies at which rolloff starts and ends (at both high and low frequency limits) are determined by impedances of the various circuit combinations, as shown by the equations of Fig. 4-9.

4-8.1 Design example

Assume that the circuit of Fig. 4-9 is to provide approximately a 20-dB minimum gain at all frequencies and approximately a 30-dB gain at the passband. Gain is to start increasing at frequencies above 10 kHz and rise to an approximate 30-dB passband at 40 kHz. The passband must extend to 200 kHz and then drop back to 20 dB at frequencies 800 kHz and above.

Note that if phase compensation is required for the basic IC, the compensation values must be based on the *minimum of* 20-*dB gain* and not on the passband gain of 30 dB.

The value of R_1 is chosen on the basis of input bias current and voltage drop, as described previously in this chapter. Assume an arbitrary value of 1 kΩ for R_1. The value of R_{OM} is then the same, or slightly less.

With a value of 1 kΩ for R_1, a value of 10 kΩ is used for R_F. With a value of 10 kΩ for R_F, the value of R_R is 330 Ω. These relationships produce gains of 20 dB and 30 dB, respectively. In practice, it may be necessary to reduce both of these trial values to get the desired gain relationship.

With a value of 330 Ω for R_R, the value of C_R is 0.012 μF.

With a value of 10 kΩ for R_F, the value of R_N is 3 kΩ.

With a value of 13 kΩ for $(R_N + R_F)$, the value of C_N is 61 pF.

With these values established, the remaining equations in Fig. 4-9 are used to confirm the four frequencies.

4-9. OP-AMP INTEGRATION AMPLIFIER

Figure 4-10 is the working schematic of an IC op-amp used as an integration amplifier (or *integrator*). Integration of various signals (usually

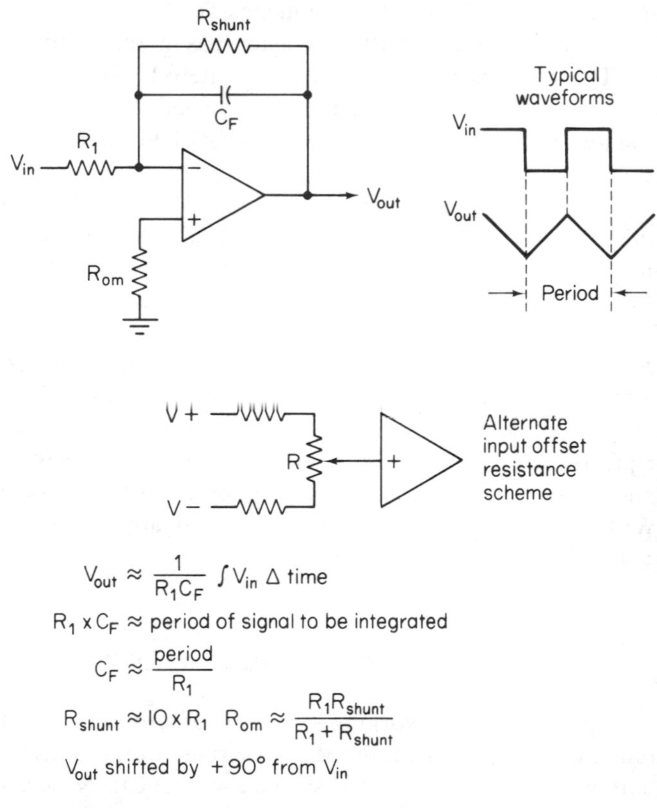

$$V_{out} \approx \frac{1}{R_1 C_F} \int V_{in}\, \Delta \text{ time}$$

$R_1 \times C_F \approx$ period of signal to be integrated

$$C_F \approx \frac{\text{period}}{R_1}$$

$$R_{shunt} \approx 10 \times R_1 \quad R_{om} \approx \frac{R_1 R_{shunt}}{R_1 + R_{shunt}}$$

V_{out} shifted by +90° from V_{in}

FIG. 4-10. Integration amplifier (integrator) using IC op-amp.

square waves) can be accomplished using this circuit. The output voltage from the amplifier is inversely proportional to the time constant of the feedback network and directly proportional to the integral of the input voltage.

4-9.1 Design example

The value of R_1 is chosen on the basis of input bias current and voltage drop, as described previously in this chapter. Assume that the input bias current is 5 μA and that the desired voltage drop across R_1 is not to exceed 180 mV (say, to provide 10 percent of a 1.8-V input signal). A value of 33 kΩ for R_1 will produce a 165-mV drop. Assume an arbitrary value of 33 kΩ for R_1.

The value of the R_1C_F time constant must be *approximately equal to the period* of the signal to be integrated.

The value of the $R_{shunt}C_F$ time constant must be *substantially larger than the period* of the signal to be integrated (approximately ten times longer). Thus, R_{shunt} is ten times R_1.

Keep in mind that R_{shunt} and C_F form an impedance that is frequency-sensitive (i.e., it is most noticed at low frequencies).

Assume that the circuit of Fig. 4-10 is to be used as an integrator for 1-kHz square waves. This requires a period of approximately 1 millisecond.

Any combination of C_F and R_1 can be used, provided the value of C_F times R_1 is approximately 0.001. Using the assumed value of 33 kΩ for R_1, the value of C_F is 0.03 μF.

The value of R_{shunt} must then be at least 330 kΩ. Note that the purpose of R_{shunt} is to provide *direct-current feedback*. This feedback is necessary so that an offset voltage can not continuously charge C_F (which can result in amplifier limiting). If the offset voltage is small or can be minimized by means of R_{OM}, it is possible to eliminate R_{shunt}.

Resistor R_{shunt} may have the effect of limiting gain at very low frequencies. However, at frequencies above about 15 Hz, the effect of R_{shunt} is negligible (because of C_F in parallel).

If greater precision is required, particularly at low frequencies, the input offset resistance R_{OM} can be replaced by the potentiometer network shown in Fig. 4-10. With this arrangement, potentiometer R is adjusted so that there is no offset voltage under no-signal conditions.

4-10. OP-AMP DIFFERENTIATION AMPLIFIER

Figure 4-11 is the working schematic of an IC op-amp used as a differentiation amplifier (or differentiator). Differentiation of various signals (usually square waves, or sawtooth and sloping waves) can be accomplished

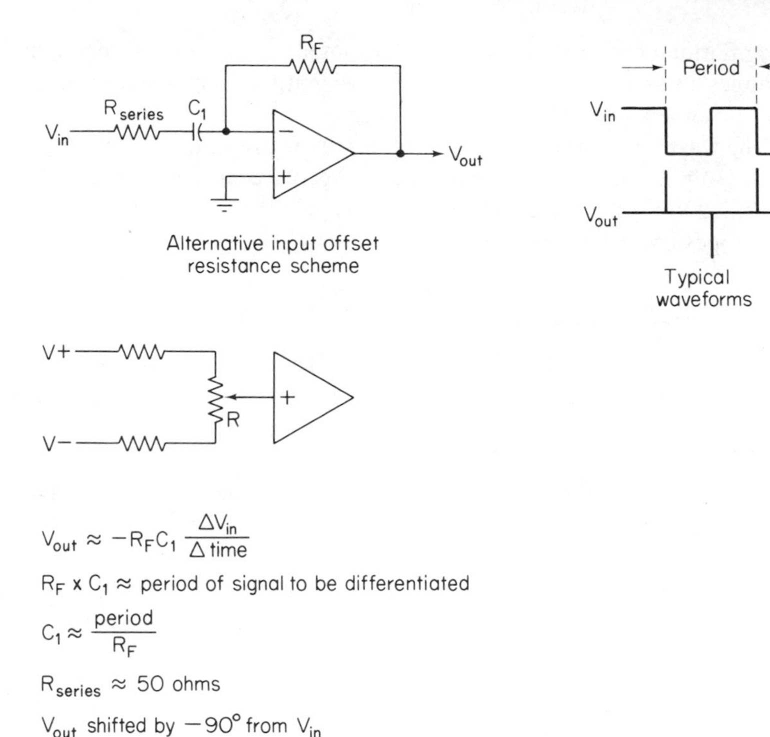

FIG. 4-11. Differentiation amplifier (differentiator) using IC op-amp.

using this circuit. The output voltage from the amplifier is inversely proportional to the feedback time constant, and is directly proportional to the time rate of change of the input voltage.

The value of the $R_F C_1$ time constant should be *approximately equal* to the period of the signal to be differentiated. In practical applications, the time constant must be chosen on a trial and error basis to obtain a reasonable output level.

The main problem in design of differentiating amplifiers is that the *gain increases with frequency*. Thus, differentiators are very susceptible to high-frequency noise. The classic remedy for this effect is to connect a small resistor (on the order of 50 Ω) in series with the input capacitor so that the high-frequency gain is decreased. The addition of the resistor results in a more realistic model of the differentiator because a resistance is always added in series with the input capacitance by the signal source impedance.

Conversely, in some applications a differentiator may be advantageously used to detect the presence of distortion or high-frequency noise in the signal. A differentiator can often detect hidden information that is not detected in the original signal.

Differentiation permits slight changes in input slope to produce very significant changes in output. An example of this feature is in determining the linearity of a sweep sawtooth waveform. Nonlinearity results from changes in slope of the waveform. Therefore, if nonlinearity is present, the differentiated waveform (amplifier output) indicates the points of nonlinearity quite clearly. (However, it should be noted that repetitive waveforms with a rise and fall of differing slopes can show erroneous waveforms.)

4-10.1 Design example

Assume that the circuit of Fig. 4-11 is to be used to differentiate 1-kHz waves. This requires a period of approximately 1mS. Any combination of C_1 and R_F can be used, provided the value of C_1 times R_F is approximately 0.001.

The values used for the integrator of Sec. 4-9 can be used as initial trial values, even though the components are interchanged. Thus, C_1 is 0.03 μF, with a 33-kΩ value for R_F.

The value of R_{Series} (if used) should be an arbitrary 50 Ω. Keep in mind that R_{Series} and C_1 form an impedance which is frequency sensitive (that will be most noticed at high frequencies).

4-11. OP-AMP VOLTAGE-TO-CURRENT CONVERTER

Figure 4-12 is the working schematic of an IC op-amp used as a voltage-to-current converter (or transadmittance amplifier). This circuit is used to supply a current (to a variable load) which is proportional to the voltage applied at the input of the amplifier (rather than proportional to the load). The current supplied to the load is *relatively independent* of the load characteristics. This circuit is essentially a *current feedback* amplifier.

Current sampling resistor R is used to provide the feedback to the positive input. When R_1, R_2, R_3 and R_4 are all the same value, the feedback maintains the voltage across R at the same value as the input voltage. When R_2 is made larger than R_1, the voltage across R remains constant at a value equal to the ratio R_2/R_1.

If a constant input voltage is applied to the amplifier, the voltage across R also remains constant, regardless of the load (with very close tolerances). If the voltage across R remains constant, the current through R must also remain constant. With R_3 and R_4 normally much larger than the load impedance, the current through the load remains nearly constant, regardless of a change in impedance.

The most satisfactory configuration for the circuit of Fig. 4-12 is where the IC is operated as a unity gain amplifier, with the values of R_1 through R_4 all

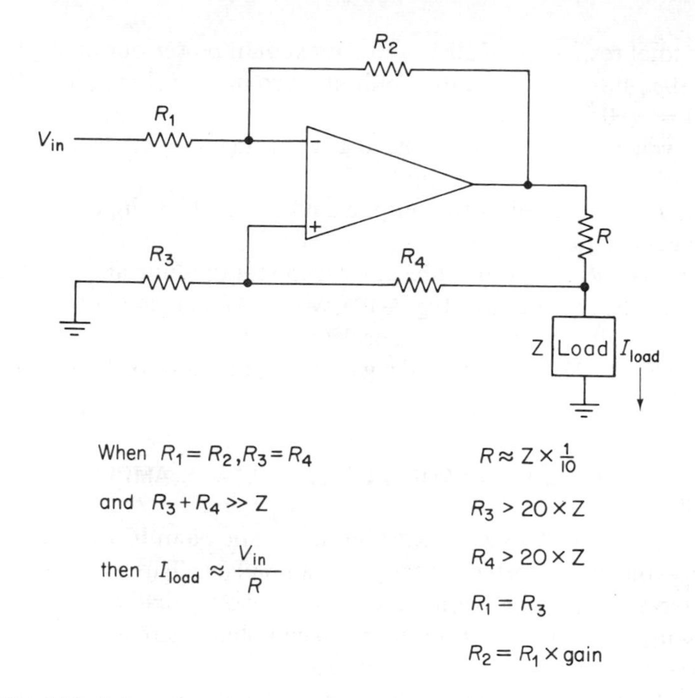

FIG. 4-12. Voltage-to-current converter (transadmittance amplifier) using IC op-amp.

the same. This requires that the input voltage be sufficient to produce the desired current (or power) for the load. If the input voltage or signal is not sufficient, the values of R_1 and R_2 must be selected to provide the necessary gain.

The value of R must be selected to limit the output power ($I^2 \times (R + \text{load})$) to a value within the capability of the IC. For example, if the IC is rated at 600 mW total dissipation, with 100 mW dissipation for the basic IC, the total output power must be limited to 500 mW. As a guideline, make the value of R approximately 1/10 of the load.

4-11.1 Design example

Assume that the circuit of Fig. 4-12 is to be used as a voltage-to-current converter. The output load is 45 Ω (nominal). The maximum power output of the IC is 500 mW. It is desired to maintain the maximum output current, regardless of variation in output load, with a constant input voltage of 5 mV.

With a 45-Ω load, the value of R is approximately 4.5 Ω. The combined resistance of R and the load is then 49.5 Ω (rounded off to 50 Ω for convenience).

With a total resistance of 50 Ω, and a maximum power output of 0.5 W for the IC, the maximum possible output current is 0.1 A ($I = \sqrt{P/R} = \sqrt{0.5/50} = \sqrt{0.01} = 0.1$ A)

With a value of 4.5 Ω for R, and 0.1 A through R, the drop across R is 450 mV.

With 450 mV required at the output, and 5 mV at the input, the required amplifier gain is 90.

The values of R_3 and R_4 must be at least 950 Ω (anything above 900 Ω) each (as shown by the equations of Fig. 4-12), with a nominal load of 45 Ω.

The value of R_1 is the same as R_3, or 950 Ω.

With a value of 950 Ω for R_1, and a gain of 90, the value of R_2 is 85.5 kΩ.

4-12. OP-AMP VOLTAGE-TO-VOLTAGE AMPLIFIER

Figure 4-13 is the working schematic of an IC op-amp used as a voltage-to-voltage converter (voltage gain amplifier). This circuit is similar to the voltage-to-current converter (Sec. 4-11) except that the load and the current sensing resistor are transposed. The voltage across the load is relatively independent of the load characteristics.

The most satisfactory configuration for the circuit of Fig. 4-13 is where the IC is operated as a unity gain amplifier, with the values of R_1 through R_4 all the same. This requires that the input voltage be sufficient to produce the desired current (or power) for the load. If the input voltage or signal is not sufficient, the values of R_1 and R_2 must be selected to provide the necessary gain.

The value of R must be selected to limit the output power ($I^2 \times (R + \text{load})$) to a value within the capability of the IC. For example, if the IC is rated at

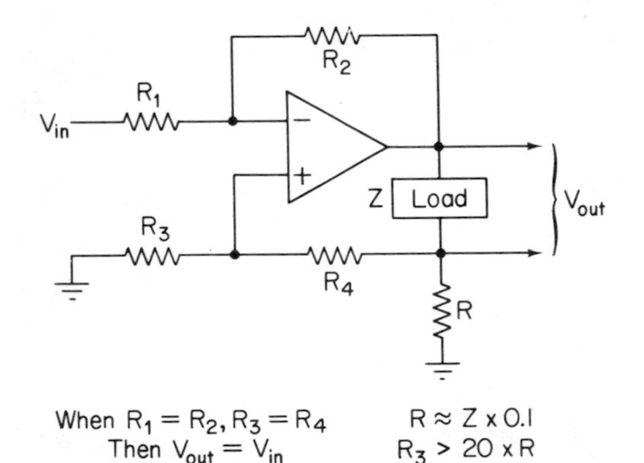

FIG. 4-13. Voltage-to-voltage converter (voltage gain amplifier) using IC op-amp.

600-mW total dissipation, with 100-mW dissipation for the basic IC, the total output power must be limited to 500 mW. As a guideline, make the value of R approximately 1/10 of the load.

4-12.1 Design example

Assume that the circuit of Fig. 4-13 is to be used as a voltage-to-voltage converter. The output load is 100 Ω (nominal). The maximum power output of the IC is 500 mW. It is desired to maintain the maximum output voltage across the load, regardless of variation in load (within the current capabilities of the IC), with a constant input voltage of 50 mV.

With a 100-Ω load, the value of R is approximately 10 Ω. The combined resistance of R and the nominal load is then 110 Ω.

With a total resistance of 110 Ω, and a maximum power output capability of 0.5 W for the IC, the maximum possible output current is 0.07 A ($I = \sqrt{P/R} = \sqrt{0.5/110} = \sqrt{0.0045} = 0.07$ A).

With a nominal value of 100 Ω for the load, and 0.07 A through the load, the maximum drop across the load is 7 V. This value may be used, provided the IC is capable of a 7-V output. If not, use a voltage output that is within the IC capabilities.

With 7 V required at the output, and 50 mV at the input, the amplifier gain must be 140.

The values of R_3 and R_4 should be at least 200 Ω each (as shown by the equations of Fig. 4-13), with a value of 10 Ω for R.

The value of R_1 should be the same as R_3, or 200 Ω.

With a value of 200 Ω for R_1 and a gain of 140, the value of R_2 should be 28 kΩ.

4-13. OP-AMP LOW-FREQUENCY SINEWAVE GENERATOR

Figure 4-14 is the working schematic of an IC op-amp used as a low-frequency sinewave generator. This circuit is a parallel-T oscillator. Feedback to the negative input becomes positive at the frequency indicated in the equation. Positive feedback is applied at all times. The amount of positive feedback (set by the ratio of R_1 to R_2) is sufficient to cause the IC amplifier to oscillate. In combination with the feedback to the negative input, feedback to the positive input can be used to stabilize the amplitude of oscillation.

The value of R_1 is approximately 10 times the value of R_2. The ratio of R_1 and R_2, as set by the adjustment of R_2, controls the amount of positive feedback. Thus, the setting of R_2 determines the stability of oscillation.

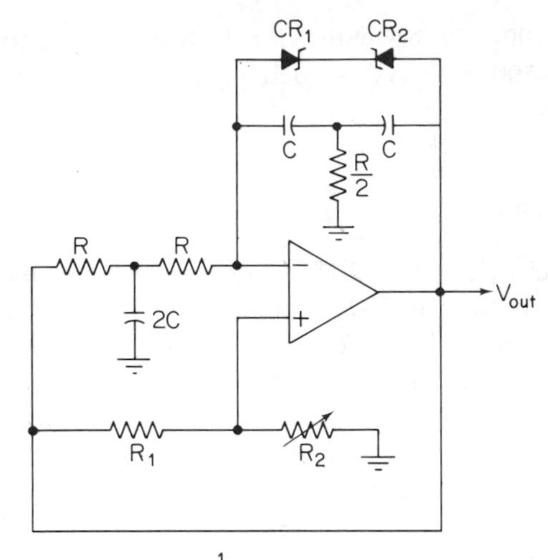

FIG. 4-14. Low-frequency sine-wave generator using IC op-amp.

$$\text{Frequency} \approx \frac{1}{6.28\,RC}$$

$$R \approx \frac{1}{6.28\,FC}$$

$$R < 2 \text{ megohms}$$

$$C \approx \frac{1}{6.28\,RF}$$

$$R_1 \approx R_2 \times 10$$

Zener point of CR_1 and $CR_2 \approx 1.5 \times V_{out}$ (peak-to-peak)

The amplitude of oscillation is determined by the peak-to-peak output capability of the IC, and the values of Zener diodes CR_1 and CR_2. As shown by the equations, the Zener voltage should be approximately 1.5 times the desired peak-to-peak output voltage. The non-linear resistance of the back-to-back Zener diodes is used to limit the output amplitude and maintain good linearity.

The frequency of oscillation is determined by the values of C and R. The upper-frequency limit is approximately equal to the bandwidth of the basic IC. That is, if the open-loop gain drops 3 dB at 100 kHz, the oscillator should provide full voltage output up to about 100 kHz.

4-13.1 Design example

Assume that the circuit of Fig. 4-14 is to provide 10-V sine-wave signals at 1 Hz.

Since R_2 is variable, its exact value is not critical. Assume a maximum value of 10 kΩ for R_2.

With a maximum value of 10 kΩ for R_2, the value of R_1 is 100 kΩ.

With a required 10 V peak-to-peak output, the values (Zener voltage) of CR_1 and CR_2 should be 15 V. It is assumed that the basic IC is capable of 10 V peak-to-peak output.

The values of R and C are related to the desired frequency of 1 Hz. Any combination of R and C can be used, provided that the combination worked out to a frequency of 1 Hz. However, for practical design, the value of R should not exceed about 2 MΩ (for a typical IC). Using a value of 1.6 MΩ for R, the value of C is 0.1 μF.

4-14. OP-AMP PARALLEL-T FILTER

Figure 4-15 is the working schematic of an IC op-amp used as a low-frequency filter. The operating principle involved is similar to that of the parallel-T oscillator described in Sec. 4-13. However, the function is that of a narrow band filter (tuned peaking amplifier) described in Sec. 4-7. The circuit described in Sec. 4-7 uses an inductance as part of the resonant circuit. At very low frequencies, the high values of inductance (and capacitance) required make a circuit similar to that of Sec. 4-7 impractical. Thus, for low frequencies, the parallel-T (or twin-T as it is sometimes called) filter is generally a better choice.

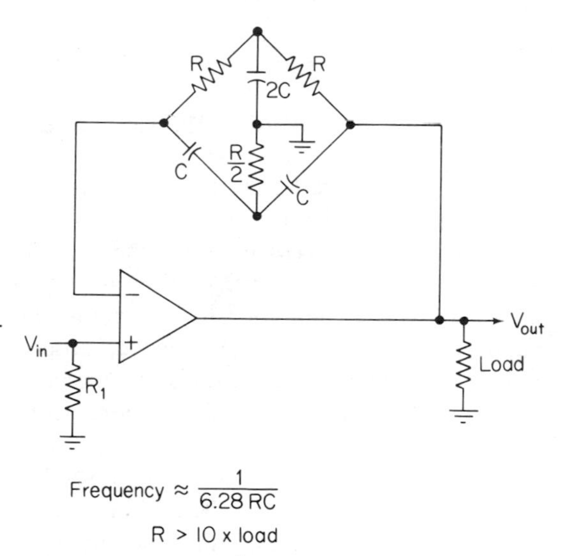

FIG. 4-15. Low frequency parallel-T (twin-T) filter using IC op-amp.

$$\text{Frequency} \approx \frac{1}{6.28\,RC}$$

$$R > 10 \times \text{load}$$

$$C \approx \frac{1}{6.28\,RF}$$

R_1 = see text

In the circuit of Fig. 4-15, gain is determined by the open-loop gain of the IC. For this reason, the parallel-T filter is somewhat less stable than the tuned amplifier.

The frequency at which maximum gain occurs (or narrow band peak) is determined by the values of R and C. Any combination of R and C can be used, provided that they work out to the desired frequency. However, the value of R should be selected on the basis of the load resistance (or load impedance). The value of R and the load are, in effect, in parallel. If the value of R is many times (at least 10) that of the load, the net parallel resistance will be just slightly less than the load. Thus, the output current requirements for the IC are increased only slightly (for a given voltage).

4-14.1 Design example

Assume that the circuit of Fig. 4-15 is to provide a peak frequency of 10 Hz, and that the load impedance is 15 kΩ.

The value of R_1 is chosen on the basis of input bias current and voltage drop, as described previously in this chapter. In some cases, R_1 is chosen to provide a given input impedance, since filters must often work between two stages which require impedance match. Assume an arbitrary value of 10 kΩ for R_1.

With a load of 15 kΩ, the value of R is at least 150 kΩ. Use an arbitrary 160 kΩ for R.

With a value of 160 kΩ for R, and a desired peak frequency of 10 Hz, the value of C is 0.1 μF.

4-15. OP-AMP MONOSTABLE MULTIVIBRATOR

Figure 4-16 is the working schematic of an IC op-amp used as a monostable (one-shot) multivibrator. Note that Fig. 4-16 also includes a timing diagram. It is essential that the user study the timing diagram, as well as the following design considerations, when using an IC as the active element in a monostable multivibrator. Although an IC op-amp can be readily adapted to multivibrator use, failure to observe certain conditions will result in improper timing, and could result in damage to the IC.

Before discussing specific design considerations, we shall describe operational theory of the basic circuit.

Basic theory. The function of a monostable MV is to produce an output in response to an input trigger. The width or duration of the output is set by the monostable circuit values, rather than by the trigger input. Thus, the output pulse can be "stretched" or set to a given width by selection of circuit values.

When op-amps are used as monostable multivibrators, the op-amp action is similar to that of a comparator. One side of the differential input stage is

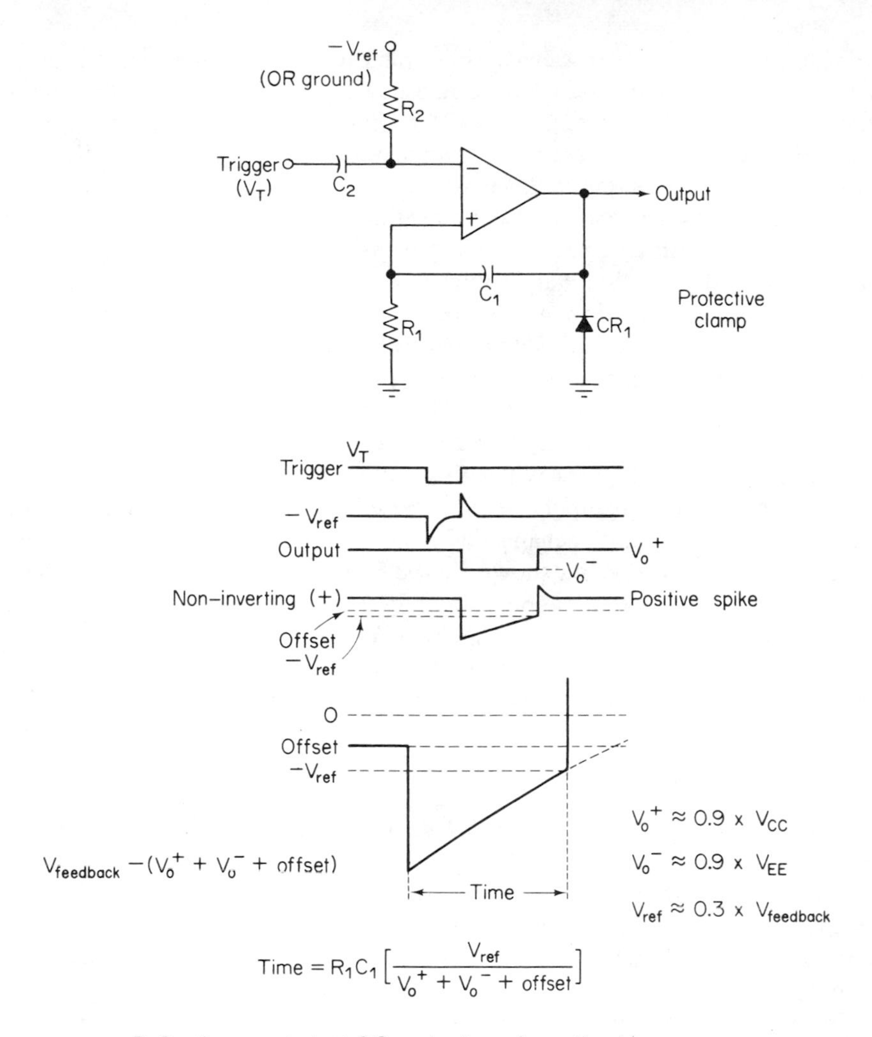

$$\text{Time} = R_1C_1\left[\frac{V_{ref}}{V_o^+ + V_o^- + \text{offset}}\right]$$

R_2C_2 time constant $\approx$ 0.2 x duration of negative trigger

FIG. 4-16. Monostable (one-shot) multivibrator using IC op-amp.

set initially "on" whereas the remaining side is "off." The output stage will either be saturated or cut off, depending upon which side of the differential input is involved. The magnitude and polarity of the differential input voltage will, therefore, control the state of the output, and can be set accordingly.

In the absence of an input trigger pulse, the non-inverting input (positive) is near ground potential, and the inverting input (negative) is at the level of the reference voltage V_{ref}. Since these two levels constitute inputs to a differential input amplifier stage, the output of the IC is at voltage level V_o+, the maximum positive output voltage swing.

When positive-going trigger V_T is coupled to the inverting input through capacitor C_2, the input base is driven in a positive direction. When the resultant base voltage becomes more positive than the non-inverting input, the input differential stage reverses state and drives the IC output to level V_o^-, the maximum negative output voltage swing.

The output voltage change is coupled through timing capacitor C_1 to the non-inverting input as a negative-going transient. Thus, the IC is effectively latched into the existing state (output at V_o^-). Timing capacitor C_1 then charges toward ground through resistor R_1, as shown in the timing diagram.

When the voltage level at the non-inverting input charges to a level equal to that at the inverting input (equal to $-V_{ref}$), differential action occurs again, and the output returns to level V_o^+. This completes the basic timing period.

The initial change in IC states (beginning of timing period) occurs as the result of a positive trigger. When the trailing edge of the trigger input is coupled to the inverting input, the input base is driven more negative, and the output of the IC remains unchanged (V_o^-).

Design considerations. As shown by the equations of Fig. 4-16, amplitude of the monostable output is approximately 90 percent of the supply voltage (V_{CC}, V_{EE}). Thus, if V_{CC} and V_{EE} are 6 V, the outputs are +5.4 V and −5.4 V (typical levels for most digital logic systems). When the output switches to a negative state (at the beginning of the timing period), a negative transient equal to the sum ($V_o^+ + V_o^-$) is coupled back to the input. Precautions must be taken to insure that the IC input is capable of accepting this input without damage.

For example, if V_o^+ and V_o^- are 5.4 V, the non-inverting input will receive about 10.8 V at the start of the timing period. If this value is too high for the IC, protective clamp diode CR_1 can be added as shown in Fig. 4-16. With CR_1 in the circuit, the IC input receives the sum of V_o^+ and the threshold voltage of CR_1 (typically 0.5 to 0.7 V). Thus, if V_o^+ is 5.4, the IC input receives about 5.9 to 6.1 V.

In addition to the negative transient, the IC input has an offset voltage developed across R_1. This voltage is determined primarily by circuits within the IC (base-emitter voltage of input stages), and is typically less than 1 V. The offset voltage must be added to the sum of the output voltages to find the actual voltage at the non-inverting input when the timing period starts (as shown by the timing diagram). This total voltage is designated as $V_{feedback}$.

Resistor R_2 is returned to a negative reference voltage, rather than to ground. This reference voltage insures complete cutoff of the IC during initial circuit conditions (before the positive trigger is applied). The reference voltage also controls the period of the output pulse. For example, as shown by the timing diagram, the output pulse period is increased if the reference voltage is decreased, and vice versa. The IC switches back to the initial state (end of the output pulse) when C_1 discharges to the level of the reference voltage.

The limits of the reference voltage are set primarily by the trigger voltage and the offset voltage across R_1. The negative reference voltage must be larger than the offset voltage (to insure cutoff), but smaller than the trigger (to insure that the trigger will switch states of the IC). As a secondary consideration, V_{ref} must be less than $V_{feedback}$. Typically V_{ref} should be no greater than about $\frac{1}{3}$ of $V_{feedback}$.

The period of the output pulse is determined by the RC time constant of R_1C_1, the ratio of V_{ref} to $V_{feedback}$, and by constants of the IC. In practical use, trial values of R and C must be selected, and the actual period of the IC output measured on an oscilloscope. However, the graph of Fig. 4-17 provides an approximate constant that can be applied for a typical IC. The curve shows a fixed $V_{feedback}$ of 6–7 V versus a V_{ref} that varies between 0.5 and 2.5 V. To find the approximate period of the output pulse with the graph, multiply the R_1C_1 time constant by the graph constant.

For example, assume a V_{ref} of -2 V and an R_1C_1 time of 100 ms. The -2 V line (vertical) intersects the curve at a constant of about 1.2. Under these conditions, the period of the output pulse is 120 μs (100×1.2).

Keep in mind that the graph of Fig. 4-17 provides approximate values that will vary from IC to IC. Actual values must be determined by test. Refer to Chapter 7. Also, note that the graph does not go below about 0.5 V (which is considered as a typical value for offset voltage across R_1) or above 2.5 V (which is greater than $\frac{1}{3}$ of $V_{feedback}$, and possibly greater than a typical trigger voltage).

The input R_2C_2 circuit must be capable of charging and discharging at a rate faster than that of the input. If not, it is possible that a slow or incomplete discharge of C_2 will affect the output timing, and could affect circuit opera-

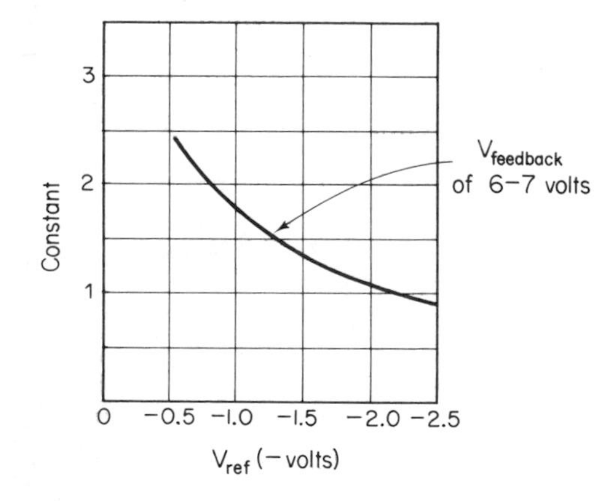

FIG. 4-17. Graph for determining period of monostable multivibrator output pulse.

tion. As a guideline, the duration of the negative trigger should be 5 times that of the R_2C_2 time constant. Often, R_2 must be chosen to match an input impedance.

In some uses, it is necessary that the output pulse not go to or below the zero level. This can be accomplished by means of the circuit in Fig. 4-18a. With this circuit, the output pulse will not go below the value of V^+. Of course, V^+ must not be greater than V_o^+. Likewise, the total output swing is equal to the difference between V_o^+ and V^+.

In other uses, it is necessary that the output pulse go to the full positive and negative extremes (V_o^+ and V_o^-). If the combination of these voltages is above the safe input level of the IC, a protective diode can be connected into the circuit as shown in Fig. 4-18b. The output then remains at full value, but $V_{feedback}$ is equal to V_o^+ plus the threshold voltage of the diode (about 0.7 V).

Note that the timing is affected by either circuit shown in Fig. 4-18. That is, the output pulse duration will not be the same as for the basic circuit of Fig. 4-16, even though identical component values are used. However, the component values shown for Fig. 4-16 can be used as trial values for the Fig. 4-18 circuits, with actual timing determined by test.

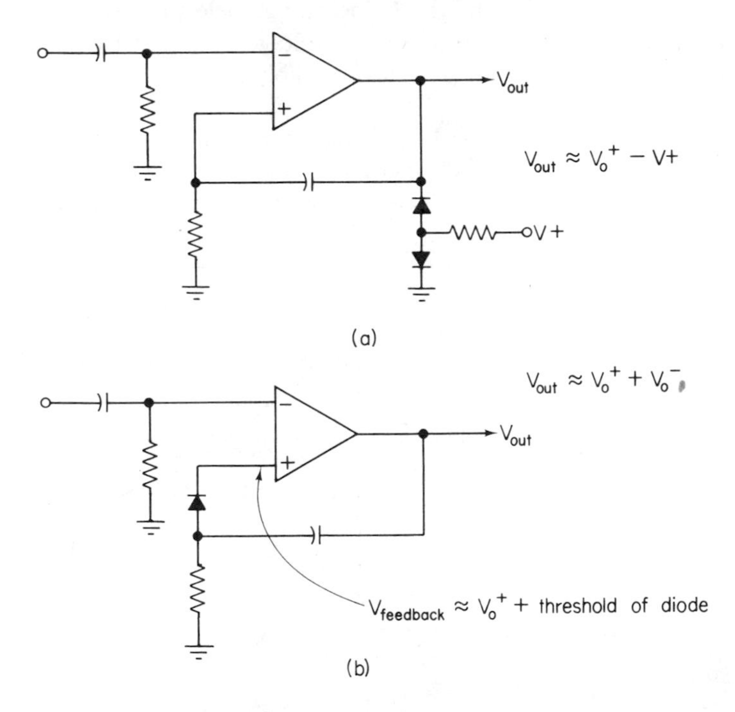

FIG. 4-18. Circuit modifications to set output level of monostable multivibrator.

4-15.1 Design example

Assume that the circuit of Fig. 4-16 is to provide an output pulse that goes from approximately +5 V to 0 V, and then return to +5 V. The required output pulse duration is 190 μs. The input is a −2-V pulse with a duration of 15 μs, from a 100-Ω impedance source. The IC op-amp is rated from a maximum 8 V input, and a maximum V_{CC}, V_{EE} of 10 V.

The value of R_2 should be 100 Ω to match the input impedance of the pulse source.

The R_2C_2 time constant should be $\frac{1}{5}$ of the input pulse duration, or 15/5 = 3 μs. With an R_2 value of 100 Ω, and a constant of 3 μs, the value of C_2 is: 3 μs/100 = 0.03 μF.

With a desired output of about 5 V, both V_{CC} and V_{EE} must be between 5.5 and 6 V. Assume a value of 6 V for both V_{CC} and V_{EE}.

With outputs (V_o^+ and V_o^-) of about 5 V, the V_{feedback} would be about 10 V. This is greater than the rated 8 V maximum input. The feedback can be reduced by means of clamp diode CR_1. The total V_{feedback} is then on the order of 6–7 V, allowing about 5 V for V_o^+, 0.7 V for the threshold of CR_1, and 0.5 V or less for the offset voltage drop across R_1.

The values of R_1 and C_1, as well as V_{ref} must be selected to get the desired 190 μs output pulse duration. The value of V_{ref} must be greater than the offset voltage of 0.5 V, but less than the trigger of +2 V. Using the graph of Fig. 4-17, note that a V_{ref} of 2 V provides a constant of about 1.07. At the other extreme, a V_{ref} of 0.5 V provides a constant of about 2.4. These extremes of V_{ref} should not be used in practical application. Instead, a safer value of V_{ref} is something between 1.0 V and 1.5 V, which provide constants of about 1.9 and 1.4, respectively.

Divide the desired output pulse duration by the constants. Note that a constant of 1.9 (V_{ref} of 1.0 V) divides equally into the 190 μs pulse duration (190/1.9 = 100 μs). Thus, R_1 times C_1 should equal 100 μs. If the constant of 1.4 (V_{ref} of 1.5 V) is used, R_1 times C_1 must produce about 135 μs (190/1.4 = 135).

Assume that the 1.9 constant is used (V_{ref} of 1.0 V). Then, any combination of R_1C_1 values can be used, provided that the product is equal to 100 μs. From a practical standpoint, a larger value of R_1 and (a corresponding value of C_1) will keep the physical size of R_1C_1 small. However, keep in mind that the input bias current of the IC will produce a voltage drop across R_1. A very large value of R_1 will produce a large drop which, in turn, could affect the output pulse timing, and could even affect circuit operation. As a guideline, use a value of R_1 that produces a drop (as a result of IC input bias current) equal to 10 percent (or less) of the offset drop (as a result of charging C_1). For example, with an assumed offset of 0.5 V, the bias current drop should be 0.05 V or less. Assuming a 5-μA input bias, the value of R_1 should be:

(0.05/0.000005 = 10 kΩ). With a value of 10 kΩ for R_1, and a required 100 μs, the value of C_1 should be: (100/10,000 = 0.01 μF.)

In practical use, the completed circuit should be checked for proper output on an oscilloscope. If necessary, adjust the value of V_{ref} to get the desire pulse duration. However, keep the value of V_{ref} well within the limits (0.5 V to 2 V). If this is not possible, then change the value of C_1.

4-16. OP-AMP COMPARATOR

Figures 4-19 and 4-20 are the working schematics of IC op-amps used as comparators. A typical use for such comparators is in digital work where the circuit must produce a logical "1" or "0" output, depending on the polarity of a differential input signal around some reference level.

As discussed, IC op-amps are high-gain differential amplifiers. Thus, without any circuit modification, an op-amp can be used as a logic comparator. For example, in Fig. 4-19, if the positive (non-inverting) input is connected to ground, and the negative input varies between +1 mV and −1 mV, an IC op-amp with an open-loop gain of 10,000 will produce an output of −10 V and +10 V respectively. If V_{CC} and V_{EE} are 10 V, or less, the IC is driven into saturation by 1-mV inputs. This is shown by the voltage transfer characteristics of Fig. 4-19b.

Note that the edge of the output waveform is not perfectly vertical. Part of this condition is the result of the IC offset voltage V_{io} which must be overcome by differential input signal. For example, if the offset is 1 mV, and a 1-mV differential is required to produce saturation, then there must be a 2-mV differential.

There is also a delay of the signal through the IC. This delay or "response time" (shown as transition voltage or width, V_W, in Fig. 4-19b) is dependent upon IC characteristics, and cannot be altered.

Although some comparators are used with dc voltages or sinewaves, most comparators are used to sense the difference voltage between a pulse waveform and a fixed reference voltage. When working with pulses, *switching times* become important, rather than V_W. This is shown in Fig. 4-19c. The delay (or response time) is the sum of the storage time and fall time. The storage time is essentially caused by saturation occurring in the latter stages of the amplifier, while the fall time is essentially controlled by the parasitic capacitance in the input stages. Both factors are characteristics of the IC.

There are two generally accepted methods to minimize delay (or to improve response time). One way is to overdrive the amplifier. For example, if a 1-mV differential produces saturation, then use 5 to 10 mV. Typically, overdrive should not exceed a factor of about 20. That is, if a 1-mV differential is

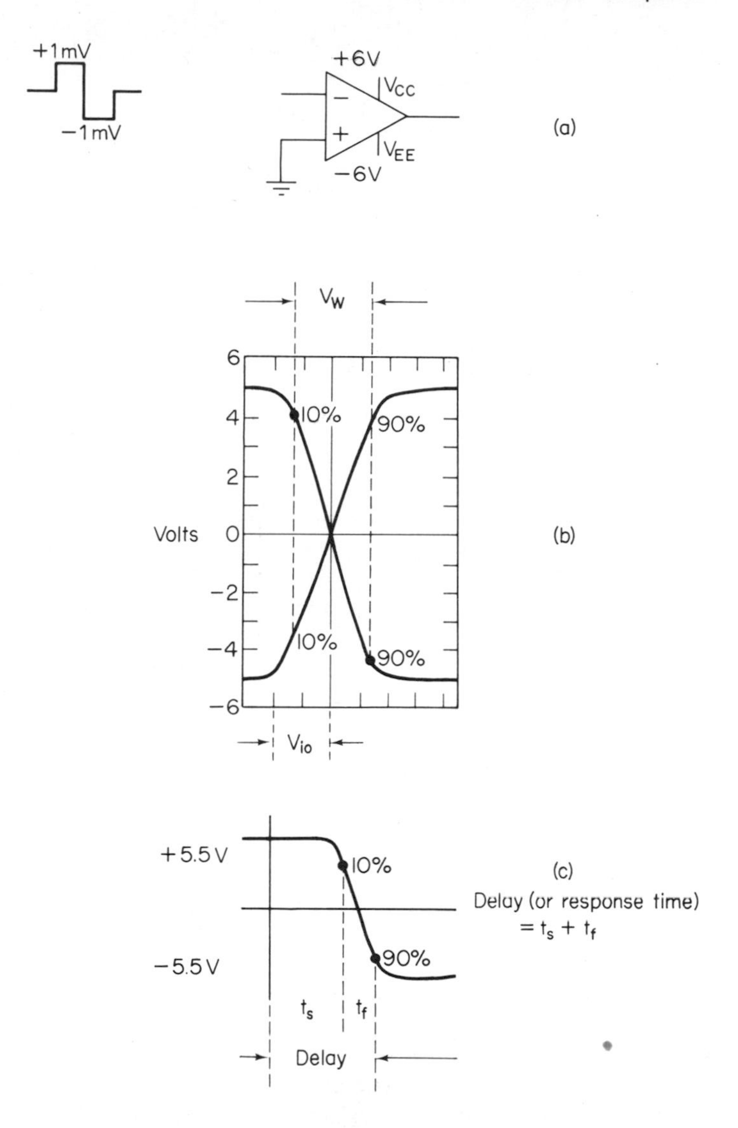

FIG. 4-19. Basic IC op-amp comparator characteristics.

needed, do not exceed 20 mV in input. Any further overdrive will probably not improve response time, but could result in malfunction.

The other method used to improve comparator response time is to use positive feedback as shown in Fig. 4-20. The positive feedback causes the circuit to switch into the desired state in a manner similar to that of the monostable multivibrator (described in Sec. 4-15). The feedback circuit helps prevent the IC from operating on any linear portion of the transfer

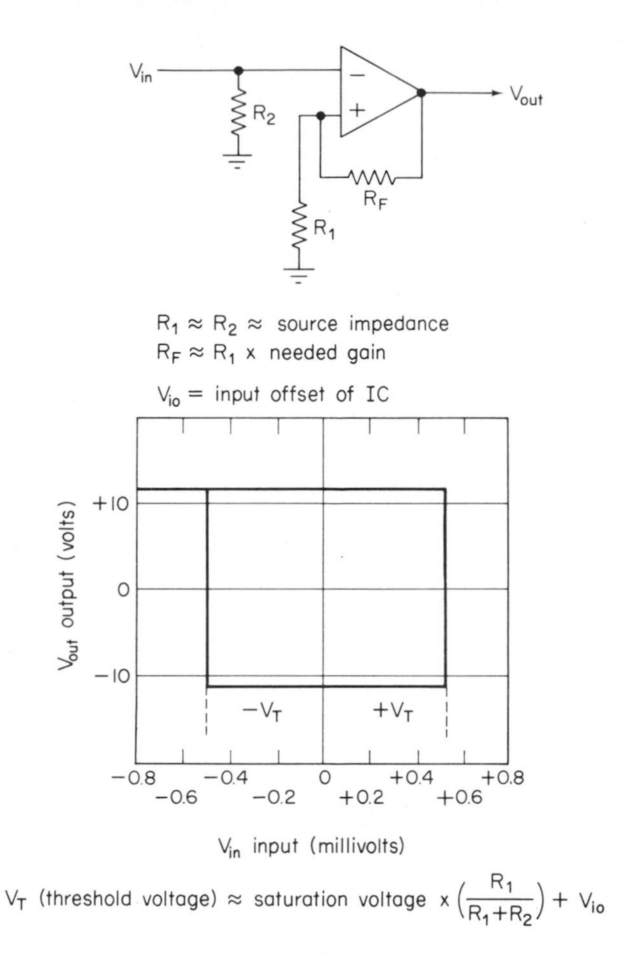

FIG. 4-20. IC op-amp comparator using feedback to improve response time.

curve. As shown by the equations, the threshold voltage V_T (voltage differential required to switch states) becomes a function of maximum output voltage, times the ratio of feedback resistors, plus any offset V_{io}. In effect V_W is reduced to zero, and is replaced by V_T. Of course, V_T can never become zero, even with an offset V_{io} of zero, since response time is also limited by the slew rate of the IC.

Some sensitivity is lost when using the positive feedback of Fig. 4-20. However, this loss of sensitivity generally improves noise immunity which may be advantages in certain applications.

Another problem with the circuit of Fig. 4-20 is that a voltage is developed across R_1 due to input bias. The drop across R_1 represents a fixed differential that must be overcome by the input voltage. This problem is offset by the addition of R_2. The values of R_1 and R_2 are approximately the same, or slightly different so as to offset any unbalance in the differential input of the IC op-amp.

With the circuit of Fig. 4-20, R_2 becomes an impedance for the input. Thus, R_2 should be chosen to match the source impedance. Generally a low value (200 Ω, or less) for the source impedance (and R_2) is best. Since the circuit is designed to saturate, the value of the source impedance (and R_2) becomes a factor in determining the response time due to the *RC* time constant of the source impedance, and the input capacitance of the input differential amplifier. A low-value source impedance will minimize this effect. Along with minimizing delay time, a small source impedance will also give a smaller input noise voltage which is always desirable.

Another method of overcoming the offset problem is to apply a fixed compensating voltage to the positive input as shown in Fig. 4-21a. The com-

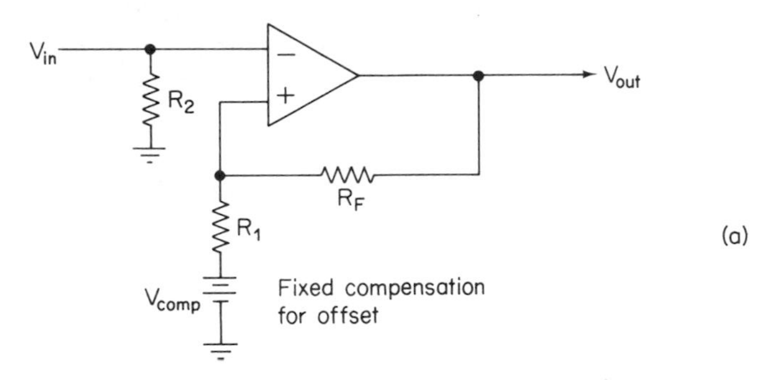

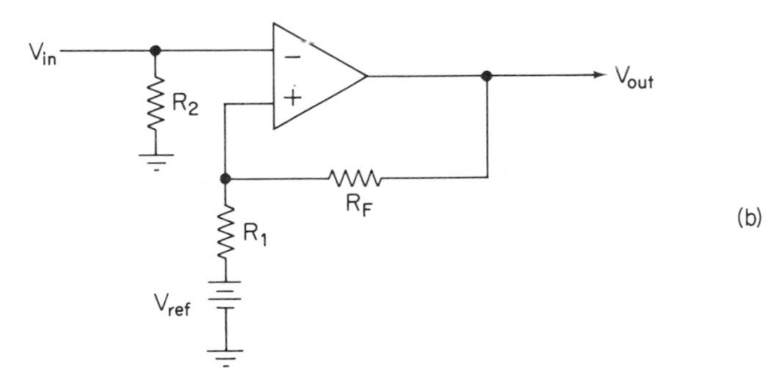

FIG. 4-21. Methods to overcome input offset, and to set comparison about a fixed reference level.

pensating voltage must be of opposite polarity, but equal to the total offset at the positive input. For example, if the normal offset of the IC is +1 mV at the positive input, and there is a +1 mV offset (caused by a differential of drops across R_1 and R_2), a −2 mV compensating voltage is required.

The circuits of Figs. 4-19 and 4-20 compare the input voltage with zero volts, but can be used to compare an input voltage with a reference voltage other than ground. However, the reference voltage can not exceed the maximum common mode range of the IC. When the input must be compared with a reference voltage, rather than ground or zero volts, the fixed reference voltage is applied to the positive input as shown in Fig. 4-21b.

4-16.1 Design example

Assume that the circuit of Fig. 4-20 is to compare an input against a fixed reference of +1 V. The output is to switch states whenever the input goes above or below 1 V by 0.5 mV. That is, any input above +1.0005 or below 0.9995 V will cause the output to change to the opposite state. The available V_{CC} and V_{EE} are about 10 V. The input source impedance is 50 Ω. The IC has an open-loop gain of about 50,000 or greater, and an input bias of 1 μA. For simplicity, assume that there is no input offset, or that the IC has provisions for neutralizing the offset. (See Sec. 3-3.10.) The IC has a common-mode input maximum of 2 V.

Since the source impedance is 50 Ω, R_2 must be 50 Ω. With a 1-μA bias, the drop across R_2 is 50 μV, which can be ignored in practical use. With R_2 at 50 Ω, R_1 must be approximately 50 Ω.

With V_{CC} and V_{EE} at 10 V, the saturation is 10 V. In practice, saturation will occur at some value slightly less than 10 V.

With 10-V saturation, and an input of 0.5 mV, the needed gain is 20,000, well below the open-loop gain of 50,000. With a needed gain of 20,000, and an R_1 of 50 Ω, R_F must be 1 MΩ (50 × 20,000 = 1 MΩ).

Using the equation of Fig. 4-20, the threshold voltage will work out to slightly less than 0.5 mV (about 0.4999 mV).

Since the comparator circuit is to operate about a +1 V point, +1 V must be applied to R_1 as shown in Fig. 4-21. The +1 V is well below the common-mode range of the IC (2 V). However, care should be taken that the input to be compared does not go below 0 V, or above +3 V. Otherwise, the common mode range will be exceeded.

4-17. OP-AMP LOG AND ANTI-LOG AMPLIFIERS

Figure 4-22 is the working schematic of an IC op-amp used as a log (logarithmic) amplifier. A log amplifier is non-linear so that a large

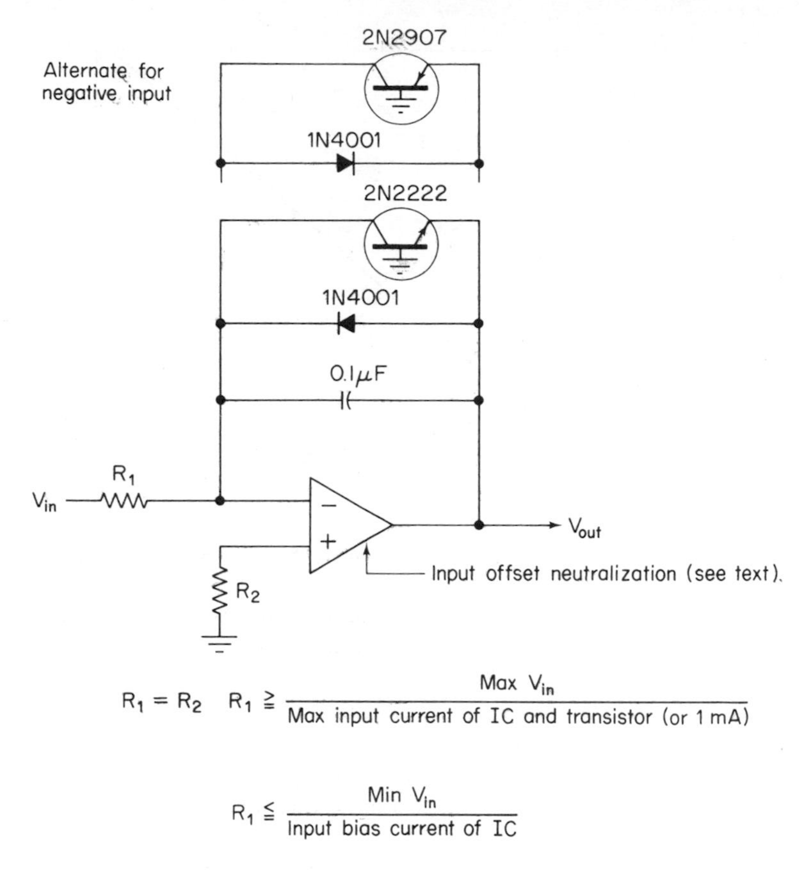

FIG. 4-22. Logarithmic amplifier using IC op-amp.

input variation produces only a small output variation. This is shown by the curve of Fig. 4-23 where the input varies from 1 mV to 100 V, but produces an output variation from about 350 mV to 640 mV (an approximate 300 mV output swing). Note that the output appears as a straight line on Fig. 4-23 since the horizontal lines represent logarithmic variations (five decades in this case).

The amplifier of Fig. 4-22 compresses the five decades of information into a small output swing. One circuit requiring such a log amplifier is a display which reads out data that spans many orders of magnitude in a single range. An NPN transistor is used when the input is positive. A negative input requires a PNP, as shown in Fig. 4-22.

The circuit uses the base-emitter junction of a transistor to provide the logarithmic response. The principle of operation relies on the exponential relationship of the base-emitter voltage and collector current of the transistor. All transistors do not exhibit good logarithmic characteristics, so care must

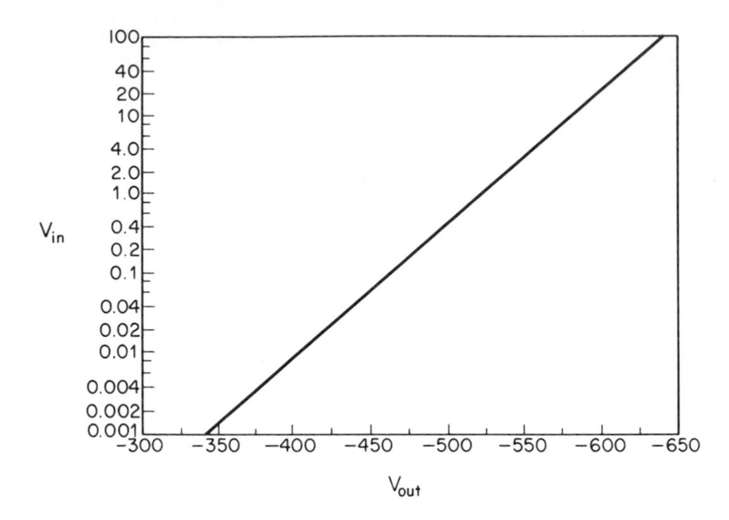

FIG. 4-23. Input versus output response of logarithmic amplifier.

be taken in selecting the proper transistor. Likewise, since the transistor junction is temperature-sensitive, the amplifier response will also be subject to temperature variations.

Capacitor C_1 across the feedback transistor is necessary to reduce the ac gain (and thus reduce noise pickup). Use a value of 0.1 μF for C_1 for frequencies up to about 10 MHz. The diode CR_1 protects the transistor against excessive reverse base-emitter voltage should the polarity of the input voltage be reversed accidently.

A log amplifier generally requires an IC that has provisions for neutralizing input offset voltage (Sec. 3-3.10). The effects of offset voltage are noticed at the low limit of the input range. This is shown in Fig. 4-24.

If an IC without input offset provisions must be used, it is possible to provide offset neutralization with the circuit of Fig. 4-25. This circuit uses the available values of V_{CC} and V_{EE}, and provides both coarse and fine neutralization of the input offset voltage.

The circuit of Fig. 4-25 also provides for neutralization or compensation of the input bias current. As a rule, an IC with the lowest possible input bias current is best for log amplifiers. This permits operation with low input signal voltages.

Referring back to the circuit of Fig. 4-22, the values of R_1 and R_2 are selected on the basis of input bias current, input voltage extremes, and the current limits of the transistor and IC, as shown by the equations.

When the input signal is at its high limit, there is a large voltage drop across resistor R_1 (almost the full value of the input signal). If a low value of R_1 is used, there is a large current flow through R_1. This current flow could exceed

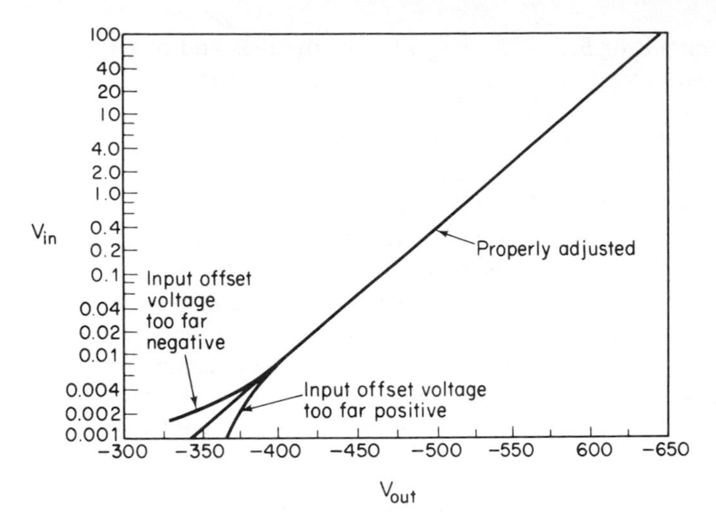

FIG. 4-24. Effects of improper offset adjustment on logarithmic amplifier.

the input current limits of the IC and the transistor. In the absence of any specific datasheet information, use a value of 1 mA for maximum input current, when calculating the value of R_1.

On the other hand, a large value of R_1 will produce a corresponding voltage drop due to the input bias current. This voltage drop must be less than the low limit of the input voltage.

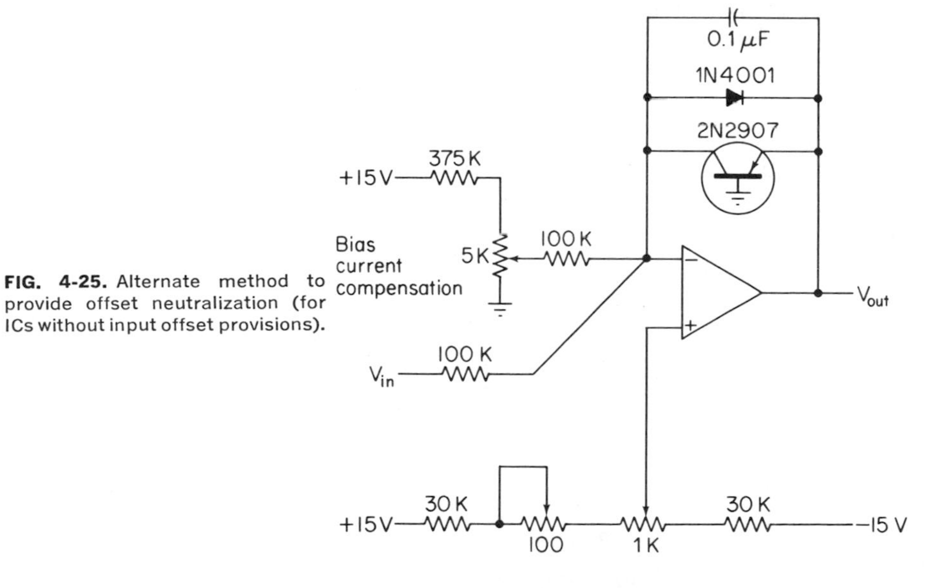

FIG. 4-25. Alternate method to provide offset neutralization (for ICs without input offset provisions).

In the circuit of Fig. 4-25, the values of the bias and offset compensating networks are based on the limits of V_{CC} and V_{EE}. The current through the compensating networks should be less than 1 mA, and preferably in the order of 0.5 mA. The value of R_1 should be selected on the same basis as R_1 in Fig. 4-22. The value of R_2 should be equal to R_1.

Figure 4-26 is the working schematic of an IC op-amp used as an anti-log amplifier. The anti-log circuit is the complement of a log amplifier (the transistor is in the input circuit rather than as the feedback element). In an anti-log circuit, larger input voltages produce progressively higher amplifier gains. Because the input is applied across the transistor base-emitter junction, input voltages are generally 1 V or less. The maximum circuit output voltage is limited by the maximum voltage swing of the IC. The values of R_1 and R_2 should be selected on the same basis as R_1 and R_2 in Fig. 4-22.

4-17.1 Design example

Assume that the circuit of Fig. 4-22 is to provide a logarithmic output response similar to that shown in Fig. 4-23. The input voltage varies five decades, from 1 mV to 100 V. Also assume that the IC has provisions for neutralizing any input offset voltage, that the input bias is 8 nA (nanoamperes), and that the maximum input current is 1 mA.

With a maximum input voltage of 100 V, and a maximum input current of 1 mA, R_1 must be 100 kΩ or larger (100 V/1 mA = 100,000).

With a minimum input voltage of 1 mV, and an input bias current of 8 nA, R_1 must be 125 kΩ or smaller (1 mV/8 nA = 125,000). Use a trial value of 100 kΩ for both R_1 and R_2.

The curve of Fig. 4-23 is the "ideal" rather than the practical curve of an IC log amplifier. In use, the circuit characteristics must be measured and plotted on a log graph. It is not practical to predict exact output for a given

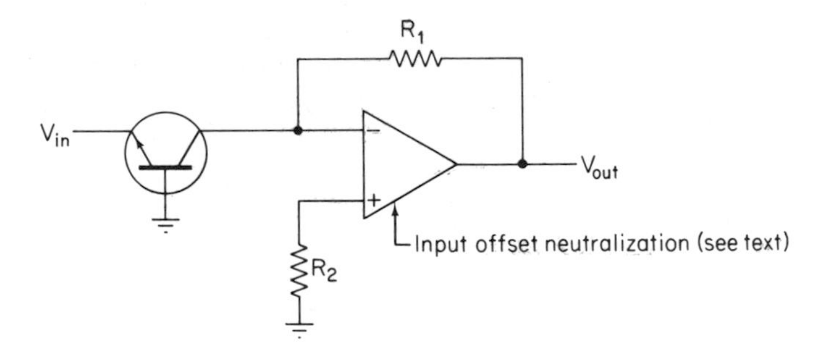

FIG. 4-26. Anti-logarithmic amplifier using IC op-amp.

input, since the final characteristics are directly dependent upon the log characteristics of the particular transistor used.

4-18. OP-AMP HIGH-IMPEDANCE BRIDGE AMPLIFIER

Figure 4-27 is the working schematic of an IC op-amp used as a high-impedance bridge amplifier. The circuit is not limited to use with a bridge. However, a bridge of any type will provide greater accuracy when its output is fed to a high-impedance amplifier. Under these conditions, little or no current is drawn from the bridge, and the bridge voltage output is amplified as necessary to provide a given reading.

Amplification for the entire circuit is provided by IC_3, and is dependent upon the ratio of R_2/R_1. IC_1 and IC_2 act as voltage followers, and provide no amplification. However, IC_1 and IC_2 do provide a high impedance to the bridge (easily 100 kΩ and above).

The value of R_1 is chosen on the basis of input bias current and voltage drop, as described previously in this chapter. The value of R_2 is selected to provide the desired gain. Resistors R_3 and R_4 should be equal to the values of R_1 and R_2, respectively.

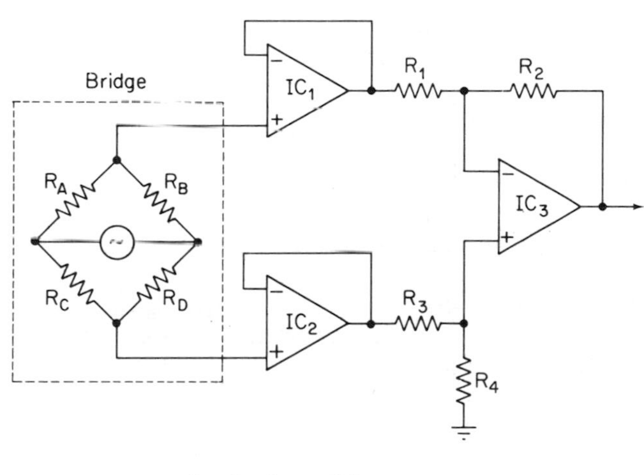

$$R_1 = R_3 \geqq \frac{\text{Input voltage} \times 0.1}{\text{Input bias of } IC_3}$$

$$R_2 = \text{Gain} \times R_1$$

$$R_4 = R_2$$

FIG. 4-27. High-impedance bridge amplifier using three IC op-amps.

4-18.1 Design example

Assume that the circuit of Fig. 4-27 is to provide about 10 V output. The bridge output is approximately 1 V. The IC_3 input bias is 10 μA, and the available $V_{CC} - V_{EE}$ voltages are 10 V.

The outputs of IC_1 and IC_2 are equal to (or slightly less) than the bridge output, since IC_1 and IC_2 are connected as voltage followers (Sec. 4-4.) Therefore, IC_3 must provide a gain of 10, for a circuit output of 10 V.

The voltage drop across R_1, due to input bias current, should be no greater than 10 percent of the input voltage. With a drop of 0.1 V across R_1, and a 10-μA input bias current, the value of R_1 is: 0.1 V/10 μA = 10,000 Ω. R_3 should also be 10,000 Ω.

With a value of 10 kΩ for R_1, and a required gain of 10, the value of R_2 is: 10,000 × 10 = 100 kΩ. R_4 should also be 100,000 Ω.

4-19. OP-AMP FOR DIFFERENTIAL INPUT/DIFFERENTIAL OUTPUT

Figure 4-28 is the working schematic of two IC op-amps connected to provide differential-in/differential-out amplification. Generally,

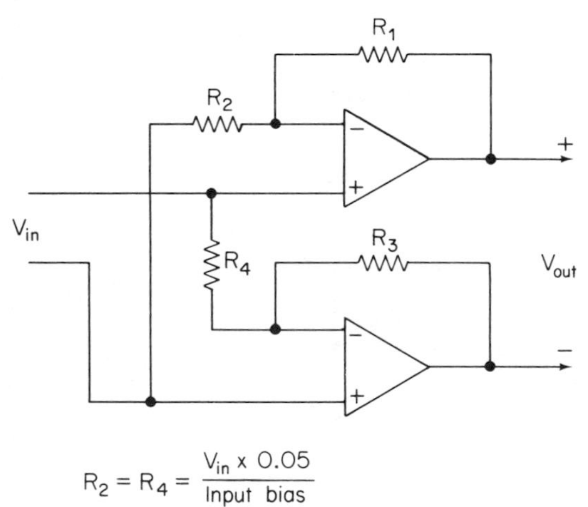

FIG. 4-28. Differential input/differential output using two IC op-amps.

$$R_2 = R_4 = \frac{V_{in} \times 0.05}{\text{Input bias}}$$

$$R_1 = R_3 = R_2 \times \text{gain of each IC}$$

$$V_{out} = V_{in}\left[\frac{R_1}{R_2} + \frac{R_3}{R_4}\right]$$

$$\frac{V_{out}}{V_{in}} = \frac{R_1}{R_2} + \frac{R_3}{R_4}$$

IC op-amps have a single-ended output. There are cases where a differential output is required. For example, the circuit of Fig. 4-28 can be used at the input of a system where there is considerable noise pickup at the input leads, but the input differential signal is small. Since the noise signals are common mode (appear on both input leads simultaneously, at the same amplitude and polarity) they will not be amplified. However, the differential input is amplified.

The circuit of Fig. 4-28 can be formed with two identical IC op-amps, or a dual-channel IC op-amp. The dual-channel IC is preferable since both channels will have identical characteristics (gain, bias, etc.). This is because both channels are fabricated on the same semiconductor chip. However, the circuit will work satisfactorily if the two ICs are carefully matched as to characteristics (particularly in regards to input bias and input offset voltage).

As shown by the equations, each IC (or each channel) provides one-half of the total differential output gain. Thus, if each IC provides a gain of 10, the differential output is 20 times the differential input.

The output voltage swing is double that of the individual ICs. Generally, the maximum output swing of a single-ended IC is slightly less than the $V_{CC} - V_{EE}$ voltage. Using the circuit of Fig. 4-28, the maximum differential output swing is twice that of the $V_{CC} - V_{EE}$ voltage. For example, if the $V_{CC} - V_{EE}$ voltage is 6 V, the maximum differential output is slightly less than 12 V.

The values of R_2 and R_4 are chosen on the basis of input bias current and voltage drop, as described in previous sections of this chapter. The values of R_1 and R_3 are chosen to provide the desired gain. Resistors R_3 and R_4 should be equal in value to R_1 and R_2, respectively.

4-19.1 Design example

Assume that the circuit of Fig. 4-28 is to provide a differential output with a swing of approximately 20 V. The available differential input is 100 mV. The IC input bias is 10 μA.

With an input of 100 mV and a desired output of 20 V, the circuit must provide a total gain of 200 (20V/0.1 V = 200). Each IC must provide a gain of 100.

The voltage drop across R_2, due to input bias current, should be no greater than 5 percent of the input voltage. With a drop of 5 mV across R_2, and a 10-μA input bias current, the value of R_2 is: 5 mV/10 μA = 500 Ω. R_4 should also be 500 Ω.

With a value of 500 Ω for R_2, and a required gain of 100, the value of R_1 is: 500 × 100 = 50,000 Ω. R_3 should also be 50,000 Ω.

With a required differential output swing of 20 V, the $V_{CC} - V_{EE}$ voltage must be about 10 V.

4-20. OP-AMP ENVELOPE DETECTOR

Figure 4-29 is the working schematic of two IC op-amps used as an envelope detector. The purpose of the circuit is to provide a visual indication (panel lamp) when an input voltage goes above or below two given reference voltages. The two ICs operate essentially as comparators (Sec. 4-16). The output of the two ICs is converted to a panel lamp indication by a logic circuit. (Operation of logic circuits is discussed in Chapters 1 and 6.)

Both ICs are connected for positive feedback so that they will switch states (similar to a comparator or multivibrator), rather than amplify any input differential as a linear output. A small differential input will drive the ICs into saturation. However, the output of both ICs is clamped to ground by CR_1 and CR_2. The negative (or low) output of each IC will not go below the normal voltage drop of CR_1 and CR_2 (typically 0.7 V for silicon diodes). On the positive saturation, the IC output is about 90 percent of V_{CC}. Thus, with a V_{CC} of 6 V, the IC outputs switch from about +5.4 V to −0.7 V. As output load increases, the +5.4 V output may drop, but should remain above +5.0 V.

Operation of the circuit can best be understood by reference to the curve of Fig. 4-30, and an example. Assume that the low reference V_1 is 2.5 V, the high reference V_2 is 3.5 V, V_{CC} is 6 V, and the logic gates require 0 V (or less) for a low (or false or "0") input, and +5 V (or more) for a high (or true

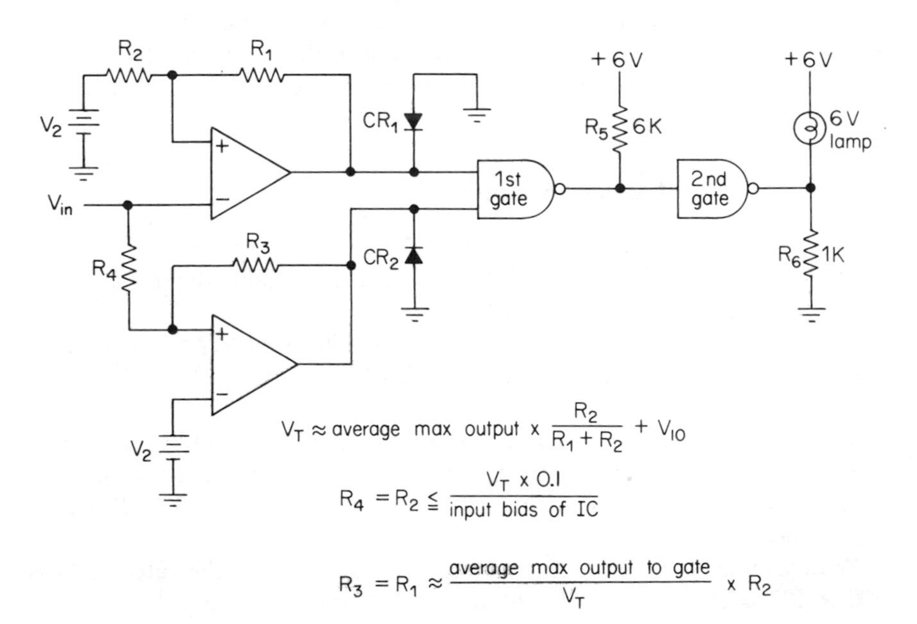

FIG. 4-29. Envelope detector using two IC op-amps.

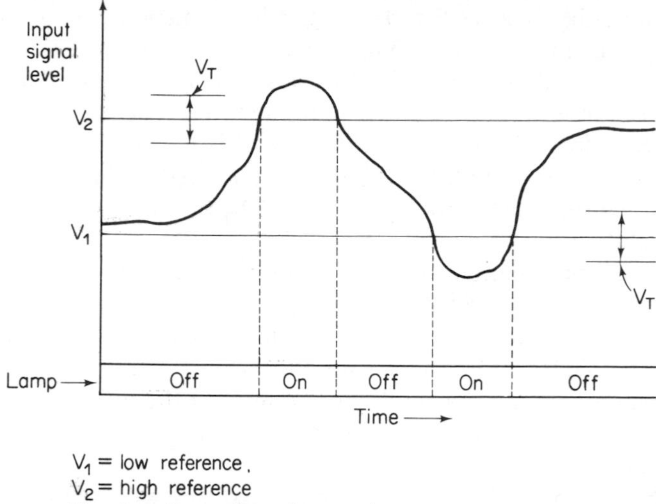

FIG. 4-30. Relationship of reference voltages, lamp indications and input voltage on IC op-amp envelope detector.

or "1") input. Now assume that the input signal varies between 2 V, 3 V and 4 V, in turn.

With a 2-V input signal, the input to IC_1 is low (2 V is less than 3.5 V), and the IC assumes the corresponding state. Since the signal input to IC_1 is at the inverting input, the output of IC_1 is high (saturated at about 5 V). This represents a high input to the first logic NAND gate. With the same 2-V signal, the input to IC_2 is low (2 V is less than 2.5 V) and, since the signal is at the non-inverting input, the output of IC_2 is low to the first NAND gate. A high and low input at a NAND gate produces a high output, which is inverted to a low by the second NAND gate. In effect, this shorts out R_6 (drops both sides of R_6 to ground or 0 V), and places the full 6 V across the lamp. Thus, the lamp turns on when the signal input (2 V) is below the low reference (2.5 V).

With a 4-V input signal, the same action occurs, but the IC's transpose states. That is, the IC_1 output is low (−0.7 V), the IC_2 output is high (+5 V), and the lamp turns on to indicate that the input has gone above the high reference.

With a 3-V input signal, the input to IC_1 is low, and the output is high. The input to IC_2 is high, as is the output. Two high inputs at a NAND gate produces a low output. This low is inverted by the second NAND gate to a high, making both sides of the lamp about 6 V. Thus, the lamp turns off when the signal input (3 V) is above the low reference (2.5 V) but below the high reference (3.5 V).

As shown by the curve of Fig. 4-30, there is a transition voltage V_T about both the high and low reference points. (This transition voltage is also known as voltage spread, dead band, lag, hysteresis, and threshold voltage.) No matter what term is used, the effect is noticed when the signal voltage crosses the reference points. For example, if the signal is moving from 3 V to 4 V, the lamp will remain off until the signal has gone slightly higher than 3.5 V. Likewise, when the signal goes from 4 V to 3 V, the lamp stays on until the signal is slightly below 3.5 V.

As shown by the equations, V_T is a function of the maximum output voltage, the ratio of feedback resistors, and any input offset voltage V_{io}, as is the case of the comparators described in Sec. 4-16. However, the maximum output voltage for the ICs of Fig. 4-29 is not the same for both states, so an average maximum output must be used. For example, if the ICs go from -0.7 V to $+5.4$ V, the average maximum output is about 3 V ($-0.7 + 5.4 = 6.1$; $6.1/2 = 3.05$).

As will be seen by the equations, a larger feedback will produce a smaller V_T and vice versa.

The values of R_2 and R_4 are chosen on the basis of input bias current and voltage drop, as described in previous sections of this chapter. However, the drop produced by bias current should be about 10 percent (or less) of V_T, rather than 10 percent of the actual input signal voltage, since the IC must operate (change states) on the small differential between signal and reference. The values of R_1 and R_3 are chosen to provide the desired gain. In the absence of some specific desired value of V_T, use a gain of about 50. Resistors R_3 and R_4 should be equal to the value of R_1 and R_2, respectively.

4-20.1 Design example

Assume that the circuit of Fig. 4-29 is to provide a V_T of about 60–70 mV, with all of the values described in the previous example: $V_{CC} =$ 6 V, $V_1 = 2.5$ V, $V_2 = 3.5$ V. Also assume that the IC has a V_{io} of 1 mV, and an input bias of 5 μA.

With a V_T of 60–70 mV, and an input bias of 5 μA, the value of R_2 should be less than 1.2 kΩ (60 mV $\times$ 0.1/5 μA = 1.2 kΩ). Use 1000 Ω for R_2 to simplify calculations. Resistor R_4 should also be 1000 Ω.

With a V_{CC} of 6 V, the maximum voltage on the positive saturation is about 5.4 V (6 $\times$ 90 percent = 5.4). The maximum negative swing is limited to about -0.7 V by diodes CR_1 and CR_2. The average maximum output voltage is about 3 V ($+5.4$ to -0.7 to 6.1; one half of 6.1 is 3.05 V). With an average maximum output voltage of 3 V, a V_T of 60 mV, and an R_2 of 1000 Ω, the value of R_1 is 50,000 Ω (3 V/60 mV = 50; 50 $\times$ 1000 = 50,000). Resistor R_3 should also be 50 kΩ.

Using the values of 3 V for average maximum voltage, 1 kΩ for R_1, 50 kΩ for R_2 and an input offset voltage V_{io} of 1 mV, find the V_T as shown by the equation on Fig. 4-29:

$$3\text{ V} \times \left(\frac{1000}{50{,}000 + 1000}\right) + 1\text{ mV} \approx 60\text{ mV}$$

If actual test of the circuit proves that V_T is below the 60-mV value, decrease the value of R_1 (and R_3). For example, a value of 47 kΩ for R_1 and R_3 should produce a V_T of 63 to 64 mV.

The values for resistors R_5 and R_6 are not critical. These values shown on Fig. 4-29 are for use with a 6-V lamp, and on the assumption that the logic circuit operates from the same 6-V source (V_{CC}) as the ICs. Refer to Chapter 6 for further data on logic circuits.

4-21. OP-AMP WITH ZERO OFFSET SUPPRESSION

Figure 4-31 is the working schematic of an IC op-amp used to provide amplification of a small signal riding on a large, fixed direct current

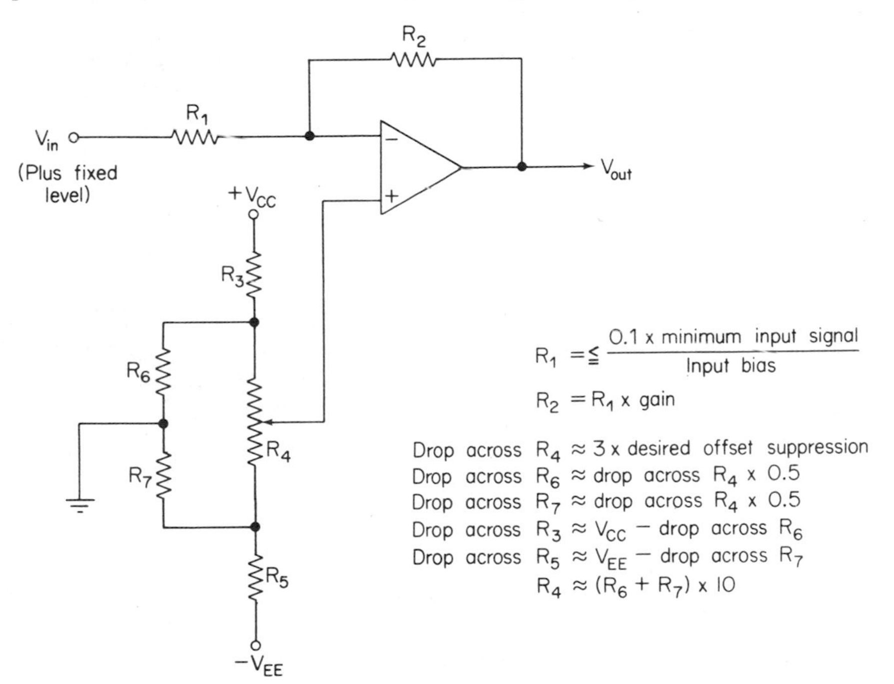

FIG. 4-31. IC op-amp with zero offset suppression to provide amplification of small signal riding on large, fixed level.

level. For example, assume that the input signal varies between 2 and 10 mV, and that the signal source never drops below +5 V. That is, the source is +5 V with no signal, and +5.002 to +5.010 V with signal. Now assume that the output is to vary between 200 mV and 1 V.

The obvious solution is to apply a fixed +5 V to the non-inverting input. This will offset the +5 V at the inverting input, and result in 0-V output (under no-signal conditions). This solution ignores the fact that the IC probably has some characteristic input offset V_{io}, or assumes that the IC has some provisions for neutralizing V_{io}. The solution also assumes that the signal is riding on exactly +5 V.

If V_{io} can not be ignored (say because it is large in relation to the signal), or if the fixed dc voltage is subject to possible change, the alternate offset circuit of Fig. 4-31 should be used. The alternate circuit is a simple resistance network that makes use of existing V_{CC} and V_{EE} voltages. In use, potentiometer R_4 is adjusted to provide zero output from the IC under no-signal conditions. That is, the 2 to 10 mV signal is removed, but the +5 V remains at the inverting input, while R_4 is adjusted for zero at the IC output.

The values for the offset network are not critical. However, the values should be selected so that a minimum of current is drawn from the $V_{CC} - V_{EE}$ supplies, and a minimum of current should flow through R_4. This is discussed further in the design example.

Although it may not be obvious, the principles shown in Fig. 4-31 can be applied to most of the IC op-amp applications described in this chapter. That is, any zero offset (at the IC input and output) can be suppressed by applying a fixed (or adjustable) voltage of correct polarity and amplitude to the opposite input.

For example, in the summing amplifier (adder) of Fig. 4-1, assume that each of the input voltages V_1 through V_4 is riding on fixed dc voltage levels. The V_1 and V_2 signals have fixed levels of +4 V each, V_3 has a level of −5 V, and V_4 has a level of +3 V. The net dc level at the inverting input is thus +6 V (+4, +4, −5, +3 = +6 V). This results in a −6 V at the output, under no-signal conditions, assuming that R_F is the same value as R_1 through R_4 and there is no gain. In effect, the output is offset from zero by −6 V.

The offset can be suppressed by application of +6 V at the non-inverting input, in place of the R_{OM} resistance (or in addition to R_{OM}). The offset suppression voltage can be fixed, or adjustable, using the circuits of Fig. 4-31, as applicable.

4-21.1 Design example

Assume that the circuit of Fig. 4-31 is to monitor a 2 to 10 mV input signal, riding on a +1 V level, and is to produce a 200 mV to 1 V output. The IC does not have provisions for V_{io} offset neutralization. It is important that the no-signal output be exactly 0 V. The IC has an input bias of 5 μA. V_{CC} and V_{EE} are 6 V.

The voltage drop across R_1, due to input bias current, should be no greater than 10 percent of the lowest input signal. With a low signal of 2 mV, and a 5-μA bias current, the value of R_1 is 40 Ω (2 mV $\times$ 0.1 = 0.2 mV; 0.2 mV/5 μA = 40).

With R_1 at 40 Ω, and a required gain of 100, the value of R_2 is 4000 Ω.

With V_{CC} and V_{EE} at +6 V and −6 V, respectively, the total drop across the offset adjustment network R_3 through R_7 is 12 V. Allowing an arbitrary 1-mA current through the network, the total resistance should be about 12 kΩ. Since the desired offset suppression is approximately 1 V (to offset the +1 V level), the drop across R_4 should be approximately 3 V. This results in a drop of about 1.5 V each across R_6 and R_7. In turn, the drop across R_3 and R_5 (each) should be about 4.5 V (6 V − 1.5 V = 4.5 V). With a desired 4.5-V drop, and approximately 1-mA current flow, the values of R_3 and R_5 should be 4500 Ω each. Likewise, the values of R_6 and R_7 should be 1500 Ω each. With R_6 and R_7 at 1500 Ω each, the value of R_4 should be 30,000 Ω ($R_6 + R_7$ = 3000; 3000 $\times$ 10 = 30,000).

4-22. OP-AMP ANGLE GENERATOR

Figure 4-32 is the working schematic of two IC op-amps connected as an angle generator. As shown by the equations, the output of IC_1 is proportional to the sine of the input phase angle, while the output of IC_2 is proportional to the cosine of the input phase angle. The ICs are connected as a Scott-T transformer into a three-wire synchro line.

When all resistor values are the same, the output of IC_2 is equal to twice the input voltage, multiplied by the cosine of the phase angle. For example, if the phase angle is 33°, and the input voltage (line-to-line three-phase input) is 1 V, the IC_1 output is: 2 $\times$ 1 = 2; 2 $\times$ 0.8387 (the cosine of 33°) = 1.6774 V.

With the same conditions (phase angle of 33°, input 1 V) the IC_1 output is: 1.732 $\times$ 0.5446 (the sine of 33°) = approximately 0.94 V.

The accuracy of the circuit in Fig. 4-32 is dependent upon matching of the ICs, as well as matching of the resistors. A dual-channel IC op-amp is ideal for the circuit since both channels are fabricated on the same chip. However, two separate ICs with closely-matched characteristics will produce satisfactory results. The resistors should have a tolerance of 1 percent (or better).

The circuit of Fig. 4-32 is most effective when the three-phase voltages are on the order of 1 V, or a fraction of 1 V (such as some analog computer servo systems), and it is desired to have output readings in the 5- to 10-V range (typical of digital logic systems).

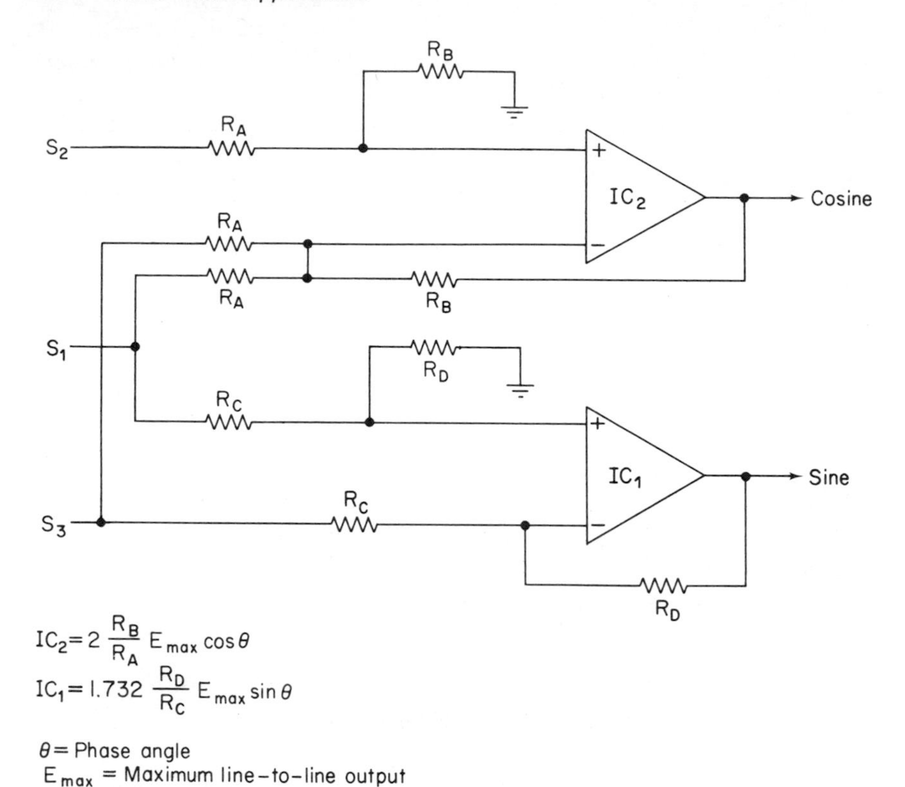

FIG. 4-32. Angle generator using IC op-amp.

The circuit of Fig. 4-32 has several advantages over direct measurement of the phase angle by meters. First, the ICs present far less loading than a meter. (Of course, this is more important with low-voltage systems than with three-phase systems in the 120-V range). Second, the circuit is independent of frequency. Most meters are for one frequency only. Third, the output voltages can be "weighted" or "scaled". For example, some design problems may require that the sine output be multiplied by 5, while the cosine output is multiplied by 10, or vice versa. This is accomplished by setting the gain of the individual ICs to different levels (by different ratios of R_B/R_A and R_D/R_C).

The values of the input resistors R_A and R_C are chosen on the basis of input bias current and voltage drop, as described in previous sections of this chapter. The same value should be used for all input resistors. The drop produced by bias current across the input resistors should be 10 percent (or less) of the line-to-line input voltage. The values of R_B and R_D are chosen on the basis of desired gain for the individual outputs. The values of both R_B resistors should be the same, as should both R_D resistors. However, R_B need not equal R_D.

4-22.1 Design example

Assume that the circuit of Fig. 4-32 is to provide outputs from a three-phase system with a maximum line-to-line output of 1 V. The IC has an input bias of 5 μA. The sine output should not exceed 2 V, while the cosine output should be 10 V or less.

With a 1-V input, and a 5-μA bias current, the value of each R_A and R_C resistor should be 20,000 Ω (1 V × 0.1 = 0.1 V; 0.1 V/5 μA = 20 kΩ).

Since the sine output is not to exceed 2 V, IC_1 should have no gain. Thus, both R_D resistors should be 20,000 Ω. Under these conditions (no gain), the output of IC_1 will go from 0 V, when the phase angle is 0°, to about 1.732 V, when the phase angle is 90°.

Since the cosine output is not to exceed 10 V, find the maximum output from IC_2 with no gain (2 V in this case), then divide the maximum output limit by the no-gain output, or: 10/2 = 5. Thus, a gain of five is required for IC_2, and the values of both R_B resistors should be 100 kΩ (20kΩ × 5 = 100 kΩ).

To provide a 10-V cosine output, V_{CC} and V_{EE} for IC_2 should be 10 V, or greater (11 or 12 V, in practical use). A lower value could be used for IC_1. However, a better match, and a more practical circuit, will be obtained if both ICs have the same $V_{CC} - V_{EE}$ voltage.

4-23. OP-AMP TRACK-AND-HOLD AMPLIFIER

Figure 4-33 is the working schematic of two IC op-amps used as a track-and-hold amplifier (also known as a sample-and-hold amplifier).

With this circuit, when voltages are applied to the gate inputs, the diode bridge conducts, and the output voltage tracks the input voltage. However, the polarity of the output signal is reversed from that of the input. If it is necessary to have an output that tracks the input directly, an IC connected as a unity-gain inverter, can be used at the output.

With conventional silicon diodes, which have a normal voltage drop of about 0.7 V, the gate voltage must be about 1.5 V, plus the maximum input voltage. When the gate voltage polarity is reversed from that shown in Fig. 4-33 by an external switching system, the diode bridge stops conducting.

Due to the charge on capacitor C_1, the value of the output voltage is equal to that of the input voltage, just prior to switching the bridge off. The output voltage remains at this value for a time period, the length of which is dependent on the value of C_1, the diode leakage current, and the input bias current of the IC, as shown by the equation. Neither of the currents can be altered. However, the time period can be set to a given value by proper selection of capacitor C_1 value.

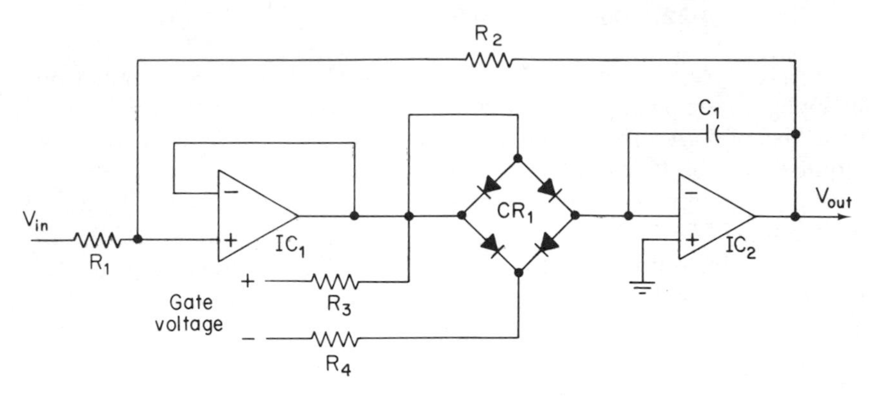

$$R_1 \geq \frac{V_{in} \times 0.1}{\text{Input bias of IC}}$$

$R_2 = R_1 \times \text{gain}$

$R_3 = R_4 = 1\text{K}$

Gate voltage $\approx 1.5 + \max V_{in}$

$$C \approx \frac{t\,(I_R + I_B)}{\Delta_{out}}$$

t = Time in seconds for no change in V_{out}

I_R = CR_1 diode leakage

I_B = IC_2 input bias

Δ_{out} = Difference in output voltage

FIG. 4-33. Track-and-hold (sample-and-hold) amplifier using IC op-amp.

The value of R_1 is chosen on the basis of input bias current and voltage drop, as described in previous sections of this chapter. Since the usual purpose of the circuit is to track the input voltage, no gain is required. Thus, R_2 should be the same value as R_1, at least as the first trial value. However, if the circuit shows some loss during test (output voltage is less than input voltage) increase the value of R_2 as necessary to produce the desired output.

The values of R_3 and R_4 are not critical. Use the 1 kΩ values for both resistors, assuming that standard silicon diodes are used, and that the gate voltage does not exceed about 7 V.

4-23.1 Design example

Assume that the circuit of Fig. 4-33 is to track and hold a signal voltage in the range of 0 to 0.5 V. The IC has an input bias of 5 μA. The diodes have a leakage of 10 μA. It is desired to hold the maximum output voltage at a 90 percent level for 2 s. For simplicity, assume that R_1 and R_2 are 10 kΩ.

Since the output must remain at 90 percent of maximum for 2 s, there must be a change no greater than 10 percent of the maximum for two seconds. Or, the output must be 0.45 V two seconds after the bridge is switched off (0.5 V × 0.1 = 0.05; 0.5 V − 0.05 V = 0.45 V). Thus, the difference in output voltage is 0.05 V.

Using the equation of Fig. 4-33, the value of C_1 is:

$$C \approx \frac{2(10 + 5) \times 10^{-6}}{0.05} \approx 600\ \mu\text{F}$$

4-24. OP-AMP PEAK DETECTOR

Figure 4-34 is the working schematic of an IC op-amp used as the major active element in a peak detector system. Where accuracy is required for peak detection, the conventional diode-capacitor detector is often inadequate because of changes in the diode's forward voltage drop (due to variations in the charging current and temperature). Ideally, if the forward

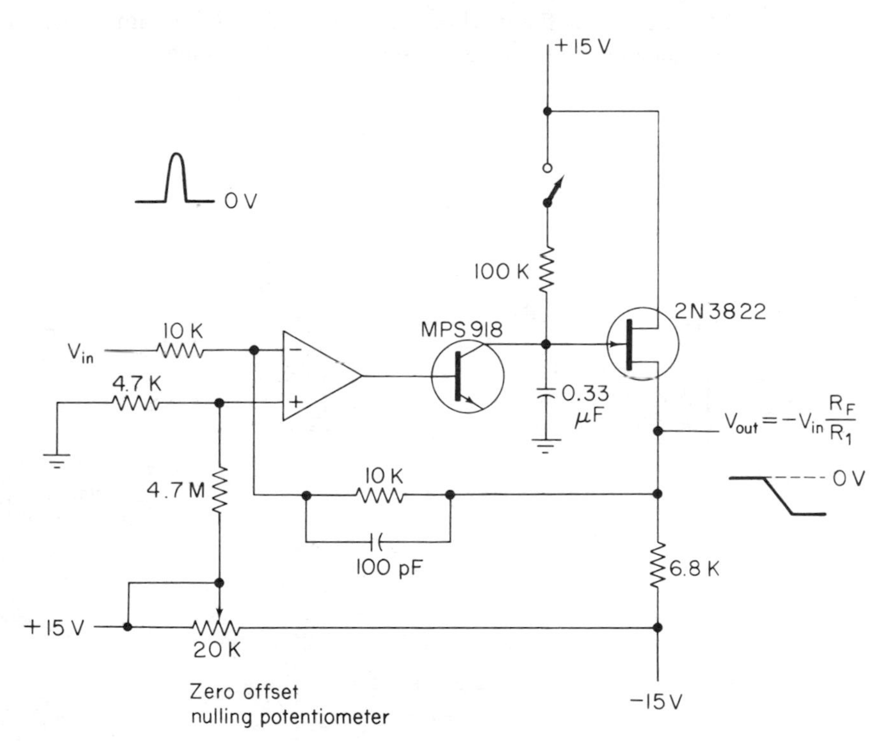

FIG. 4-34. Peak detector using IC op-amp.

drop of the diode can be made negligible, the peak value is the absolute value of the peak input, and will not be diminished by the diode-voltage drop (typically 0.7 V for a silicon diode). This can be accomplished by means of an IC op-amp.

The circuit of Fig. 4-34 uses the base-collector junction of a transistor as the detecting diode. The transistor (acting as a diode) is contained within the feedback loop. This reduces the effective forward diode drop by the loop gain. Any transistor can be used provided the leakage is low (preferably 10 nA, or less, at 15 V).

The storage time of the circuit in Fig. 4-34 is dependent upon the leakage of the diode (transistor) and the FET, as well as the value of C_1.

Note that the values shown on Fig. 4-34 apply to a typical IC that requires a $V_{CC} - V_{EE}$ of 15 V. These values can be used as a starting point for design.

In use, the output is adjusted to zero offset (zero volts output with no signal input) by closing the nulling switch and adjusting the potentiometer.

4-25. OP-AMP TEMPERATURE SENSOR

Figure 4-35 is the working schematic of an IC op-amp used as the major active element in a temperature sensor circuit. Temperature sensing

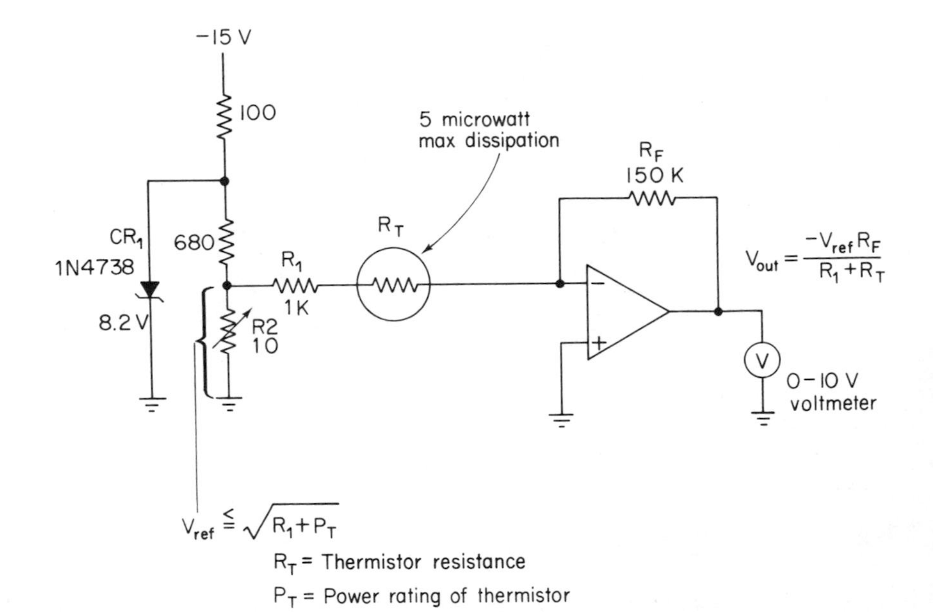

FIG. 4-35. Temperature sensor using IC op-amp.

is popular because thermistors are inexpensive and easy to use. However, thermistors are non-linear and can supply only very small output signals (generally a few microwatts).

The circuit of Fig. 4-35 overcomes these limitations. The output voltage is relatively linear over the temperature range of the IC. The output is exactly linear about the temperature where the thermistor resistance equals the resistance of R_1. Thus, R_1 should be equal to R_t (the thermistor resistance) at the center of the desired temperature range.

The reference voltage V_{ref} is obtained from the Zener diode CR_1 and voltage divider. The upper limit is a value determined by the power rating (P_t) of the thermistor, as shown by the equation. With the values shown, R_2 is adjusted so that V_{ref} is -0.067 V. Under these conditions, the maximum power dissipated by the thermistor is about 2.5 μW, well under the 5-μW rating.

4-26. OP-AMP VOLTAGE REGULATOR

In this section, we shall discuss IC op-amps used as voltage regulators. These are not to be confused with IC packages that function as complete regulator circuits. Here, the discussion is limited to using an IC op-amp as the active element in a voltage regulator circuit. Complete IC regulators are discussed in Chapter 5.

Regulators using IC op-amps as the gain elements usually show better regulation than packaged IC regulators. This is primarily due to the higher available loop gain in the op-amp. Voltage drops of less than 0.01 percent over the entire load range are commonplace, and with care 0.001 percent is possible, using op-amps.

4-26.1 Basic regulator theory

Regulators can best be analyzed as a feedback system such as shown in Fig. 4-36. The output voltage and reference voltage in this theoretical system can be calculated using the equations shown. However, these equations do not take into account the output impedance of the op-amp, or the effects of the load.

The circuit of Fig. 4-37 is somewhat more practical. Here, the op-amp is followed by an emitter-follower stage which provides necessary current gain (for a useful output current), as well as low output impedance. As shown by the equation, the output impedance of the regulator is approximately equal to the impedance seen at the base of Q_1, divided by the beta of Q_1. In turn, the base impedance is essentially the open-loop output impedance of the op-amp. Thus, the open-loop output impedance of the regulator is equal to the op-amp output impedance divided by the beta of Q_1.

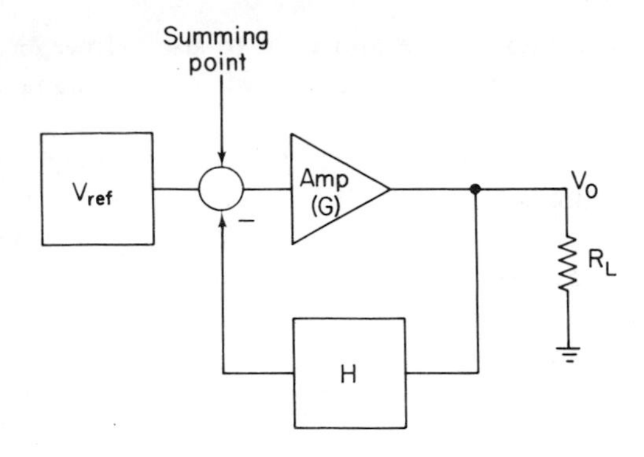

FIG. 4-36. Theoretical feedback amplifier form of voltage regulator.

$$V_O = V_{ref}\,\frac{G}{1+GH}$$

G = Amplifier gain (A_{vol})
H = Fraction of V_O fed back to summing point

$$V_{ref} - V_O = V_{ref}\,\frac{1}{1+G} \quad \text{for } H=1$$

When feedback is added, the regulator output impedance drops (by an amount proportional to open-loop gain), as shown by the closed-loop gain equation.

Using either circuit (Fig. 4-36 or 4-37), any variations in output voltage (due to changes in load), or variations in input voltage (changes in V_{ref}) cause a corresponding change in feedback voltage. In turn, this causes the amplifier gain to change in a direction to oppose the initial change. For example, if

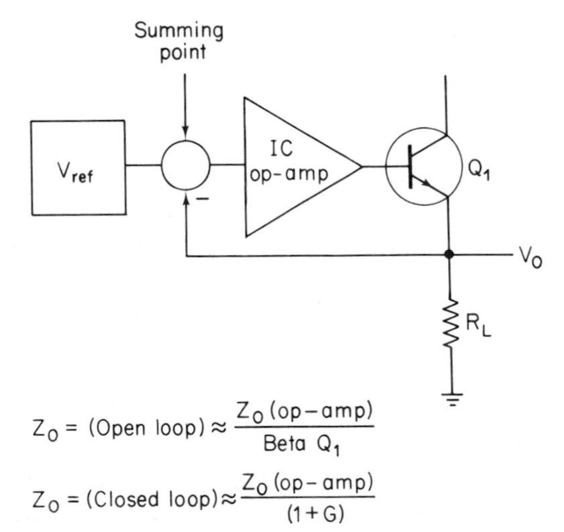

FIG. 4-37. Theoretical feedback regulator with improved current capability.

V_{ref} is increased, V_o starts to increase as does feedback. The increase in negative feedback offsets the initial increase in V_{ref}, so the net effect is that V_o remains constant.

In a practical circuit, V_o is never constant. There is always some variation in output, which is expressed as a percentage of regulation. This is discussed in the next paragraph and, as is shown, the percentage of regulation decreases (regulation improves) when the closed-loop output impedance of the circuit decreases (and when op-amp gain increases).

4-26.2 Calculating percentage of regulation

The circuit of Fig. 4-38 can be used to calculate the theoretical percentage of regulation for any feedback regulator. Figure 4-38 shows the regulator as a voltage source V', with a finite output impedance R_o. One equation for percentage of regulation involves the voltage source V' and output voltage V_o. The alternate equation involves source V' and the term $I_L R_o$, which is the load current through the regulator output impedance. The $I_L R_o$ term is equivalent to the voltage drop across R_o, since $E = IR$. The term $I_L R_o$ also equals the difference between V' and V_o $(V' - V_o)$.

No matter which equation is used, the percentage of regulation is decreased (regulation is improved) when R_o is decreased. A decrease in R_o indicates a lower regulator system output impedance, a lower $I_L R_o$ term, a lower voltage drop across R_o, and a smaller difference between V' and V_o, all of which lower the percentage of regulation.

Now let us calculate the percentage of regulation using the equations established thus far.

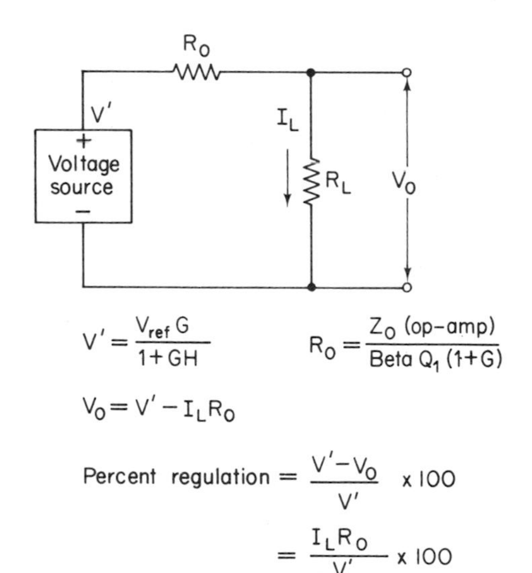

Ro
V'
+
Voltage source
−
IL
RL
Vo
$V' = \frac{V_{ref} G}{1+GH}$
$R_O = \frac{Z_O \text{ (op-amp)}}{\text{Beta } Q_1 (1+G)}$
$V_O = V' - I_L R_O$
$\text{Percent regulation} = \frac{V' - V_O}{V'} \times 100$
$= \frac{I_L R_O}{V'} \times 100$

FIG. 4-38. Theoretical regulator model.

Using the theoretical circuit of Fig. 4-37, assume that the IC op-amp has an open-loop gain of 100,000, an output impedance of 1000 Ω, that the emitter-follower has a beta of 20, and that V_{ref} is 10 V.

Using the equation of Fig. 4-38, find the value of R_o:

$$\frac{1000}{20 \times (1 + 100{,}000)} \approx 0.5 \text{ m}\Omega$$

Assuming a load current I_L of 100 mA, and using the $I_L R_o$ equation of Fig. 4-38, find the percentage of regulation:

$$\frac{100 \text{ mA} \times 0.5 \text{ m}\Omega}{10 \text{ V}} \times 100 = 0.0005 \text{ percent}$$

Using the $V' - V_o$ equation of Fig. 4-38, find the same percentage of regulation:

$$V_o = V' - I_L R_o = 10 \text{ V} - 0.00005 = 9.99995$$

$$\frac{V' - V_o}{V'} = \frac{10 \text{ V} - 9.99995}{10 \text{ V}} \times 100 = 0.0005 \text{ percent}$$

From a user's standpoint, a 0.0005 percent value for regulation is not practical. There are several factors which prevent such a low percentage. The following paragraphs describe the major problems found during practical design.

4-26.3 Effects of voltage offset on regulator action

Using the theoretical values of the previous example, the output voltage is supposed to follow the input or reference voltage within about 500 μV. This value can easily be masked by the input offset voltage V_{io} of a typical IC op-amp (1 to 10 mV). Thus, unless some provisions are made to neutralize V_{io}, the output voltage will always be removed from the input voltage by an amount equal to V_{io}. For example, assume that the uncorrected V_{io} is 1 mV, and that the same reference (10 V) as the previous example is used to find percentage of regulation:

$$\frac{10 \text{ V} - 9.999 \text{ V}}{10 \text{ V}} \times 100 = 0.01 \text{ percent}$$

Another effect of V_{io} is output drift with temperature and loading. Typical V_{io} drift with temperature is 5 μV/ °C to 20 μV/ °C. This not only shows up as added shift between output voltage and reference voltage over a temperature range, but with variations in load as well. For example, assume that the temperature increases 50 °C for a given load change. If the drift is 20 μV/ °C, V_{io} will change by a value of 1 mV (50 × 20 μV). Even if V_{io} is neutralized to zero at the lower temperature, the output will be offset by 1 mV from the input, resulting in a 0.01 percent of regulation (assuming the values of the previous example).

In a practical IC op-amp regulator circuit, the output voltage will appear to drift after the load is applied, or the load is changed. Then, after some time, the output voltage will settle to a constant value. The regulation figure using the equations of Fig. 4-38 will be of little value if the V_{io} drift completely covers the short-term regulation. A typical IC requires about 1 to 2 minutes to settle after a change in temperature. This can easily interfere for good regulation for rapid load changes.

4-26.4 Effects of common mode rejection and power supply sensitivity on regulator action

If the IC op-amp receives its power directly from the regulator circuit input voltage as shown in Fig. 4-39, any variation in input voltage can produce some change in input offset voltage V_{io}. The amount of V_{io} change for a given power supply change is a function of power supply sensitivity (Sec. 3-3.11). It is obvious that the percentage of regulation can be no better than the power supply sensitivity. For example, assuming the same 10-V reference, a change in V_{io} of 1 mV (due to power supply variation) produces 0.01 percent of regulation.

Another effect of power supply variation is that a common mode signal (Sec. 3-3.7) or error voltage is generated when V_{ref} remains constant (as it should), but the power supply voltages change. (The error is common

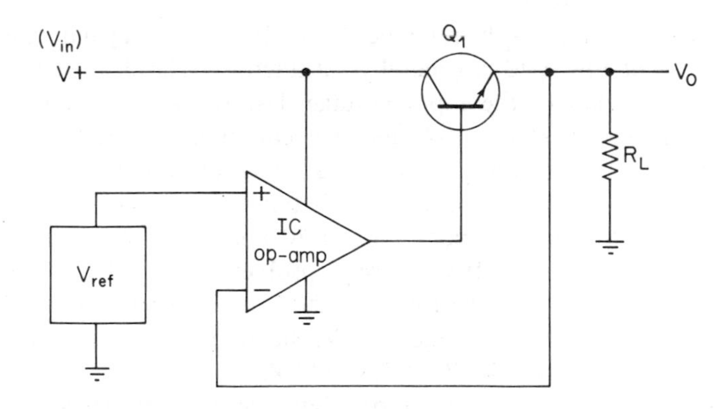

FIG. 4-39. IC op-amp regulator powered by line voltage.

mode since both power supplies $V_{CC} - V_{EE}$ or $V+$ and $V-$ change by the same amount.)

When V_{ref} is made the same as $V+$ and $V-$, the common mode error signal is equal to the difference between V_{ref} and the average of $V+$ and $V-$, or:

$$\text{common mode signal} = \left(\frac{V+ + V-}{2}\right) - V_{ref}$$

As a guideline, any IC op-amp used as the active element in a regulator should have a common mode rejection ratio of at least 90 dB, and preferably 100 dB.

4-26.5 Effects of voltage reference on regulator action

No matter what regulator circuit is used, output voltage stability over a period of time, and with temperature variations, depends for the most part on the *quality of the voltage reference*. An IC op-amp, even with a very low temperature drift characteristic, can not guarantee stable regulation. Thus, considerable care must be used to select a good voltage reference.

4-26.5.1 Zener diode voltage reference

The most obvious solution is to use a Zener diode which has a low temperature coefficient by nature, or a Zener that has been compensated by adding forward diodes. Zener diodes have many problems. For example, low-voltage Zeners are difficult to use when the regulator must provide a high-voltage output. High-voltage compensated Zeners are expensive. Also, when the reference voltage is equal to the output voltage, current from the input voltage must be used to operate the Zener. This can cause some line ripple to feed through to the output.

If a Zener is to be used as the reference for an IC op-amp regulator, a circuit similar to that of Fig. 4-40 is generally satisfactory. The drop across series resistance R_S is equal to the input voltage, less the Zener voltage (which becomes V_{ref}). Note that Fig. 4-40 includes equations for trial values of the reference circuit. The following summarizes the use of these equations.

The Zener voltage should be about 0.7 times the lowest input voltage. Thus, if the input voltage varies between 10 and 15 V, the Zener voltage should be 10×0.7 or 7 V. If the design problem is stated in reverse, the minimum input voltage must be 1.4 times the desired Zener voltage. Thus, if the required Zener (or V_{ref}) voltage is 10 V, the minimum input voltage from the basic power supply must be 10×1.4 or 14 V.

Because the Zener voltage must be about 30 percent below the input voltage (to insure that the Zener will go into avalanche condition), there is a

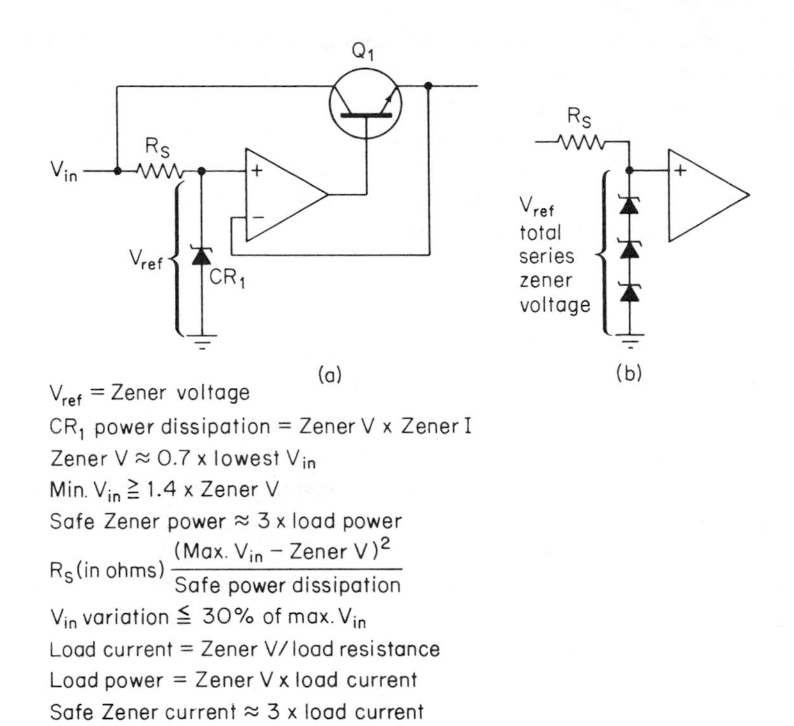

Q_1
R_S
V_{in}
V_{ref}
CR_1
R_S
V_{ref} total series zener voltage
(a)
(b)
V_{ref} = Zener voltage
CR_1 power dissipation = Zener V x Zener I
Zener V ≈ 0.7 x lowest V_{in}
Min. V_{in} ≧ 1.4 x Zener V
Safe Zener power ≈ 3 x load power
R_S (in ohms) $\dfrac{(\text{Max. } V_{in} - \text{Zener V})^2}{\text{Safe power dissipation}}$
V_{in} variation ≦ 30% of max. V_{in}
Load current = Zener V/ load resistance
Load power = Zener V x load current
Safe Zener current ≈ 3 x load current

FIG. 4-40. Basic Zener voltage reference for IC op-amp regulator.

corresponding drop across the series transistor Q_1 when the Zener is used as V_{ref} (and the regulator output must equal V_{ref}). A large drop across Q_1 can create a power dissipation problem for Q_1.

Note that when a series of Zener diodes are used to achieve a given V_{ref} (Fig. 4-40b), the total series Zener voltage should be about 0.7 times the lowest input voltage.

A safe power dissipation rating for the Zener diode is 3 times the load power. Thus, with a 250-mW load, a Zener capable of disispating 750 mW should be satisfactory. In the case of a regulator, the input resistance of the IC represents the load. Generally, IC op-amp resistance is quite high in relation to the voltage involved. Thus, the power dissipating rating of the Zener is low. Typically, a Zener (or series of Zeners) used as V_{ref} in an IC op-amp regulator requires a 0.5-W power dissipating rating.

The value of R_S is found using the maximum input voltage, the desired Zener voltage, and the safe power dissipation. Using the same 10- to 15-V input, a desired Zener voltage (V_{ref}) of 7 V, and a safe power dissipation of 0.5 W, the first trial value of R_S is: $(15 - 7)^2/0.5 = 128\ \Omega$. (In practical design, use the next highest value of 130 Ω.)

An alternate method of using a Zener as V_{ref} is shown in Fig. 4-41. Here, a FET is used in place of R_S. Note that the FET gate and source terminals are shorted together. With this arrangement, the FET will conduct when the drop across the FET exceeds the "pinch off" voltage. Typically, the pinch off voltage is on the order of 1 to 2 V (or less). Thus, the input voltage need only be 1 to 2 V higher than the desired Zener (or V_{ref}) voltage. This provides a minimum drop across the series transistor Q_1.

4-26.5.2 Variable voltage reference using a FET

One problem in using a Zener as V_{ref} is that the reference voltage is fixed. Thus, the voltage is dependent upon available Zener values (either a single Zener or a series of Zeners). Likewise, the reference voltage cannot be adjusted. Often, it is more desirable to have a regulator with a more flexible output. In a typical solid-state system, the power supply output (regulator output, in this case) should be capable of at least ±5 percent variation (preferably ±10 percent).

One inexpensive solution to this is shown in Fig. 4-42. Here, the voltage reference is maintained at a selected value by series resistors and a FET. The values shown are typical for input voltages up to about 20 V, and possibly higher. The input voltage must be higher than the highest desired V_{ref} by about 1 to 2 V (or less). This provides a minimum drop across the series transistor Q_1.

In use, potentiometer R_2 is adjusted for the desired V_{ref}. Any variation from this voltage is applied to the FET gate, and results in a corresponding variation in voltage drop across R_1. For example, if the input voltage goes up, V_{ref} goes up, as does the FET gate voltage. The FET draws more current, and the drop across R_1 increases to offset the initial change in V_{ref}.

Needless to say, the FET circuit of Fig. 4-42 can not provide as stable a V_{ref} as the Zener circuits of Figs. 4-40 and 4-41. However, when adjustable V_{ref} is

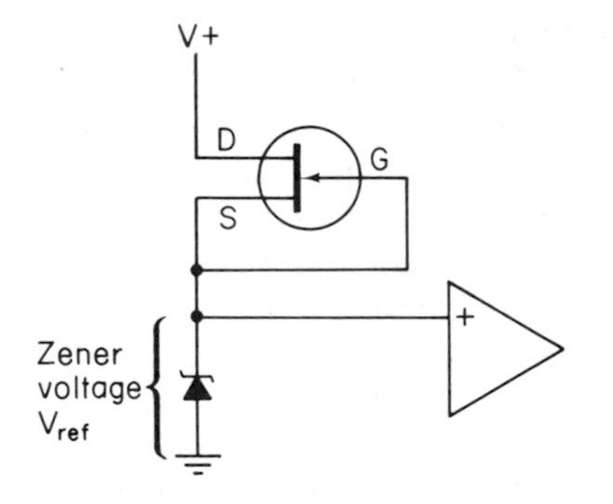

FIG. 4-41. Zener voltage reference using FET in place of series resistance.

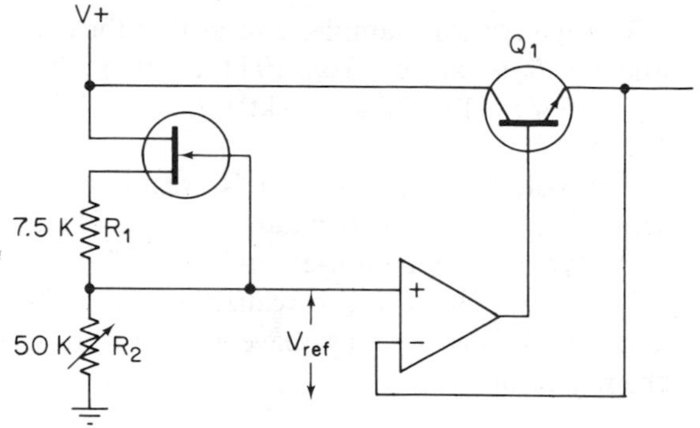

FIG. 4-42. FET variable voltage reference for IC op-amp regulator.

of greater importance, and cost must be kept to a minimum, the circuit of Fig. 4-42 will provide satisfactory results.

4-26.5.3. IC regulators as a voltage reference

When V_{ref} must be adjustable, and temperature-stable, the packaged IC regulators described in Chapter 5, provide the most satisfactory results. The basic circuit for using an IC regulator as V_{ref} in an op-amp regulator is shown in Fig. 4-43. Here, the V_{ref} is set by the ratio of R_1/R_2, and a constant which is a characteristic of the IC regulator package.

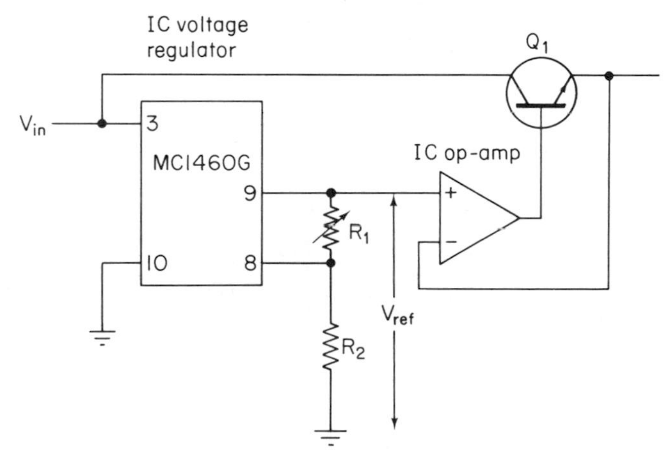

$$V_{ref} = \left(1 + \frac{R_1}{R_2}\right) 3.5\ \text{V}$$

3.5 V = constant of IC voltage regulator

FIG. 4-43. Basic IC op-amp voltage regulator using IC regulator as voltage reference.

Using a typical example, assume that the constant is 3.5 V, that R_2 is 10 kΩ, and that R_1 is variable from 10 kΩ to 40 kΩ. The V_{ref} is then variable from 7 V to 17.5 V: $(1 + 10\text{ k}\Omega/10\text{ k}\Omega) \times 3.5 = 7$ V; $(1 + 40\text{ k}\Omega/10\text{ k}\Omega) \times 3.5 = 17.5$ V.

The voltage range of such IC regulators is typically up to about 18 V. However, there are IC regulators capable of handling up to about 37 V.

A typical temperature drift specification for the IC regulators is 0.002%/° C. However, to realize the full value from the IC regulator, both resistors R_1 and R_2 must have temperature coefficients that match that of the regulator.

4-26.6 Operating at multiples of V_{ref}

When the regulator output must be considerably higher than the available V_{ref}, it is possible to operate at a multiple of V_{ref}, using the circuit of Fig. 4-44. As shown, the regulator output voltage V_o is set by the ratio of R_2/R_1, plus 1, times V_{ref}. Thus, if R_2 is four times the value of R_1, V_o is five times V_{ref}.

Of course, regulation for this circuit is worse than for a circuit where V_{ref} and V_o are equal, because of the loop attenuation introduced by R_1 and R_2. As shown by the equations, regulator output impedance Z_o, common mode rejection, and power supply sensitivity are all degraded.

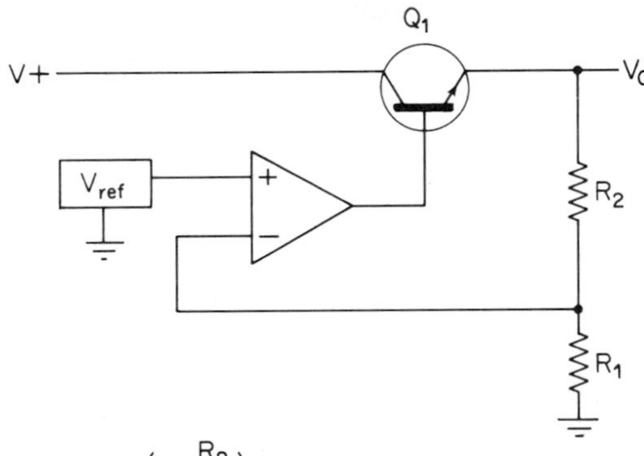

$$V_O \approx V_{REF}\left(1+\frac{R_2}{R_1}\right)$$

$$Z_O \text{ (closed loop)} \approx \frac{Z_O(\text{op-amp})}{\text{Beta } Q_1\left[1+\frac{A_{vol}\,R_1}{R_1+R_2}\right]}$$

$$\text{CMRR}' \approx \text{CMRR} \mid \text{dB} - \left(1+\frac{R_2}{R_1}\right) \mid \text{dB}$$

$$\text{Power supply sensitivity}' \approx \text{PSS}\left[1+\frac{R_2}{R_1}\right]$$

FIG. 4-44. IC op-amp voltage regulator operating at multiples of V_{ref}.

4-26.7 Floating IC op-amp regulators

IC op-amp regulators are often used for stabilizing voltages considerably higher than the ratings of the IC. This control is possible because the entire circuit is not referenced to ground, but rather "floated" between ground and the supply voltage. Such a circuit is shown in Fig. 4-45. The circuit responds as a Zener "multiplier". The output V_o is a multiple of the reference voltage (across CR_3), and set by the ratio of R_1/R_2. Zener CR_1 insures that the IC positive supply is greater than the required output swing. Zener CR_2 maintains a constant supply voltage for the unit. The voltage

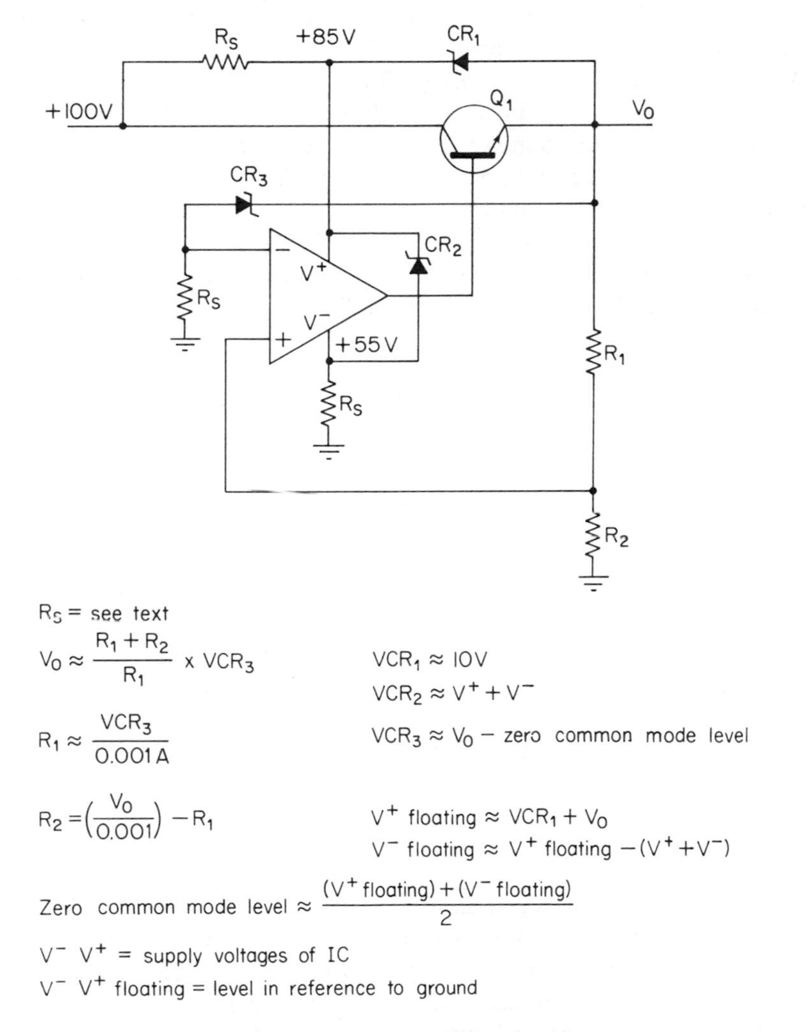

FIG. 4-45. Floating voltage regulator using IC op-amp.

across CR_2 is the sum of the positive and negative supply voltages. Under these conditions, the voltage across the IC is tied to the output voltage by fixed constants. The voltage across CR_3 is equal to the output voltage, less the zero common mode level (which is the average of the $V+$ and $V-$ voltage levels).

Typically, the voltage across CR_1 is 10 V, while the voltage across CR_2 is equal to the sum of the supply voltages (if V^+ and V^- are 15 V, CR_2 must be 30 V). Using these same values, and assuming a desired V_o of +75 V, the remaining values can be found using the equations of Fig. 4-45 as follows.

With a V_o of 75 V and a V_{CR1} of 10 V, the value of V^+ is floated at +85 V, while the V^- floating point is +55 V (30 V, or the sum of the positive and negative supply voltages, below the V^+ floating point).

With the V^+ and V^- voltage levels (floating points) at +85 V and +55 V, respectively, the zero common mode level is: 70 V; $(85 + 55)/2 = 70$ V. Thus, both the inverting and non-inverting inputs to the IC are floated at 70 V.

With V_o at +75 V, and the zero common mode level of +70 V, the voltage across CR_3 is 5 V $(75 - 70 = 5)$.

Assume an arbitrary 1-mA current through R_1 and R_2, or a total of 75 kΩ for $R_1 + R_2$ (75 V/0.001 A = 75 kΩ). Using the equations, $R_1 = V_{CR3}/0.001$ A, then $R_1 = 5$ kΩ (5/0.001 = 5 kΩ). With $R_1 + R_2$ at 75 kΩ, and R_1 at 5 kΩ, $R_2 = 70$ kΩ.

The output V_o should then be:

$$V_o \approx \frac{R_1 + R_2}{R_1} \times V_{CR3} \frac{5\ \text{k}\Omega + 70\ \text{k}\Omega}{5\ \text{k}\Omega} \times 5\ \text{V} = +75\ \text{V}$$

The value of R_S is selected on the basis of voltage drop and power dissipation, as described for Zener diode references (Sec. 4-26.5.1).

One limitation of the circuit in Fig. 4-45 is that regulation decreases (percentage of regulation increases) for increasing output voltage (all other factors being equal). Other limitations include the fact that there is a large voltage drop across R_2 and R_S in series with CR_3, as well as across the series transistor Q_1. However, the circuit can provide reasonably good regulation at voltages between about 100 and 250 V.

4-26.8 Effects of ground loops

Thus far, we have discussed the effects of devices (op-amp, Zener, etc.) on overall regulator performance. In practical use, a well-designed regulator can show very poor performance if layout is poor, and little consideration is given to current paths. Typically, the regulator circuit elements will be mounted on the same board or card as the power supply. However, the regulator output will just as typically be delivered to a load at some remote location. In this sense, the term "remote" can mean another board or card only a few feet (or even a few inches) away from the regulator. In any event, the load current must pass through wires which will produce some voltage drop.

A regulator diagram is given in Fig. 4-46 with critical load path wire resistances shown as R_{W_1} and R_{W_2}. These resistances are the most important simply because they carry the most current, and thus drop the most voltage.

Even if the output voltage V_o equals V_{ref}, and the IC op-amp gain is infinite (neither of these conditions is practical), the actual voltage across the load V_L will still be a fraction of V_o. In effect:

$$V_L = V_o \times \frac{R_L}{R_{W_1} + R_{W_2} + R_L}$$

This may seem insignificant, but consider that No. 20 wire shows about 10 mΩ of resistance per foot. This means about 1 mV per 100 mA of load current per foot. Assuming a 100-mA load (which would be quite small for the typical power transistor used as series transistor Q_1), a V_{ref} of 10 V, and a distance of 6 inches from the regulator card to the load card (12 inches total), the actual V_o will be offset by 1 mA from V_{ref}. This results in a best possible 0.01 percent regulation.

An additional problem arises if the regulator output is connected to the load by binding posts or plug-in terminals, rather than being soldered. Even solder joints can result in loss of a few millivolts if not properly made.

All of these conditions can be minimized (but not eliminated) by "remote sensing" as shown in Fig. 4-47. Assuming that the "sense" lines are the same length as the load lines, the voltage across the inverting and non-inverting inputs to the IC are as shown by the equations on Fig. 4-47. In effect,

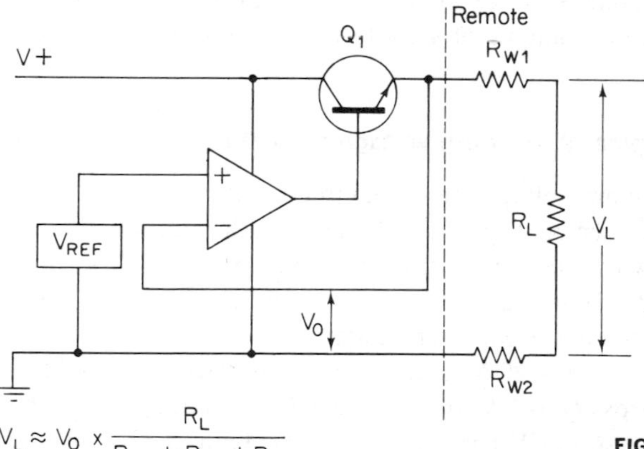

FIG. 4-46. Basic ground loop model of IC op-amp regulator.

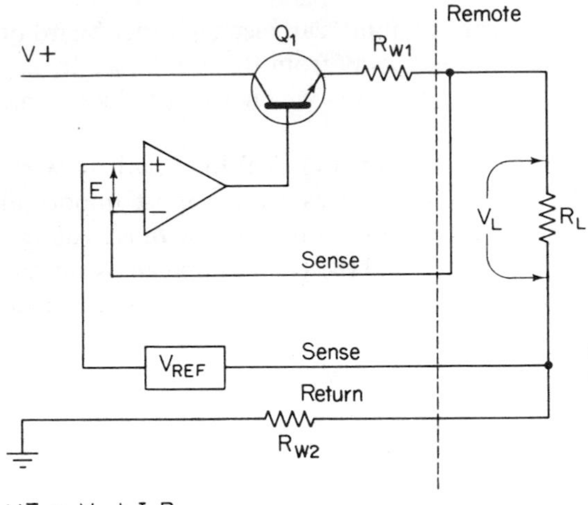

FIG. 4-47. Basic IC op-amp voltage regulator with remote sensing.

$$V_E^- = V_L + I_L R_{W2}$$

$$V_E^+ = V_{REF} + I_L R_{W2}$$

$$V_E^+ - V_E^- = E = V_{REF} - V_L, \text{ and}$$

$$\frac{V_E^+ + V_E^-}{2} = \frac{V_L + V_{REF}}{2} + I_L R_{W2}$$

both inputs are offset by the same amount. That is, the common mode voltage is increased by the drop across R_{W_2}.

If the common mode rejection of the IC is good (at least 90 dB), the added common mode voltage will not significantly affect performance of the regulator.

The sense lines can be very small gauge wire since they carry very little current. The sense lines should be No. 20 wire or smaller. Resistance R_{W1} increases the open-loop output impedance of the regulator. However, the additional 10 to 20 mΩ (for a load wire length of 1 foot) will not significantly affect performance.

4-26.9 A typical IC op-amp voltage regulator

A typical op-amp voltage regulator circuit is shown in Fig. 4-48, with a regulation curve shown in Fig. 4-49. Note that the output voltage drops less than 0.25 mV, with a load variation from 0 to 300 mA. Also note that the circuit components are identified as to type number or value. These values can be used as a "starting point" for design.

The regulator operates from a +20-V source, and is adjusted for a +15-V output by R_1. The drop across Q_1 is 5 V and, with a 300-mA load, Q_1 must be capable of dissipating at least 1.5 W (preferably more). Beta of Q_1 must be at least 20.

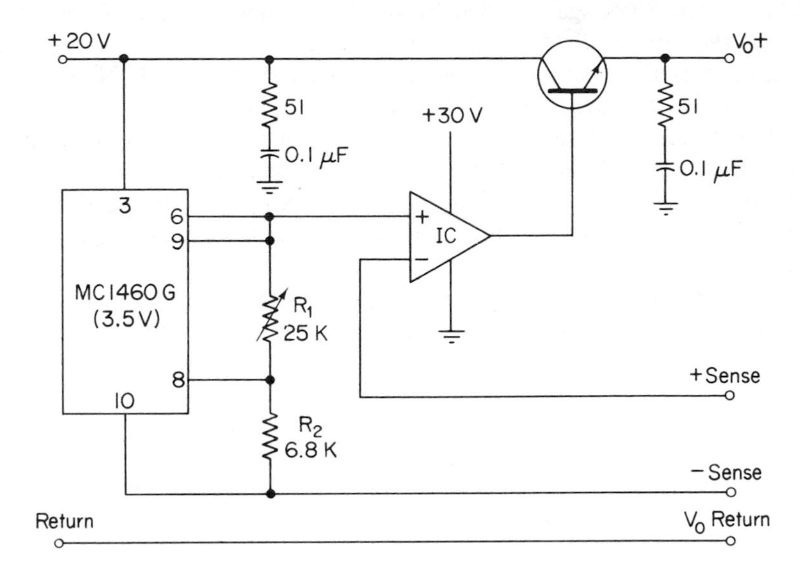

IC op-amp = MC 1539 G, or following characteristics
A_{Vol} = 50,000 (min)
Z_0 = 4 K
CMRR = 100 dB
Offset voltage = −4 mV max
TC_{Vio} = 5 μV/°C
Power supply sensitivity = 150 μV/V (max)
Q_1 = 2N4921 with beta of 20 (min)

FIG. 4-48. Typical IC op-amp voltage regulator (Motorola).

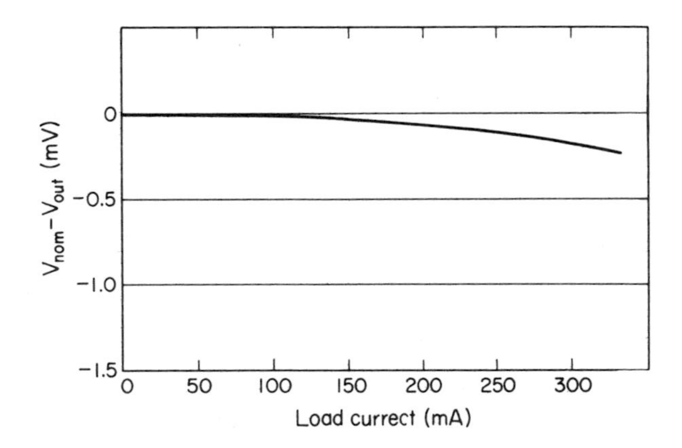

FIG. 4-49. Typical load regulation curve for IC op-amp regulator.

The reference voltage supplied by the IC regulator to the IC op-amp is equal to the ratio of R_1/R_2 times a constant of 3.5.

The characteristics of the IC op-amp are listed on Fig. 4-48. Note that the IC op-amp is operated from a single supply (+30 V).

For best results, R_1, R_2 and Q_1 should be mounted close together so that they will all be at the same temperature. This will minimize the change in percentage of regulation with changes in temperature. R_1 and R_2 should also be matched in temperature characteristics to the IC regulator, and to each other.

In the final analysis, regulation will be no better than reference voltage stability which, in turn is dependent upon temperature characteristics.

5. LINEAR INTEGRATED CIRCUIT PACKAGES

In this chapter, we shall discuss linear ICs that are not op-amps. There are two basic types of such ICs. First, there are the general-purpose ICs. These include diode arrays, transistor arrays, Darlington-pair arrays, and similar groups of solid-state elements fabricated on a single semiconductor chip. Second, there are the special purpose ICs that can replace all of the active elements in a solid-state circuit. These include regulators, IF amplifiers, amplifier-phase detectors, comparators, and so on. Both types of ICs are discussed from the user's standpoint.

Unlike Chapter 4 where the applications or uses applied to many ICs, the values and design examples in this chapter are based on the characteristics of the particular ICs discussed. Although these values and examples are "typical", they do not necessarily apply to all ICs of a given type. Thus, the datasheet must be consulted (or the IC must be tested as described in Chapter 7), to determine the IC characteristics and values. However, the values given here can be used as a "starting point" for design.

5-1. IC DUAL DIFFERENTIAL COMPARATOR

The dual differential comparator performs the same function as the op-amp comparator described in Sec. 4-16. However, no op-amps are needed when an IC differential comparator is used. In effect, the circuit is complete, with the possible exception of a few resistors and external power.

The IC differential comparator is similar to an op-amp. This is shown in Fig. 5-1 which is the schematic of a Motorola MC1514 comparator. In general, op-amps are designed for analog applications, and are not easily

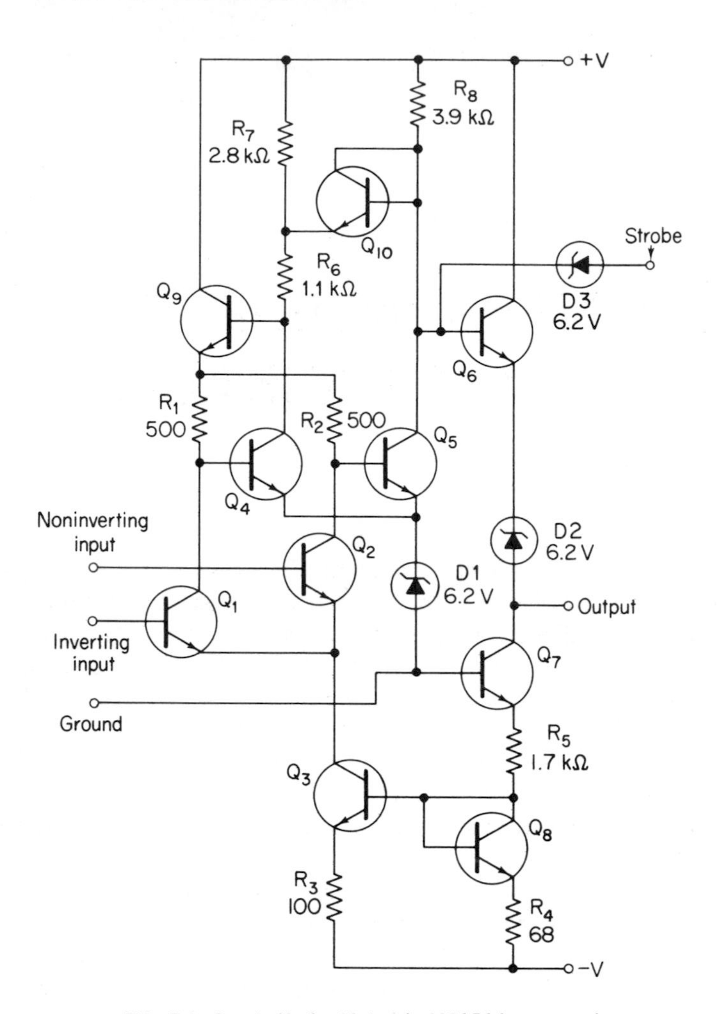

FIG. 5-1. One-half of a Motorola MC1514 comparator.

adapted to digital use. For example, the output voltage swing is not usually compatible with the various logic families. Likewise, there is a long propagation time and slow recovery time in most op-amps, making them unsuited for high-speed digital use.

The IC differential comparator was developed to provide a circuit that is easily interfaced with digital logic (in regard to propagation time, recovery time, and output swing), but still retains some op-amp characteristics. For example, as shown in Fig. 5-1, the comparator has two inputs (inverting and non-inverting, a single-ended output, dual power supplies, plus a "strobe" to enable the output. The strobe feature is essential in digital use where data must be held until called for at a specific time.

5-1.1 Circuit description

Basically, the comparator consists of two amplifier stages, output level shifting, and strobe capability. (See Fig. 5-1.)

The balanced input differential amplifier stage consists of Q_1 and Q_2 with their emitters tied to the current source Q_3 and its associated biasing circuitry. The current source improves the common mode rejection of the input amplifiers by making the collector currents insensitive to common mode input voltages.

The differential output of the first stage feeds a single-ended amplifier stage consisting of Q_4 and Q_5. Q_4 provides biasing for Q_5 as well as unity gain inversion so that the full differential output from the input stage is delivered to the base of Q_5. The biasing scheme used in the second stage renders the output of Q_5 insensitive to positive power supply variations. This is due to the excellent matching of R_1, R_2, Q_4 and Q_5 made possible by monolithic processing techniques. A change in the positive power supply voltage will change the collector currents in both Q_4 and Q_5 equally, thereby holding the output voltage of Q_5 constant. Emitter-follower Q_9 isolates the collector currents of the first and second stage amplifiers. The combined voltage gain of the two amplifier stages is typically 1700.

The single-ended output of the second stage is referenced several volts positive with respect to ground. In order to make the output voltage swing compatible with the popular TTL (Chapter 6) logic levels, the output level is shifted down by the emitter follower Q_6 and the 6.2-V Zener diode D_2.

When using the recommended power supply voltages of +12 V and −6 V, the typical output voltage swing is from −0.5 V to +3 V. The emitter follower output also supplies a high output current capability to the load. Zener diode D_1 permits a large input voltage swing. The maximum input positive voltage is limited to the voltage on the base of Q_4 or Q_5 which is +7 V. Q_7 isolates the output from the diode compensating bias driver network of the current source for Q_1 and Q_2. Also Q_7 has sufficient current sink capability to drive one DTL or TTL load.

The diode-connected transistor Q_8 compensates for the base-emitter voltage drift of Q_3 with temperature, and helps reduce voltage gain variations over temperature. The diode-connected transistor Q_{10} limits the output positive voltage swing. External clamping of the output emitter follower Q_6 is accomplished by the strobe input, acting through Zener diode D_3.

The strobe circuit clamps the output one base-emitter voltage drop below the strobe input potential. The strobe can be driven with standard DTL or TTL logic. The strobe adds considerable flexibility to the comparator, especially when used in noisy environments. Typically, minimum strobe width is 30 ns, with a strobe load of less than 1 mA. Delay time between strobe input and comparator output is 4 to 12 ns, depending on temperature.

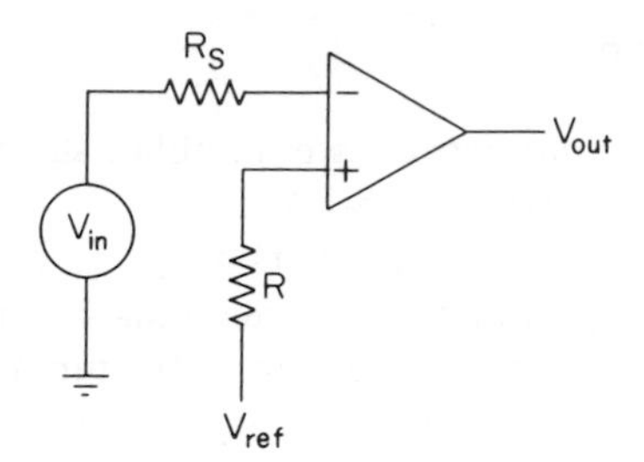

FIG. 5-2. Basic level detector using IC comparator.

5-1.2 Application as level detector

Figure 5-2 shows the IC comparator connected as a basic level detector. A reference voltage between ± 5 V can be applied to one input, and the signal to be detected is applied to the opposite input. When the input signal exceeds the reference voltage, the comparator switches state. The voltage transfer function for a positive reference voltage is shown in Fig. 5-3. If a negative reference voltage is applied to the inverting input, then the transfer curve is inverted about the vertical axis.

In order to minimize the turn-on error voltage of the circuit in Fig. 5-2, the dc resistance of the reference voltage source R should equal the source resistance R_S of the signal source. Minimum input offset voltage is generated when the two resistances are equal, because the input bias currents to the comparator produce nearly equal voltages across each source resistance (as is the case with an op-amp). Also, a low source resistance will minimize the effects of the input offset current. From a practical standpoint, R and R_S should be selected for minimum resistance, and to match any differences between source and reference resistances.

The circuit of Fig. 5-2 can be used for a variety of threshold detector applications. Other applications include a voltage comparator in analog-to-digital converters, and a high noise immunity buffer circuit.

5-1.3 Application as level detector with hysteresis

Figure 5-4 shows the IC comparator connected as a level detector with hysteresis. In some slow-speed switching applications, noise can

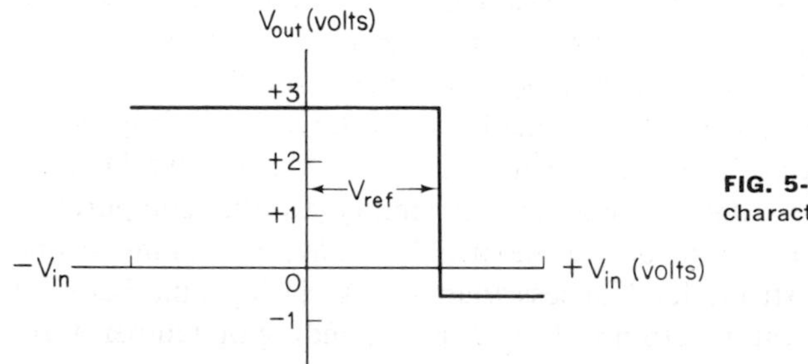

FIG. 5-3. Level detector transfer characteristics.

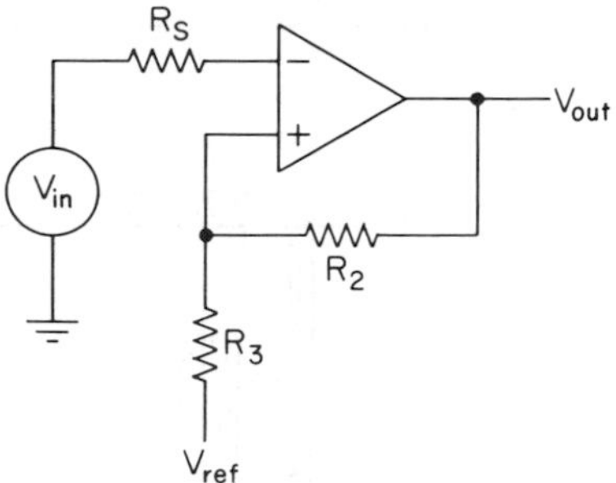

FIG. 5-4. Basic level detector with hysteresis.

$$V_{high} = V_{ref} + \frac{R_3\left[V_{out}(\mathrm{max}) - V_{ref}\right]}{R_2 + R_3}$$

$$V_{low} = V_{ref} + \frac{R_3\left[V_{out}(\mathrm{min}) - V_{ref}\right]}{R_2 + R_3}$$

$$V_H\,(\text{hysteresis loop}) = V_{high} - V_{low}$$

$$V_H = \frac{R_3}{R_2 + R_3}\left[V_{out}(\mathrm{max}) - V_{out}(\mathrm{min})\right]$$

cause oscillation during the time the input signal is passing through the transition region of the comparator. This problem can be greatly reduced, or eliminated, by using hysteresis in the form of external feedback as shown. The turn-on and turn-off points of the circuit are then controlled by the ratio of the feedback resistances (all other factors being equal). This is shown by the equations of Fig. 5-4 and the graph of Fig. 5-5.

If possible, the source resistance should be equal to the parallel resistance of R_2 and R_3. This will minimize input offset voltage. Hysteresis curves for various values of feedback resistors are given in Fig. 5-5. There is one precaution that should be observed with the feedback resistors. Oscillation can occur if the ratio of the feedback resistors is greater than the small-signal gain of the comparator.

Note that the circuit of Fig. 5-4 could be used as a variable-threshold Schmitt trigger.

5-1.4 Application as a double-ended limit detector

Figure 5-6 shows two IC comparators connected as a double-ended limit detector. The circuit switches the output of the comparators to the high state whenever the input voltage is outside the bounds of the reference voltages. This circuit could be used in test equipment to test for some set limits. The input-output transfer function is also given in Fig. 5-6. Note that the strobe input could be used to disable the circuit when not in use.

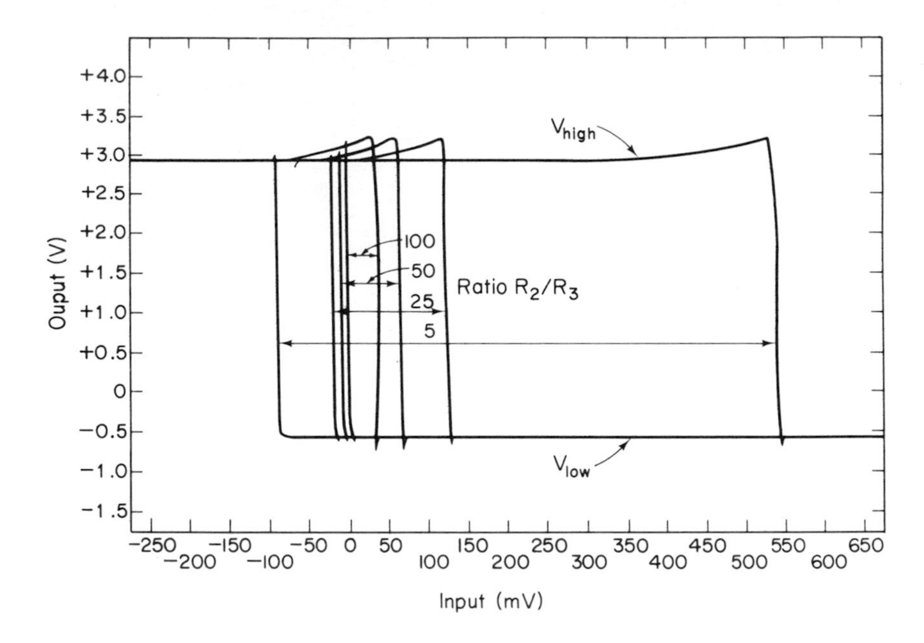

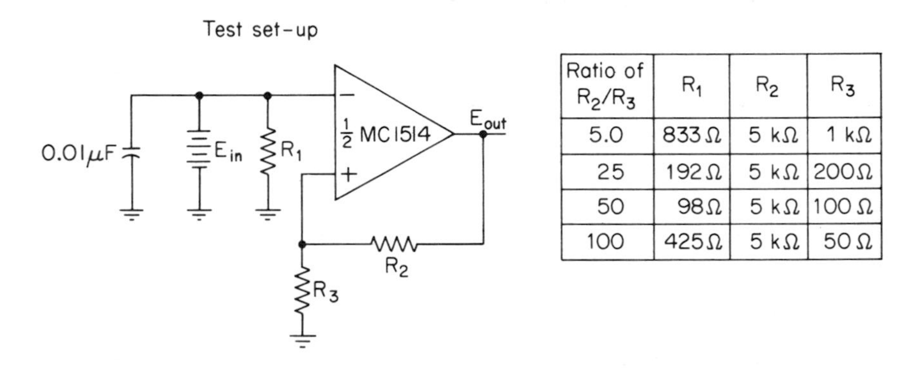

Ratio of R_2/R_3	R_1	R_2	R_3
5.0	833 Ω	5 kΩ	1 kΩ
25	192 Ω	5 kΩ	200 Ω
50	98 Ω	5 kΩ	100 Ω
100	425 Ω	5 kΩ	50 Ω

FIG. 5-5. Motorola MC1514 comparator hysteresis curves.

5-1.5 Application as a monostable multivibrator

Figure 5-7 shows two IC comparators connected as a monostable multivibrator with a fixed threshold. Figure 5-8 shows the connections for an adjustable threshold monostable MV. In either circuit, the monostable multivibrator action is accomplished with the two cross-coupled comparators. Both circuits are switched from the stable state to the quasi-stable state with a positive-going signal.

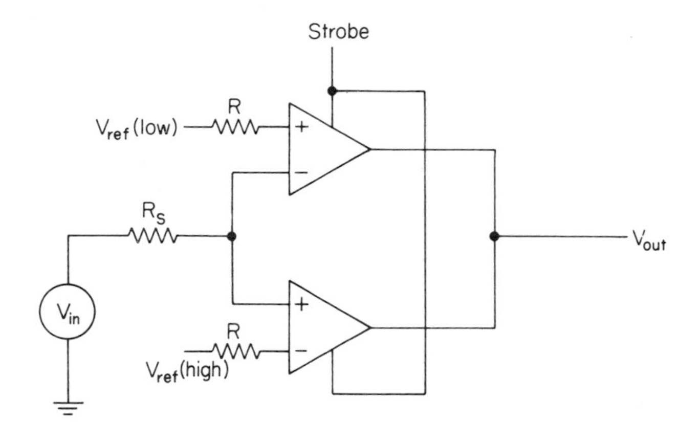

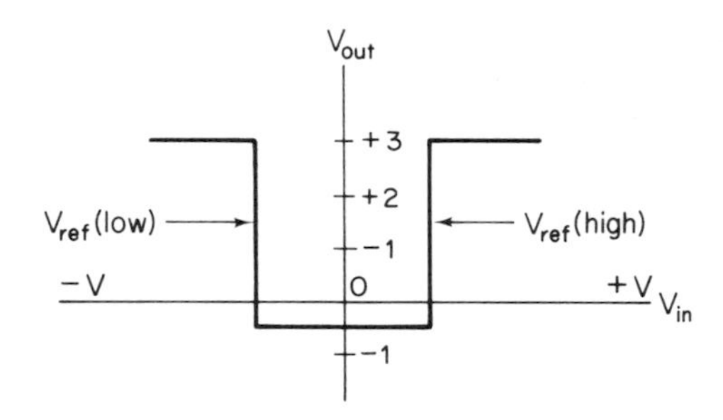

FIG. 5-6. Double-ended limit detector using IC comparator.

In the circuit of Fig. 5-7, the threshold or triggering point is determined solely by the comparator output in the high state (V_{OH}). In the circuit of Fig. 5-8, the trigger point is determined by V_{OH} and the ratio of R_1 and R_2, as shown by the equations.

In either circuit, the amount of time the circuit remains in the quasi-stable state (duration of the monostable or one-shot pulse) is determined by the *RC* time constant, and the ratio of ΔV_{out} and ΔV_{ref}, as shown by the equations. For example, assume an *RC* time constant of 1000 ns (say an *R* of 1 kΩ and a *C* of 1000 pF), and a ratio of 3 (ΔV_{out} of 3 and a V_{ref} of 1). This will produce an approximate 1300 ns duration of the monostable pulse (1000 ns × 1.3 = 1300).

A positive reference voltage is required to hold the comparator output V_{out} high during the time the circuit is in the stable state (no input pulse). Secondly, the reference voltage can be used to vary the output pulse. For example, using the previous values, if the reference voltage is increased to 1.5 V, the

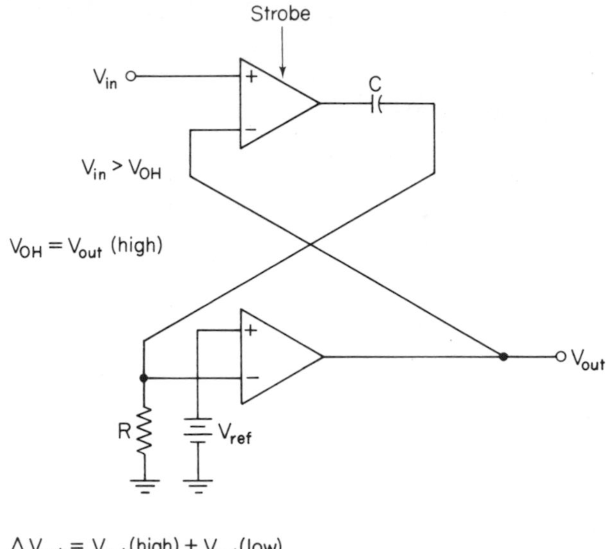

$\Delta V_{out} = V_{out}(\text{high}) + V_{out}(\text{low})$

$T \approx R_C \times \left[1 + \frac{\Delta V_{out}}{V_{ref}}\right]$

T = Duration of monostable output pulse

FIG. 5-7. Monostable multivibrator with fixed reference using IC comparator.

ratio is 2 (V_{out} of 3 and V_{ref} of 1.5 = 2). This will produce an approximate 1200 ns duration of the pulse.

The strobe feature of the comparator can provide additional flexibility. For example, the strobe could be used to gate output pulses of different time duration. This can be accomplished since the strobe and input signals are essentially an AND function. If desired, the strobe can be used to inhibit selected input pulses.

In the circuit of Fig. 5-8, the input voltage need not be greater than V_{OH}, as is the case with the Fig. 5-7 circuit. As shown, the threshold (trigger point) of the circuit is set by the values of R_1 and R_2, for a given value of V_{OH}. For example, assume that V_{OH} is 3 V, R_1 is 500 Ω, and R_2 is 50 Ω. Under these conditions, the circuit should trigger with an approximate 0.27-V input pulse:

$$V_{in} = \frac{50 \times 3}{500 + 50} = \frac{150}{550} \approx 0.27 \text{ V}$$

5-1.6 Application as an astable multivibrator

Figure 5-9 shows the IC comparator connected as an astable multivibrator. As shown by the equations, the period of the output pulse

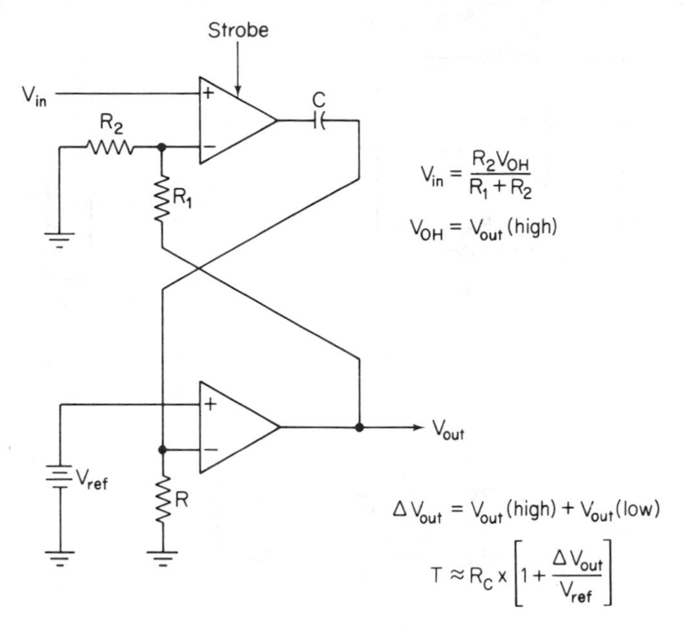

FIG. 5-8. Monostable multivibrator with adjustable reference using IC comparator.

waveform (and hence the frequency) is determined by the *RC* time constant, the ratio of feedback resistors R_2 and R_3, the ratio of high and low output voltages, and the level of the negative power supply voltage.

Resistors R, R_1, R_2 and R_3 bias the comparator in the active region to ensure self-starting. In practical applications, R_1 can be made variable to insure turn-on of the circuit, and to adjust the frequency (within limits set by the *RC* values).

The capacitor *C* produces a net positive feedback at high frequencies which insures oscillation. The resistance value of *R* should be limited from 1 kΩ to 20 kΩ. If *R* is less than 1 kΩ, the current capability of the comparator may not be sufficient to allow the output to switch fully to the low state. For values of *R* larger than 20 kΩ, the input offset current can affect the output pulse width.

For high frequency operation, the value of *C* has to be small. If the value of the *C* calculated to generate a desired frequency approaches the stray capacitance of the physical layout, the frequency generated may vary considerably from the calculated value. To minimize this effect, a minimum *C* of 100 pF is recommended. With these values of *R* and *C*, operation is limited to frequencies below about 5 MHz.

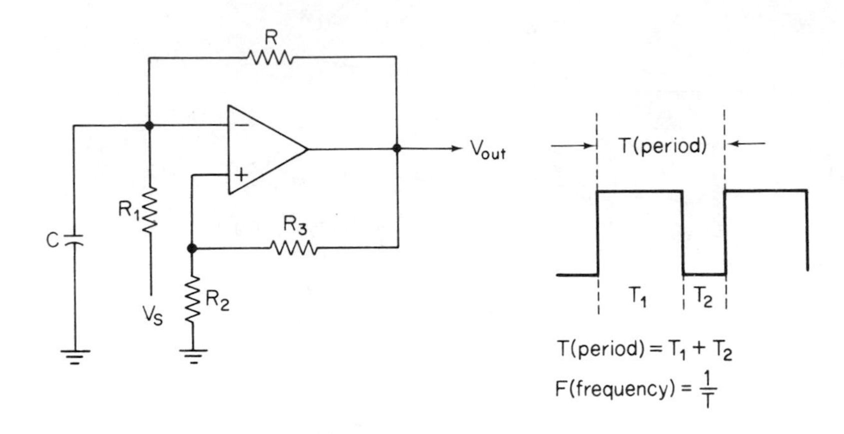

$$T_1 \approx \frac{R_1RC}{R+R_1} \ln \left[\frac{R_1V_{OH} - RV_S + \dfrac{R_2}{R_2+R_3}(R_1+R)V_{OL}}{R_1V_{OH} - RV_S - \dfrac{R_2}{R_2+R_3}(R_1+R)V_{OH}} \right]$$

$$T_2 \approx \frac{R_1RC}{R+R_1} \ln \left[\frac{R_1V_{OL} + RV_S + \dfrac{R_2}{R_2+R_3}(R_1+R)V_{OH}}{R_1V_{OL} + RV_S - \dfrac{R_2}{R_2+R_3}(R_1+R)V_{OL}} \right]$$

ΔV_{out} = the total output voltage swing
V_{OH} = total voltage in the high state
V_{OL} = total voltage in the low state
V_S = negative power supply voltage

FIG. 5-9. Astable multivibrator using IC comparator.

In order to minimize the effects of offset current and voltage, the parallel combination of R_2 and R_3 should be equal to the parallel combination of R_1 and R.

5-1.7 Application as a zero-crossing pulse generator

Figure 5-10 shows two IC comparators connected as a zero-crossing pulse generator. This circuit is a special application of a double-ended limit detector, and is designed to generate a digital output pulse at the zero crossover points of a sinewave input signal. This is shown in Fig. 5-11.

Two comparators are needed. One comparator is set to detect a minus voltage, and the other comparator is set to detect a positive voltage. The output of the two wire-ORed comparators is low only during the time the input

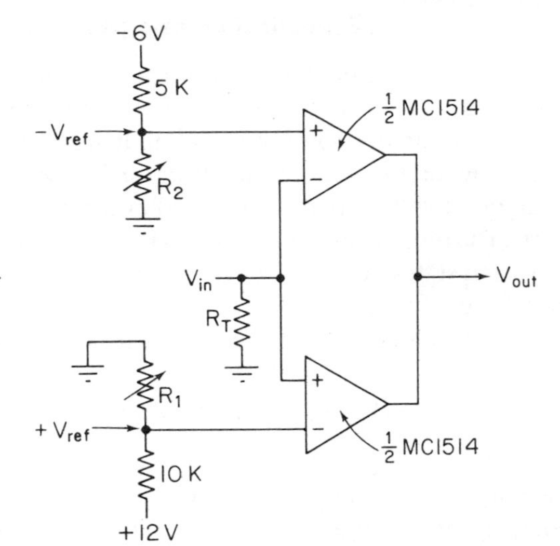

FIG. 5-10. Zero-crossing pulse generator using IC comparator (Motorola).

signal is between $-V_{ref}$ and $+V_{ref}$. Otherwise, the comparator output is high as shown in Fig. 5-11. The circuit is most accurate when the reference voltages are as close to the zero as is possible.

As shown by the equations, the duration of the output pulse is determined by the total reference voltage, the amplitude (RMS) of the sinewave, and the sinewave frequency. For example, assume a total reference voltage of 20 mV (+10 mV and −10 mV), a frequency of 10 kHz, and a sinewave of 2.5 V. The duration of the output is approximately:

$$\frac{20 \text{ mV}}{2.5 \times 6.28 \times 10 \text{ kHz}} = 127 \text{ ns}$$

The value of R_T should be chosen to minimize input offset voltage.

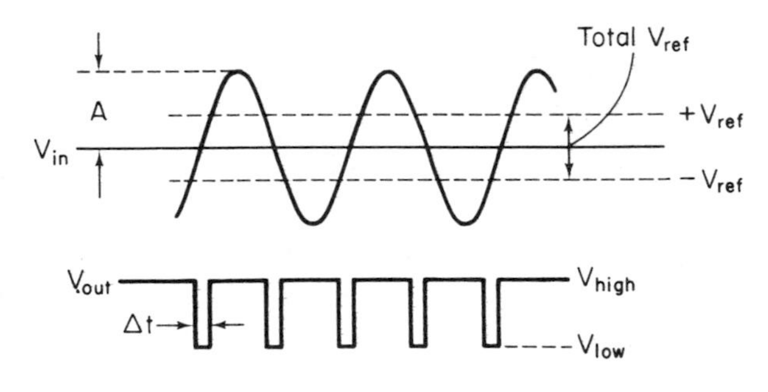

FIG. 5-11. Input-output waveforms for zero-crossing pulse generator (Motorola).

5-1.8 Application as a peak voltage detector

Figure 5-12 shows the IC comparator connected as a peak voltage detector. With the *RC* network connected at the comparator input, the voltage at the non-inverting input will always lag V_{in} (at the inverting input) by an amount dependent upon frequency. This lag in voltage results in an input voltage polarity reversal every time the input signal changes direction. The reversal in input voltage is sufficient to cause the comparator output to switch states. Capacitor C_1 is used to filter out possible oscillations during switching.

Typical values and waveforms are shown on Fig. 5-12.

5-1.9 Application as a core memory sense amplifier

Figure 5-13 shows two IC comparators connected as a core memory sense amplifier. This was one of the original applications for the comparator. A sense amplifier is essentially an interface between the memory elements and the logic elements of a digital computer. The signals generated by the tiny ferrite cores of the memory are of insufficient amplitude to drive the logic circuits directly. The function of the sense amplifier is to convert the relatively small signals from the core memory into logic levels such as DTL and TTL. Not only must the sense amplifier detect small signals, but it also has to distinguish between signals which differ by only a few millivolts in amplitude. The problem is further complicated by the fact that noise and relatively large common mode signals are present during most of the computer's memory cycle.

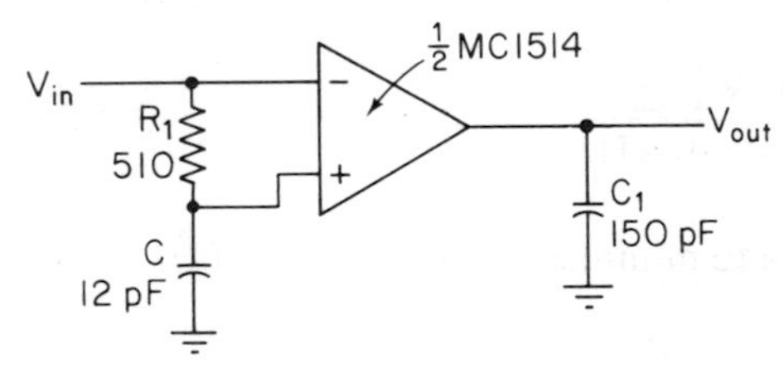

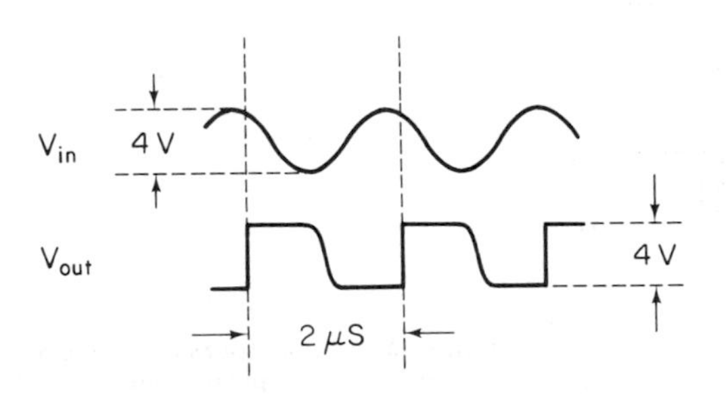

FIG. 5-12. Peak voltage detector using IC comparator (Motorola).

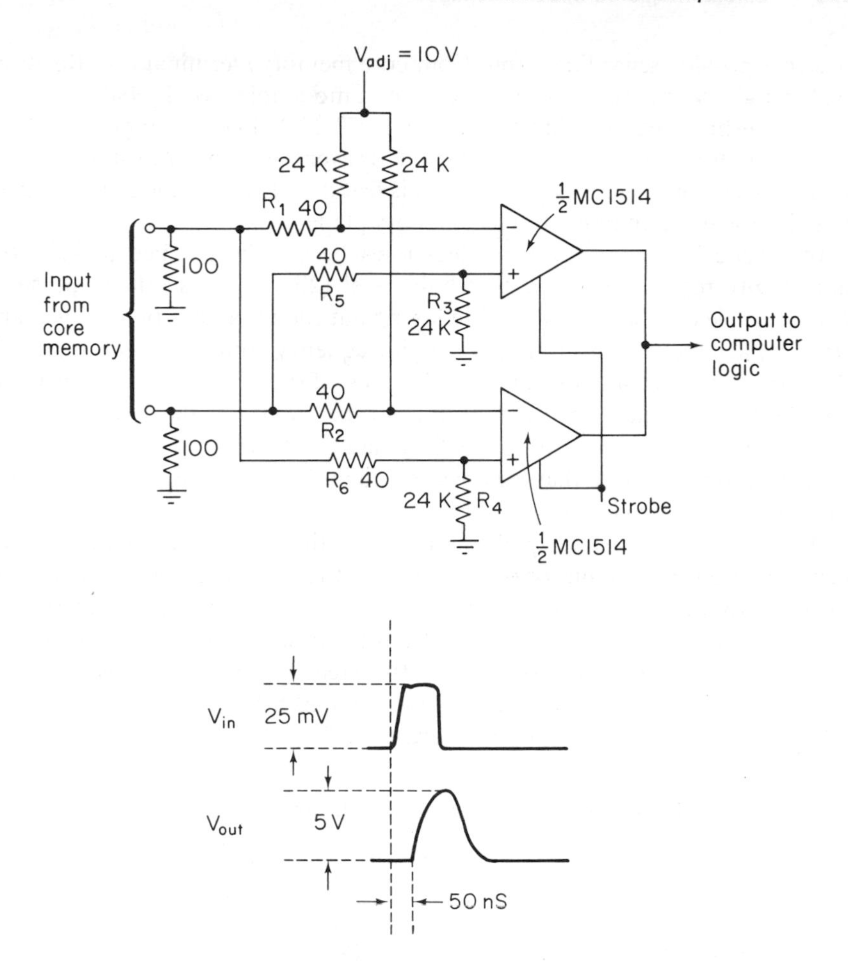

FIG. 5-13. Core memory sense amplifier using IC comparator.

For example, a typical core memory produces outputs of less than 50 mV, while a typical logic level is +5 V. This requires a gain of 100. To make the problem more complex, a logic 1 output from a core could be represented by a 15 to 50 mV signal, whereas a logic 0 output might be a 0 to 10 mV signal. The problem can be solved by setting the comparator threshold so that the output (to the computer logic) goes to a logic 1 (+5 V) whenever the input (from the core memory) is above about 20 mV. The comparator output switches to a logic 0 (0 V) with inputs below 20 mV.

In the circuit of Fig. 5-13, the threshold is set by the resistor divider network. The two comparators with their outputs wire-ORed together are required to respond to both positive and negative input signals. The threshold is essentially equal to the voltage across resistors R_1 and R_2. The 100-Ω input resistors provide sense line (input from core memory) termination. Resistors R_3 through R_6 are used to increase common mode rejection by balancing the input threshold circuit. With V_{adj} tied to the +12 V line, the input threshold of the resistor divider is set at 20 mV. However, the input threshold can be varied by changing the V_{adj} voltage. This feature allows the memory designer to tailor the sense amplifier to each memory plane.

In practical use, the resistor values must be closely matched to keep the uncertainty region about the threshold level (20 mV) as small as possible. Considerable care should be used in laying out the resistor network since any stray capacitance can result in poor high-frequency common mode rejection. Since the input threshold voltage is a function of the V_{adj} voltage, it must also be tightly controlled once set. If the strobe input to the comparator is grounded, the comparator output will be clamped to about −0.5 V. This will prevent undesired noise that may pass through the comparator from entering the computer logic circuits.

All other factors being equal, the response time of the comparator is dependent upon the amount of overdrive from the core memory. As discussed in Sec. 4-16, the more the threshold level is exceeded by the input voltage, the faster the comparator will respond. As in the case of any logic element, the faster (or shorter) the response time, the greater the system speed. In the circuit of Fig. 5-13, with a 25-mV input and a 20-mV threshold (5-mV overdrive), the response time of the comparator is about 50 ns.

5-1.10 Miscellaneous applications

In addition to the applications described thus far, the comparator can be used whenever it is necessary to produce a pulsed or uniform level-changing output (suitable for digital circuits) in response to various input levels and waveforms. The comparator switches output states in a uniform manner (1 or 0, high or low, etc.) in response to a differential input. This input can be pulsed, sinewave, or simply a difference in dc levels (or a combination). No matter what input is used, the output is a pulse or level change that can be adapted to most of the existing logic families. In this regard, the comparator is a pulse restorer.

As an example, the comparator can be used as a line receiver where digital pulse data must be transmitted through long lines. Assume that the digital logic levels are 0 V and +5 V (representing 0 and 1), and that the pulses are attenuated to about 3 V by the transmission line. Also assume that there is a 1-V noise level on the line. If the line is fed directly to the digital circuit, the 3-V pulse might not be sufficient to trigger the digital circuit elements, unless gain is introduced. On the other hand, a high gain could cause the 1-V noise signals to trigger the digital circuits. This problem can be solved by a comparator where the input threshold is set above 1 V, but below 3 V (say 2 V), and the output is set for 0 V and +5 V.

5-2. IC DIFFERENTIAL AMPLIFIER

As discussed in Sec. 4-19, there are many applications where an amplifier with differential input and differential output is required. A laboratory differential voltmeter is a classic example. Such meters must amplify very small differential signals in the presence of large common mode signals. The output of the meter amplifier is fed to a meter movement where both the amplitude and polarity of the differential signal is indicated, but common mode signals are rejected.

Also as discussed in Sec. 4-19, a differential output amplifier can be made up using two op-amps. To be really effective, both op-amps must be identical in characteristics (gain, temperature drift, etc.). This is best accomplished using a dual-channel op-amp. Also available are differential amplifiers in IC package form. These differential amplifiers are similar to dual-channel op-amps, except that both channels are already interconnected on the semiconductor chip.

This is shown in Fig. 5-14 which is the schematic of a Motorola MC1520 differential amplifier.

5-2.1 Circuit description

Transistors Q_2 and Q_4 differentially amplify the input signal applied between pin 9 and pin 10. The emitter follower input buffering provided by Q_1 at the non-inverting input and Q_3 at the inverting input is used to obtain the high input impedance, to minimize the input capacitance, to reduce the dependence of offset and impedance variations upon device current gains, and to reduce the susceptibility of "latch up".

Transistor Q_5 serves as a temperature compensated current source used primarily to provide a high common mode rejection ratio. A common mode feedback loop is incorporated around the input stage to provide quiescent operating stability for the input devices, and thus insure a more predictable input common mode voltage swing. The base of Q_7 and collector of Q_8 are wire bonded to pins 1 and 7 (as are the base of Q_9 and the collector of Q_{10} to pins 2 and 6) to provide for open-loop frequency compensation capacitors.

A resistive level shifter follows the input stage to maintain the desired voltage levels at the overall device output. The second differential gain stage also contains emitter-follower buffering to minimize losses in the resistive level shifter and loading of the first stage collectors.

Low output impedance and a large output swing are provided at the output by differential emitter followers Q_{15} and Q_{17}. To improve peak negative swing of the amplifier, a current source Q_{18} (rather than the usual emitter resistor) is used to bias the emitter-follower. Stabilization of output quiescent operating point is obtained by using a local common mode feedback loop incorporating a separate differential amplifier stage (Q_{13} and Q_{14}). This circuit

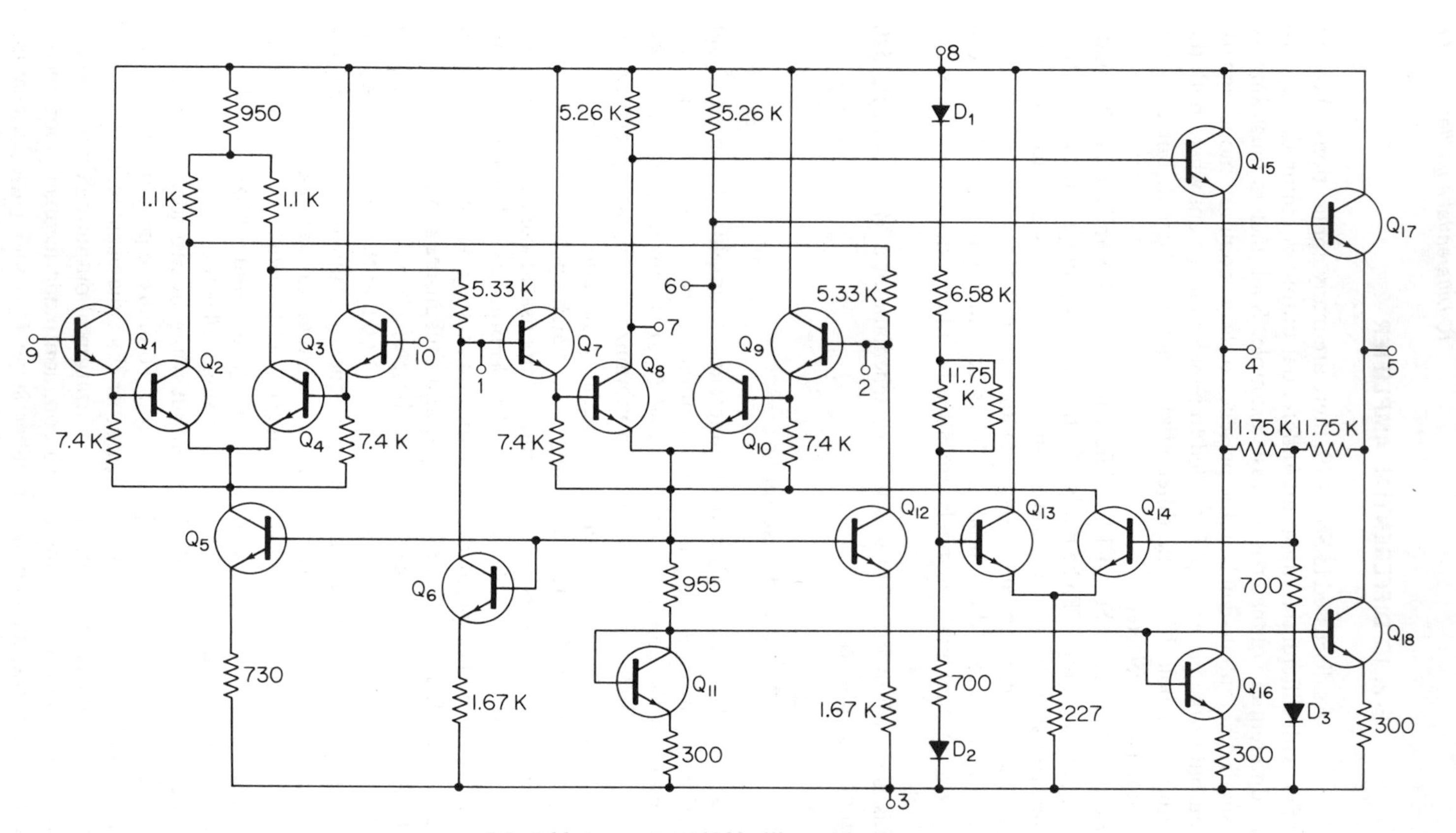

FIG. 5-14. Motorola MC1520 differential output op-amp.

holds average voltage output at ground potential for split supply operation, and at $V_{CC}/2$ for single supply operation. Changes in output quiescent point (also known as no-signal drift or zero drift) are minimized by using temperature compensation diodes D_2 and D_3 in base circuits of the feedback differential amplifier.

5-2.2 Typical connections and applications

The IC shown in Fig. 5-14 can be used wherever a high-gain, wide frequency range differential amplifier is required. Figure 5-15 shows the connections for both balanced-input and unbalanced-input operation. The recommended values for the external resistors and capacitors, together with corresponding gain and frequency ranges, are given on the datasheet. The IC can also be used as a single-ended op-amp (both inverting and non-inverting) if desired.

5-3. IC VIDEO, IF, AND RF AMPLIFIERS

There are a great number of linear amplifiers in IC package form suitable for use with video, intermediate and radio frequencies. Some of these packages include provisions for automatic gain control (AGC) or similar features. Likewise, there is an almost infinite variety of applications for these amplifiers. It would be impossible to discuss all of the ICs and

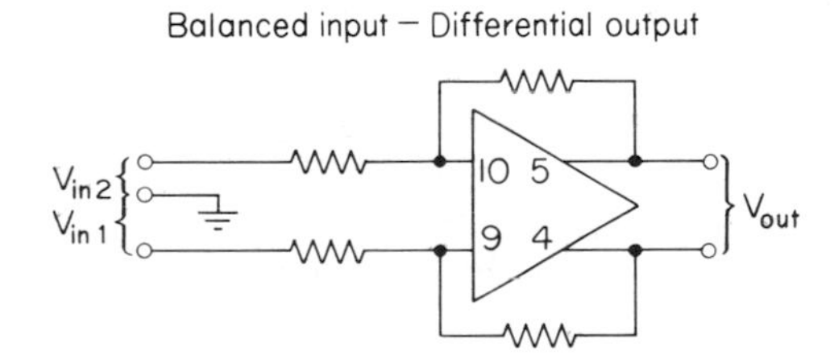

FIG. 5-15. Typical connections for differential output op-amp.

Single-ended input — Differential output

V_{in}

10 5

9 4

V_{out}

applications in any detail here. (Such discussions are available in the manufacturer's datasheets and application notes.) However, the following paragraphs provide a cross-section of IC amplifier packages and typical applications. Where practical, both the external circuit values and the performance characteristics are given. Needless to say, an ingenious experimenter can often adapt the application circuits to other ICs of similar design.

5-3.1 Motorola MC1550 amplifier

Figure 5-16 shows the full schematic and simplified schematic of the amplifier. As shown, the input is applied to the base of Q_1, and the output

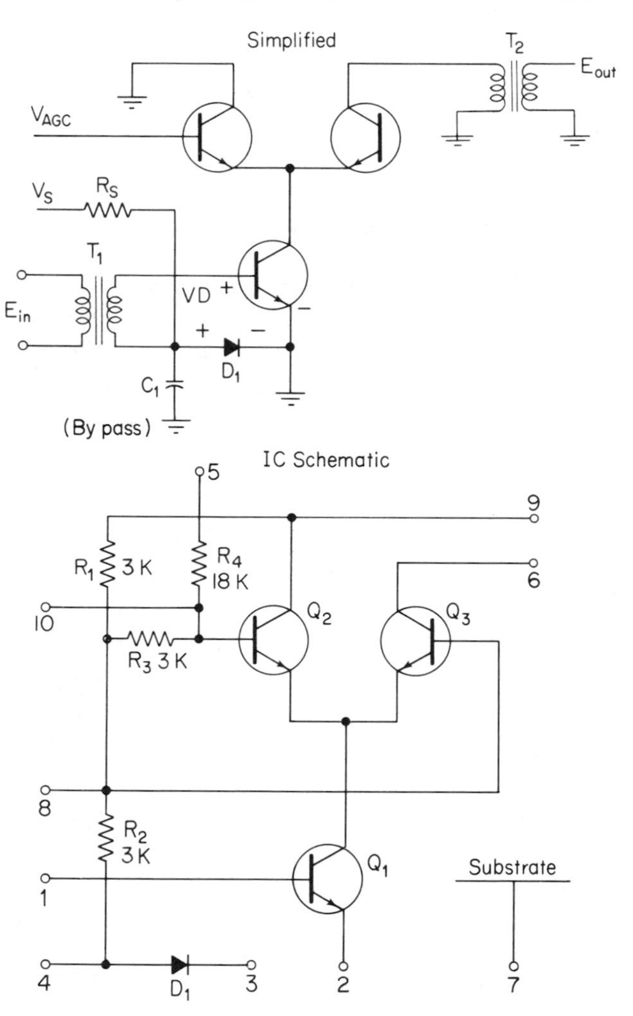

FIG. 5-16. Motorola MC1550 RF-IF amplifier.

is taken from the collector of Q_3. The combination of Q_1 and Q_3 acts as a common-emitter, common-base pair. This combination reduces internal feedback. AGC is introduced at Q_2, with a reference voltage V_R at Q_3. The amount of AGC is determined by the difference between V_{AGC} and V_R. When V_{AGC} is at least 114 mV greater than V_R, Q_3 is turned off and ac gain is at a minimum. If V_{AGC} is less than V_R by 114 mV or more, Q_3 is turned on, and ac gain is maximum. Changing the AGC voltage has little effect on the Q_1 operating point, and the input impedance of Q_1 remains constant. Voltage V_S and resistance R_S establish the current in diode D_1 which is on the same silicon die as Q_1. The emitter current of Q_1 remains within 5 percent of the diode current. This biasing technique takes advantage of the matching characteristics that are available with ICs. Resistors R_1 and R_2 bias the diode D_1 and also establish a base voltage for transistor Q_3. Resistors R_3 and R_4 serve to widen the AGC voltage from 114 mV to about 0.86 V. This is necessary in some applications so that the AGC line will be less susceptible to external noise.

5-3.1.1 Narrow band tuned amplifier

Figure 5-17 shows the MC1550 used as a tuned, narrow-band amplifier. Note that with an AGC of 0 V, and a 30-dB power gain, the bandwidth is about 0.5 MHz, centered on 60 MHz.

5-3.1.2 Wide band tuned amplifier

Figure 5-18 shows the MC1550 used as a tuned, wide-band amplifier. Both the external circuit values and the performance characteristics are given. Note that with an AGC of 0 V, and a 30-dB power gain, the bandwidth is about 15 MHz, centered on 45 MHz.

5-3.1.3 Oscillator

Figure 5-19 shows the MC1550 used as a 5–10 MHz oscillator. Note that the frequency can be increased to about 13 MHz when C_2 is reduced to about 80 pF, and can be decreased to about 3 MHZ when C_2 is increased to about 2000 pF. Using the 3-kΩ load (R_L) shown, the output voltage will be about 5 V for a V_{CC} of 6 V, and about 10 V for a V_{CC} of 12 V. If the load R_L is increased to about 7–8 kΩ, the output voltage will be approximately equal to V_{CC}.

5-3.1.4 Video amplifier

Figure 5-20 shows the MC1550 used as a video amplifier. Note that the voltage gain remains essentially flat to about 3MHz, and remains above the 3 dB point to about 20 MHz.

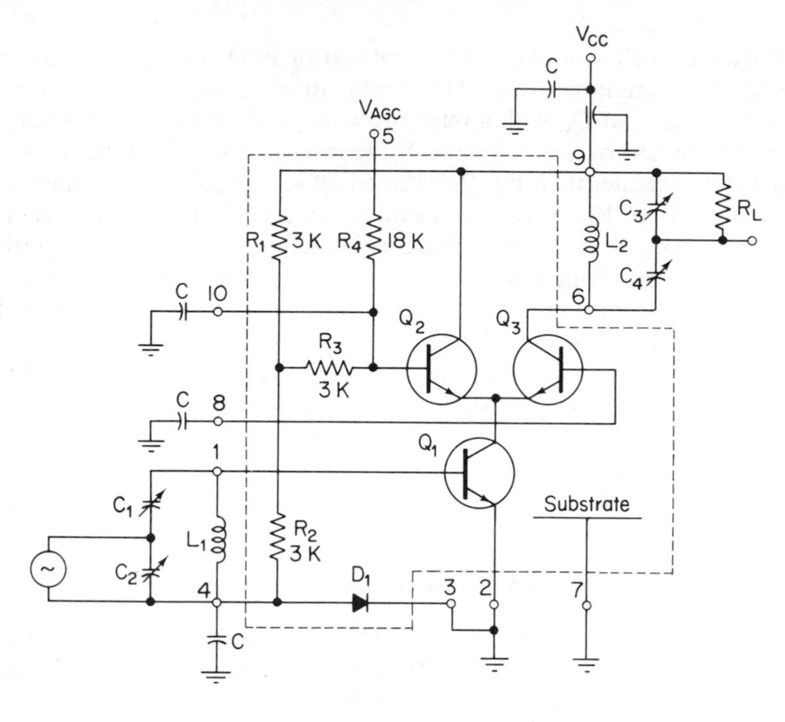

$L_1 = 0.23\ \mu H$
$C_1 = 36$ pF
$C_2 = 65$ pF
C (by pass) = 1000 pF

$L_2 = 0.26\ \mu H$
$C_3 = 630$ pF
$C_4 = 29$ pF
$R_L = 50\ \Omega$

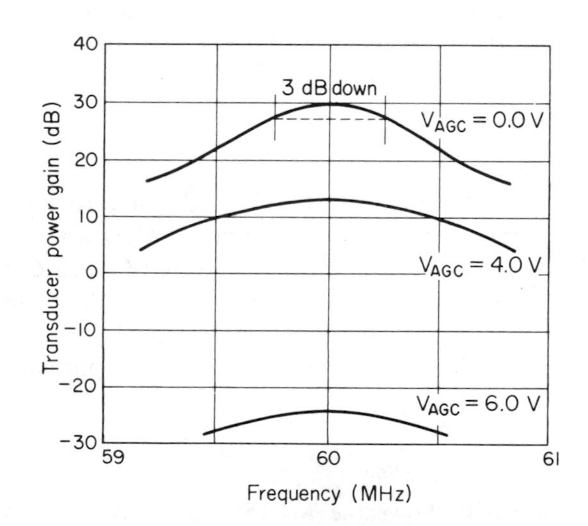

FIG. 5-17. 60-MHz tuned amplifier (Motorola).

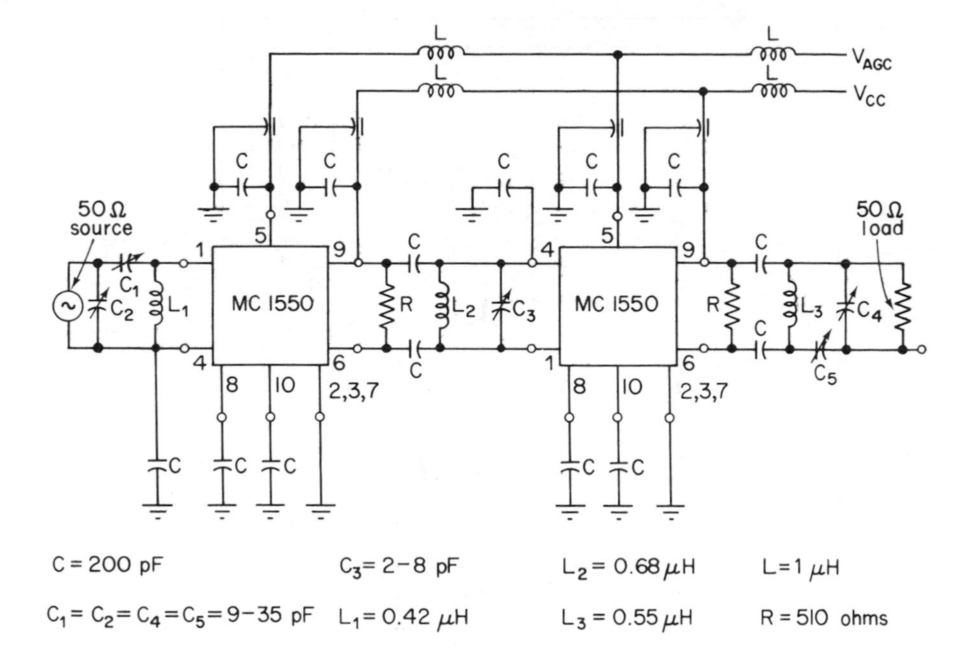

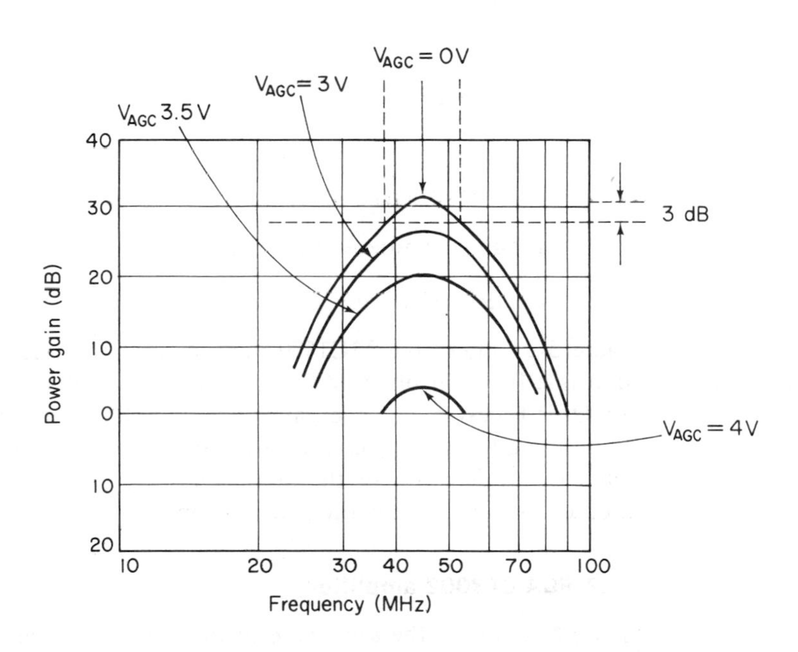

FIG. 5-18. Wide band tuned amplifier (Motorola).

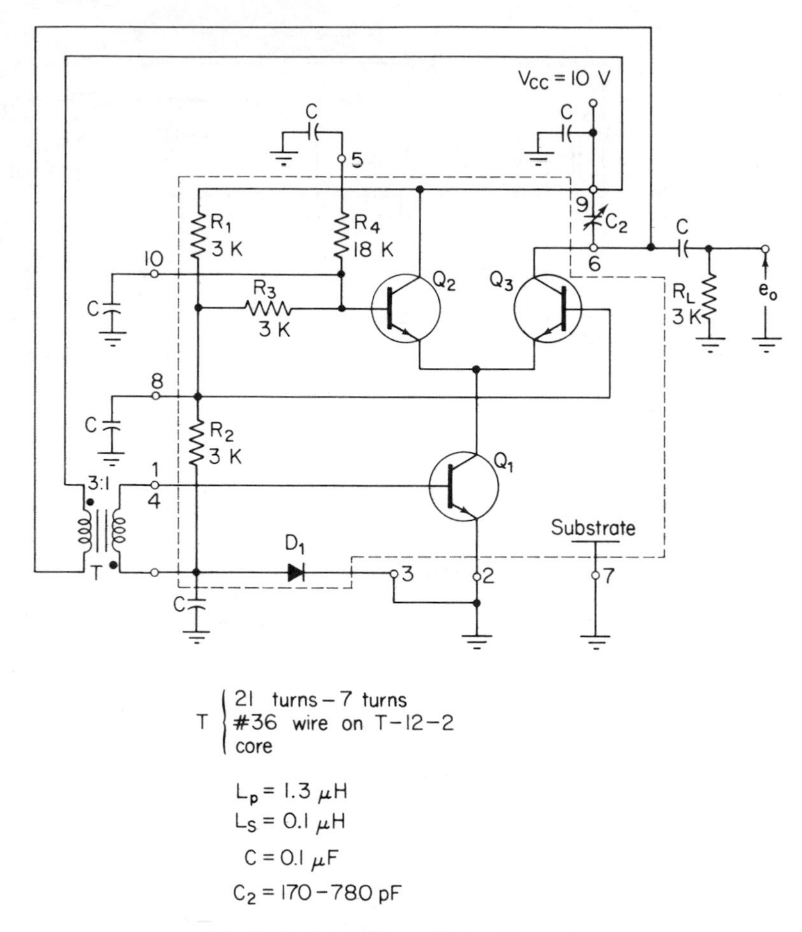

FIG. 5-19. 5–10-MHz oscillator (Motorola).

5-3.1.5 Amplitude modulator

Figure 5-21 shows the MC1550 used as an amplitude modulator. Modulation is a function of the AGC circuit. If the voltage at the base of transistor Q_2 is time-varying, the gain of the amplifier will also be time-varying. By injecting an audio signal into the base of Q_2, the RF carrier will be amplified as a function of the audio input. Approximately 90 percent modulation can be expected without distortion.

5-3.2 RCA CA3002 amplifier

Figure 5-22 shows the schematic of the amplifier. The circuit is basically a single-stage balanced differential amplifier (Q_2 and Q_4) with

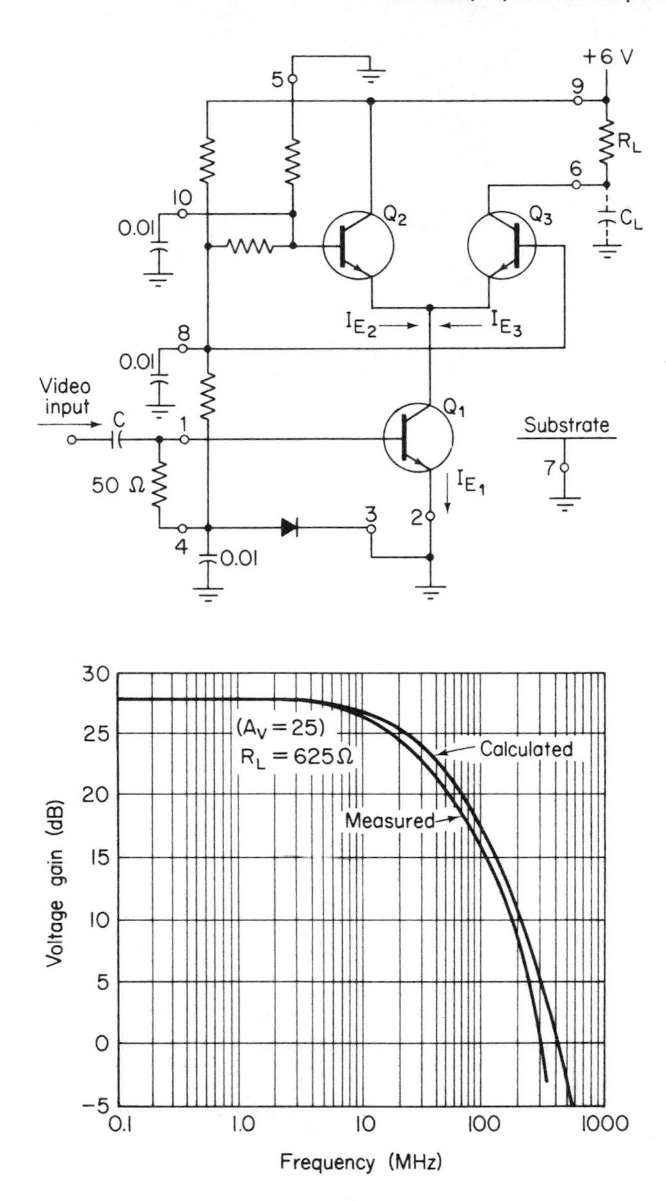

FIG. 5-20. Video amplifier (Motorola).

input emitter-followers (Q_1 and Q_5), a constant-current sink (Q_3) in the emitter-coupled leg, and an output emitter-follower Q_6. A single-ended input is connected to terminal 10, or a push-pull input to terminals 10 and 5. A single-ended output is direct-coupled at terminal 8, or capacitively-coupled at terminal 6 (an internal connection). The emitters of differential pair (Q_2 and

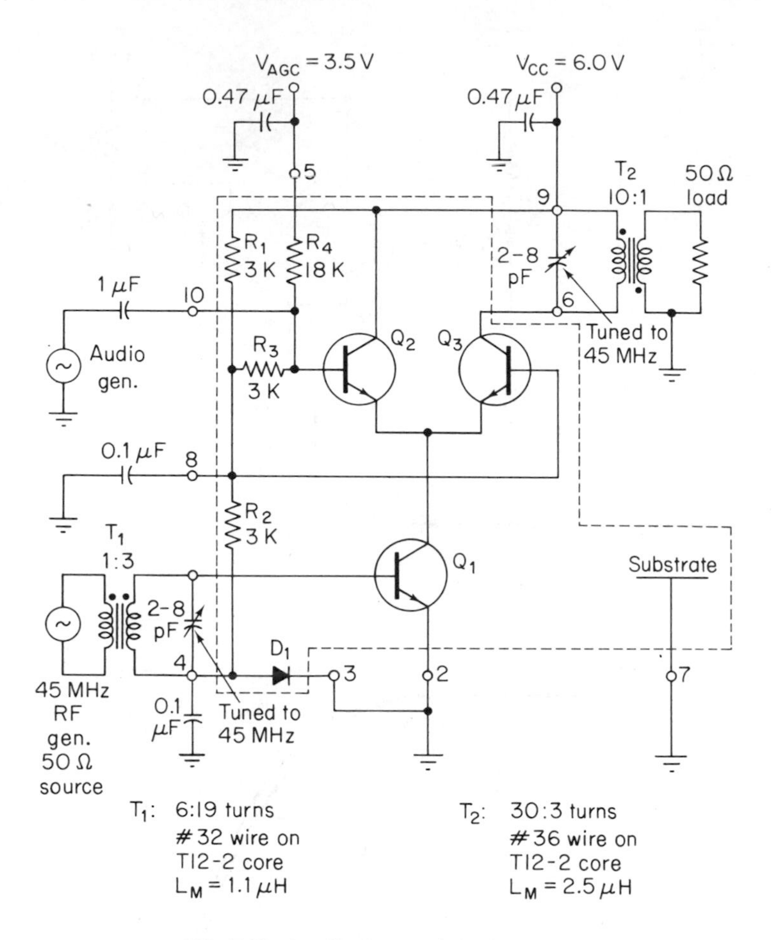

FIG. 5-21. Amplitude modulator (Motorola).

Q_4 are connected through degenerative resistors R_3 and R_4 to the transistor current source Q_3. The use of these resistors improves the linearity of the transfer characteristic and increases the signal-handling capability.

Transistor Q_1 provides a high input impedance. Transistor Q_5 preserves the circuit symmetry, and also partially bypasses the base of Q_4. Additional bypassing can be obtained by connection of an external capacitor between terminal 5 and ground. The emitter-follower transistor Q_6 provides a direct-coupled output impedance of less than 100 Ω.

5-3.2.1 Product detector

Figure 5-23 shows the CA3002 used as a product detector. A differential pair driven by a constant-current transistor can be used as a prod-

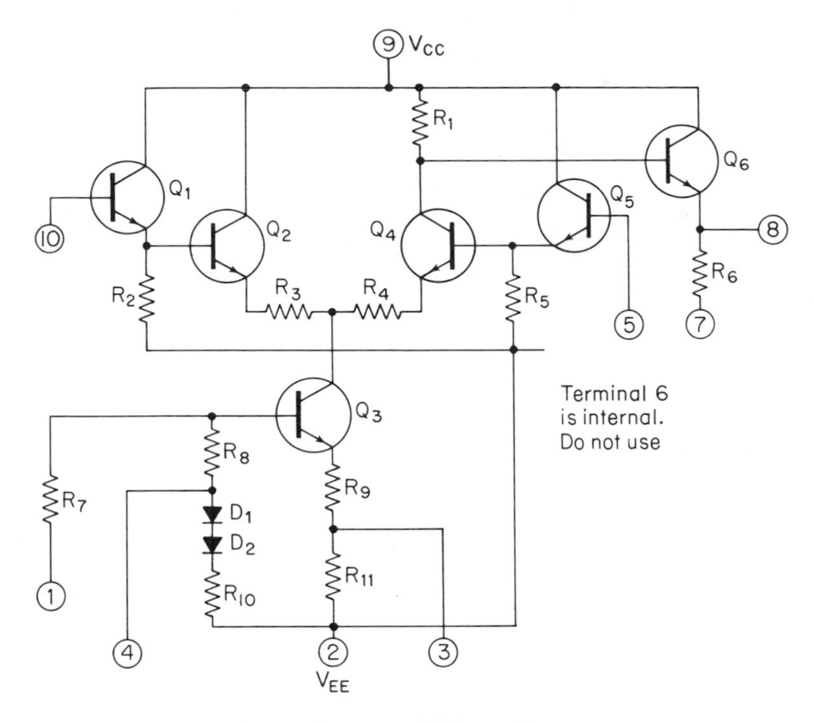

FIG. 5-22. RCA CA3002 amplifier.

uct detector if a suppressed-carrier signal is applied to the differential pair, and the regenerated carrier is applied to the constant-current transistor. Of course, the circuit must be operated in a linear region, and the current from the constant-current transistor must be linear with respect to the reinserted carrier voltage.

In the circuit of Fig. 5-23, a double-sideband suppressed-carrier signal is applied at the differential amplifier Q_1, with the opposite differential input at ground. The 1.7-MHz carrier is inserted at the current transistor Q_3 input. Because of the single-ended output, a high-frequency bypass capacitor is connected between the output and ground. This provides filtering for the high-frequency components of the oscillator signal at the output.

5-3.3 Motorola MC1590 amplifier

Figure 5-24 shows the schematic of the amplifier. The circuit consists of a common-emitter differential amplifier input stage (Q_1, Q_2) which drives a pair of common base differential amplifiers (Q_3 through Q_6). The output of the common-base stage is fed into a common-collector (Q_7, Q_{10}), common-emitter (Q_8, Q_9) cascaded, differential pair. The unnumbered transistors shown on the schematic are for dc biasing.

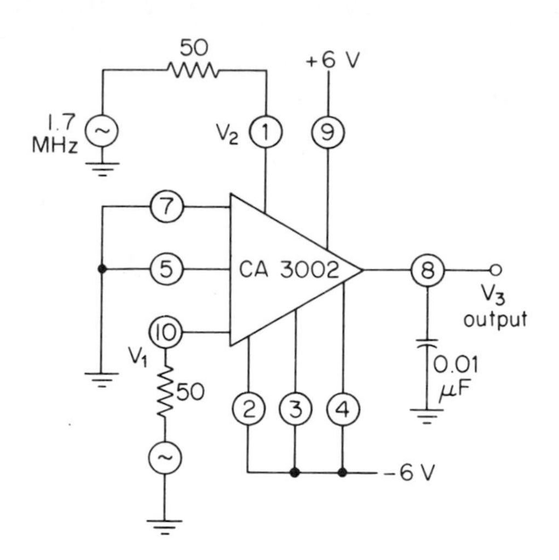

Typical characteristics

V_1 (mV)	V_2 (V)	V_3 (mV)	Conversion voltage gain (dB)	dB down from fundamental (2nd harmonic)
1	1.7	12.5	21.9	60
32	1.7	310	19.8	32
4	0.25	22	15.6	15
4	3.0	15	11.4	42

FIG. 5-23. Product detector (RCA).

Automatic gain control of the amplifier is accomplished by an increase in the dc voltage at pin 2 above a nominal voltage of 5 V (assuming +12-V power source). As the voltage on pin 2 is raised above 5 V, transistors Q_4 and Q_5 which are normally off, begin to turn on. With transistors Q_4 and Q_5 on, a portion of the signal current is diverted from Q_3 and Q_6.

A simplified version of the signal path is also shown in Fig. 5-24.

5-3.3.1 100-MHz mixer

Figure 5-25 shows the MC1590 used as a 100-MHz mixer. When pin 2 is biased near the center of the linear portion of the AGC characteristic, the MC1590 can be used as a frequency converter. When the local oscillator signal is applied to the AGC pin, and the signal applied between the input pins, the familiar sum and difference frequencies are produced at the output. In Fig. 5-25, the 30-MHz difference frequency is filtered at the output.

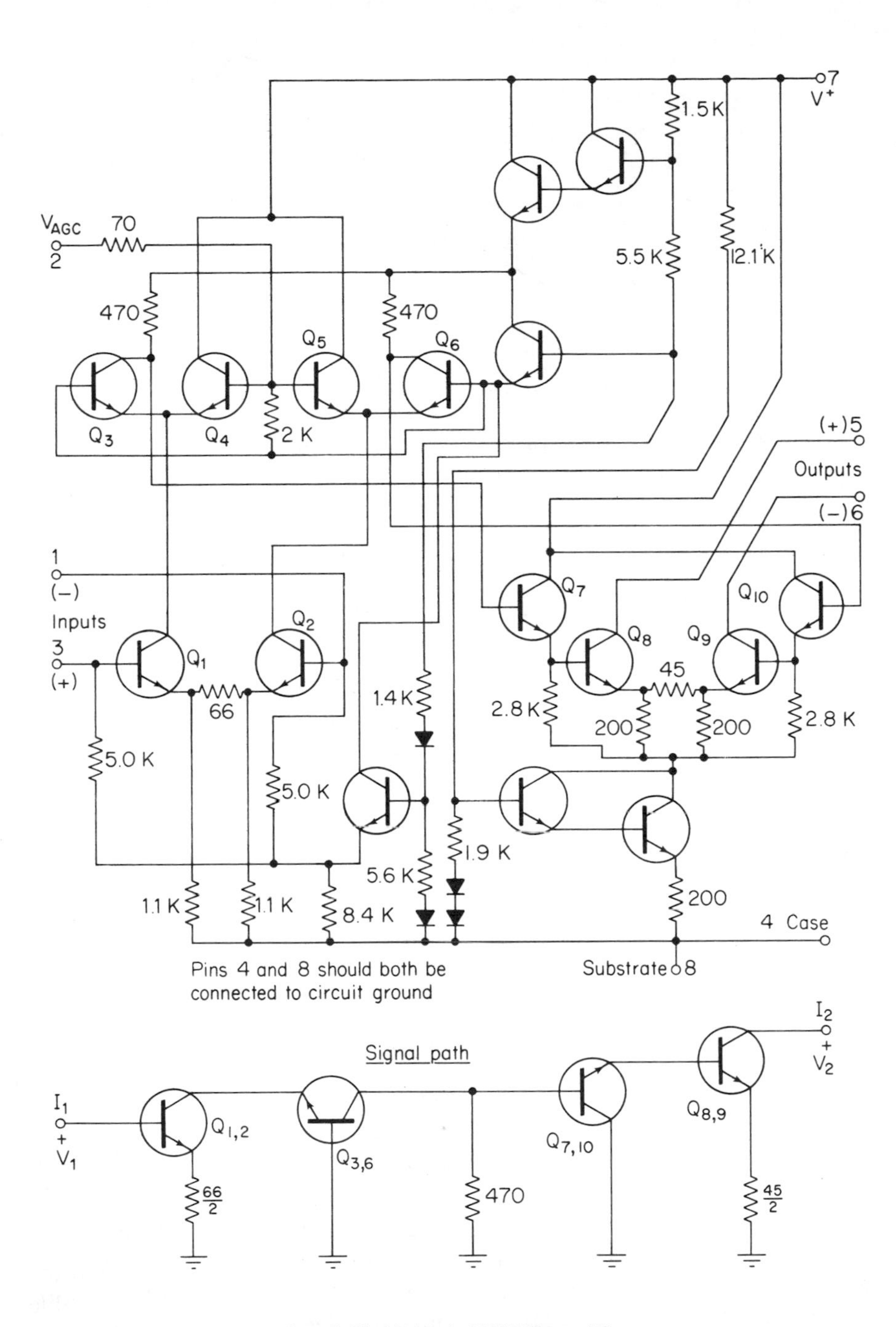

FIG. 5-24. Motorola MC1590 amplifier.

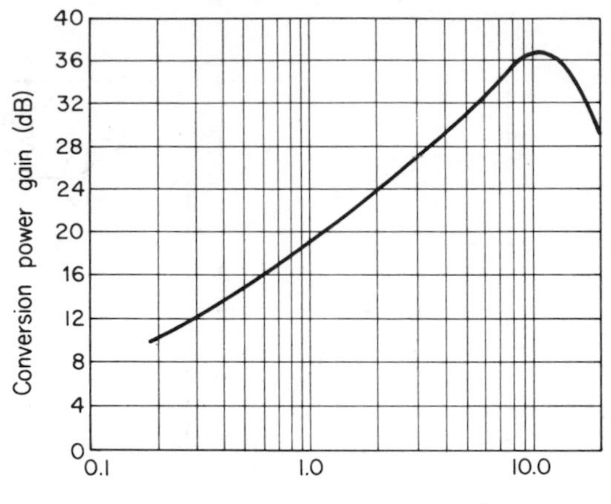

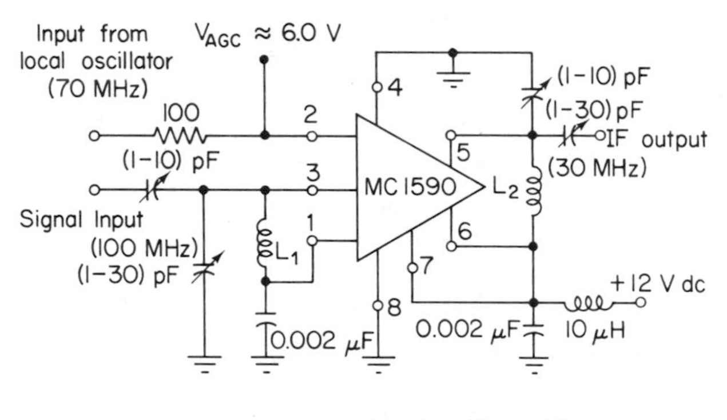

FIG. 5-25. 100-MHz mixer (Motorola).

5-3.4 RCA CA3028A amplifier

Figure 5-26 shows the schematic of the amplifier. The circuit consists of a balanced differential amplifier that is driven from a controlled, constant-current source. No emitter resistors are provided for the differential pair of transistors. As a result, the circuit has a narrow dynamic range, but higher gain (since there is no emitter feedback). Also, without emitter resistors, the transistors can be driven into saturation, thus making the IC suitable as a limiter. Exceptional versatility is made possible by the availability of in-

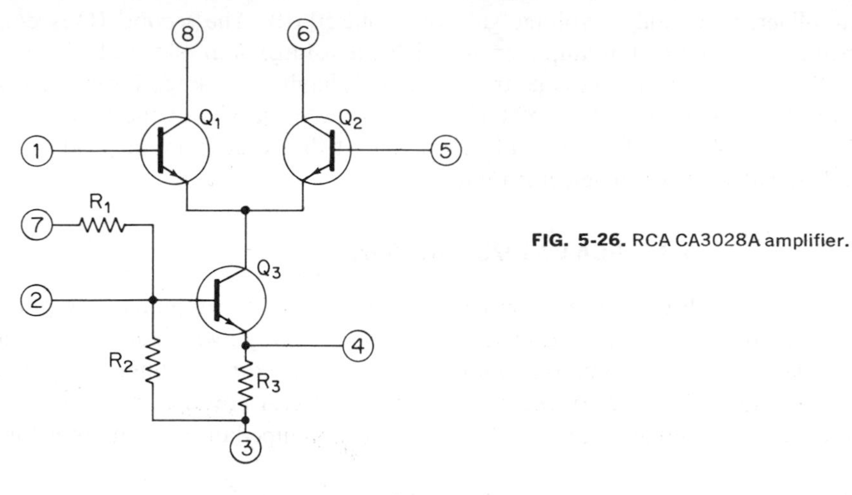

FIG. 5-26. RCA CA3028A amplifier.

ternal circuit points to which external circuit elements may be connected. The circuit can be operated as push-pull amplifiers, cascode amplifiers, single amplifiers in cascade, or parallel channels.

5-3.4.1 10.7-MHz IF strip

Figure 5-27 shows two CA3028A amplifiers used as a 10.7-MHz IF strip for an FM receiver. The first amplifier is connected as a cascode

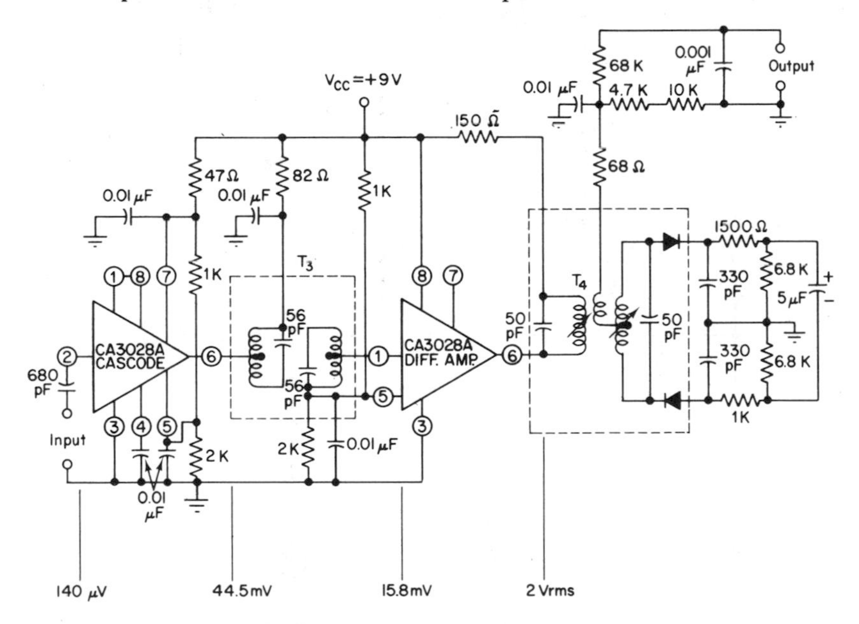

FIG. 5-27. 10.7-MHz IF strip (RCA).

amplifier, and yields a voltage gain of about 50 dB. The second IC is connected as a differential amplifier and yields a voltage gain of about 42 dB.

When a practical interstage transformer T_3 having a voltage insertion loss of 9 dB is used, over-all gain is 83 dB, and the sensitivity at the base of the first IC is 140 μV. That is, a 140-μV signal at the IF strip input produces a 2-V signal at the ratio detector input.

5-3.5 RCA CA3023 amplifier

Figure 5-28 shows the schematic of the amplifier. Amplifier gain is obtained by use of transistors Q_1, Q_3, Q_4 and Q_6, which are connected as two dc-coupled common-emitter/common-collector amplifiers having a voltage gain of about 60 dB. The common-collector configuration provides the necessary impedance transformation (high-impedance input and low-

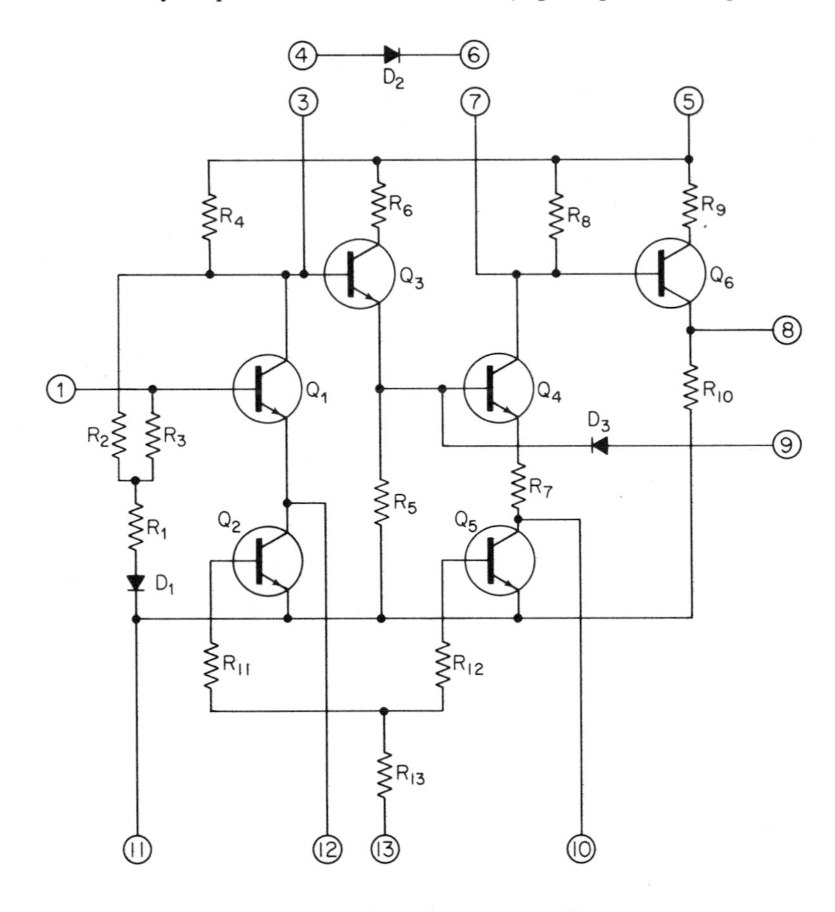

FIG. 5-28. RCA CA3023 amplifier.

impedance output) for wide bandwidth. The output transistor Q_6 provides the low output impedance. The circuit must be capacitively coupled, and should have a low-impedance source for best operation.

Figure 5-29 shows typical connections for wide-band and bandpass applications, with and without AGC, and for limiter applications. An external feedback resistor R_F, or a tuned circuit can be added between terminals 3 and 7 for desired bandwidth and gain performance. Linear operating conditions are maintained by the bias applied between the collector and base of Q_1 by the resistor-diode network R_1, R_2, R_3 and D_1. Because the collector of Q_1 is held at a fixed potential that is relatively independent of supply, device characteristics, and temperature, dc coupling to the remainder of the circuit can be used.

For applications in which gain control is desired, terminals 10 and 12 are left floating and AGC is applied to terminal 2. For maximum gain, terminal 2 is operated at a positive voltage not larger than the supply voltage applied to terminal 5. In the positive voltage condition, transistors Q_2 and Q_5 are saturated and the impedance in the emitters of Q_1 and Q_4 is low. When the gain-control voltage becomes negative, Q_2 and Q_5 come out of saturation and provide high emitter feedback to reduce the gain.

In limiting applications, diodes D_2 and D_3 are connected in the feedback loops. The diodes provide clamping for sufficient input-signal swing. Limiting can be achieved with input signal swings up to 2.5 V.

5-3.5.1 28-MHz limiting amplifier

Figure 5-30 shows two CA3023 amplifiers used as a 28-MHz limiter amplifier. Terminals 3 and 7 are connected to terminals 4 and 6, respectively. Terminal 8 is connected to terminal 9 to provide limiting action. A self-resonant coil in parallel with a 2-kΩ resistor is inserted in the feedback loop of each amplifier to provide gain and stability. The bandwidth of the system, before limiting, is 3.8 MHz, and the effective Q is 7.35. The total gain is 61 dB (30.5 dB per stage), and the power dissipation is 66 mW. Full limiting occurs at an input of about 300 μV.

5-3.6 Motorola MC1552 and MC1553 Amplifiers

Figures 5-31 and 5-32 show the schematics of the MC1552 and MC1553 amplifiers respectively. Both devices are video (or wideband) amplifiers designed for single-ended input and single-ended output operation from one power supply.

Each IC uses a three-stage direct-coupled common emitter cascade (Q_1, Q_2, Q_3) with series-series negative feedback from the third stage emitter to the first stage emitter. The main difference between devices is the magnitude of

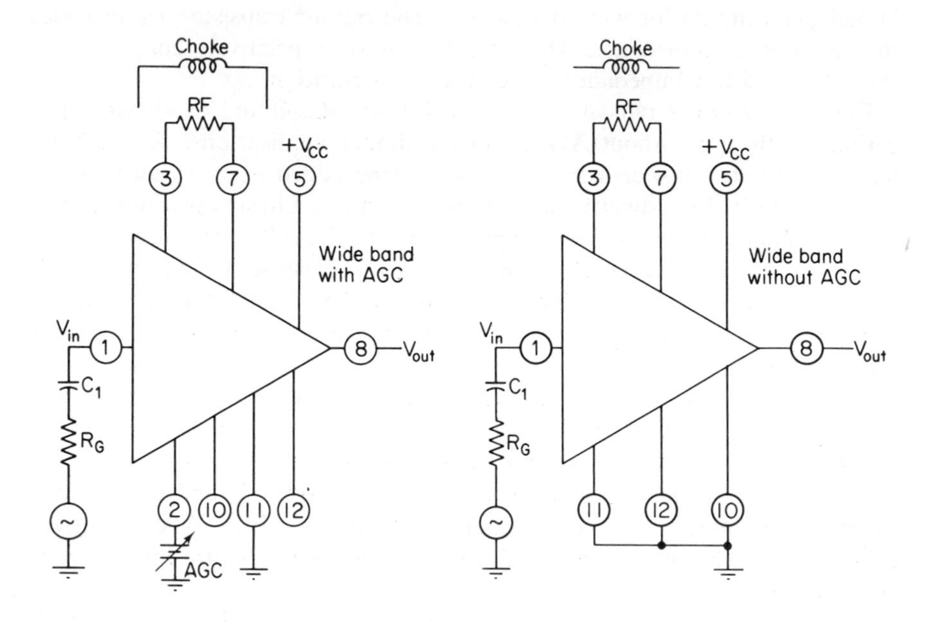

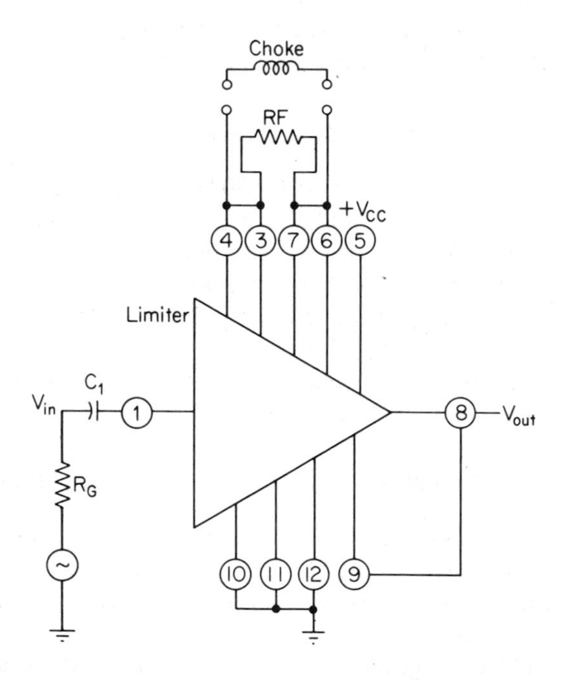

FIG. 5-29. Wide band and limiter amplifiers (RCA).

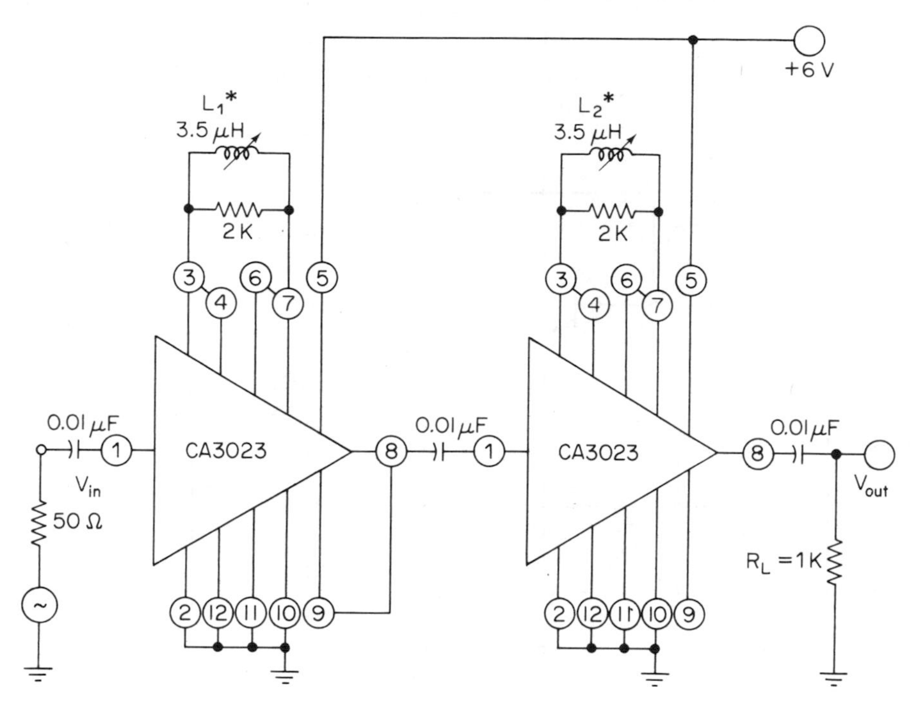

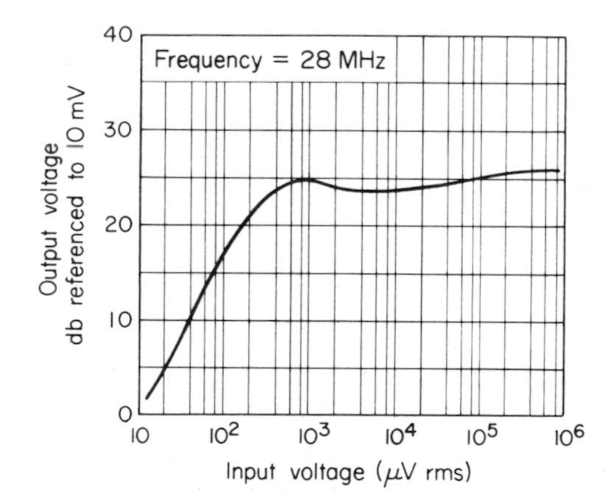

FIG. 5-30. 28-MHz limiting amplifier.

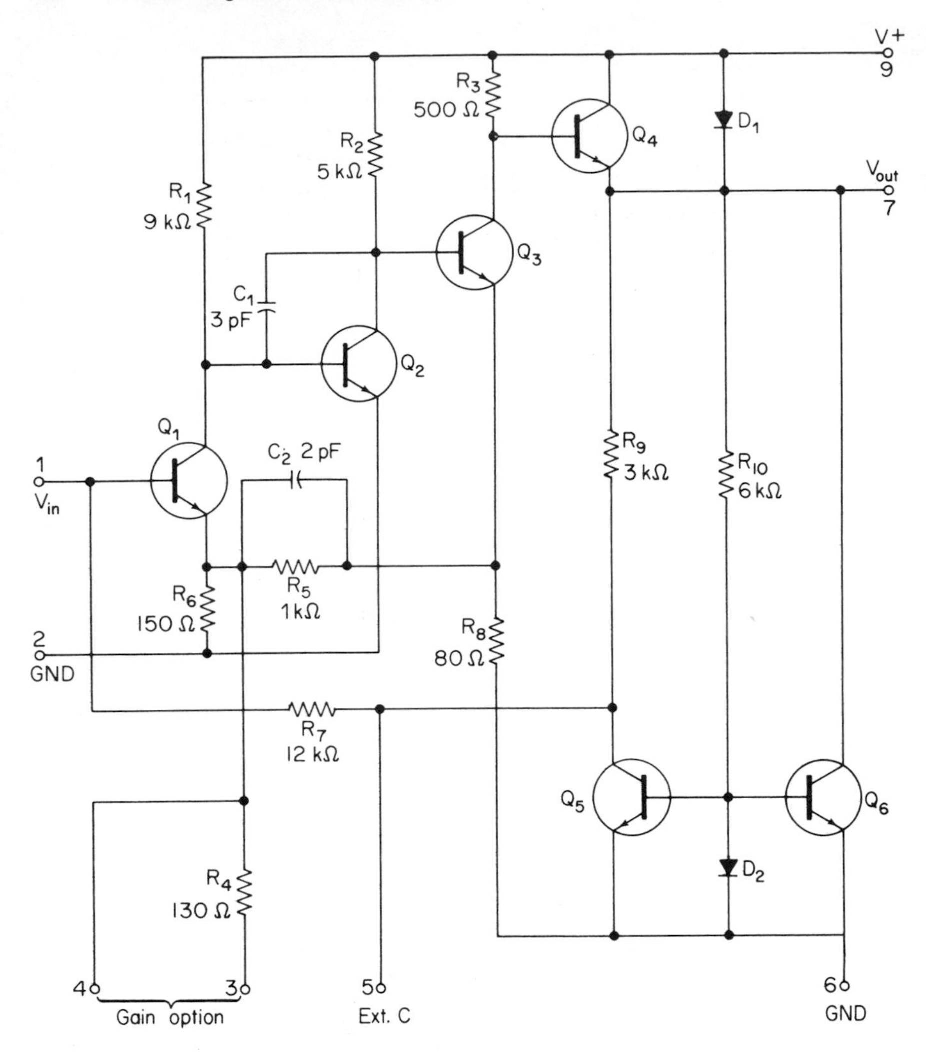

FIG. 5-31. Motorola MC1552 video amplifier.

the feedback and emitter resistances used to set up the voltage gain levels (the MC1552 is low gain; whereas the MC1553 is high gain).

An emitter follower Q_4 is used at the output of the basic feedback triple to provide low output impedance. A separate dc feedback loop, which is normally decoupled with respect to ac signals, consisting of emitter follower resistor R_9 and shunt feedback resistor R_7, extends from the input to the output to set the no-signal output voltage. Current extracted from this loop by Q_5 provides close temperature compensation of the no-signal point.

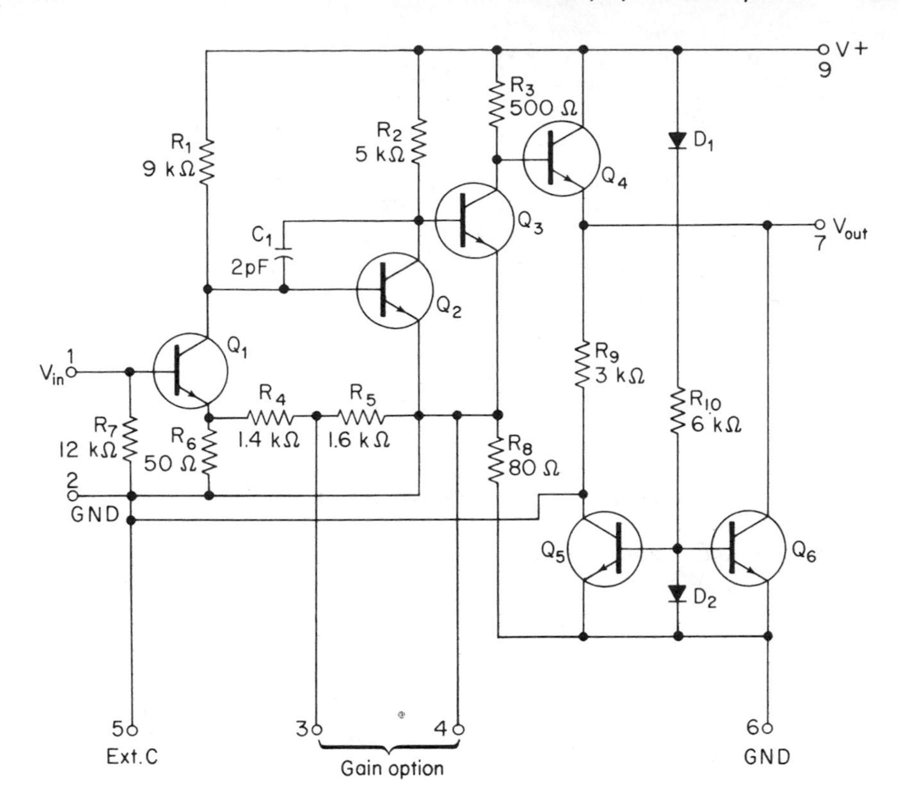

FIG. 5-32. Motorola MC1553 video amplifier.

The diode bleeder network consisting of D_1, R_{10}, and D_2 sets the level of current flowing in current sources Q_5 and Q_6. When D_2 and Q_5–Q_6 are physically close on the IC chip, the diode voltage-current relationships are essentially the same as the base-emitter diode of Q_5 and Q_6. Thus, when a finite direct current is passed through D_2, an identical direct current emitter current flows through the transistors to establish bias conditions.

The output no-signal level of the ICs is maintained at one-half the operating supply voltage. Current source Q_6 is used to establish symmetrical positive and negative load current excursions, regardless of emitter resistance magnitude, power supply voltage, or ambient temperature.

5-3.6.1 Pulse amplifier

Figure 5-33 shows the basic connections for both the MC1552 and MC1553 as pulse amplifiers. Note that terminals 3 and 4 can be used as low-gain outputs if desired. Figure 5-34 shows the MC1552 connected as a

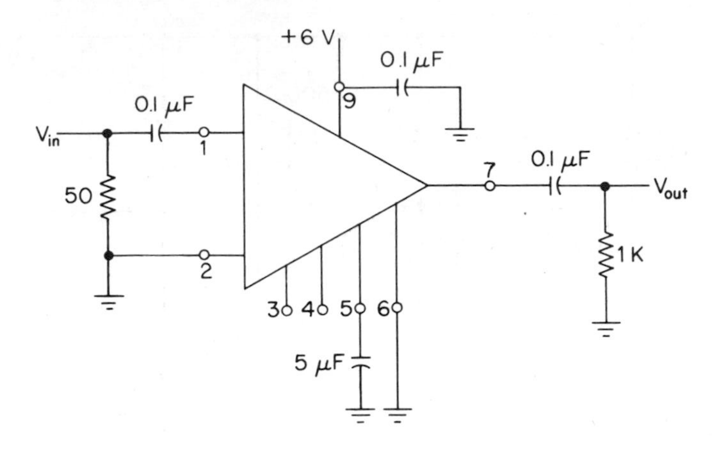

FIG. 5-33. Basic pulse amplifier (Motorola).

non-inverting pulse amplifier. Here, the input transistor Q_1 is bypassed. Pulse voltage gains of up to 3000 are possible with proper selection of values.

5-3.6.2 Summing/scaling amplifier

Figure 5-35 shows the MC1552 connected as a summing/scaling amplifier. In this non-inverting configuration, the summation of input signal currents is accomplished at the summing point, pin 4, through

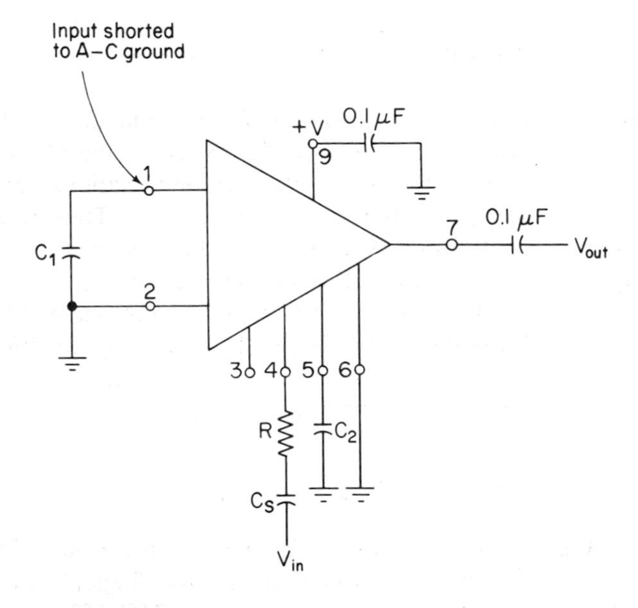

FIG. 5-34. Non-inverting pulse amplifier (Motorola)

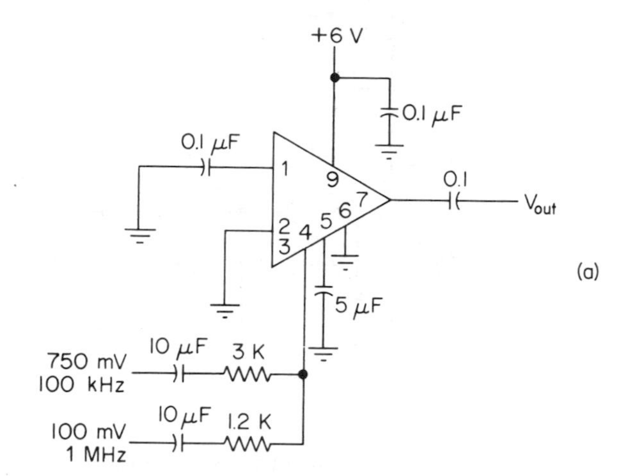

FIG. 5-35. Summing/scaling amplifier (Motorola).

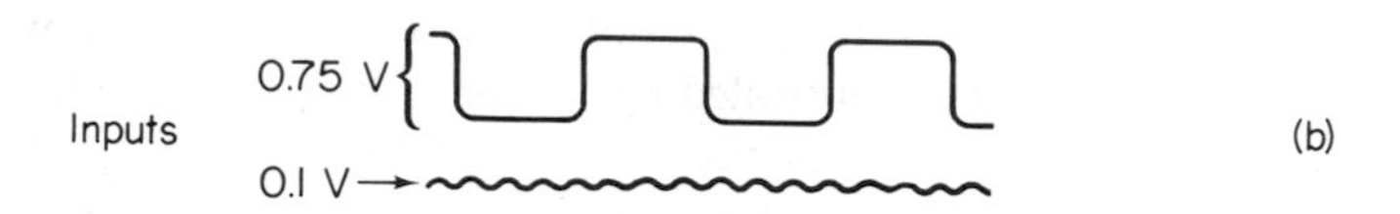

Composite 2 V (c)

the input resistors. Scale factor considerations are accomplished by selection of the design values of the input summing resistors. (Refer to Sec. 4-2.)

Using the resistor values indicated in Fig. 5-35a to establish the scale factor levels, and the input signals of Fig. 5-35b, produces the resultant composite output signal shown in Fig. 5-35c.

5-3.6.3 Oscillator

Figure 5-36 shows the MC1553 connected as an oscillator. The wide bandwidth and output swing capability of the IC makes it suitable for high frequency master clocks or local oscillators in many system designs. In this instance, positive feedback is injected through the 1-MHz crystal to input pin 1. The bias decoupling capacitor normally on pin 5 is omitted to insure that the crystal operates into a relatively low impedance.

The output of the oscillator is taken from pin 7 which is buffered from the oscillator circuit proper by a stage of gain and an emitter-follower. No

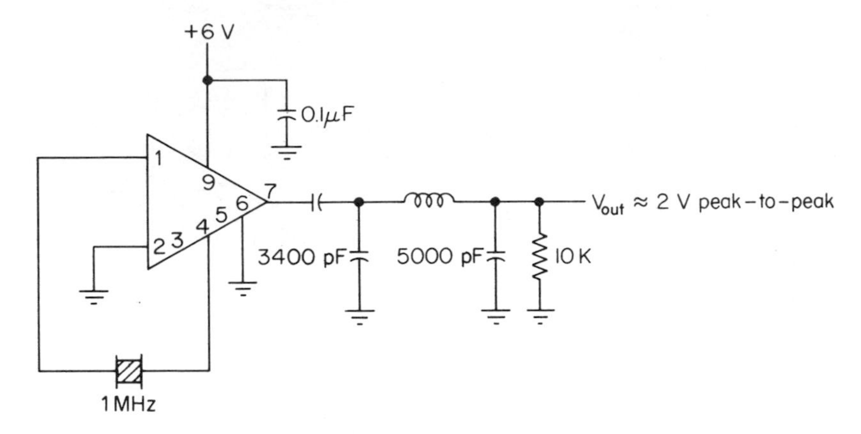

FIG. 5-36. 1-MHz oscillator (Motorola).

provisions are made to control the loop gain or the amplitude of oscillation. As shown, an approximate 2-V output is obtained with a power source of 6 V. Because gain is not controlled by feedback, there is a high harmonic content at the output signal. For this reason, a "brute force" pi-filter is inserted at the output to extract the fundamental frequency. (The values shown are for 1 MHz.)

5-3.6.4 Tuned amplifier

Figure 5-37 shows the MC1552 connected as a tuned amplifier. With the values shown, the circuit is tuned for a center frequency of 30 MHz. Two responses are presented in the graph of Fig. 5-37. The first uses the input impedance of the IC as the transformer secondary termination (shown as R = infinity). The second response uses a 1-kΩ damping resistor on the secondary. Insertion of the damping resistor broadens the response at the expense of gain, but reduces loading variations on the source which can become substantial with this particular type of tuning.

5-3.6.5 Amplifier with AGC

Figures 5-38 and 5-39 show two automatic gain control (AGC) configurations for the ICs. In many amplifier applications, it is necessary to include AGC capabilities to accommodate signal levels with wide dynamic ranges.

In the circuit of Fig. 5-38, the amplifier gain is a function of diode control current. Gain increases as control current increases. Thus, the AGC feedback must be negative to produce the required results. The 1N914 diodes are chosen solely for convenience and can be replaced by any other similar silicon diode.

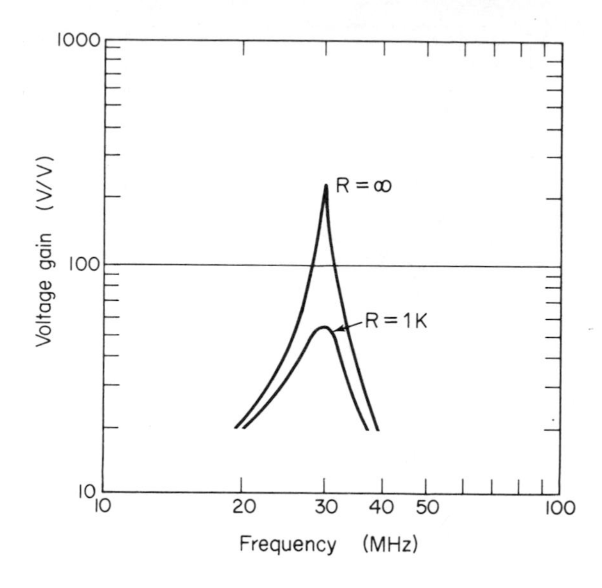

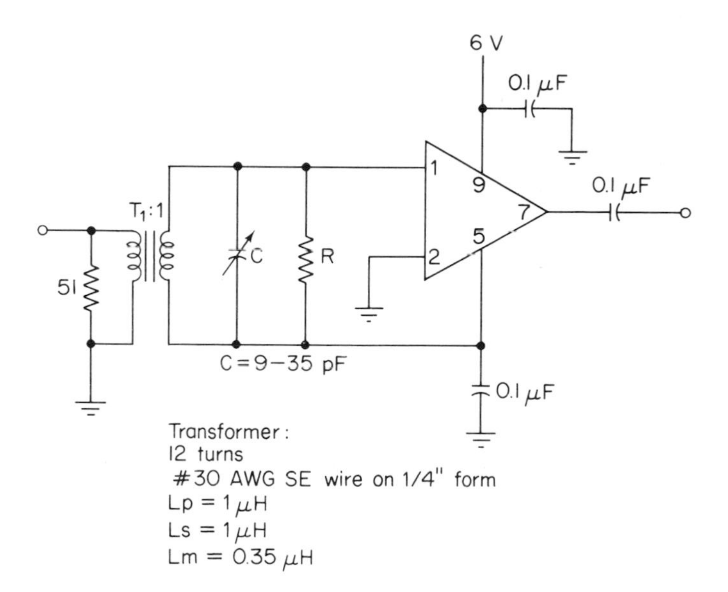

FIG. 5-37. 30-MHz tuned amplifier (Motorola).

The circuit of Fig. 5-38 has limitations in that the lowest level of gain that can be obtained is the normal unmodified gain of the amplifier. For lower gain level control, the circuit of Fig. 5-39 is used. In this circuit, the diode is used simply as a variable impedance in a voltage divider network.

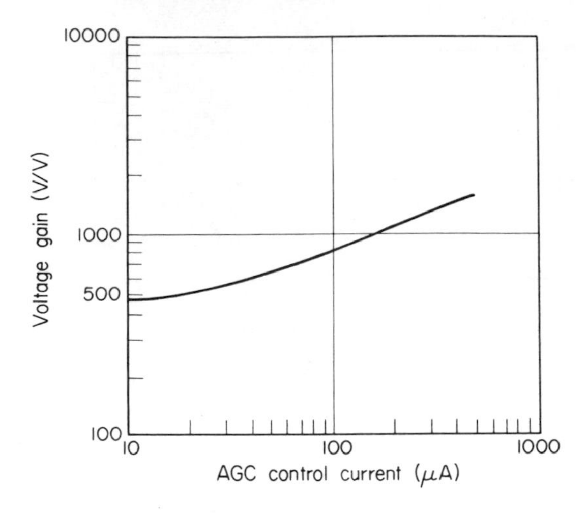

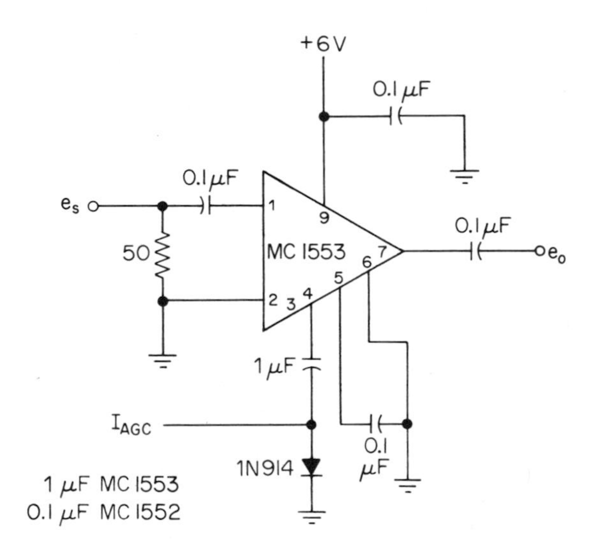

FIG. 5-38. Amplifier with AGC, for high gain control (Motorola).

5-3.7 Motorola MC1545 amplifier

Figure 5-40 shows the schematic of the MC1545, which is a gated video amplifier. The IC is designed as a wideband video amplifier. However, because of the gated, two-channel, differential design, a number of other applications are possible.

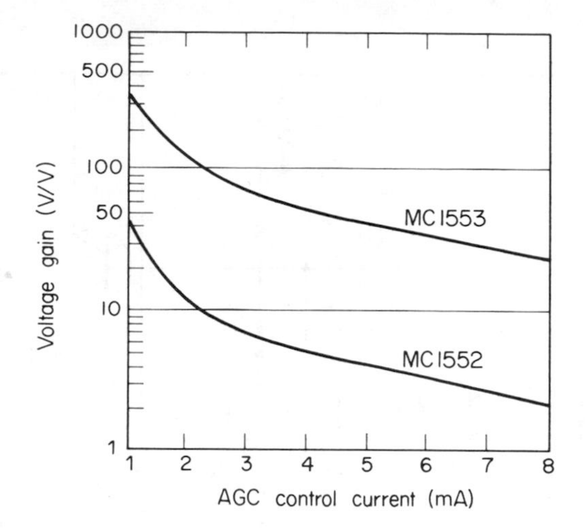

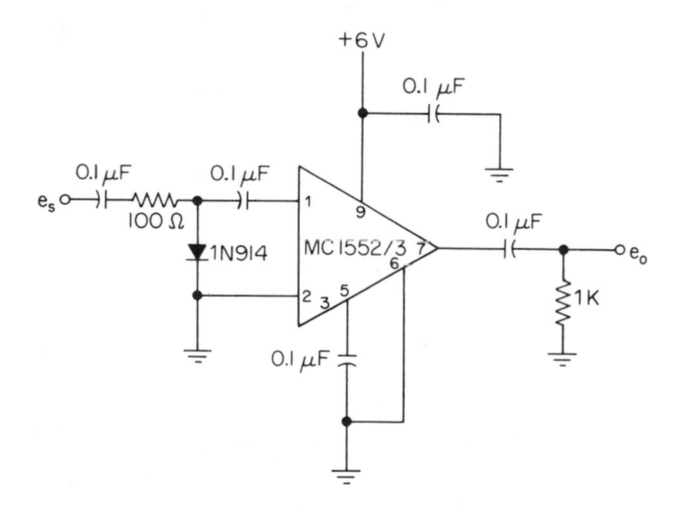

FIG. 5-39. Amplifier with AGC, for low gain control (Motorola).

The IC is basically a dual differential amplifier with controlled current source switching, using a single temperature-compensated current source. The differential amplifier pairs are connected to common load resistors which minimizes output level changes due to circuit imbalances.

Channel switching is accomplished by using a differential switching amplifier (transistors Q_5 and Q_6). With the gate (pin 1) open, or at a high level

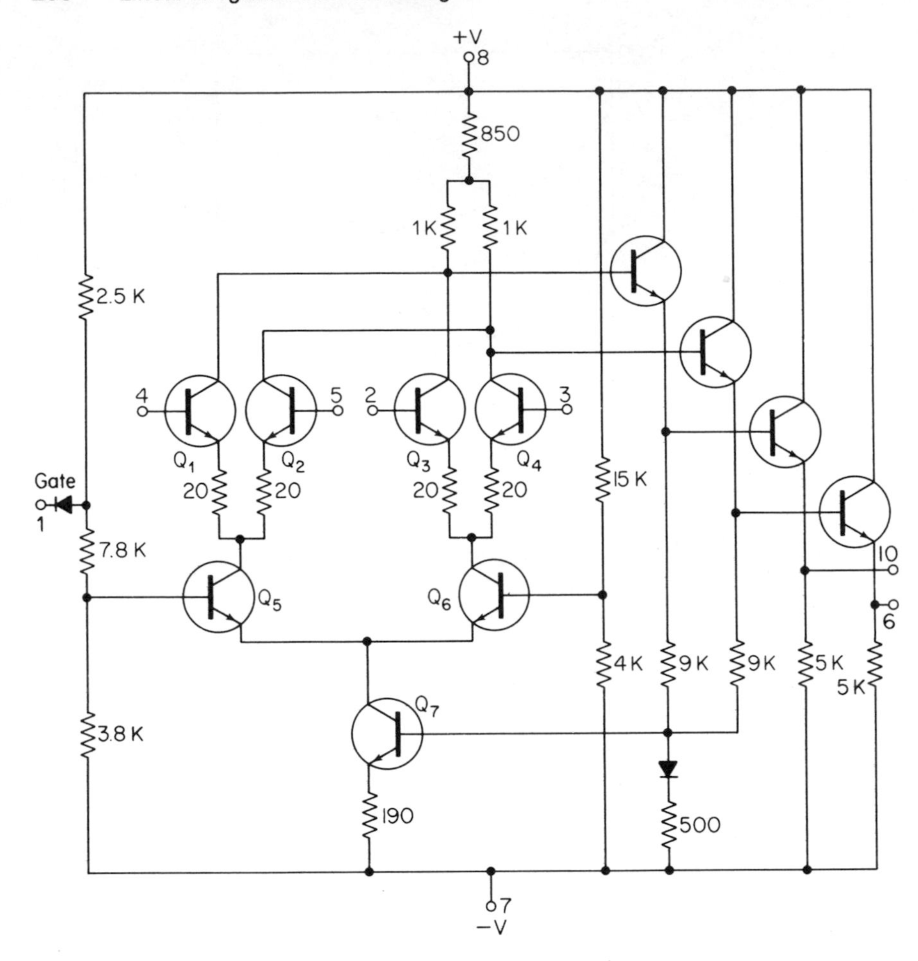

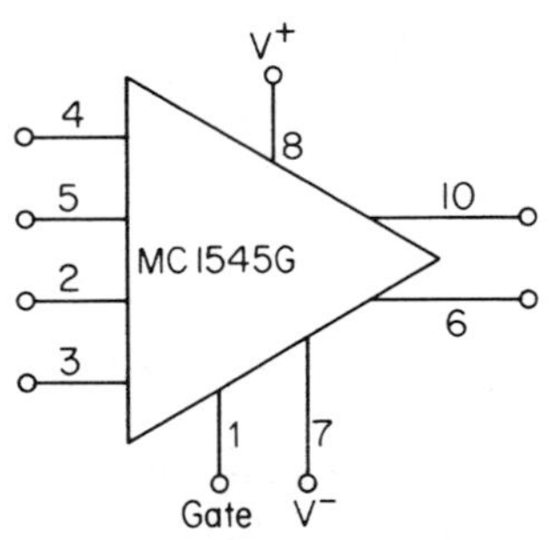

Pin numbers are for G package (TO−5)
Refer to data sheet for F & L pinouts.
On TO−5: Pin 10 under tab
Pin 7 connected to case

FIG. 5-40. Motorola MC1545 amplifier.

(greater than 1.5 V for ±5 V supplies) the base of Q_6 will be at a lower potential than Q_5. Then Q_5 will be on and Q_6 off. The current to the current source Q_7 is then provided entirely by Q_5 and the Q_1-Q_2 differential amplifier. The input to this amplifier is amplified and passed to the output.

When the gate is grounded or at a low potential (less than 0.4 V) Q_6 is on and Q_5 is off, and the transistor pair Q_3–Q_4 is operating. The signal applied to this input will appear at the output. The gate drive switching characteristic is compatible with the output of the conventional saturated logic devices (TTL, DTL), when used with symmetrical ±5-V supplies.

5-3.7.1 *Basic video switch*

The basic configuration for the IC used as an open-loop, two-channel, wideband amplifier or video switch is shown in Fig. 5-41. The single-ended gain in this configuration is flat at about 18 dB to 10 or 11 MHz, with a 3-dB drop at about 13 MHz.

The IC can be used as a closed-loop amplifier. However, there is little advantage in most wideband applications.

5-3.7.2 *Temperature compensated gate*

The amount of gate voltage required to switch the IC varies with temperature. This presents no problem with digital logic circuits where the gate signal is a pulse, and the gate is either "full on" or "full off".

Typically, the required gate switching voltages varies about 2 mV/°C. A digital logic pulse of one or two volts can easily overcome this variation. However, when the IC is used with an analog voltage in linear applications, some form of temperature compensation may be required.

A quick method of temperature compensation is to use another silicon diode in the opposite direction in series with the gate input diode, as shown in Fig. 5-42. This provides forward-biased, back-to-back diode compensa-

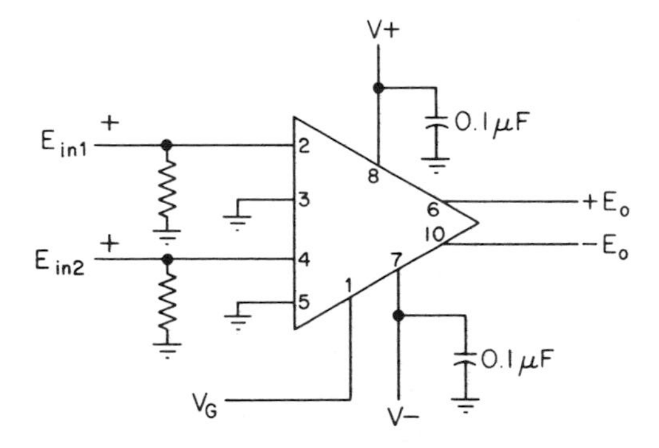

FIG. 5-41. Basic open-loop, two channel wideband amplifier or video switch (Motorola).

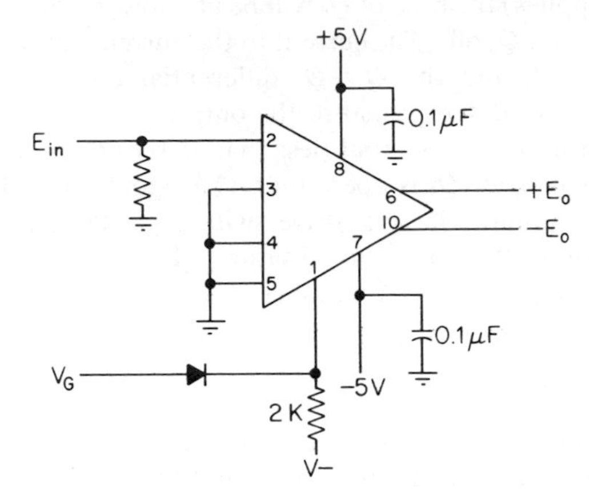

FIG. 5-42. Amplifier with temperature compensated gate (Motorola).

tion so that the temperature drift of the two diodes will (ideally) cancel. Typically, the temperature is about 0.2 mV/°C with the two-diode scheme. Of course, the gate characteristics are shifted higher about 0.6 or 0.7 V (or the equivalent of the normal voltage drop across a silicon diode).

5-3.7.3 Video amplifier with AGC

The gate may be used in an active mode to provide AGC action. Figure 5-43 shows the connections using the gate to provide AGC for one amplifier channnel. Also shown on Fig. 5-43 is a gain versus gate voltage curve.

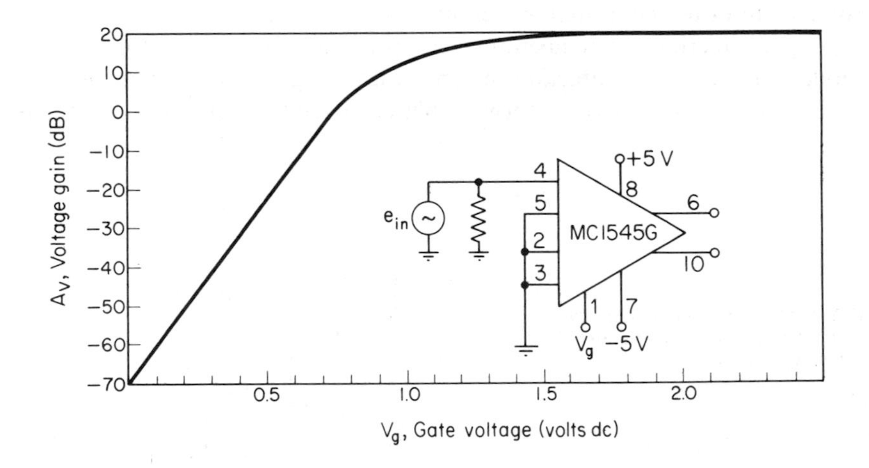

FIG. 5-43. Video amplifier with AGC.

5-3.7.4 Amplitude modulation

The gate may also be used to provide amplitude modulation, as shown in Fig. 5-44. The carrier signal is introduced into the input of one channel (pin 4). The remaining channels are disabled by connecting the inputs to ground. The modulation signal is introduced into the gate through a coupling capacitor. The gate is also connected to ground through a 5-kΩ potentiometer which sets a bias level on the gate. By using a potentiometer to bias the gate at the midpoint of the channel linear region (with a typical gate voltage of 1.2 to 1.4 V), a symmetrically-modulated result is obtained. Of course, the modulation voltage must not exceed the value that would operate the amplifier out of its linear range. Typically, the input carrier voltage is limited to 150 mV, while the input modulation voltage is approximately 400 mV (peak-to-peak).

5-3.7.5 Balanced modulator

When balanced modulation is required, the IC is connected in the circuit of Fig. 5-45. Note that the circuit of Fig. 5-45 is similar to that of Fig. 5-44, except for connections at the input. In the balanced circuit of Fig. 5-45, the internal differential amplifiers have been connected in a manner which cross-couples the collectors.

Assuming that the carrier level is adequate to switch the cross-coupled pair of differential amplifiers, the modulating signal at the gate is switched between collector loads at the carrier rate. This results in multiplying the

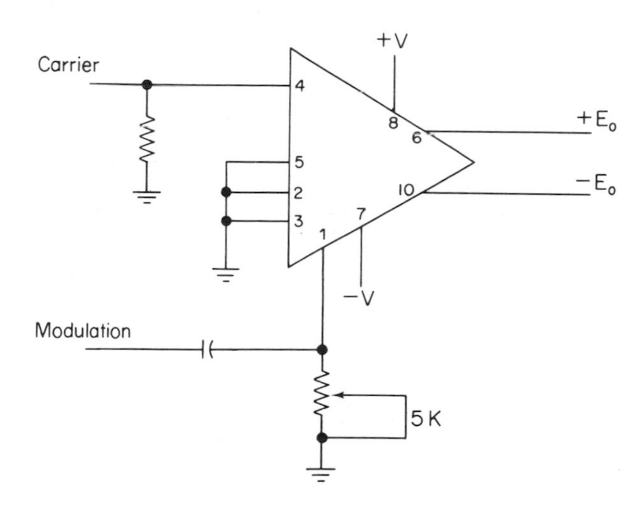

FIG. 5-44. Amplitude modulator (Motorola).

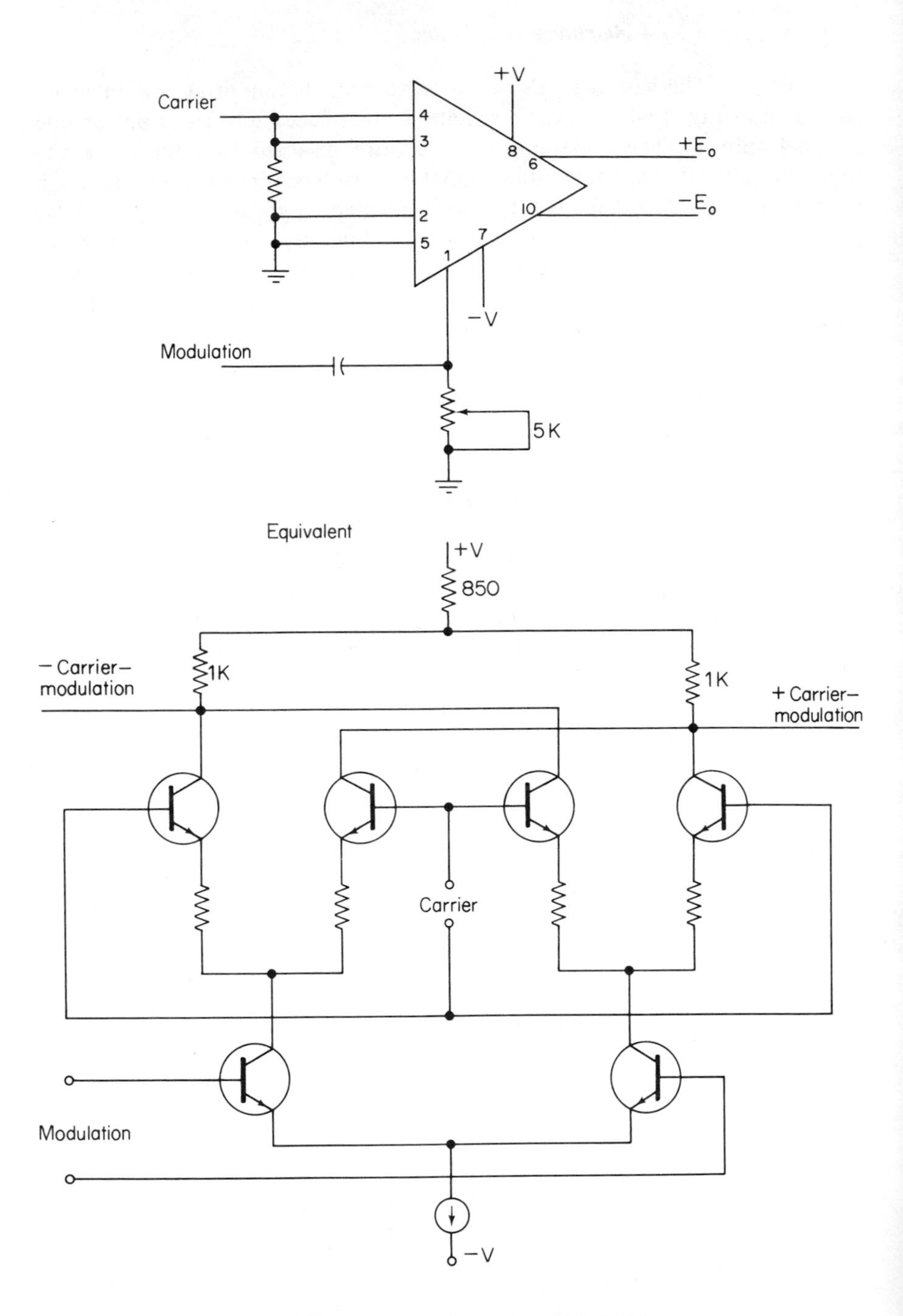

FIG. 5-45. Balanced modulator (Motorola).

modulation by a symmetrical switching function. Care must be exercised not to overdrive the modulation input (typically modulation is limited to 200 mV peak-to-peak). This will insure good carrier suppression.

To balance the IC when acting as a modulator, equal gain must be given to the two differential channels. This is done by setting the potentiometer for a proper balance. In theory, the potentiometer is adjusted to set both channels at the exact center of their linear operating range. However, in practice, the center point may not be exactly the same for both channels, so the potentiometer must be set to some point of equal gain near the center.

5-3.7.6 Gated astable multivibrator

Figure 5-46 shows two amplifiers connected as a gated astable multivibrator. The output is a square wave of approximately 2 V peak-to-peak (assuming power supply voltages of 5–6 V). As shown by the equations, the output frequency is useful up to about 2 MHz, and is set by the values of R, C, R_1 and R_2. Rise times and fall times of the square wave can be improved by means of the "speedup" capacitor C_S across R_1, as shown.

The circuit is free-running (or self-generating) so long as the gates are open. Thus, the circuit can be turned on or off by means of a bias at the gates.

5-3.7.7 Gated oscillator with level control

Figure 5-47 shows the IC used as a gated oscillator with an adjustable level control. The output is a sinewave of approximately 2 V peak-to-peak (with power supply voltages of 5–6 V). The useful frequency range is 1 kHz to 10 MHz. As shown by the equations, the frequency is set by values of R and C (both R values and both C values must be the same). The output level is set by the potentiometer.

Rise times of the circuit are dependent upon the RC network, and vary from 1 to several cycles of the oscillator frequency. At 160 kHz, a typical rise time is 30 μs. At 8 MHz, the rise time decreases to something near 100 nS.

The circuit is normally free-running, and can be turned on or off by means of a bias at the gate.

5-3.7.8 Pulse width modulator

Figure 5-48 shows the IC used as a pulse width modulator. Note that the gate is left open, and the IC is used as a differential amplifier.

A pulsewidth modulator is essentially a comparator. When the amplitude of the modulation signal is greater than that of the carrier signal, the output

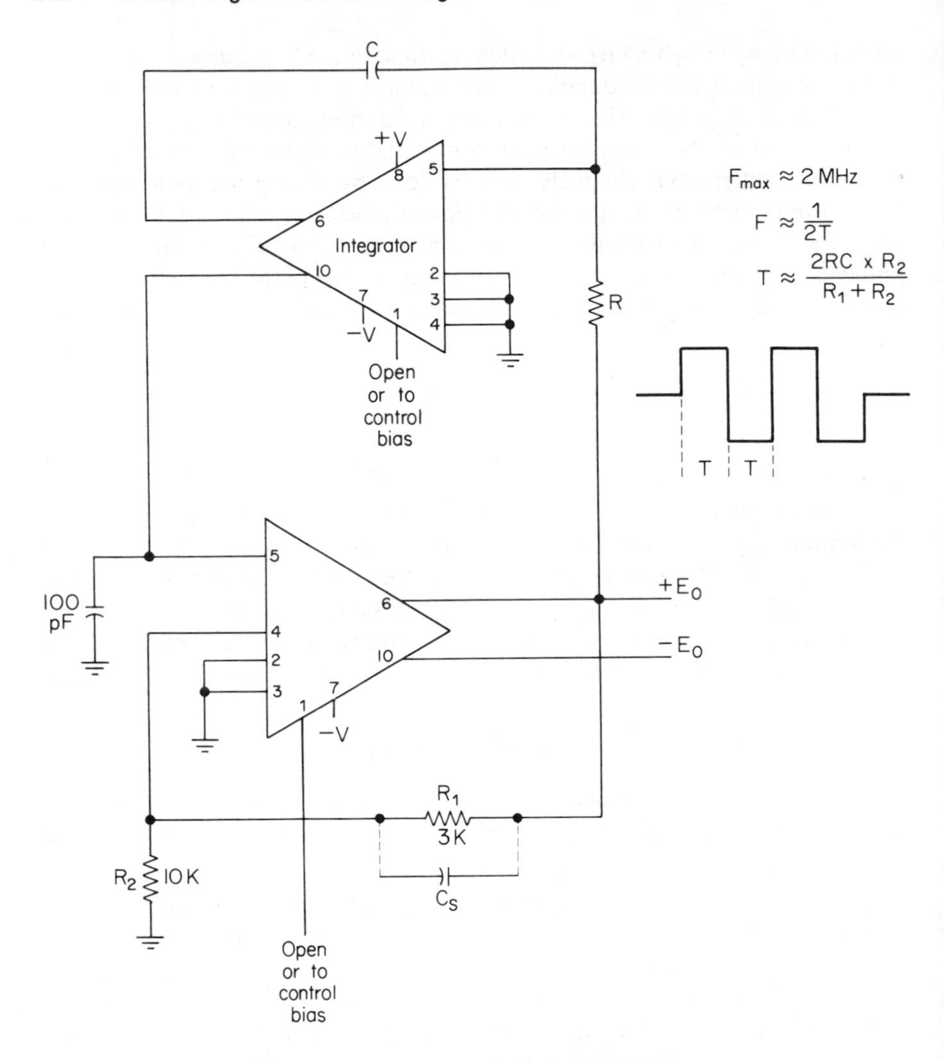

FIG. 5-46. Astable multivibrator (Motorola).

will be driven to one extreme (full on or full off), assuming that amplifier gain is sufficient to produce saturation with the given input level.

A sawtooth signal is used as the carrier in the circuit of Fig. 5-48. One edge of the output pulse is always at the trailing edge of the sawtooth, with the other edge of the output derived from the modulation signal. Thus, the rise time of the output pulses is approximately the time it takes the input signals (carrier and modulation) to give a differential voltage large enough to drive the output to full on.

Assuming that the carrier frequency is made much higher than the modulation frequency, the modulation signal level will be approximately constant during the switching time (that is, approximately constant during each saw-

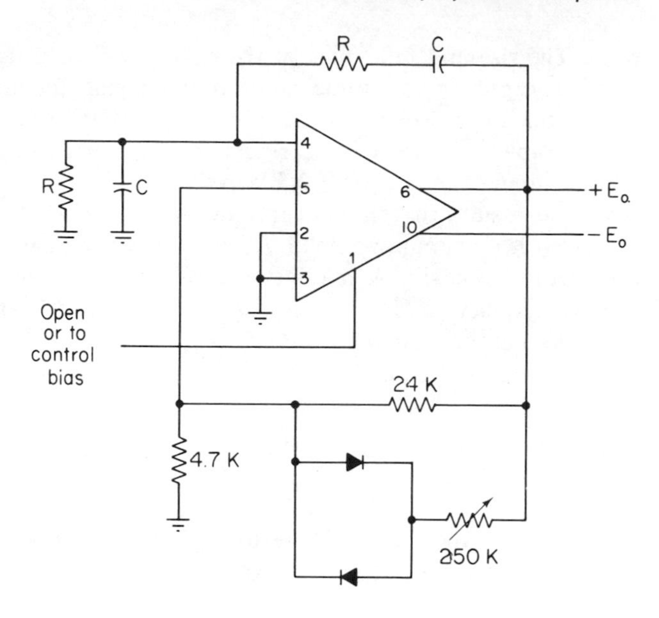

FIG. 5-47. Gated oscillator with level control.

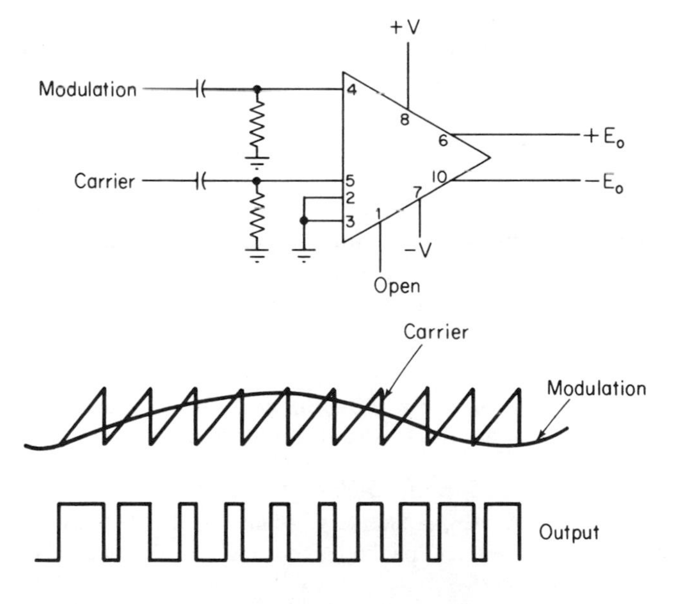

FIG. 5-48. Pulse width modulator (Motorola).

tooth cycle). The rise and fall times of the output will then depend on the carrier frequency, carrier amplitude, and amplifier gain (assuming the rise and fall times do not exceed the capability of the IC). For maximum rise time, it is desirable to have a large signal for the carrier (limited by the common mode input of the IC, or ± 2.5 V in this case.)

As an example, assume that the carrier frequency is 250 kHz (a 4-μs period), the carrier signal is maximum at ± 2.5 V (or 5 V peak-to-peak), that the fall time of the sawtooth is negligible in relation to the rise time (which is normally the case), that amplifier voltage gain is 10 (or 10 V/V as it is listed on some datasheets), and that the desired output is 2 V peak-to-peak.

The slope of the sawtooth is then:

$$\frac{5\ \text{V}}{4\ \mu\text{s}} = 1.25\ \text{V}/\mu\text{s}$$

With a gain of 10, the input change to give full output is 200 mV. The approximate rise time of the output is then:

$$1\ \mu\text{s}/1.25\ \text{V} = 0.8\ \mu\text{s per volt};\ 0.8\ \mu\text{s} \times 0.2\ \text{V} = 0.16\ \mu\text{s (or 160 ns)}$$

In addition to rise and fall times, another characteristic that indicates the quality of a pulse width modulator is the ratio of largest output pulse width to the smallest pulse width. A number of factors or variables are involved in this ratio, including pulse rise times, amplifier gain, non-linearity, input differential mode breakdown, and input loading considerations. No matter what characteristics are involved, it is obvious that a large ratio will permit more information to be transmitted by means of pulse width than a small ratio, other factors being equal. That is, pulse width modulation with a ratio of 10 is capable of passing more information than an identical modulation system with a ratio of 5.

From a practical standpoint, using the circuit of Fig. 5-48, the modulation ratio can be increased by decreasing the carrier amplitude, while maintaining all other factors (modulation signal, supply voltage, etc.) constant. Of course, a decrease in carrier amplitude will increase rise and fall times of the output pulses.

5-3.8 Motorola MC1350 and MC1330 television video amplifier and detector system

Figures 5-49 and 5-50 show the circuits of the MC1350 amplifier and MC1330 detector, respectively. (The MC1330 detector is shown in simplified form.) These IC packages form a complete video IF and detector system.

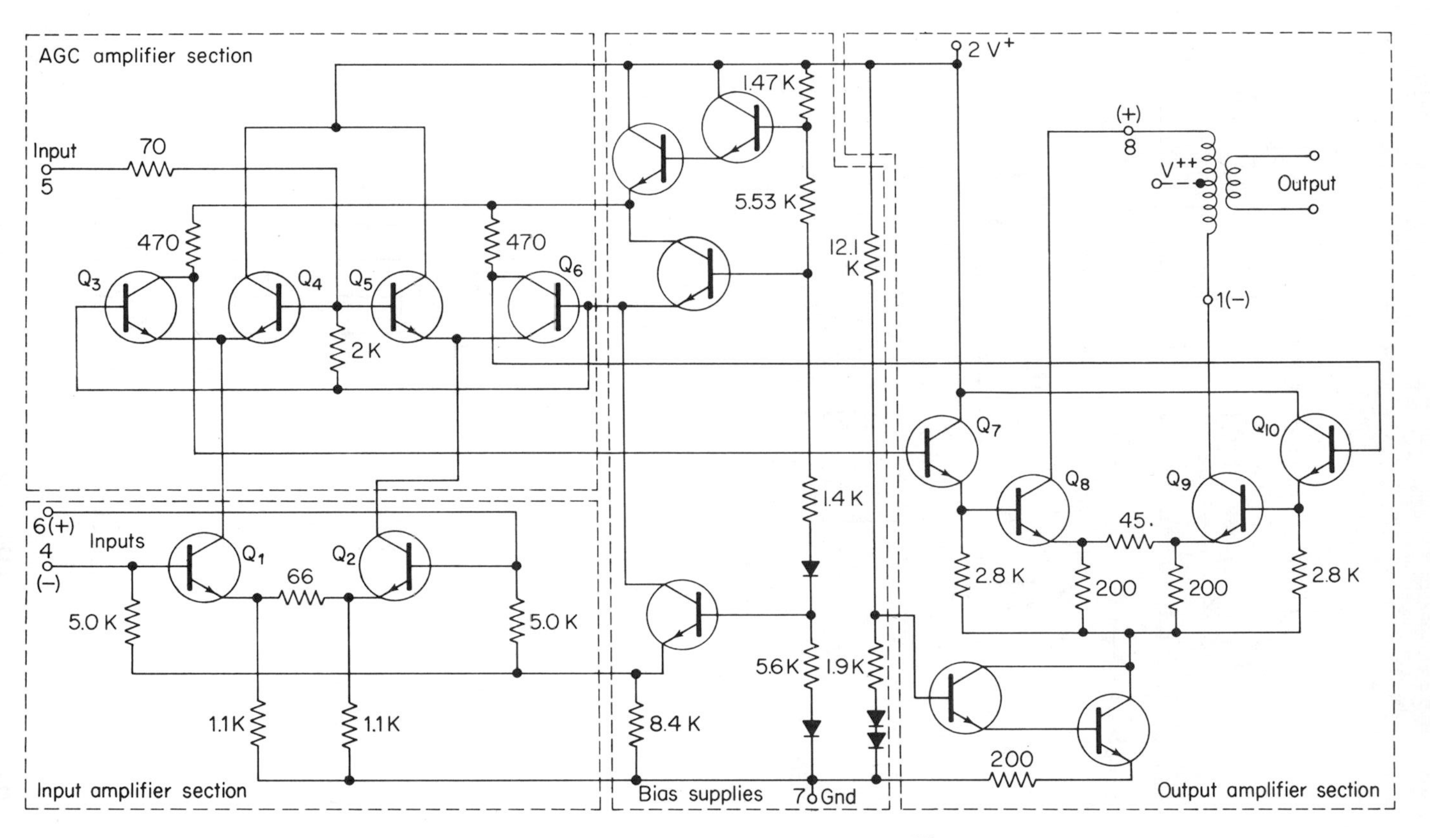

FIG. 5-49. Motorola MC1350 video IF amplifier.

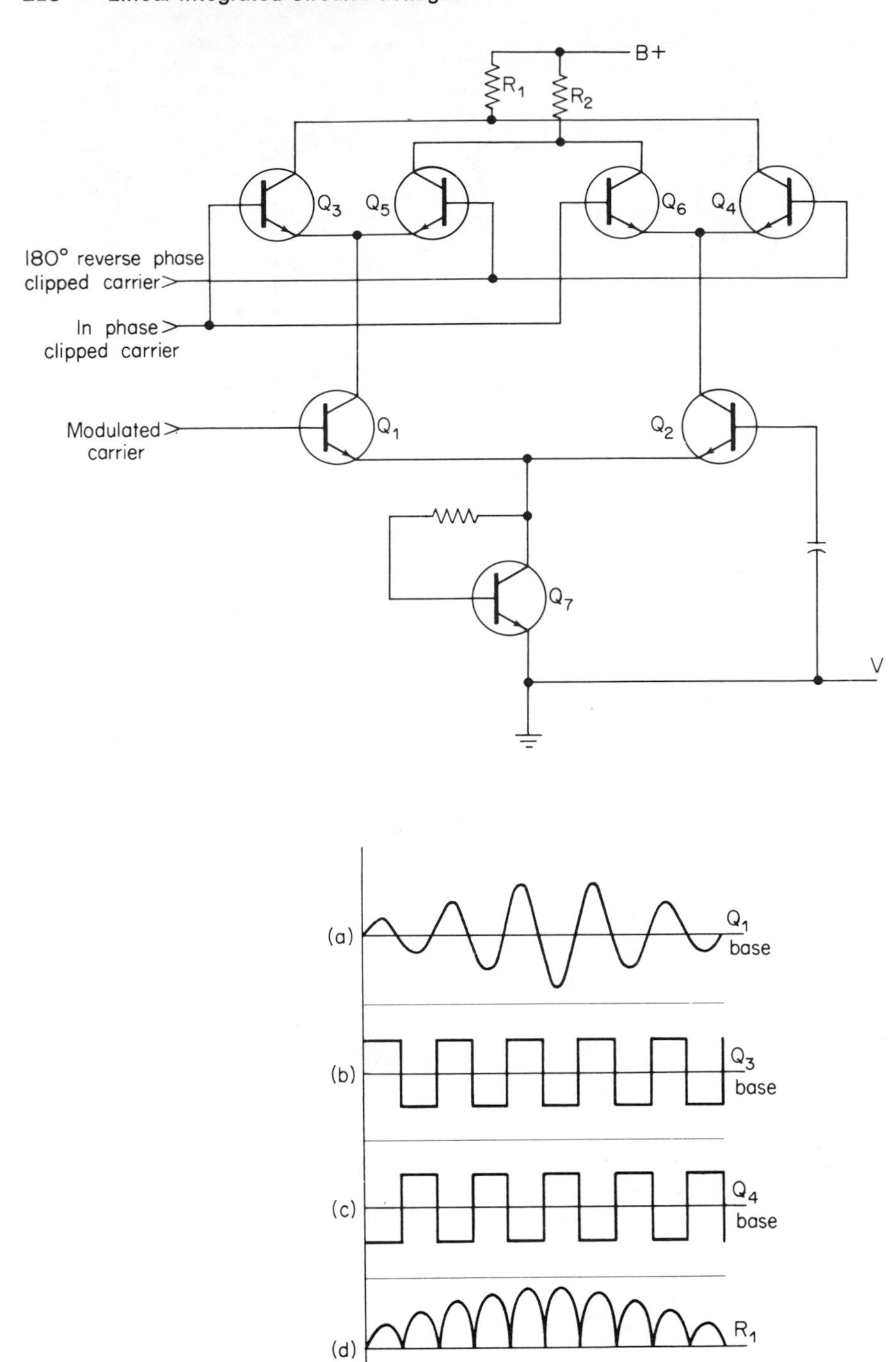

FIG. 5-50. Motorola MC1330 low level video detector simplified schematic.

Before going into circuit operation we shall consider the over-all requirements of a television video system. Figure 5-51 shows the signal levels and degree of AGC required if a television receiver is to function correctly throughout the range of input signal conditions commonly encountered. In some locations, all TV channels may provide high level signals, or conversely, all channels appear as low level signals. However, in most practical situations, each channel has its own amplitude. Some signals are nearly lost in noise, while others approach overload strength. This range of field intensity at the antenna required AGC compensation in the television receiver over a dynamic range greater than 90 dB. Some of this control can be accomplished in the tuner, since a good solid-state TV tuner has an AGC reduction capability usually greater than 36 dB. The difference of at least 60 dB must be provided by the video amplifier.

The detected video output level will depend on the video amplifier and picture tube drive requirements. In the extreme case of a single-stage tube video amplifier, as used in inexpensive monochrome receivers, the level could be as high as 6 V. But in most hybrid and all solid-state receivers, a 1- to 2-V composite video and sync signal is sufficient.

Figure 5-52 shows the two ICs and associated circuit parts connected to form a video amplifier and detector system for a television receiver. This circuit gives the required AGC gain reduction, the required IF gain, and a detected composite video output signal level of up to about 6 or 7 V. Typical voltage gain is 84 dB, and typical AGC range is 80 dB.

The detector uses a single tuned circuit, L_3 and C_{10}. Coupling between the two ICs is obtained by a double-tuned transformer L_1 and L_2. No filters or traps are included in the circuit, since different television manufacturers may have their own preferences and rejection requirements. The sound intercarrier information may be taken from the detected video output.

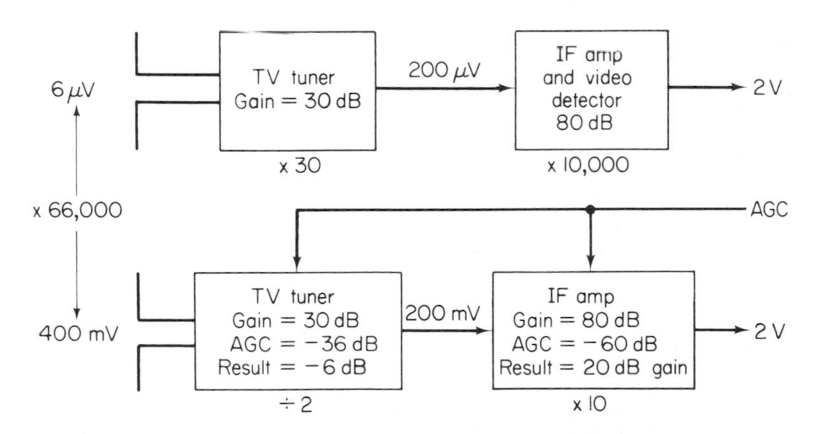

FIG. 5-51. Signal levels and AGC reduction requirements for a typical TV receiver (Motorola)

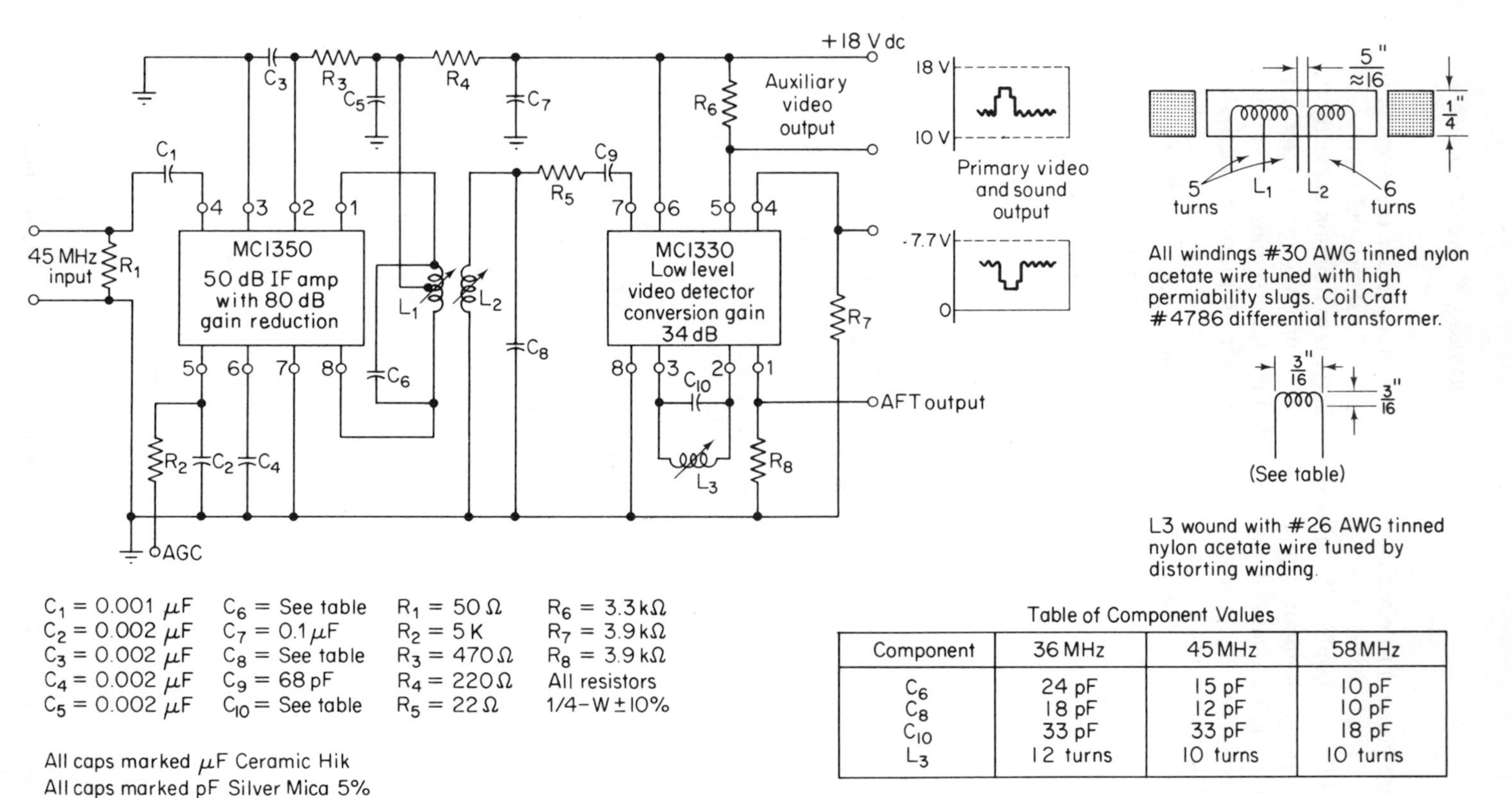

Table of Component Values

Component	36 MHz	45 MHz	58 MHz
C_6	24 pF	15 pF	10 pF
C_8	18 pF	12 pF	10 pF
C_{10}	33 pF	33 pF	18 pF
L_3	12 turns	10 turns	10 turns

FIG. 5-52. TV video amplifier and detector system (Motorola)

5-3.8.1 Video amplifier

Figure 5-49 is the schematic diagram of the MC1350 video amplifier. Q_1, Q_2, Q_3 and Q_6 form a differential cascade amplifier. When Q_4 and Q_5 are not conducting, the amplifier gain is at maximum. With a positive AGC bias voltage applied to the bases of Q_4 and Q_5, they will conduct and shunt away the signal current of Q_3 and Q_6. This will attenuate the gain of the amplifier, although the collector currents of Q_1 and Q_2 will remain constant, preventing a large input impedance change.

The output amplifiers Q_7 through Q_{10}, are supplied from an active current source that maintains a constant no-signal bias, keeping the output admittance nearly constant over the AGC range. The differential output is taken from the collectors of Q_8 and Q_9. However, single-ended output may be taken from either collector, provided the unused collector is connected to the positive supply (V^+). Operation in this latter mode reduces the circuit gain. Either differential or single-ended inputs can be applied to Q_1 and Q_2. For single-ended input, there is no loss in gain, provided the unused input is grounded through a capacitor.

5-3.8.2 Video detector

Figure 5-50 is a simplified schematic of the MC1330 video detector. The circuit is a low level detector (LLD) using the multiplier principle. (Multipliers are discussed further in Sec. 5-9.) The MC1330 is a doubly-balanced, full-wave, synchronous detector with very linear detection and excellent frequency response. Carrier rejection for this type of detector is typically 60 dB. The synchronous detector functions by multiplying the signal to be detected by the same signal which has been amplified and limited.

Q_7 is a constant current source. Q_1 and Q_2 form a differential amplifier, while Q_3 through Q_6 are carrier-operated switches. When positive half cycles of the amplitude modulated carrier appear at the base of Q_1, it begins to conduct. The in-phase, clipped carrier signal turns Q_3 on, causing current flow through R_1 to increase. No current will flow through Q_4 because it is switched off.

When negative half-cycles appear at the base of Q_1, Q_2 conducts through differential action. The reverse-phase carrier pulse turns Q_4 on, causing the current through R_1 to increase. No current will flow through Q_3, since the in-phase carrier pulse is negative at this time. The current flow through R_1 increases for either positive or negative half cycles of the carrier, producing a negative voltage charge at the collectors of Q_3 and Q_4.

The reverse action takes place in R_2 due to Q_5 and Q_6. Figure 5-50a shows the amplitude modulated carrier appearing at the base of Q_1. An identical, but inverted, waveform appears at Q_2. The two clipped carrier waveforms required as switching pulses are shown in Figures 5-50b and 5-50c. The detector is switched at twice carrier frequency, with Q_3 conducting on the positive half cycles, and Q_4 conducting on the negative half cycles. Figure 5-50d is the voltage waveform across R_1. Note that the original carrier no longer exists and that the detected modulation is constructed of pulses of double the carrier frequency.

The stages which follow the basic detector have limited frequency response, amplifying only lower frequency components (the modulation), and making the detector self-filtering for the high frequency products. Thus, shielding the detector to prevent spurious radiation of the IF (and harmonics) is not necessary.

5-4. IC AUDIO AMPLIFIER

Many of the linear amplifiers discussed in this chapter can be used at audio frequencies. Also, there are ICs specifically designed as audio amplifiers. In the past, and for the most part today, IC audio amplifiers are low power (1 or 2 W). These ICs are used as microphone amplifiers, drivers, pre-amplifiers, etc., to drive discrete component power amplifiers. However, the future trend is toward medium and high power IC audio amplifier packages. Of course, such ICs require a large heat sink capability. The following discussion is limited to the low-power ICs which are typical of present-day units.

5-4.1 RCA CA3007 audio amplifier

Figure 5-53 shows the schematic of the amplifier. As shown, the circuit is a balanced differential configuration with either a single-ended or a differential input, and two push-pull emitter-follower outputs. The circuit, is intended for use as a direct-coupled (no coupling capacitors) driver in a class B audio amplifier with squelch provisions.

The input stage consists of a differential pair of transistors Q_1 and Q_2, operating as a phase splitter with gain. The two output signals from the phase splitter, which are 180° out-of-phase, are direct-coupled through two emitter-followers, Q_4 and Q_5. The emitters of the differential pair of transistors are connected to the transistor constant-current sink Q_3.

The diodes D_1 and D_2 make the emitter current of Q_3 essentially dependent on the temperature coefficient of the diffused emitter resistor R_3. Because diffused collector resistors R_{15} and R_{16} should have identical temperature coefficients, constant collector-voltage operating points should result at the collectors of Q_1 and Q_2. However, the no-signal operating voltages at the output terminals 8 and 10 increase as temperature increases because the base-

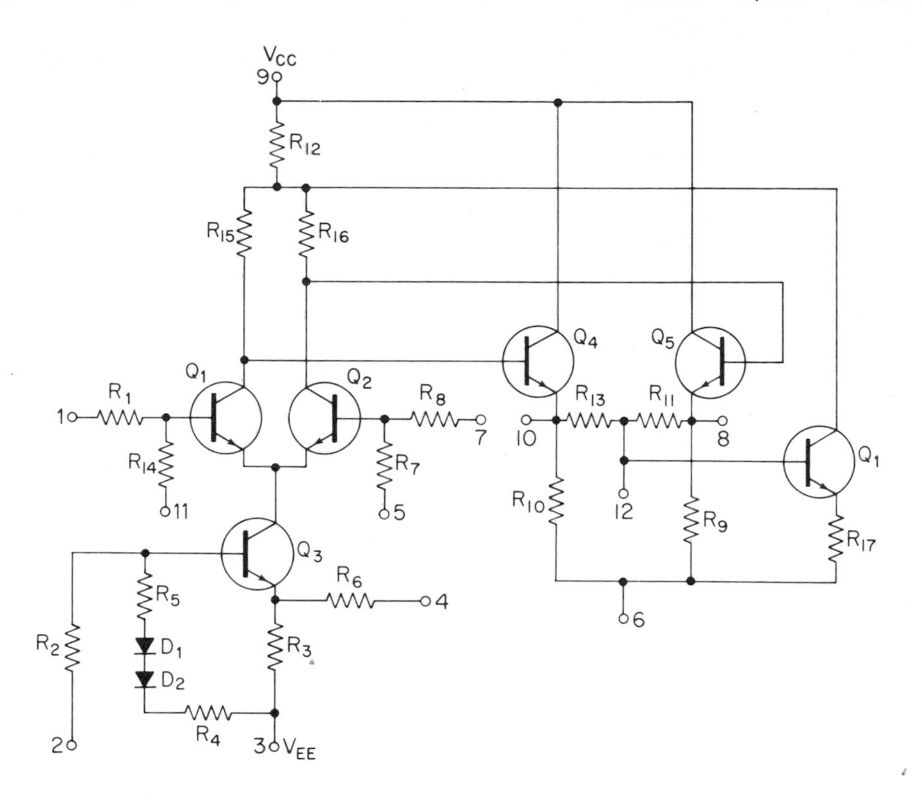

FIG. 5-53. RCA CA3007 audio amplifier.

emitter voltage drops of transistors Q_4 and Q_5 decrease as temperature increases. This small variation in output no-signal operating voltage is sufficient to cause a large variation in the standby current of a class B push-pull output stage when the IC and output stage are direct-coupled. Resistors R_{11}, R_{12}, R_{13} and R_{17} and transistor Q_6 form a dc feedback loop which stabilizes the no-signal operating voltage at output terminals 8 and 10 for both temperature and power-supply variations so that variations in the output operating points are negligible.

Resistors R_1, R_7, R_8 and R_{14} form the input circuit. A double-ended input is applied to terminals 1 and 5, and a single-ended input is applied to either terminal 1 or terminal 5, with the other terminal returned to ground. The IC must be ac coupled to the input source. That is, a coupling capacitor must be used. Also, any dc resistance between terminal 1 and ground should be added between terminals 5 and ground. Output power-gain stabilization for a direct-coupled driver and output stage is accomplished by means of an ac feedback loop that connects terminals 7 and 11 to the proper emitters of the push-pull output stage.

Connections of power supplies to the IC requires that the most positive voltage be connected at terminal 9, and the most negative voltage to terminal 3 (which is internally connected to the substrate and case). The IC can be operated from various supplies, and at various levels. Either single or dual supply operation is possible. For dual-supply operation, symmetrical supplies must be used if the IC is to be direct-coupled to the audio power stage. For single-supply operation, the IC must be ac coupled to the audio output stage, and the number of external components required increases.

For operation from either single or dual supplies, the operating current in transistor Q_3 is determined by the bias voltage between terminals 2 and 3. The more negative terminal of this bias voltage must be connected to terminal 3. For dual-supply systems, terminal 2 is either grounded or connected to a trigger circuit for audio-squelching purposes.

5-4.1.1 Dual-supply, direct-coupled audio driver

Figure 5-54 shows the IC used as a dual-supply audio driver in a direct-coupled audio amplifier. This amplifier provides a power output of 300 mW for an audio input of 0.3 V (RMS), with V_{CC} at +6 V, V_{EE} at −6 V, and V at 30 V.

The external resistor R connected between terminals 3 and 4 is used to set the class B output-stage standby current as required for a particular application. If the standby current is too low, crossover distortion will result. If the standby current is too high, standby power drain is excessive. Decreasing

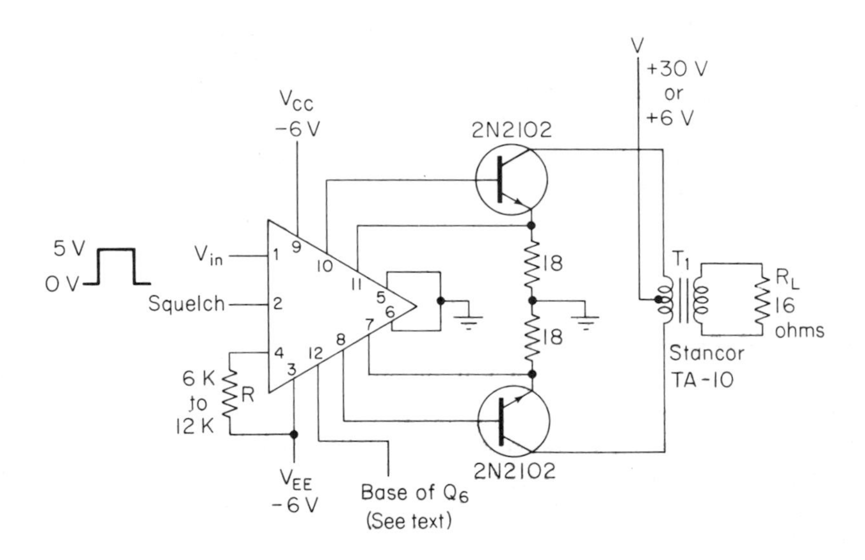

FIG. 5-54. Dual-supply, direct-coupled audio driver (RCA).

the value of resistor R reduces the standby current. For a standby current of 10 mA, resistor R is typically 1 kΩ.

Terminal 2 must be grounded or, if an audio squelch is desired, must be connected to a positive voltage supply of 5 V minimum. When terminal 2 is near ground, the IC functions normally. When terminal 2 is at 5 V, the differential pair of the audio driver saturates, and the push-pull output stage is cut off. The squelch source must be capable of supplying a current of 1.5 mA in the 5-V condition, and 0.75 mA in the near-ground condition.

For a symmetrical audio driver, there is no ac signal present at the base of Q_6. However, unbalances between the two halves of the circuit may require that the base of Q_6 be bypassed for proper operation. The base of Q_6 may be bypassed by connection of an external capacitor (typically 50 μF, 6 V) from terminal 12 to ground. Bypassing is usually not used, unless high undistorted power outputs are required over the complete temperature range of the IC.

5-4.1.2 Single-supply, capacitor-coupled audio driver

Figure 5-55 shows the IC used as a single-supply audio driver in a capacitor-coupled audio amplifier. The connection still represents a

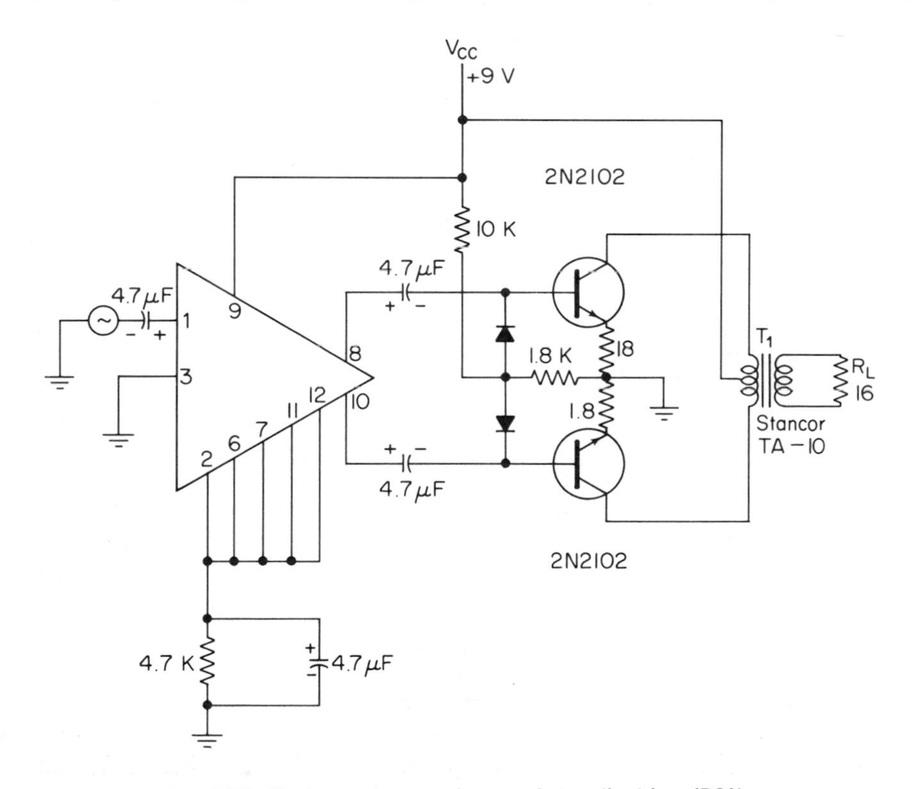

FIG. 5-55. Single-supply, capacitor-coupled audio driver (RCA).

differential-pair phase splitter fed from a constant-current transistor. The two output signals from the phase splitter are direct-coupled through two emitter-followers which are capacitor-coupled to the push-pull output stage. Because of the ac coupling, there is no longer a dc dependence between the driver and the output stage, and any desired audio output design or drive source can be used.

As a single stage, the IC provides a voltage gain of 24 dB for a dc power dissipation of 20 mW, with the harmonic distortion reaching 3 percent for outputs of 0.6 V (RMS) at terminals 8 and 10 (without feedback).

Both dc and ac feedback loops are eliminated in the circuit of Fig. 5-55. Although the dc feedback loop is no longer required because of the ac coupling, removal of the ac feedback loop causes the output power gain to decrease about 1 dB for a 50 °C rise in temperature.

5-4.2 Motorola MC1554 audio power amplifier

Figure 5-56 shows the schematic of the amplifier. The output stage is quasi-class B, obtained by use of diodes CR_2, CR_3, and CR_4. Diodes CR_2 and CR_4 are conducting for most of the complete cycle. This fixes the voltage across the input of Q_5, Q_6 and the diode CR_3 at two diode drops. Thus, Q_6 and CR_3 can not be conducting simultaneously. This prevents excessive current flow from one power source to another through the output circuit elements. The resulting dead-band or cross-over distortion is easily removed from the output signal by the action of the overall feedback loop.

Operation of the output stage can be seen by noting that V_{out} depends upon the current flow in Q_7. A rising output voltage is obtained by decreasing current in Q_7 which reduces the voltage drop across R_3, and makes the base of Q_5 rise toward $+V_{CC}$. This action turns both Q_5 and Q_6 on to supply current to the load for the positive output swing. For the negative swing Q_7 is driven to a larger value of collector current which increases the drop across R_3 and turns both Q_5 and Q_6 off.

5-4.2.1 Basic non-inverting power amplifier

Figure 5-57 shows the IC used as a basic non-inverting power amplifier. Both single-supply and dual-supply connections are shown. Note that the output must be capacitively coupled to the load when the single-supply is used. A dual-supply (or split-supply as it is sometimes called) eliminates the need for capacitive coupling.

With either supply system, the IC package dissipates about 1.5 W (with about 1-W output). If the IC case is mounted directly against a metal chassis, a heat sink is usually not necessary. If the IC is mounted on a typical composition printed-circuit board, a heat sink is required. The manufacturer recom-

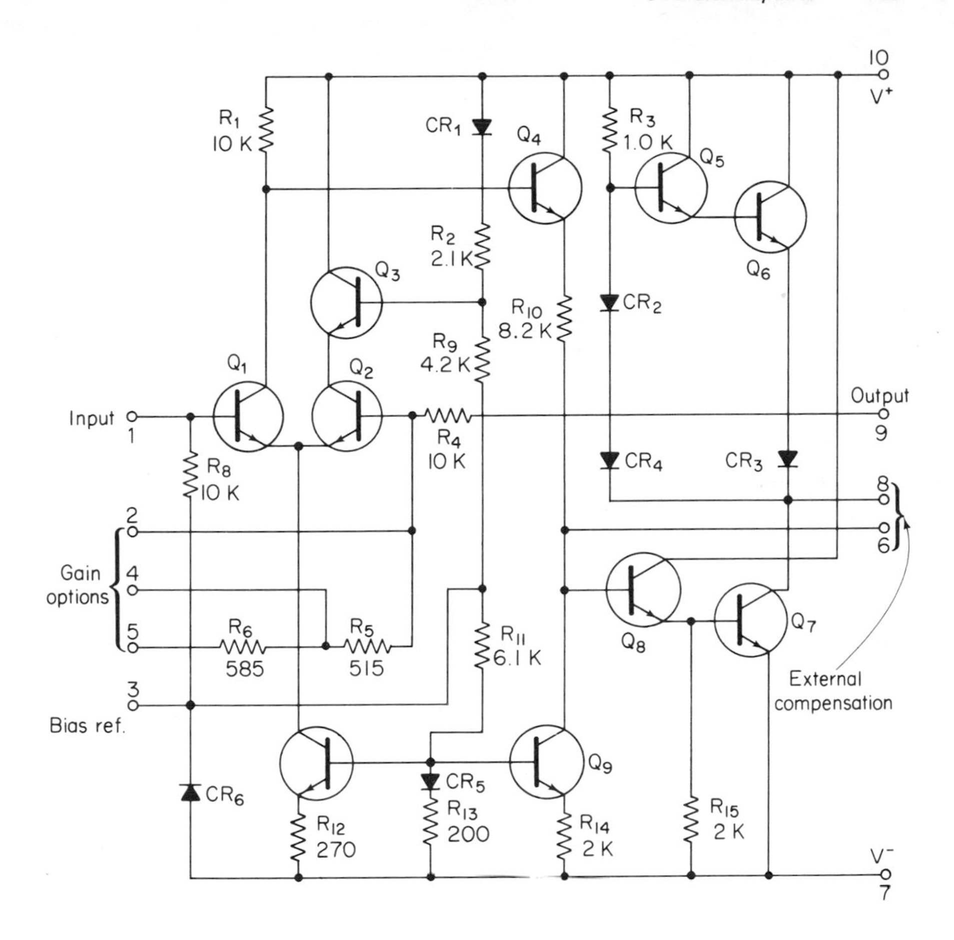

FIG. 5-56. Motorola MC1554 audio power amplifier.

mends a $2 \times 2 \times \frac{1}{8}$ inch aluminum plate with a center hole drilled and reamed such that the heat sink fits snugly over the top of the TO-5 style package, using some type of thermal conducting grease.

In the dual-supply circuit, note that a 10-μF electrolytic capacitor C_2 is connected to $-V_{EE}$. This keeps a fixed bias on the capacitor. Also, an extra bypass capacitor C_6 is required for dual-supply operation. However, the need for this extra capacitor is usually offset by the improved low-frequency response. With the values shown, typical low-frequency response of the dual-supply connection is 40 Hz, while the single-supply low end is 200 Hz. Another advantage of the dual-supply is elimination of the "thump" caused in a loudspeaker by the turn-on charging transient of C_3. On the other hand, direct coupling to the dual-supply circuit input will introduce an offset at the output. This may be objectionable. For example, if the output load is a

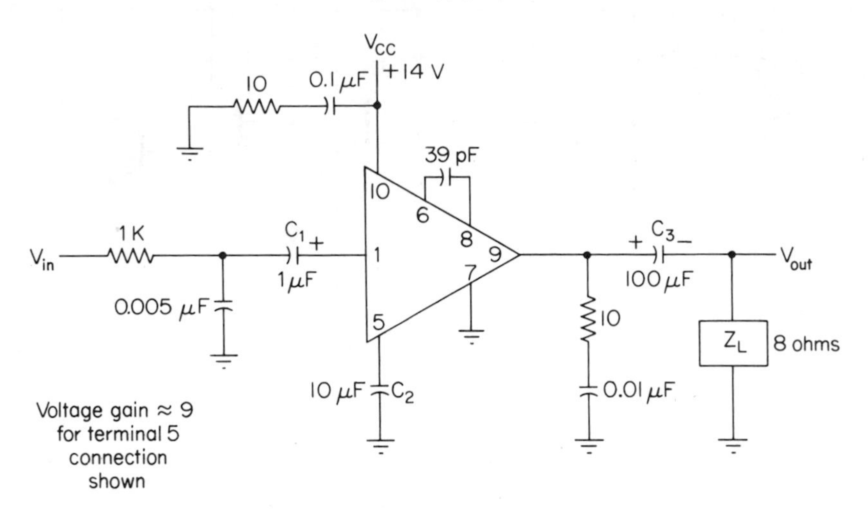

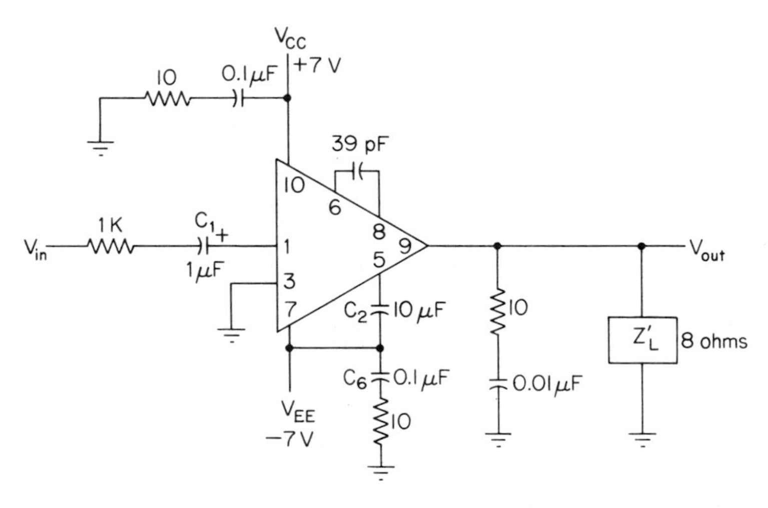

FIG. 5-57. Basic non-inverting power amplifier (Motorola).

loudspeaker, an offset voltage at the output will apply a constant current through the loudspeaker winding.

The high-frequency limit of both circuits is about 22 kHz (well beyond the normal audio range). This limit is set by the input *RC* filter network (1-kΩ resistor and 0.005-μF capacitor). The *RC* filter can be eliminated, thus extending the high-frequency limit. The IC will provide 20-dB gain up to about 2 or 3 MHz, and over 30-dB gain up to about 500 kHz. However, at higher frequencies, the harmonic distortion increases. At frequencies below 22 kHz, and with 20-dB gain, the total harmonic distortion is less than 1 percent (typically 0.75 percent).

The low-frequency limit of the dual-supply circuit is set by the value of C_1 and the input impedance of the IC (typically 10 kΩ), as well as the feedback capacitor C_2. The single-supply low-frequency limit is also affected by the output capacitor C_3 and the impedance of the load. As in the case with discrete-component circuits, the coupling capacitors and their related impedances form *RC* low-pass filters.

As a rule-of-thumb, to find the low-frequency cutoff limit (about 3 dB down) of any *RC* filter, the equation is: $1/(6.28\ RC)$. For example, assume that the load impedance is 8 Ω, and the value of C_3 is 100 μF, as shown. Then the approximate low frequency limit is: $1/(6.28 \times 8 \times 100\ \mu F) \approx 200$ Hz. Thus, if the capacitor value is doubled to 200 μF, the low frequency limit is reduced to about 100 Hz.

Since C_3 is eliminated in the dual-supply circuit, the value of C_1 and the IC input impedance set the low-frequency limit. Using the same equation, the approximate low frequency limit for dual-supply operation is: $1/(6.28 \times 1\ \mu F \times 10\ k\Omega) \approx 16$ Hz. The fact that this is lower than the actual low limit of about 40 Hz is caused by the effect of feedback capacitor C_2 on low-frequency operation.

The connection of capacitor C_2 sets the amount of feedback from output to input, and thus fixes the IC gain. Capacitor C_2 can be connected to pins 2, 4 or 5 to produce fixed gains of approximately 32, 25 and 20 dB, respectively. The opposite side of C_2 is connected to ground for the single-supply, or to $-V_{EE}$ for the dual-supply. Capacitor C_2 thus forms a low-pass filter (for the feedback voltage) in conjunction with internal IC resistors.

Note that an *RC* bypass network (10 Ω and 0.01 μF) is shown at the output of both circuits. This network is to prevent oscillation at radio frequencies. The network is not always required, but should be included as a precaution in any power IC where high frequency operation is possible. For example, the IC here is designed to provide some gain up to about 10 MHz, but will operate at frequencies up to 50–150 MHz. If there is any feedback (say harmonics) due to stray lead inductance, stray capacitance, etc., the IC can oscillate at these high frequencies. This will pass unnoticed in normal audio operation. Likewise, the bandwidth of most oscilloscopes will not pass such oscillations. But these undetected oscillations can cause the IC to over-heat, and possibly burn out.

With single-supply operation, there is a small output offset (in the order of 20 mV) caused by an unbalance in the input differential stage. However, since single-supply operation requires capacitor coupling at the output, the offset should have little effect. No current is drawn through the load due to this offset, except the initial charging current of the coupling capacitor.

With dual-supply operation, where no coupling capacitor is used and the output must have no offset, terminal 3 of the IC is grounded. This removes the base resistance unbalance and establishes a zero-volt reference at the input. Thus, the no-signal level of the output is zero volts.

5-4.2.2 Inverting power amplifier

Figure 5-58 shows the IC used as an inverting power amplifier. The dual-supply connections are used. The output and supply connections are essentially the same as for the non-inverting circuit. However, the input connections are entirely different. The input signal is introduced at the opposite (inverting) side of the differential amplifier at pin 4. The non-inverting input, pin 1, is returned to ground through a 0.1 μF capacitor. Terminals 2 and 5 are shorted together. This arrangement provides the full gain of 32 dB. However, the input impedance (at pin 4) is reduced to about 250 Ω. As a result, a large value input coupling capacitor is needed to pass low frequencies. For example, using the equation and the 40-Hz input described in Sec. 5-4.2.1, the value of input capacitor C is about 16 μF.

The input can be direct coupled from the source. However, the output will then be offset by any no-signal dc that appears at the input, and by any drop across the approximate 250-Ω input impedance. This offset can be compensated for by properly biasing pin 1, or terminating pin 1 with approximately 250 Ω.

With the values shown, the IC delivers about 1-W output, depending upon the value of the load. The available peak-to-peak output voltage swing is

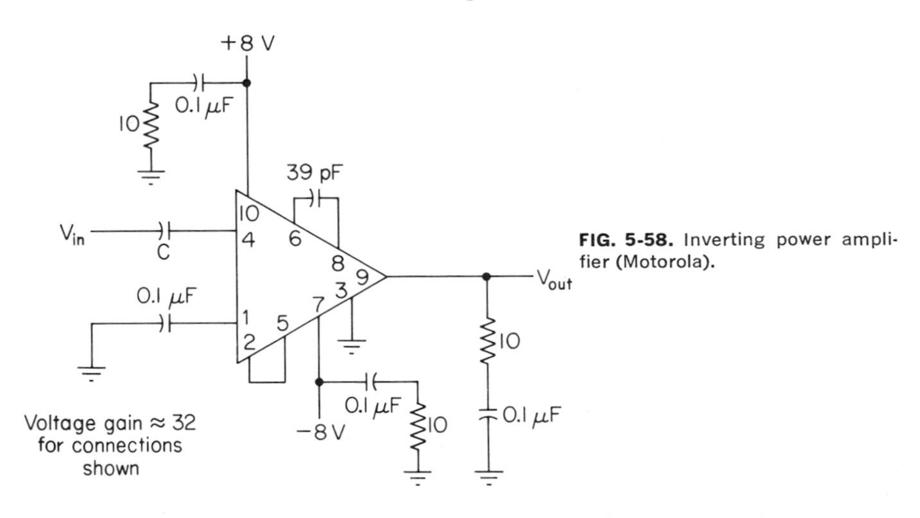

FIG. 5-58. Inverting power amplifier (Motorola).

about 12 V into a 12-Ω load. If the load resistance is lower than 12 Ω, the peak-to-peak voltage swing must be reduced (by reducing the input signal) to comply with the absolute maximum peak current rating of 500 mA.

5-4.2.3 Pulse power amplifier

Figure 5-59 shows the IC used as a pulse power amplifier. The single-supply connections are used, with the gain set for approximately 25 dB (pin 4 connected to ground through the 10-μF capacitor). Because the output is a pulse, and not a continuous linear signal, the normal power dissipation of about 1.5 W can be increased so as to produce a peak power output of about 3 W. This is based on the assumption that the pulse duty cycle is 50 to 75 percent.

In addition to not exceeding the peak power, care must be taken not to exceed peak output current (which in this case is 500 mA). For example, assume that the load must be 10 Ω. Then the peak output voltage must be limited to 5 V ($E = IR$; $0.5 \times 10 = 5$). Since the gain is set for approximately 25 dB, or a voltage gain of about 20, the input pulse must be limited to a peak of 0.25 V.

5-4.2.4 Differential output power amplifier

Figure 5-60 shows two ICs connected as a differential output power amplifier. The single-supply connections are used for both ICs. Both

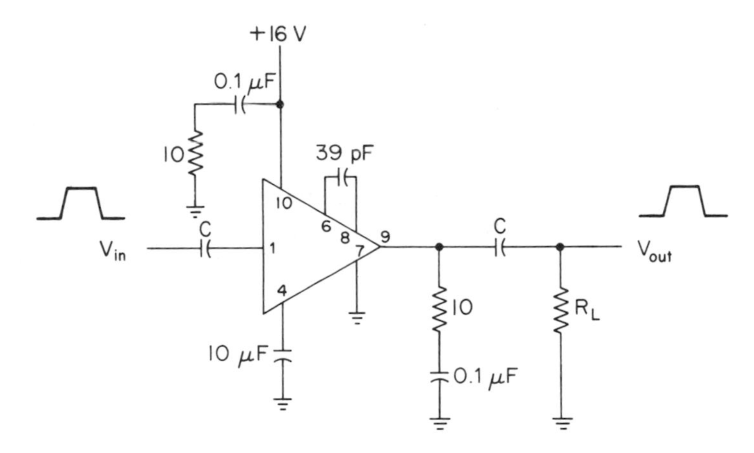

FIG. 5-59. Pulse power amplifier (Motorola).

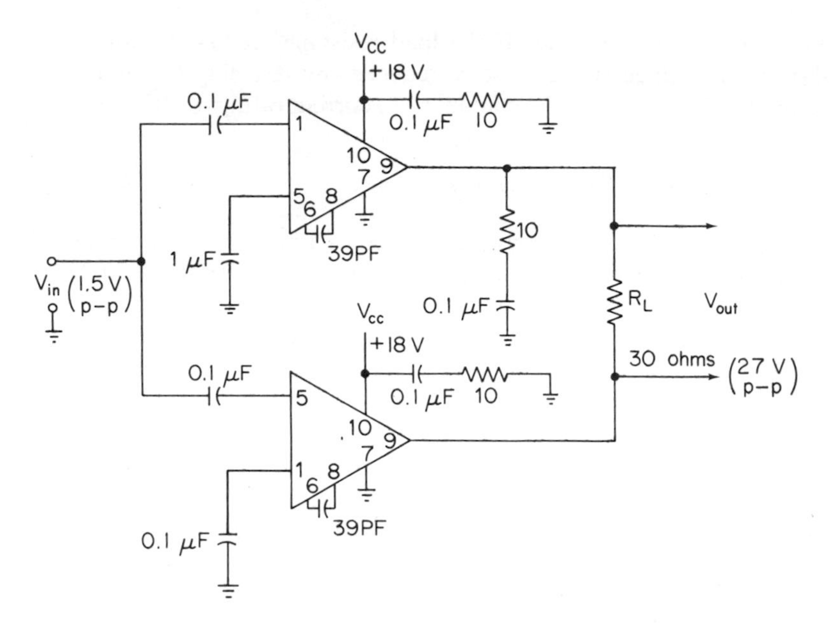

FIG. 5-60. Differential output power amplifier (Motorola).

amplifiers are connected for a 20-dB gain. However, the top amplifier is non-inverting, while the bottom amplifier is inverting. This results in an effective overall voltage gain of twice that of either IC operating alone. That is, if the voltage gain of each IC is 10, the effective voltage gain of the circuit is 20.

The input impedance of the upper amplifier is 10 kΩ, while that of the lower amplifier is 1 kΩ. This unequal input impedance requires unequal capacitance values for the input coupling capacitors (to produce the same low-frequency cutoff point).

Because of the differential output connections, no output coupling capacitor is required, even though the single-supply connections are used. Also, the peak-to-peak output voltage swing capabilities of the amplifier can and do exceed the supply voltage. As a rule of thumb, the maximum possible peak-to-peak output voltage is about 75 percent of the sum of the two supply voltages (in this case $18 + 18 = 36$; $36 \times 0.75 = 27$ V).

5-5. IC FM SYSTEMS

There is an increasing use of ICs in FM systems. The IC is particularly useful in stereo systems since two channels of amplification can be fabricated on one semiconductor chip. Thus, both channels should have identical characteristics.

At present, ICs are used to replace the IF strip in FM receivers, as preamplifiers in stereo audio sections, and as complete FM decoder systems. The ICs used to replace IF strips in FM receivers are similar to the IF ICs used in TV receivers. Such units are discussed in Sec. 5-3.8, and will not be repeated here. Instead, we will concentrate on FM stereo preamplifiers and FM decoder systems.

5-5.1 Motorola MC1303P dual preamplifier

Figure 5-61 shows the schematic of the amplifier. Note that there are two identical channels of amplification. Thus, the IC is ideally suited for preamplification of two stereo channels.

One channel of a typical stereo preamplifier is shown in Fig. 5-62. Two dual-channel ICs are used. The input IC provides compensated gain for phonograph or tape playback. The output IC provides broadband (flat) gain for the system. The circuits between the two ICs provide bass, treble, balance, and volume controls, as well as an emitter-follower to provide low input impedance for the output IC.

The values shown in Fig. 5-62 are selected to give desired playback and tone characteristics for the system. Although the values are selected on the basis of using the MC1303, similar (or possibly identical) values can be used for other dual-channel ICs. Thus, the user will find the following paragraphs particularly helpful when designing any audio system. Before going into the procedures for selecting external components, we shall discuss the basic amplifier characteristics.

5-5.1.1 Basic amplifier characteristics

As shown in Fig. 5-61, each channel of the amplifier has a differential input amplifier, followed by a second differential stage with single-ended output, and two emitter-follower stages.

The input differential amplifier is fed from a constant-current source in the emitters which, in turn, is biased from a voltage divider in the emitter circuit of the second-stage differential amplifier. This arrangement is used to provide common-mode negative feedback, thus increasing rejection of the common-mode signal. The input transistors are biased at approximately 250 μA to provide low-noise operation. By cascading the two differential amplifiers, low drift, dc bias stability and temperature stability are obtained.

The second stage differential amplifier drives an emitter-follower which, in turn, drives a composite PNP output stage. In this stage, the voltage at the base and the resistor in the emitter of the PNP transistor control the current in both the PNP and NPN devices. Thus, the two transistors operate as a single PNP unit. This combination yields the required current gain and volt-

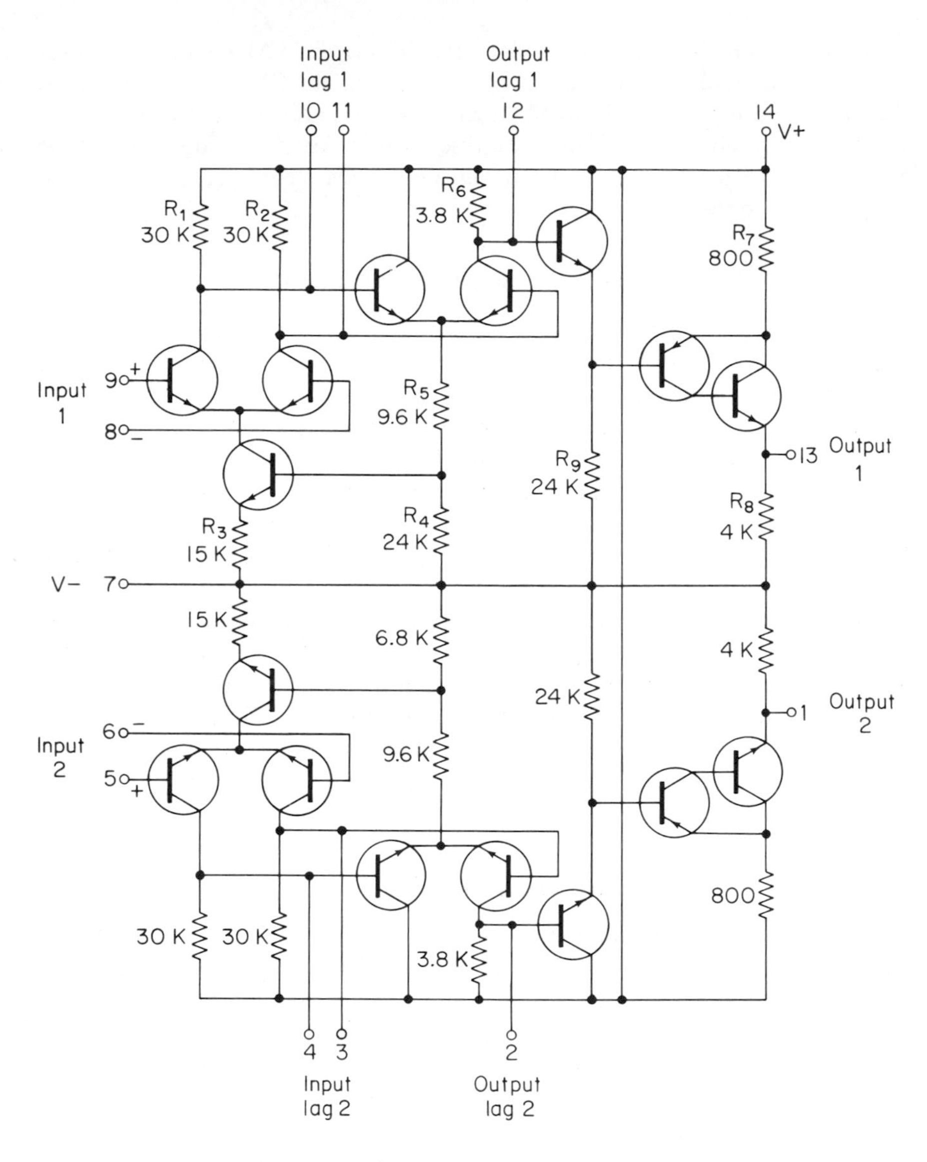

FIG. 5-61. Motorola MC1303 dual preamplifier.

age gain. The output operating level is set by making the voltage across R_8 equal to the $-V_{EE}$ supply voltage.

5-5.1.2 Playback amplifier section

The closed-loop voltage gain of the playback amplifier section is set by the ratio of the feedback network to resistor R_2. The feedback (or

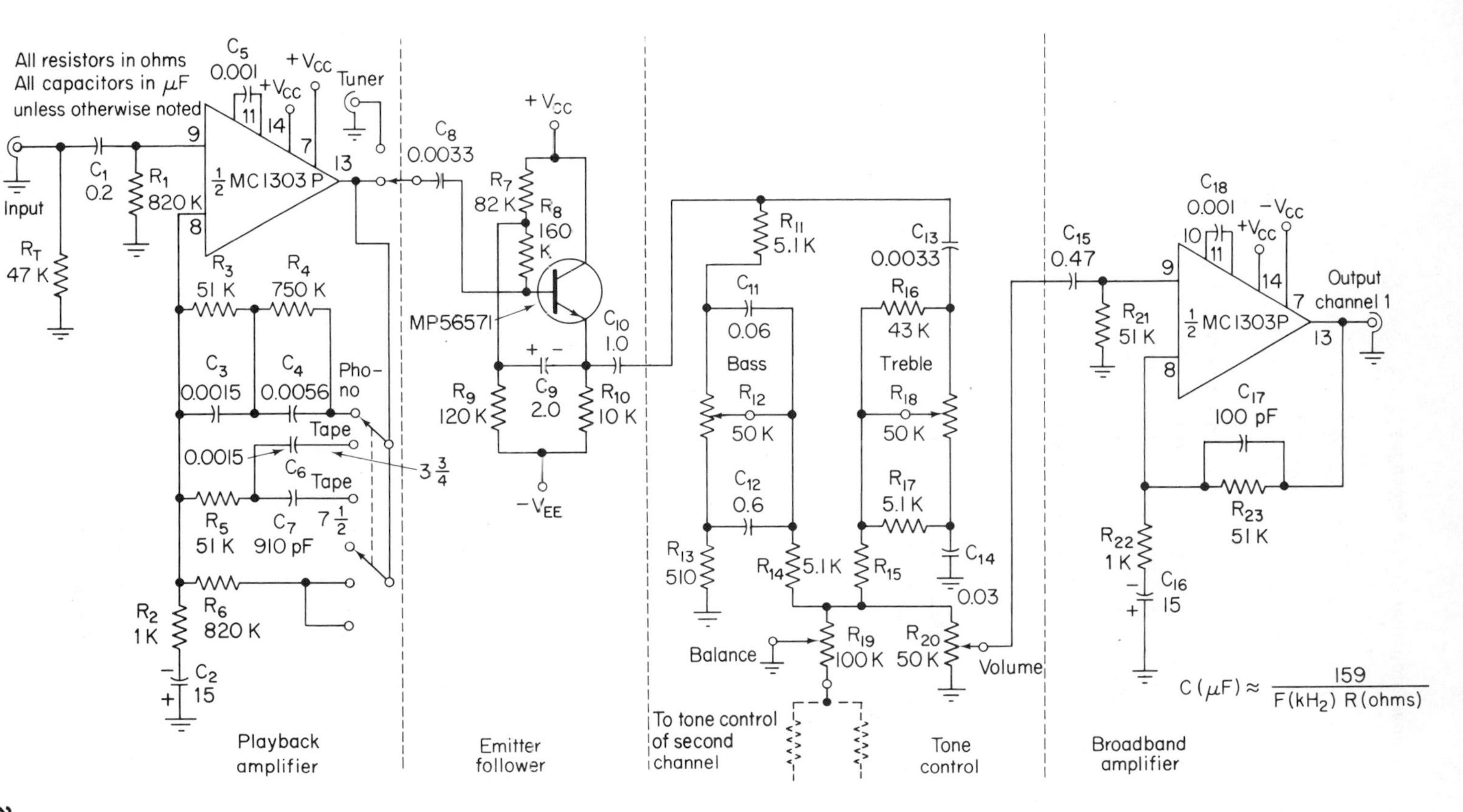

FIG. 5-62. One channel of stereo preamplifier (Motorola).

compensation) network for phonograph use is composed of C_3, C_4, R_3, R_4, while the tape network is composed of R_5, C_6, and C_7.

R_1 is approximately equal to the resistance of the compensation network. This minimizes input offset current in the usual manner. The input offset is also minimized by the addition of C_2 in series with R_2. The time constant of C_2R_2 is selected to give a 3-dB roll-off at the lowest frequency to be considered. The time constant of C_1 and the input impedance should also be selected for a 3-dB roll-off at this frequency.

The input impedance of the amplifier will be approximately equal to R_1. If the amplifier is to be used for a magnetic phonograph cartridge, the value of R_1 is generally much larger than the required 50 kΩ of a typical magnetic cartridge. To overcome this problem, resistor R_T is added to properly terminate the cartridge impedance.

Input lag compensation is accomplished with C_5. According to the datasheet, the value of C_5 can range from 680 pF to about 0.002 μF. As described throughout Chapters 3 and 4, the slew rate and high-frequency response are partly governed by the value of C_5. A 0.001-μF capacitor is a good starting point for a trial capacitance.

RIAA *playback equalization* is used for the phonograph network. The standard RIAA equalization curve is shown in Fig. 5-63. The recording curve is the inverse of the playback curve so that addition of the two gives a net flat frequency-versus-amplitude response. In recording, the high frequencies are emphasized to reduce effects of noise and low inertia of the cutting stylus. The low frequencies are attenuated to prevent large excursion of

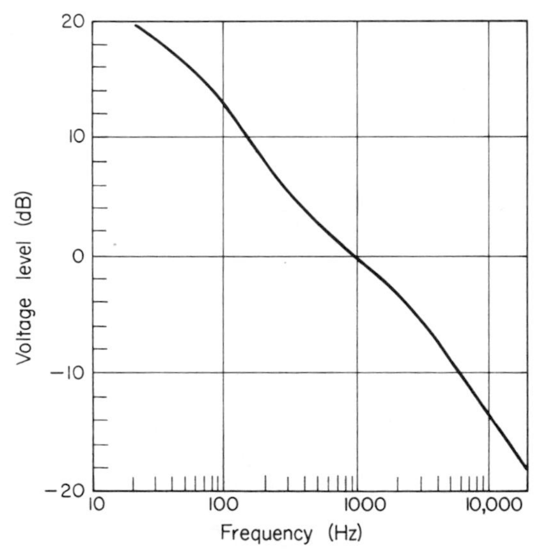

FIG. 5-63. RIAA playback equalization curve (Motorola).

the cutting stylus. It is the job of the frequency selective feedback network to accomplish the addition of the recording and playback responses.

It is impossible to have the playback network be the exact inverse of the recording compensation, since each recording system is slightly different. However, there are guidelines that can be applied. A typical audio range is from 20 Hz to 20–24 kHz. Thus, there is roll-off at both the low and high ends. At the low end, the roll-off should start at some point between 10 and 20 Hz. This can be accomplished by making the 10-Hz point about 3 dB down from the 20 Hz point. As frequency increases from 20 Hz, there must be an almost linear (hopefully) roll-off. Ideally, the voltage gain at 20 Hz should be 100 times the gain at 20 kHz, and 10 times the gain at 1 kHz. This will produce the approximate RIAA curve of Fig. 5-63.

In the circuit of Fig. 5-62, the linear roll-off is accomplished by dividing the playback network into three parts. The R_2C_2 section sets the 10-Hz point at 3 dB down from the 20-Hz point, the R_4C_4 section covers frequencies up to about 1 kHz, and the R_3C_3 network covers higher frequencies. The value of R_3 should be 1000 times the desired voltage gain at 1 kHz, while the value of R_4 should be 15 times that of R_3. The value of R_2 is also based on the value of R_3, and 3 is selected to provide the desired 1-kHz voltage gain (of 50 in this case). That is, the R_3/R_2 ratio sets the 1-kHz voltage gain.

The playback amplifier of Fig. 5-62 produces an arbitrary minimum voltage gain of 5 at the highest frequencies (20–24 kHz). Most of the gain for the complete system is provided by the broadband IC amplifier. Using a minimum gain of 5 at the highest frequency, the gain at 20 Hz must be 100 times that amount, or 500. Likewise, the gain at 1 kHz must be 50. Using these desired gains, the value of R_3 is: $1000 \times 50 = 50$ kΩ (use a 51 kΩ standard). With R_3 at 50 kΩ, the value of R_4 is: 15×50 kΩ $= 750$ kΩ; and the value of R_2 is: 50 kΩ/50 = 1 kΩ.

At low frequencies, the predominant impedance of the compensation feedback network is that of R_4. As frequency increases from about 50 Hz, the reactance of capacitor C_4 in parallel with R_4 begins to decrease the impedance of the C_4R_4 section. The reactance of C_4 is made to equal R_4 at about 35–50 Hz. At about 1 kHz, the net impedance of C_4R_4 is low compared to R_3, and R_3 sets the mid-band gain. As frequency increases to about 2 kHz, the parallel impedance of capacitor C_3 begins to shunt R_3, decreasing the impedance of the C_3R_3 section. The reactance of C_3 is made equal to R_3 at about 2.1 kHz.

The value of R_1 is made equal to the combined values of R_3 and R_4, or about 800 kΩ (use an 820 kΩ standard). The reactance of C_1 is made equal to R_1 at about 1–2 Hz. The reactance of C_2 is made equal to R_2 at about 10 Hz. If the phonograph cartridge used has an impedance close to that of R_1, then R_T can be omitted. This is generally not the case, since a typical magnetic cartridge has an impedance of about 50 kΩ. The value of R_T is made equal to the impedance of the cartridge. The 47 kΩ value shown for R_T is generally satisfactory.

Note that the same equation, based on frequency breakpoint and value of corresponding resistor, is used to find the values of C_1, C_2, C_3, and C_4.

NAB playback equalization is used for the tape network. The standard NAB equalization curve is shown in Fig. 5-64. Again, the recording curve is the inverse of the playback curve so that addition of the two gives a flat response. Likewise, the high frequencies are emphasized and the lows are attenuated. However, unlike phonograph playback, tape playback tends to flatten out after about 3–4 kHz. Also, a different response is required for different tape speeds. The playback response curves for both $3\frac{3}{4}$ and $7\frac{1}{2}$ ips (inches per second) are shown in Fig. 5-64. Up to about 1 kHz, the curves are almost identical. Because there is only one frequency breakpoint (where the curve must start to flatten) for each tape speed, a simple *RC* series compensation network is all that is required (instead of the multi-section network used for phonograph playback).

The breakpoint for $3\frac{3}{4}$ ips occurs at about 1.85 kHz. The mid-band frequency gain is still 50, so the value of R_2 remains at 1 kΩ, and R_5 is made equal to R_3, or 51 kΩ. The reactance of C_6 is made equal to 51 kΩ (R_5) at 1.85 kHz (the nearest standard value is 0.0015 μF).

The breakpoint for $7\frac{1}{2}$ ips is at about 3.2 kHz so that C_7 must have a reactance of 51 kΩ at this frequency. A C_7 capacitor value of 910 pF is the nearest standard.

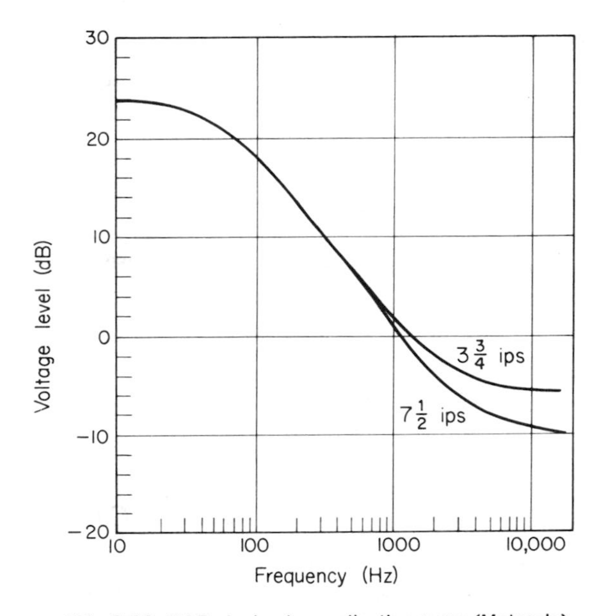

FIG. 5-64. NAB playback equalization curve (Motorola).

Because C_6 and C_7 block the dc path for the IC feedback input, resistor R_6 is added when the phono-tape switch is in either tape position. The use of R_6 prevents realization of a full 20-dB bass boost because of the shunting action across the tape compensation network. However, the network does provide about 15 dB of boost, which is generally satisfactory. The emitter-follower stage compensation network does allow the full 20 dB of bass boost to be realized.

It should be noted that the accuracy of both the RIAA and NAB compensation will be only as good as the components used. It is recommended that 5 percent tolerance resistors and capacitors be used. Likewise, it may be necessary to "trim" the values to get an "exact" performance curve.

5-5.1.3 Emitter-follower stage

The passive tone control selected for this preamplifier gives a constant slope, variable turnover characteristic which is desirable from the listener's standpoint. Since the tone control establishes the driving impedance for the output (broadband) pre-amplifier stage, it is desirable to keep the impedance low. This low impedance could load the output of the first stage, so an emitter-follower stage is placed between the input preamplifier and the tone control.

The emitter-follower stage is "bootstrapped" to provide a higher input impedance and will also allow some low-frequency compensation. By proper selection of the coupling and bootstrap capacitors (C_8 and C_9) a 12 dB per octave roll-off at the low cutoff frequency can be obtained. This greatly reduces the effects of excess noise which occurs at very low frequencies in semiconductor devices. This noise, known as "one over eff noise" or "flicker" noise, occurs more noticeably at frequencies below 10 Hz. By setting the 12 dB roll-off to start at about 20 Hz, an approximate 24-dB roll-off occurs at frequencies below 10 Hz. This arrangement also gives about 5 dB of bass boost due to the "resonant rise". This effect is produced because the so-called "bass" frequencies of both RIAA and NAB curves are above the 10 Hz point. The combined resonant rise, and the 15-dB gain of the input preamplifier stages, results in a full 20 dB of bass boost in both the RIAA and NAB positions of the preamplifier.

At frequencies where C_9 is a low impedance, the input impedance of the circuit is approximately R_{10} times the transistor gain (plus 1). However, at low frequencies when the reactance of C_9 becomes large, less signal voltage is developed across R_8. As frequency decreases further, the input impedance decreases at a rate corresponding to a 6 dB per octave slope. If C_8 is chosen to establish a breakpoint at this same frequency, the net effect is a 12 dB per octave roll-off at the low frequency cutoff point.

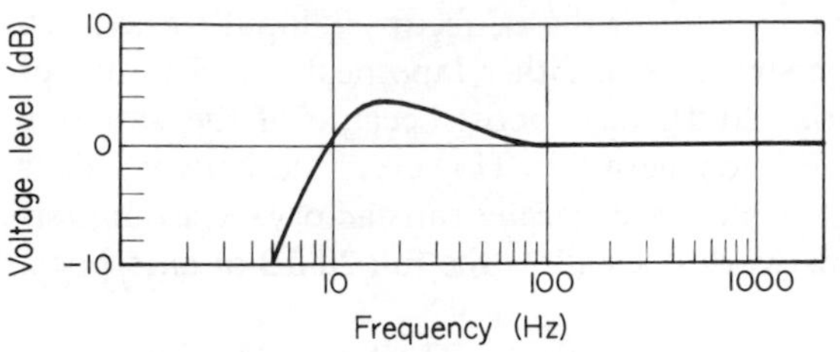

FIG. 5-65. Frequency response of emitter-follower (Motorola).

The reactance of C_9 at the low frequency cutoff point (20 Hz) is made equal to 10 percent of the parallel combination of R_7 and R_9 (approximately 5 kΩ). Thus: $159/5000 \times 0.02 = 1.59\ \mu F$. (Use a 1.5 or 2 μF standard.)

The reactance of C_8 at the 20-Hz point is made equal to an estimated 2.5 MΩ input impedance. Thus, $159/2.5\ M\Omega \times 0.02 = 0.0032\ \mu F$ (use 0.0033 μF standard). The response curve of the emitter follower is shown in Fig. 5-65.

5-5.1.4 Tone controls

The bass and treble tone controls are standard audio taper potentiometers. At 50 percent rotation, the resistance is split, 90 percent on one side of the wiper and 10 percent on the other side. The relationship between wiper position and resistance is shown in Fig. 5-66.

In the bass control circuit, when the control is in the center position, the frequency response is flat from about 50 Hz to 20 kHz. This is shown in Fig. 5-67. The reactance of C_{11} is made equal to the 45-kΩ portion of R_{12} at 50–60 Hz, and the reactance of C_{12} is made equal to the 5-kΩ portion of R_{12} at 50–60 Hz. As frequency increases from 50 Hz, C_{11} couples more signal to the output, while C_{12} shunts more signal to ground through R_{13}. The net effect is a flat response from about 50 Hz to 20 kHz with a 20-dB insertion loss.

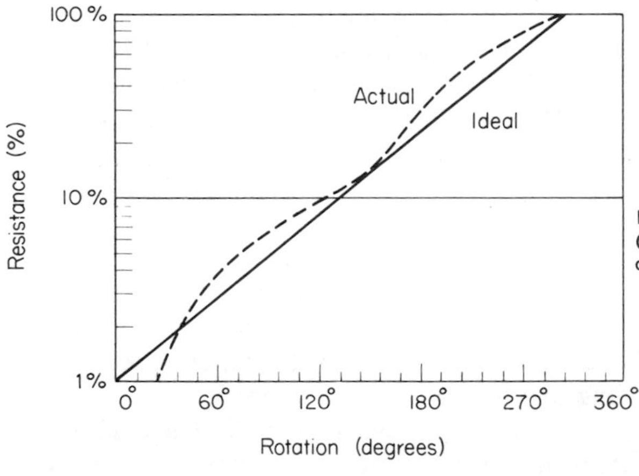

FIG. 5-66. Relationship of position (rotation) and resistance in tone controls (Motorola).

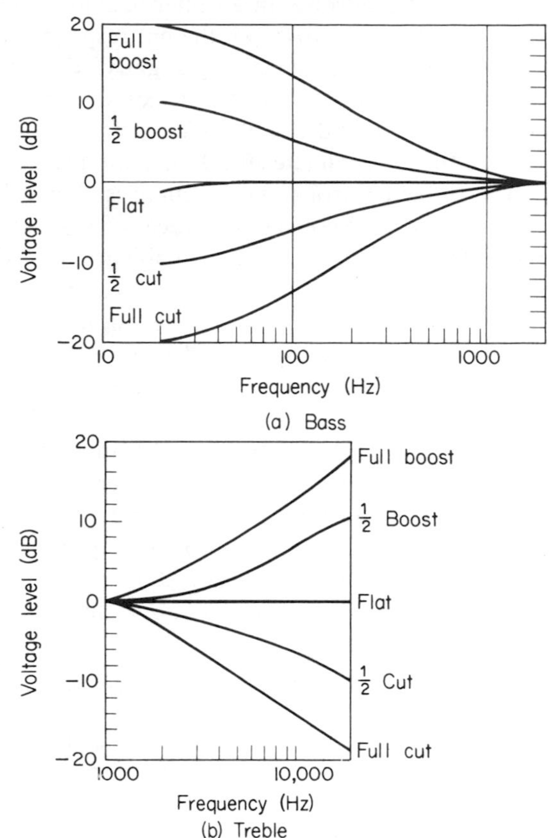

FIG. 5-67. Normalized tone control frequency response curves (Motorola).

When the wiper is in the boost position, C_{12} with a reactance $\frac{1}{10}$ the resistance of R_{12} at 50–60 Hz, effectively shunts R_{12} out of the circuit, making R_{11} and C_{12} the dominant frequency response shaping components. Ideally, the full bass boost position will supply an output voltage (at about 50 Hz) that is 20 dB greater than the center position (flat response).

The full boost position represents zero attenuation in the tone control of the bass frequencies. The amplitude of the output will decrease at a 6 dB per octave rate to the frequency where the reactance of C_{12} is negligible. The output amplitude is then determined by the ratio of R_{11} to R_{13}.

When the wiper is in the "full cut" position, the output amplitude at about 50 Hz is determined by the ratio of C_{11} reactance to R_{13} resistance, and is about 40 dB below the input voltage. As frequency is increased, the reactance of C_{11} decreases until it is equal to the resistance of R_{13}, again making the output amplitude dependent upon the ratio of R_{11} to R_{13}.

When the control is in an intermediate position, the frequency at which roll-off begins (±3 dB from the flat response curve) will vary, but the slope of the roll-off will change only slightly. Figure 5-67a shows the response of the bass control. The boost-cut axis uses the flat response position as the reference point or 0 dB, although in fact, it is 20 dB below the input signal (due to the approximate 20-dB insertion loss).

The treble control response curve is shown in Fig. 5-67b. At frequencies below about 2.1 kHz, the reactances of C_{13} and C_{14} become small when compared to the parallel divider combination of the control R_{18} and fixed resistance R_{16} and R_{17}. The resistive divider then provides the 10-to-1 voltage division to maintain the 20-dB insertion loss for the high frequencies. The net result is a 20-dB loss that is flat from about 20 Hz to 20 kHz.

The reactance of C_{13} should be about one half the resistance value of R_{18} at a frequency of 2.1 kHz (or about 25 kΩ). Thus: $159/25\ \text{k}\Omega \times 2.1 = 0.003$. The value of C_{14} should be about 10 times the value of C_{13}, or $10 \times 0.003 = 0.03$, to maintain the 10-to-1 voltage division.

The resistance of R_{16} is approximately $\frac{1}{10}$ of the control (R_{18}) resistance, while the R_{17} resistance is approximately 80 percent of R_{18}. Resistors R_{14} and R_{15} are isolation resistors made equal to 10 percent of the resistance of the respective control potentiometers R_{12} and R_{18}.

5-5.1.5 Broadband amplifier stage

The broadband stage is designed exactly like the playback amplifier except that the compensation network is replaced with a 51-kΩ resistor (R_{23}) in parallel with a 100-pF capacitor C_{17}. The capacitor is used to reduce the mid and high frequency noise of the amplifier. The input resistor R_{21} is also 51 kΩ (to minimize input offset). Typical broadband response is shown in Fig. 5-68.

5-5.2 Motorola MC1304 FM stereo decoder system

Figure 5-69 shows the schematic of the IC. Although only one package is used, the IC consists of two different functional blocks, one con-

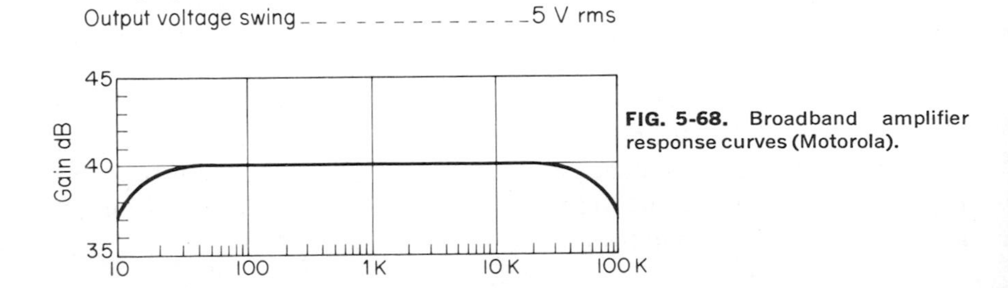

FIG. 5-68. Broadband amplifier response curves (Motorola).

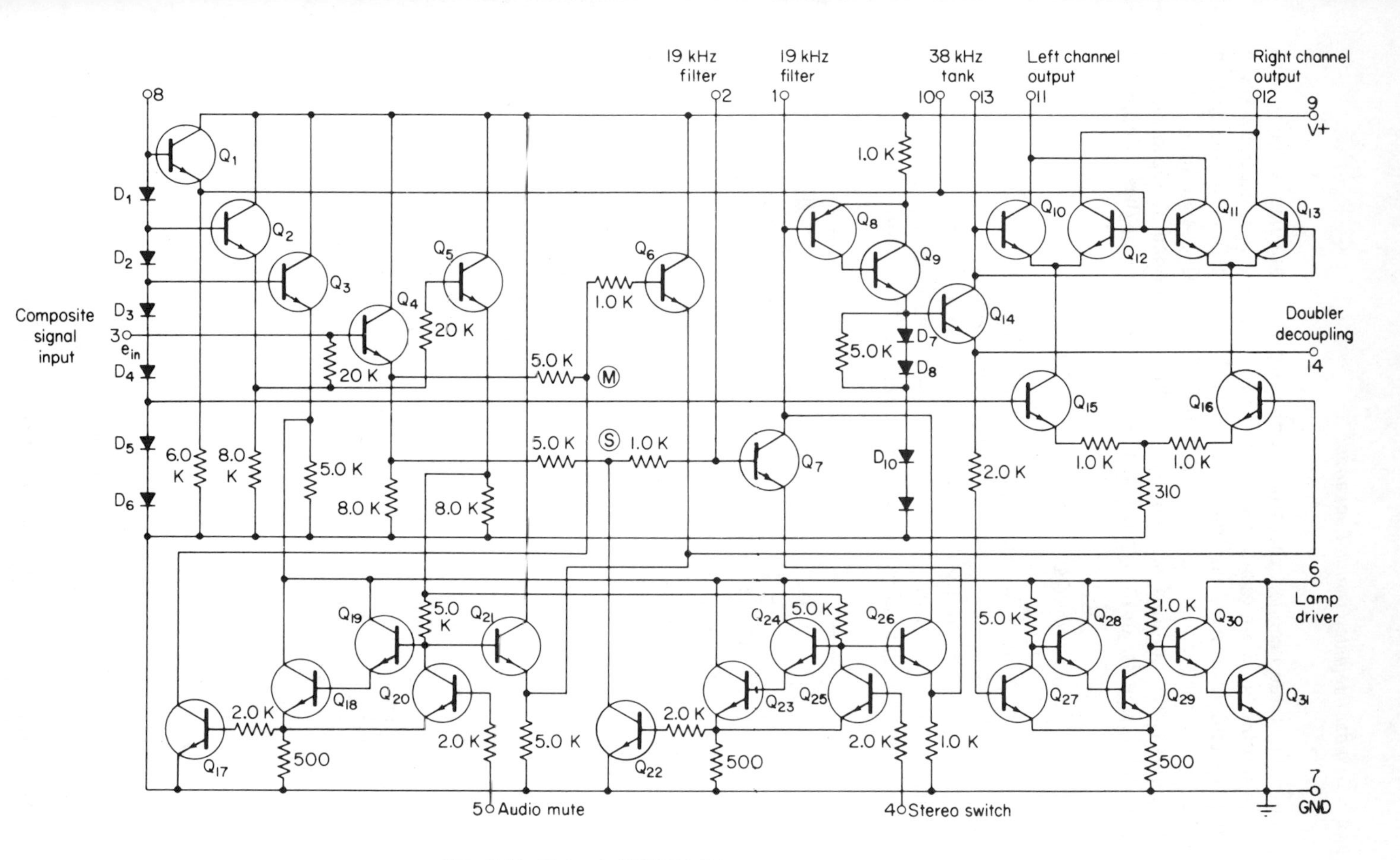

FIG. 5-69. Motorola MC1304 FM stereo decoder system.

taining the stereo decoder, the other containing three auxiliary circuits providing muting, stereo-monaural selection, and stereo lamp indication. These auxiliary circuits can be combined in several ways, giving the designer a choice of several combinations which meet personal design preferences. Muting between stations, automatic switching to monaural reception for weak stereo signals, and "stereo select" action, are some of these options.

5-5.2.1 Decoder section

Figure 5-70 shows the IC connected in a typical circuit. The following is a description of the decoder section operation.

The composite signal from the FM detector is coupled to the base of Q_4 via C_1. This external capacitor, in conjunction with the input impedance of the IC, give an *RC* time constant which determines the minimum frequency

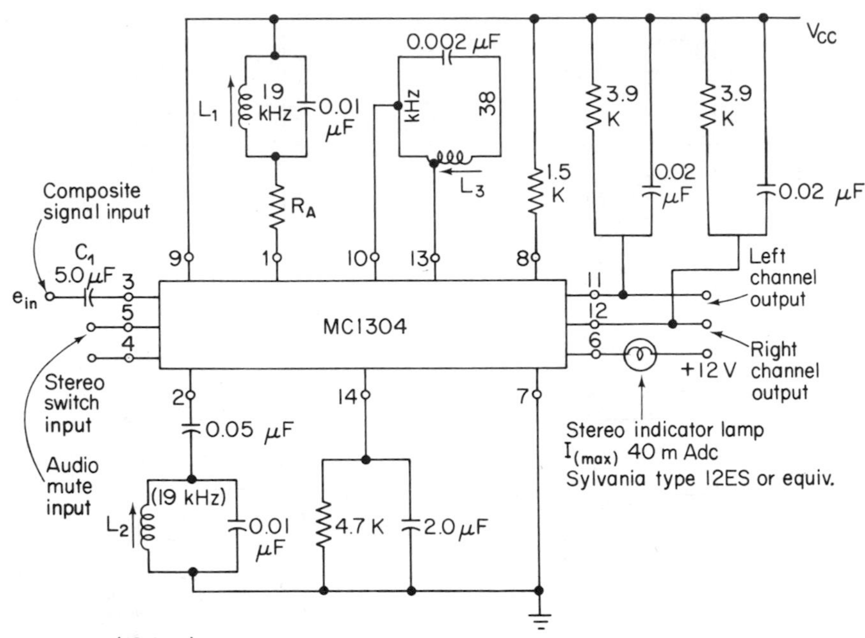

L_1, L_2 (19 kHz)
Nominal inductance: 8.0 mH
Unloaded Q: 120 (measured at 50 kHz using nominal inductance value)
Miller # 1361 or equivalent

L_3 (38 kHz)
Nominal inductance: 8.0 mH
Unload Q: 125 (Measured at 50 kHz using nominal inductance value)
Miller # 1362 or eqivalent

FIG. 5-70. Typical circuit configuration for FM stereo decoder system (Motorola).

response in the circuit and, consequently, the highest channel separation obtainable at the low end of the audio range.

In receivers where a strong high frequency roll-off of the composite signal is expected, parallel *RC* compensation could be used in series with the input to improve channel separation at the high frequency end.

Transistor Q_4 is used as an emitter-follower to provide the high input impedance level needed in the IC to avoid any serious loading of the FM detector. Due to the high degree of feedback provided by the associated emitter resistance, the input impedance is determined primarily by the 20-kΩ resistor connected between the Q_4 base and the bias network. The output of Q_4 is fed to the detector, via Q_6, and to Q_7, the 10-kHz amplifier.

A parallel tuned circuit is used across the input of Q_7 to recover the 19-kHz pilot signal. A second high-Q resonant circuit is used at the output of Q_7 to increase overall selectivity.

The combination Q_8–Q_9 acts as a high gain PNP transistor which couples the 19-kHz signal into the base of Q_{14}, the frequency doubler, which has a high-Q parallel-tuned circuit (38 kHz) for the collector load. Coil L_3 of the tuned circuit is tapped down to reflect a low impedance across the output of Q_{14}. A 10:1 transformation sets an output impedance no higher than about 2 kΩ, avoiding saturation of Q_{14}. If higher impedances are used, the time symmetry of the 38-kHz signal could be lost.

A synchronous detector (Q_{10} through Q_{16}) is used to demodulate the stereo information. When a stereo signal reaches the detector, the 38-kHz voltage developed across the doubler load will switch the transistors of the two upper differential pairs (Q_{10} through Q_{13}) on and off, according to the polarity of the driving signal. A time multiplexing of the composite stereo signal takes place at a 38-kHz rate, permitting separation of the left and right channel information. The detector can also handle a monaural signal.

5-5.2.2 Auxiliary circuits

Three auxiliary circuits are incorporated in the IC. As shown in Figs. 5-69 and 5-70, each consists of an electronic switch associated with a Schmitt trigger. This provides fast and positive switching after the respective thresholds have been reached.

Muting and stereo/monaural switches. Both the muting and stereo/monaural switches are "normally closed", and will open only if a minimum dc voltage is supplied to the proper terminal.

If no voltage is supplied to these circuits, the muting switch and the stereo/monaural switch will ground points M and S (Fig. 5-69) through Q_{17} and Q_{22}, respectively. To avoid overloading in these switching transistors, points M and S each have a 5-kΩ resistor in series with the driving source.

An effective ac grounding of the signal at M makes it possible to obtain a high level of audio attenuation at the output of the decoder. This muting action can be achieved automatically. If this is the case, the dc voltages needed to turn the switch on and off become a function of the signal-to-noise level of the incoming signal.

The stereo lamp indicator. This device is connected internally and represents a "normally open" switch in series with the bulb and the power supply.

When a stereo signal is received, the dc voltage on the emitter of Q_{14} rises, triggering the "on" switch.

The 19-kHz pilot signal level needed at the input of the IC to turn the stereo lamp indicator on (or the pilot sensitivity) can be adjusted, within a certain range, by adding a resistor R_A in series with pin 1. This resistor will change the threshold level of the combination Q_8–Q_9. In a typical system, the pilot sensitivity is increased by a factor of 2.1 by changing the value of the series resistance from 0 to 240.

The dc voltages required to turn this switch on and off differ by 2 mV on the average. This differential assures a definite on condition once the threshold level of the swtich has been reached.

Functional arrangements of the auxiliary circuits. The three auxiliary circuits can be combined in various ways to form different functional arrangements. Specific circuit considerations are as follows.

If desired, it is possible to mute particular monaural stations, as well as interstation noise (conventional audio mute) by combining the function of two switches. For example, the audio mute switch can be driven using the dc voltage developed across the 38-kHz emitter decoupling network (pin 14). The mute driving voltage is then low enough, in the absence of a valid stereo signal, to keep the switch in the "on" condition.

The short circuit between point M and ground decreases the effective emitter resistance in series with Q_4, but still provides a high enough load to develop a substantial 19-kHz signal at the input of Q_7 (pin 2) if a stereo signal is fed to the input. Accordingly, the dc voltage at pin 14 increases, triggering the mute switch off. The final result of this combination is an FM receiver with a "stereo select" action, able to reject any monaural program automatically. When this combination is being used, the use of a stereo lamp indicator becomes redundant.

Stereo reception is poor when the signal-to-noise ratio is low. The same signal, unless it is very poor, can be detected monaurally. This can be done by making the dc driving voltage applied to the stereo/monaural switch (pin 4) a fraction of that applied to the mute switch (pin 5). The voltage divider must be designed in such a way that any RF level below a predetermined point considered acceptable, will not provide enough dc voltage to switch the stereo/monaural switch into the stereo mode.

5-6. IC DIRECT-COUPLED (DC) AMPLIFIER

There are an infinite number of ICs used as dc amplifiers. Typically, such ICs use some form of differential amplifier, with a *temperature-compensation constant-current source*. Typically, all current for the differential amplifier is fed through an NPN transistor (usually connected between the emitters of the differential amplifier and V_{EE}). If there is an increase in current, a larger voltage is developed across the current-source transistor emitter resistor. This larger voltage acts to reverse bias the base-emitter, thus reducing current through the transistor. Since all current for the differential amplifier is passed through the current-source transistor, current to the amplifier is also reduced. If there is a decrease in current, the opposite occurs, and the amplifier current increases. Thus the differential amplifier is maintained at a constant current level. The current-source transistor is temperature compensated by diodes connected in the base-emitter bias network. These diodes have the same (approximate) temperature characteristics as the base-emitter junction, and offset any change in base-emitter current flow that results from temperature change.

The following is a description of a typical IC direct-coupled amplifier.

5-6.1 RCA CA3000 amplifier

Figure 5-71 shows the schematic of the amplifier. The current-stabilized, temperature-compensated differential amplifier provides push-pull outputs, high-impedance (100-kΩ) inputs, and gain of approximately 30 dB at frequencies up to about 1 MHz. The useful frequency response can be increased by means of external resistors and coils.

The circuit is basically a single-stage differential amplifier (Q_2 and Q_4) with input emitter-followers (Q_1 and Q_5) and a constant-current source Q_3 in the emitter-coupled leg. Push-pull input and output capabilities are inherent in the differential configuration.

The use of degenerative feedback resistors R_4 and R_5 in the emitter-coupled pair of transistors increases the linearity of the circuit. The low-frequency output impedance between each output (terminals 8 and 10) and ground is essentially the value of the collector resistors R_1 and R_2 in the differential stage.

5-6.1.1 Crystal oscillator

The IC can be used as a crystal oscillator at frequencies up to 1 MHz by connection of a crystal between terminals 8 and 1, and use of two external resistors, as shown in Fig. 5-72. The output is taken from the collector that is not connected to the crystal (in this case, terminal 10). If a variable-

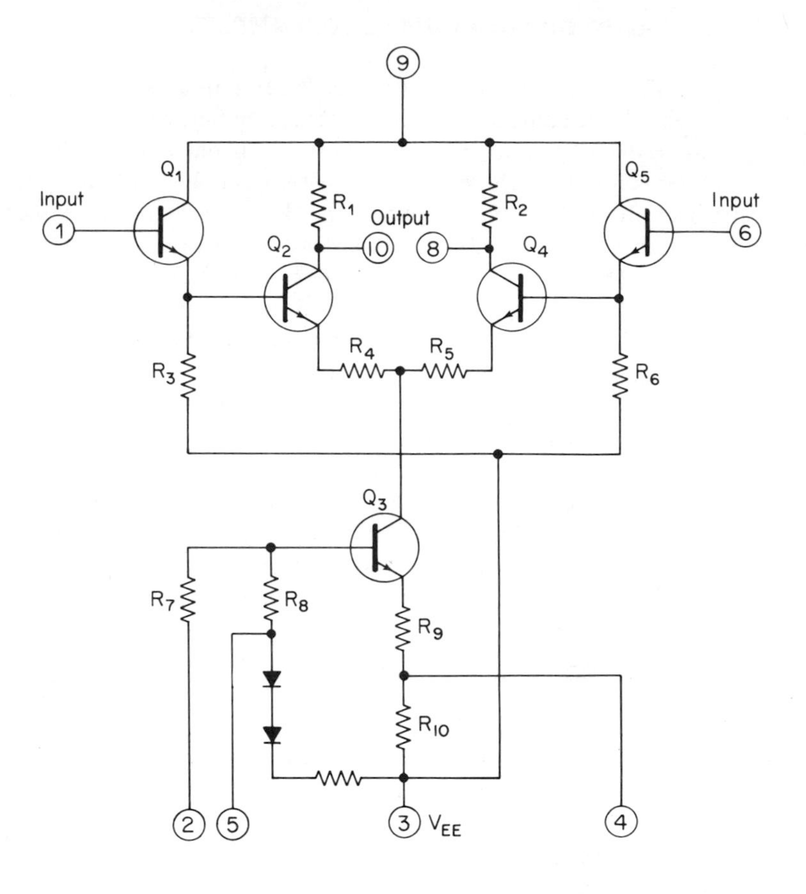

FIG. 5-71. RCA CA3000 amplifier.

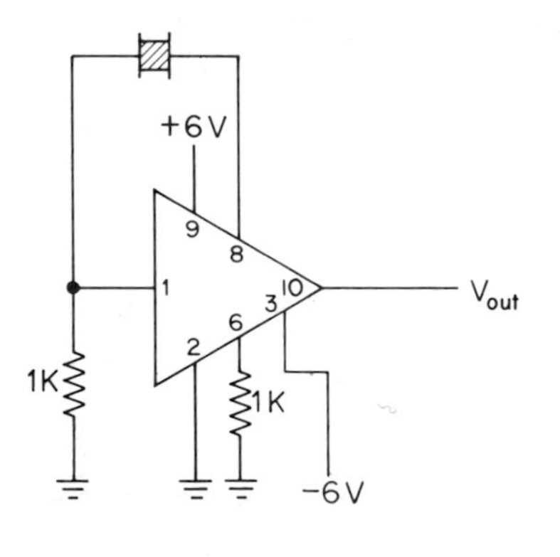

FIG. 5-72. Crystal oscillator with fixed feedback (RCA).

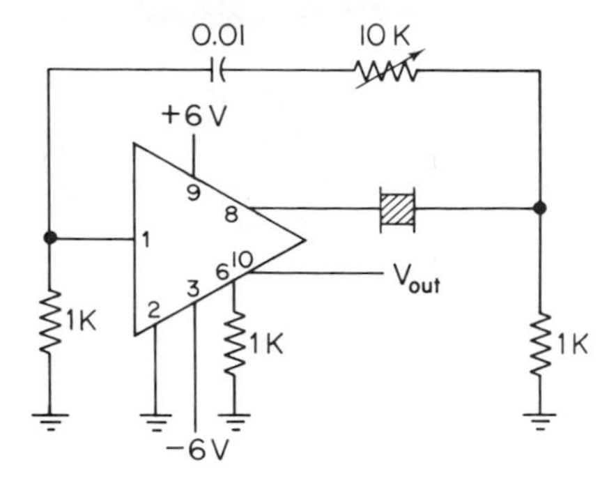

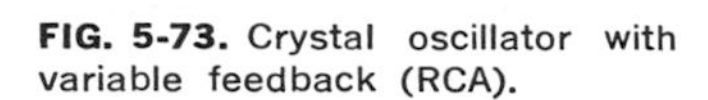
FIG. 5-73. Crystal oscillator with variable feedback (RCA).

feedback ratio network is used, as shown in Fig. 5-73, the feedback may be adjusted to provide a sinewave output. Output waveforms for both circuits are also shown. The frequency in each circuit is 455 kHz, as determined by the crystal. The range of the crystal oscillators can be extended to frequencies of 10 MHz or more by use of collector tuning.

5-6.1.2 Modulated oscillator

If a low-frequency signal is connected to terminal 2, as shown in Fig. 5-74, the IC can function as an oscillator, and produce an amplitude-modulated signal. The waveform in Fig. 5-74 shows the modulated signal output produced by the circuit when a 1-kHz signal is introduced at terminal 2 and a high-pass filter is used at the output.

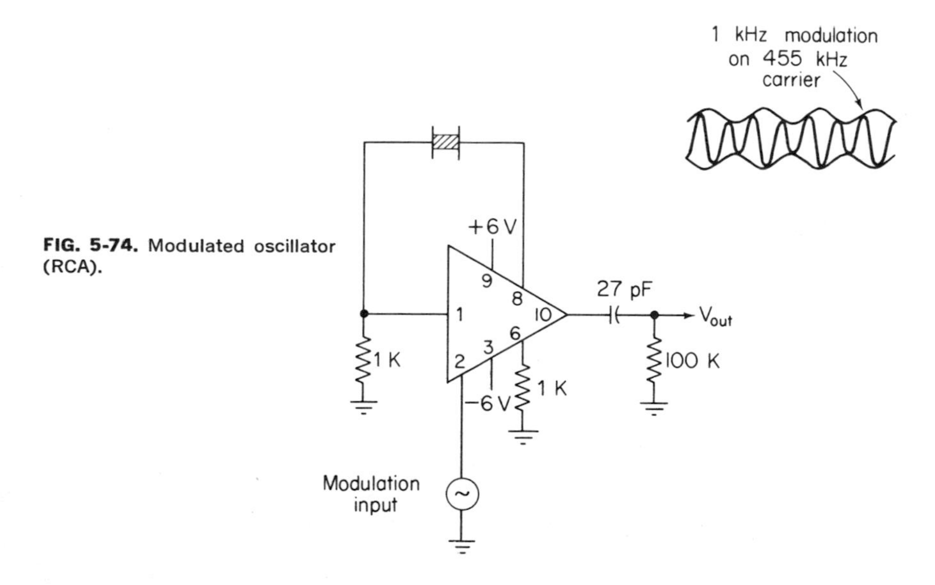

FIG. 5-74. Modulated oscillator (RCA).

5-6.1.3 Low-frequency mixer

In a configuration similar to that used in modulated-oscillator applications (Fig. 5-74), the IC can be used as a mixer by connection of a carrier signal at the base input of either differential-pair transistor (terminal 1 or 6) and connection of a modulating signal to terminal 2 or 5.

5-6.1.4 Cascaded RC-coupled feedback amplifier

The two-stage feedback cascade amplifier shown in Fig. 5-75 produces a typical open-loop mid-band gain of 63 dB. This circuit uses a 100-pF capacitor C_1 to shunt the differential outputs of the first stage. This capacitor staggers the high-frequency roll-offs of the amplifier and thus improves stability (prevents oscillation).

The low-end roll-off of the amplifier is determined by the interstage coupling. Because AGC may be applied to the first stage (terminal 2 at the base of the current-source transistor), the amplifier of Fig. 5-75 can be used in high-gain video-AGC applications under open-loop conditions. If feedback is used to control the gain, AGC may still be applied successfully.

If three or more ICs are cascaded, the low-frequency roll-offs must be staggered (by using different values of interstage coupling capacitors), as well as those at the high end, to prevent oscillation. Three cascaded stages produce about 94-dB gain.

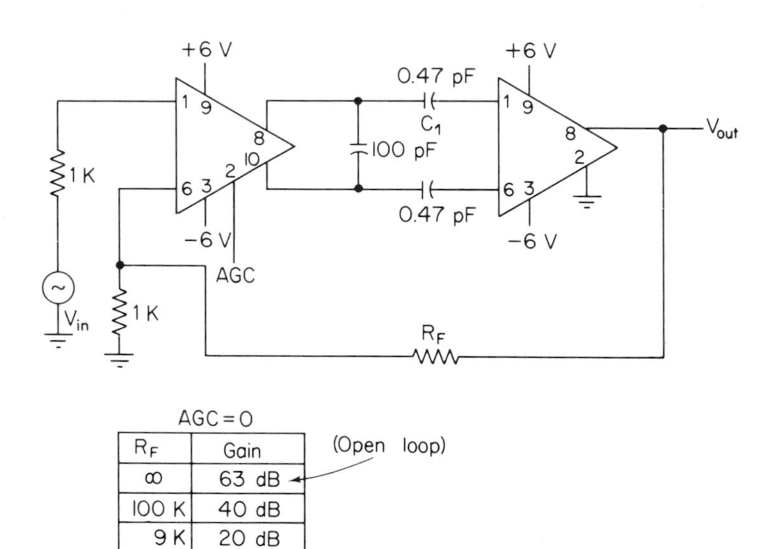

R_F	Gain
∞	63 dB
100 K	40 dB
9 K	20 dB

FIG. 5-75. Cascaded, *RC*-coupled, feedback amplifier (RCA).

5-6.1.5 Narrow-band tuned amplifier

Because of its high input and output impedances, the IC is suitable for use in parallel tuned-input and tuned-output applications. There is a comparative freedom in selection of circuit Q because the differential amplifier exhibits inherently low feedback qualities, provided the following conditions are met: (1) the collector of the driven transistor is returned to ac ground, and the output is taken from the non-driven side, and (2) the input is adequately shielded from the output by a ground plane.

The CA3000 has an output capacitance of approximately 9 pF at a frequency of 10 MHz. This capacitance will resonant at 28-μH coil at this frequency, and give a minimum Q of 4.55 when the collector load resistor is the only significant load. With this low Q, stagger tuning may be unnecessary for many broad-band applications.

Figure 5-76 shows the IC in a narrow-band, tuned-input, tuned-output configuration for operation at 10 MHz with an input Q of 26 and an output Q of 25. A typical response curve is also shown. The 10-MHz voltage gain is about 30 dB, and the total effective Q is 37. The CA3000 can be used in tuned-amplifier applications at frequencies up to the 30-MHz range.

5-7. IC VOLTAGE REGULATORS

Voltage regulators, using an IC op-amp as the active element, are discussed in Sec. 4-26 of Chapter 4. Here, we shall discuss IC packages that contain a complete regulator circuit. By their very nature, IC voltage regulators are "special purpose" devices. That is, their function is to provide a dc voltage that remains constant (within certain limits) in spite of changes in load, input voltage, power supply voltage, etc. It is not likely that the characteristics of one IC voltage regulator could be adapted to another. Thus, from the user's standpoint, there is little value in describing the many possible external connections for the variety of IC regulators available today. (Such data is found on the IC datasheets.) However, all IC voltage regulators have some points in common. These are discussed in the following paragraphs.

5-7.1 The basic regulator

Figure 5-77 shows a combined block diagram and simplified schematic of a basic IC voltage regulator. As shown, the basic regulator is divided into four parts. The control section provides a means of starting and stopping regulator action. Not all IC regulators are provided with a control section. This makes for a less expensive unit, but somewhat limits the regula-

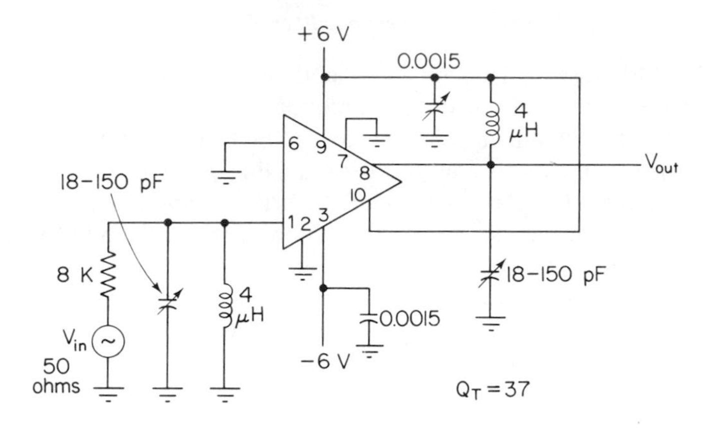

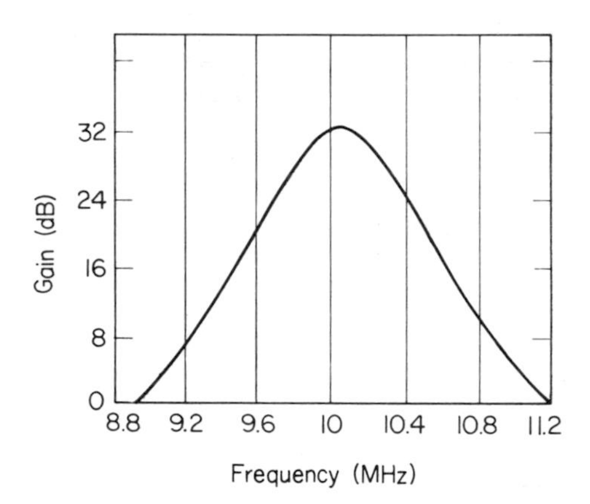

FIG. 5-76. Narrow-band tuned amplifier (RCA).

tor's usefulness in today's sophisticated equipment. The bias section provides a fixed, temperature-compensated reference voltage (V_{ref}). Typically, this reference voltage is in the order of 3–4 V. Since the desired output voltage is rarely some exact value between 3–4 V, it is necessary to "multiply" the reference voltage to some higher exact value. This is done in the dc level shift section. The available output reference voltage $V_{O(ref)}$ is determined by the ratio of resistors R_1 and R_2, which set the feedback (and thus the gain) of the reference amplifier. The $V_{O(ref)}$ can be higher or lower than the fixed V_{ref} supplied by the bias section.

The output section is essentially a unity-gain, differential amplifier, followed by a pair of NPN transistors. The differential amplifier inputs consist of $V_{O(ref)}$ and the output voltage V_O. These voltages are the same when the

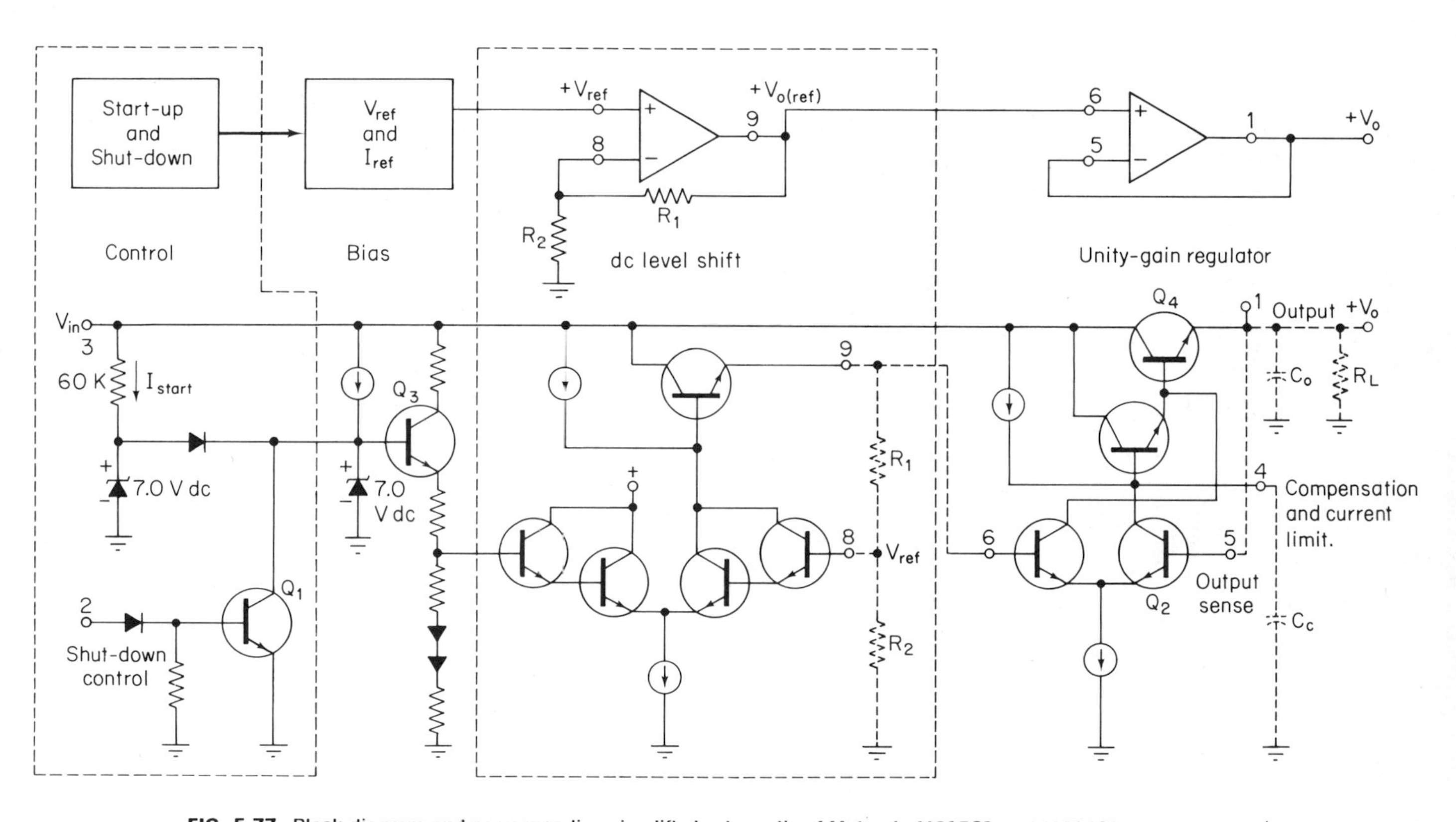

FIG. 5-77. Block diagram and corresponding simplified schematic of Motorola MC1569 and MC1469 voltage regulators.

regulator is functioning normally, and is at the selected voltage. Any change in output voltage unbalances the differential amplifier, and changes the bias on Q_1. The bias change increases or decreases the current through Q_1 as necessary to offset the initial change in output voltage. The maximum output voltage is determined by minimum input-output differential; that is, the smallest amount of voltage across Q_1 which allows normal operation.

Note that the differential amplifier has a frequency compensation capacitor connected at the collector. This external capacitor serves the same purpose as the frequency compensation capacitors in an operational amplifier. (Refer to Chapter 3.) The value of the compensating capacitor is specified on the data-sheet, and is on the order of 0.001 to 0.1 μF. Also note that the basic regulator circuit is provided with an external capacitor at the output. This capacitor is a conventional filter capacitor, and is typically on the order of 10 μF.

5-7.2 Short-circuit and overload protection

Short-circuit current limiting is provided by either an external transistor or a diode string. This is shown in Fig. 5-78. In some IC regulators the diode string is internal. Either way, the diode string (or transistor) is forward-biased when the load current creates a drop across the series resistor equal to the total diode voltage drops (or the base-emitter drop of the transistor). When the series resistor is of the proper value, the drop across the resistor will equal the total diode voltage drops only if the regulator's maximum current limit is reached (or is approached).

From the user's standpoint, external off-the-chip short-circuit diodes (or transistors) offer some advantages. If either the diodes or transistors are on the chip, the threshold voltage is altered by IC temperature rise. That is, if the temperature rises, the diodes or transistors will require a lower voltage

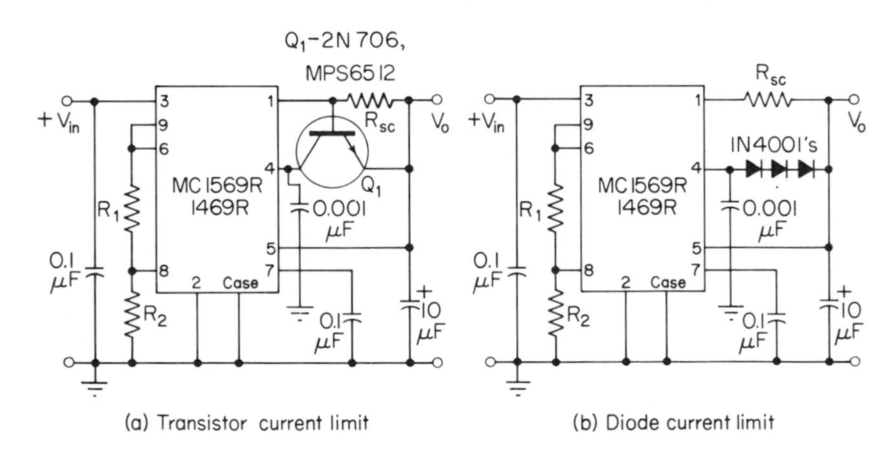

FIG. 5-78. Short-circuit protection methods (Motorola).

drop before they pass current. As more load current is drawn, even normal current, IC temperature can rise quite rapidly (particularly if the case is not on a heat sink) and current limiting may set in long before the design value is reached.

No matter what system is used, when the threshold point is reached, the diodes (or transistor) draw base current away from the NPN transistor pair, and therefore limit maximum output current.

5-7.3 Current boosting

Current boosting is often used for higher load current or better regulation. Even though the basic IC may be capable of delivering the design current, boosting will reduce junction temperature rise, since the current-carrying transistor (which dissipates the most heat) is separate from the IC. Thus, any temperature-related problems such as reference voltage drift are minimized.

A typical NPN boosted current circuit is shown in Fig. 5-79. Here, load current is passed through a series-connected power transistor which, in turn, is controlled by the IC regulator output voltage. This type of circuit can also be used to increase efficiency at low output voltages since an external series pass element requires only a volt or so to remain active.

5-7.4 Current regulator

IC voltage regulators can also be used as current regulators. The basic circuit is shown in Fig. 5-80a. The output current passed through the load R_L also passes through R_1, and produces a corresponding drop

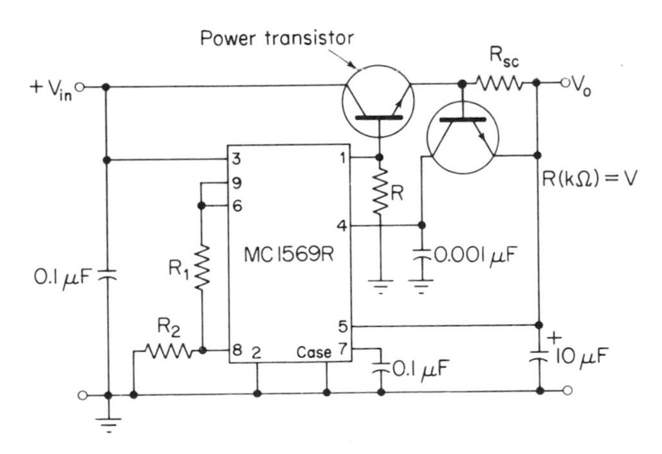

FIG. 5-79. NPN current boosting (Motorola).

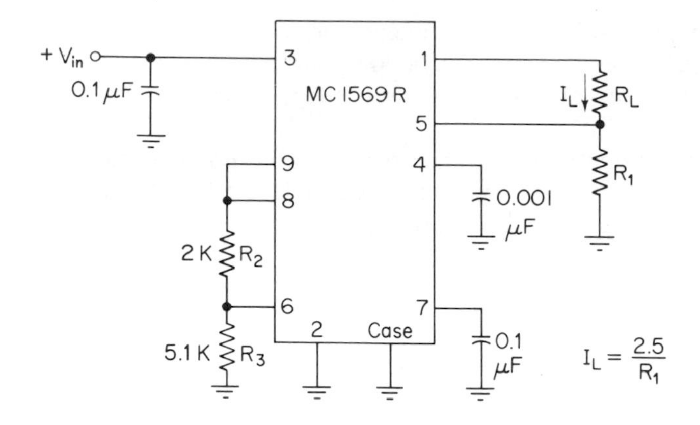

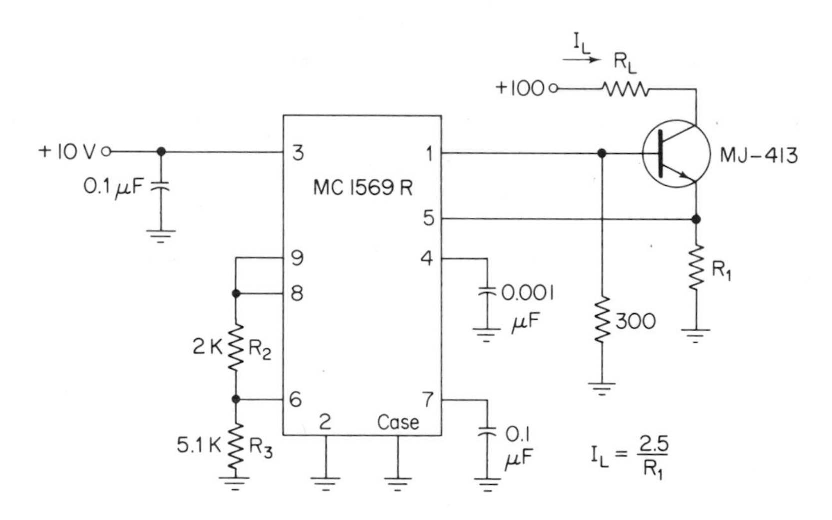

FIG. 5-80. Current regulators (Motorola).

across R_1. The output current level sensed across R_1 is compared with the reference voltage (set by the ratio of R_2/R_3). Any deviations are amplified and corrected to maintain load current constant to within a fraction of one percent (typically 0.05 percent).

Both load current and voltage may be extended by using an external power transistor as shown in Fig. 5-80b.

5-7.5 Shutdown circuits

As discussed in Sec. 5-7.1 and shown in Fig. 5-77, a well-designed IC regulator will have some means for shutdown and start-up. In the circuit of Fig. 5-77, control of the IC is accomplished by transistor Q_2 and the associated circuitry. In normal operation, when the IC regulator is functional, pin 2 (the shutdown control pin) is open, or connected to a voltage close to ground. In this condition, transistor Q_2 is non-conducting and its presence has no effect on the normal operation of the IC.

Transistor Q_1 shunts the internal reference zener Z_2. Should Q_1 be turned on, and saturated, the internal reference Z_2 is shorted to ground. Since all internal current sources are biased to the internal reference, they would not function, and the regulator would stop operation (Q_3 will be shut off). Shut down of the IC is thus accomplished by applying a voltage to pin 2 of sufficient value to turn on and saturate Q_1. The regulator will return to operation once this potential is removed.

There are several methods for applying the control voltage. The following paragraphs describe a few of the most common methods.

5-7.5.1 Logic control

Figure 5-81 illustrates how the state of the IC regulator can be controlled by a logic gate. Here, it is assumed that the regulator is operating in its normal mode—as a positive regulator referenced to ground, and that the logic gate is of the saturating type, operating from a $+V$ supply to ground. The gate can be of the DTL, RTL, or TTL type, where the output stage uses an active pull-up resistor. (Refer to Chapter 6.)

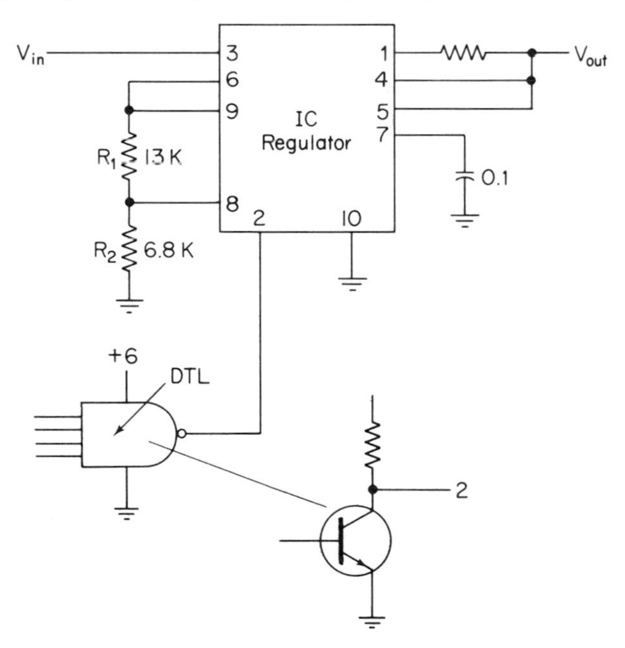

FIG. 5-81. Logic control (shutdown) of IC regulator (Motorola).

One advantage of this technique is for remote systems where power is to be conserved until a particular subsystem is needed to function. At such a time, the regulator is turned on via a computer control, it performs its task, data is collected and stored, and the subsystem is shut down.

5-7.5.2 Junction temperature control

An IC regulator has one problem not found in the discrete component regulators. Most of an IC regulator chip is occupied by the series pass transistor. The temperature of the entire IC is thus set by the series pass transistor. Variations in load cause the transistor temperature to change, thus changing the temperature of remaining components on the IC. This "thermal feedback" or "temperature feedback" can cause "thermal runaway", as described in Sec. 2-5.4. It is possible to use the thermal feedback to control the IC, and to initiate shutdown if an unsafe temperature is reached. Several methods are illustrated in Fig. 5-82.

Figure 5-82a shows a method of shutdown which is controlled by junction temperature (T_J). For practical purposes, the temperature of the D_2 and Q_2

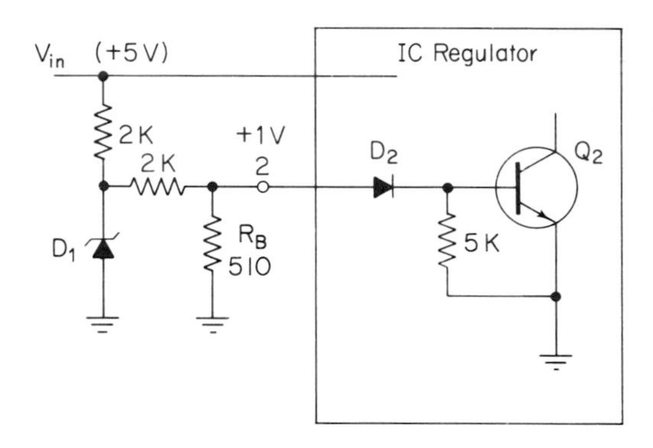

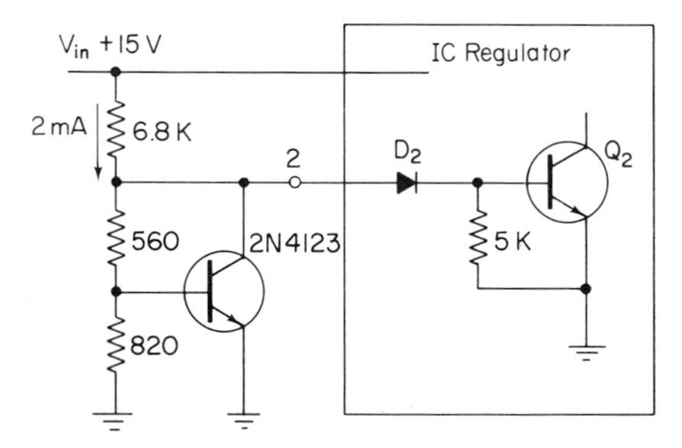

FIG. 5-82. Junction temperature control (shutdown) of IC regulator (Motorola).

junctions can be considered the same as the output series transistor. As temperature increases, the voltage required for turn-on of D_2 and Q_2 (regulator turn-off) drops. Thus, if a fixed voltage is applied, D_2 and Q_2 will turn on when a predetermined temperature is reached. The Zener diode provides the fixed dc voltage which can be varied by varying the value of R_B.

Figure 5-82b shows a shutdown method which is controlled by ambient temperature (T_A). Here, the fixed reference control voltage is set by the drop across the junction of an external transistor. Since this drop is controlled by ambient temperature, the control voltage (and turn-on point) is set by the ambient temperature.

In the circuit of Fig. 5-82a, the Zener diode is usually mounted away from the IC so that heat from the IC will not affect the Zener. On the other hand, the transistor of Fig. 5-82b is usually mounted near the IC, so that a large increase in IC heat will change the ambient temperature around the transistor.

These temperature control systems are often used in conjunction with the short circuit and overload circuits described in Sec. 5-7.2. When a short circuit or overload occurs, increased current is still drawn through IC components, and the IC can heat, possibly to a dangerous level. By using both overload and temperature control, the IC will turn off when the temperature reaches a predetermined point during overload. When the IC is cool enough, the regulator returns to the "on" condition. If the short or overload still exists, the heating process begins again, and the IC will continue in a cycling mode as long as the output is shorted. This technique provides control of both the maximum short-circuit current, and the maximum junction temperature. This complete protection is often considered superior to the "current foldback" system described in Sec. 5-7.6.

5-7.5.3 Current boost temperature control

The thermal shutdown techniques can be used when the IC is operating in a current boost circuit as described in Sec. 5-7.3. The external series pass transistor must be mounted on a common heat sink with a monitor transistor, as shown in Fig. 5-83. In this configuration, Q_2 heats when the output is shorted or there is an overload, while Q_1 monitors the heat sink temperature. Q_1 conducts for a predetermined temperature increase above the normal ambient, as set by the values of R_2 and R_3. Resistor R_4 is used to limit the drive current into pin 2.

5-7.5.4 Output short circuit shutdown

Not all shutdown systems are based on temperature. A standard circuit that does not use T_J or T_A control for shutdown is shown in Fig. 5-84. In this circuit, Q_1 is normally saturated because of the base drive

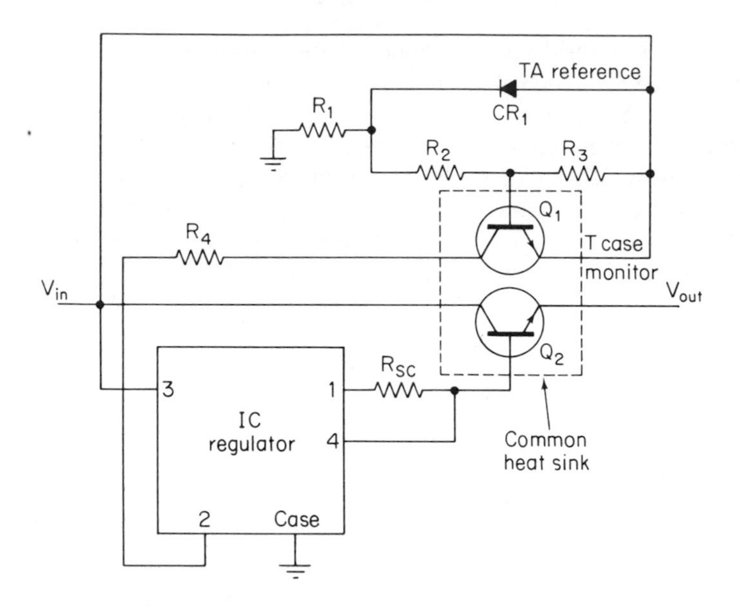

FIG. 5-83. Thermal shutdown using external phase transistor (Motorola).

supplied through the resistor divider R_1 and R_2. With Q_1 in saturation, the voltage on pin 2 is below the threshold of the IC control circuit. If the output of the IC is short circuited, Q_1 turns off, the voltage at pin 2 rises, and shutdown of the regulator occurs. R_1 and R_2 can be chosen to saturate Q_1 as well

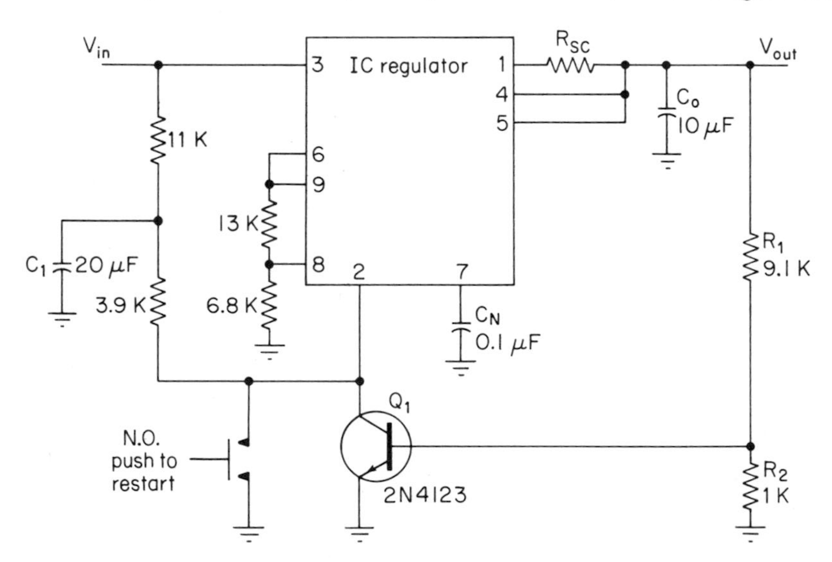

FIG. 5-84. Output short circuit shutdown (Motorola).

as to sink the minimum regulator load current (typically about 1 mA). Capacitor C_1 is necessary to provide an RC time constant which prevents shutdown when V_{in} is initially applied (before Q_1 can saturate). When the short circuit is removed, the regulator must be manually reset before it will return to full regulation.

5-7.6 Current foldback technique

It is not absolutely necessary for an IC regulator to have a shutdown or control section. There are circuits which will, in effect, shut the regulator off should an overload or short circuit occur. These circuits reduce the regulator output voltage and current to zero (or near zero) when the load current reaches a certain limit. Such a circuit, and corresponding current-voltage graph, are shown in Fig. 5-85. Note that when output current reaches a certain limit, both the voltage and current drop back (or "foldback") to some low (safe) level.

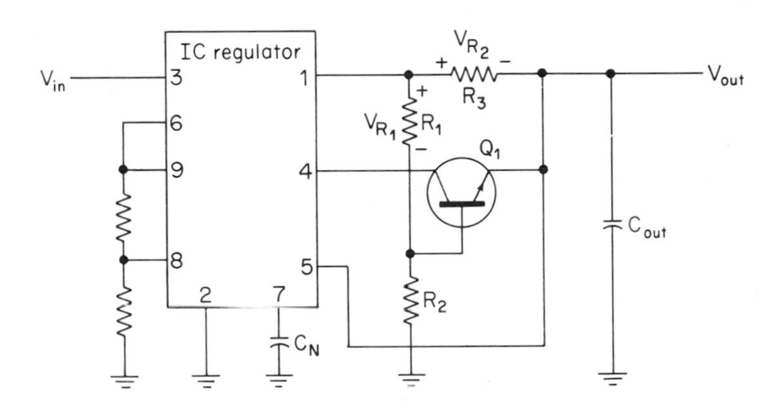

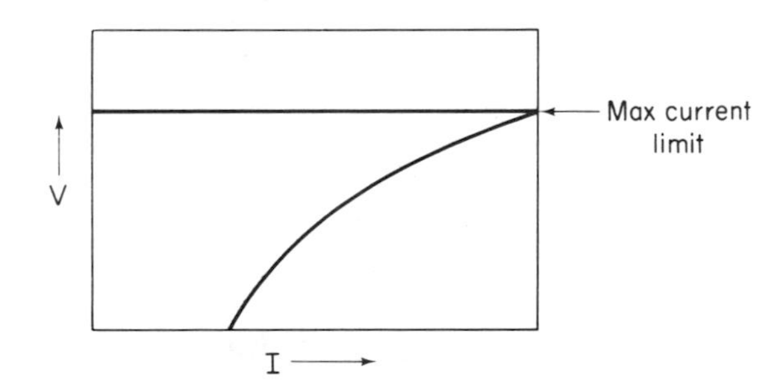

FIG. 5-85. Short circuit current foldback (Motorola).

In the circuit of Fig. 5-85, the voltage drop across R_1 is set to about twice the base-emitter drop of Q_1 (or about 1.4 V) when the output voltage is normal. For Q_1 to conduct, the voltage across R_3 (which senses the output load) must be about equal to the drop across R_1, plus the base-emitter voltage. By proper selection of resistance values, this will occur only if the current limit has been reached. When Q_1 conducts, the collector connected to pin 4 diverts current drive from the series pass elements (Fig. 5-77), producing current foldback.

5-7.7 Specifying an IC voltage regulator

Unfortunately, the specifications of IC regulators from all manufacturers are not consistent. The following is a list of the terms used in Motorola datasheets, with a brief explanation for each. If the user fully understands these terms, he should have no difficulty in specifying an IC regulator, or understanding the datasheets of other manufacturers.

Output voltage range. This is the range of output voltages over which the specifications apply. A particular output voltage is established by a user-selected external resistive divider network (R_1 and R_2 of Fig. 5-77).

Output current. All IC regulators are capable of supplying a certain amount of current to a load without the use of external transistors. The magnitude of current is specified on the datasheet as a maximum and a minimum for each device. For example, a particular IC regulator might be capable of supplying 500 mA for any output voltage from 2.5 to 37 V. However, the same regulator must supply at least 1 mA to maintain its regulator characteristics. For this reason, the IC datasheets often specify that a resistance be placed across the output to draw a minimum of 1 mA at the selected voltage.

Input-Output voltage differential. The input-output voltage differential ($V_{in} - V_O$) is the voltage required to bias the circuitry that supplies the drive current to the series pass transistor. As the input voltage drops in magnitude and approaches the output voltage dc level, the regulator drops out of regulation at the minimum value of this specification.

Input voltage range. The input voltage range is specified on the datasheet as a minimum and a maximum. This indicates that at least the minimum voltage must be present to properly bias the Zener diode (or other reference) on the IC. This input voltage must also exceed the output voltage by at least the specified input-output voltage differential. For example, if the $V_{in} - V_O$ is 2 V minimum and 3 V maximum, and the desired output is 10 V, the input voltage must be at least 12 V, and possibly 13 V.

Load regulation and output impedance. Load regulation is the percentage change in output voltage for a dc steady-state change in load current from the minimum to maximum value specified. This is expressed as:

$$\text{load regulation} = \frac{V_{NL} - V_{FL}}{V_{NL}} \times 100 = (\%V_O)$$

where V_{NL} is the output voltage with minimum load current,
and V_{FL} is the output voltage when the full load current is being drawn, and the units are a percent of V_O ($\%V_O$).

Output impedance Z_O is a small-signal ac parameter. It indicates the ability of the regulator to prevent common power supply voltage changes created by fluctuating currents drawn at signal frequencies by circuit load. High values of Z_O can create undesirable coupling between circuits powered by a common voltage regulator, and thereby cause system oscillations. Typical output impedances for an IC regulator are less than 1 Ω, and often as low as 0.025 Ω.

Input regulation. Input regulation is the percentage change in output voltage per volt change in the input voltage and is expressed as:

$$\text{input regulation} = \frac{\Delta V_O}{V_O(\Delta V_{\text{in}})} \times 100\ (\%/V_{\text{in}})$$

where ΔV_O is the change in the output voltage V_O for the input change ΔV_{in}, and $\%/V_{\text{in}}$ are the units.

Completely packaged, line-operated voltage regulators usually use the term "line regulation" to show the dependence of the output voltage on the ac power line variations. As an IC does not operate directly from the line, the term "input regulation" has been used to be more accurate.

Temperature coefficient. This is the stability of the output voltage over a change in operating temperature.

5-8. IC BALANCED MODULATORS

The main application for an IC balanced modulator is a "building block" for high frequency communications equipment. The unit functions as a broad band, double-sideband, suppressed-carrier balanced modulator without transformers or tuned circuits. The IC can also be used as an SSB product detector, AM modulator/detector, FM detector, mixer, frequency doubler, and phase detector.

5-8.1 Basic modulator

Figure 5-86 is the schematic of the basic IC balanced modulator. The circuit consists of differential amplifier Q_5–Q_6 driving a dual differential

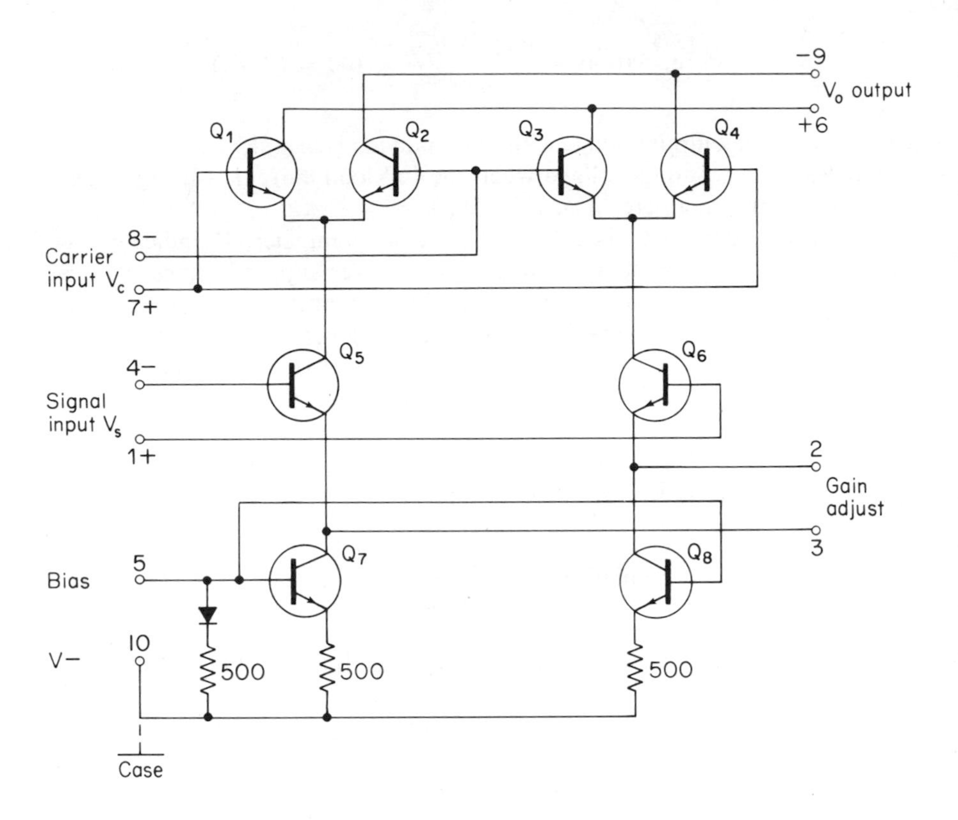

FIG. 5-86. Motorola MC1596 balanced modulator.

amplifier composed of transistors Q_1 through Q_4. Transistors Q_7 and Q_8 form constant current sources for the lower differential amplifiers Q_5–Q_6.

In operation, a high level input signal is applied to the carrier input, and a low level input is applied to the signal input. This results in saturated switching operation of carrier dual differential amplifiers (Q_1–Q_4), and linear operation of the modulating differential amplifier (Q_5–Q_6).

The resulting output signal contains only the sum and difference frequency components and amplitude information of the modulating signal. This is the desired condition for most balanced modulator applications.

Saturated operation of the carrier-input dual differential amplifiers also generates harmonics. Reducing the carrier input amplitude to its linear range greatly reduces these harmonics in the output signal. However, it has the disadvantages of reducing gain, thus causing the output signal to contain carrier signal amplitude variations.

The carrier input differential amplifiers have no emitter feedback. Therefore, the carrier input levels for linear and saturated operation are readily calculated. The crossover point is typically in the range of 15–20 mV, with linear operation below this level, and saturated operation above it.

The modulating-signal differential amplifier has its emitters brought out to pins 2 and 3. This permits the designer to select his own value of emitter feedback resistance, and thus tailor the linear dynamic range of the modulating signal input to a particular requirement. The resistor also determines device gain.

5-8.2 Balanced modulator

Figure 5-87 shows a typical balanced modulator circuit. Typical input signal levels are 60 mV for the carrier, and 300 mV for the modulation input.

The modulating signal must be kept at a level to insure linear operation of Q_5–Q_6. If the signal level is too high, harmonics of the modulating signal are generated and appear in the output as spurious sidebands of the suppressed carrier.

Operating with a high level carrier input has the advantages of maximum gain and insuring that any amplitude variations present on the carrier do not

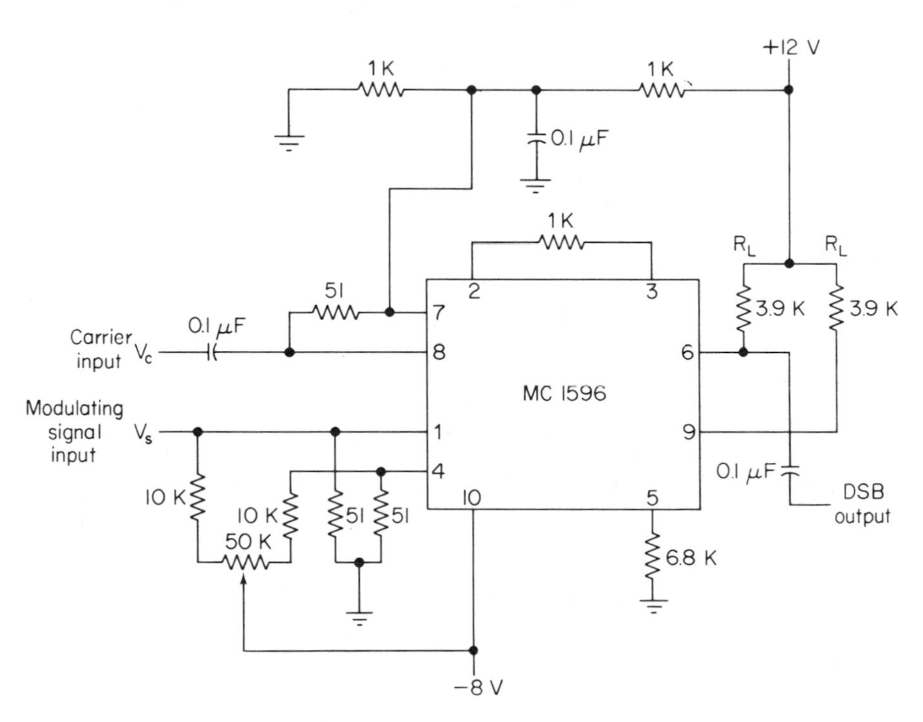

FIG. 5-87. Typical balanced modulator circuit for double sideband output (Motorola).

appear on the output sidebands. It has the disadvantage of increasing some of the spurious signals.

The decision to operate with a low or high level carrier input will, of course, depend on the application. For a typical filter-type SSB generator, the filter will remove all spurious outputs except some spurious sidebands of the carrier. For this reason, operation with a high level carrier will probably be selected for maximum gain and to insure that the desired sideband does not contain any spurious amplitude variations present on the carrier input signal.

On the other hand, in a low frequency broadband balanced modulator, spurious outputs at any frequency may be undesirable, and low level carrier operation may be the best choice.

5-8.3 Amplitude modulator

Figure 5-88 shows the balanced modulator used as an amplitude modulator. Modulation for any percentage from zero to over 100 is possible. The circuit operates by unbalancing the carrier null to insert the proper amount of carrier into the output signal. Note that the circuit for amplitude modulation is essentially the same as for balanced modulation,

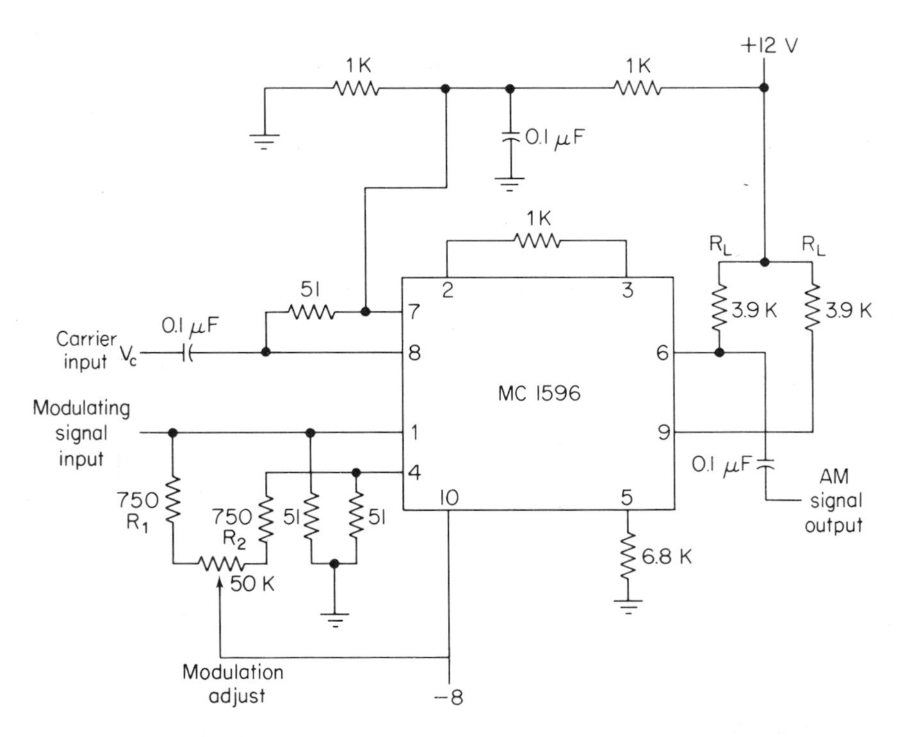

FIG. 5-88. Amplitude modulator using balanced modulator (Motorola).

except for the values of resistors R_1 and R_2. These resistors are of lower value for the amplitude modulator, thus permitting a wider range for the modulation adjust potentiometer. This increase in range permits the circuit to be unbalanced to the point where some carrier appears at the output. In use, the potentiometer is adjusted until the desired percentage of modulation is measured at the output.

5-8.4 Product detector

Figure 5-89 shows the IC in an SSB product detector configuration. For this application, all frequencies except the desired demodulated audio are in the RF spectrum, and can be easily filtered at the output. As a result, the usual carrier null adjustment need not be included.

Upper differential amplifiers Q_1–Q_4 are again driven with a high level signal. Since carrier output level is not important in this application (carrier is filtered at the output) carrier input level is not critical. A high level carrier input is desirable for maximum gain, and to remove any carrier amplitude variations from the output. Typical carrier inputs are from 100 to 500 mV.

The modulated signal (single-sideband, suppressed-carrier) input level to differential amplifier pair Q_5–Q_6 is maintained within the limits of linear operation. Typically, the modulated input is less than 100 mV. Again, no transformers or tuned circuits are required for excellent product detector performance from very low frequencies up to 100 MHz.

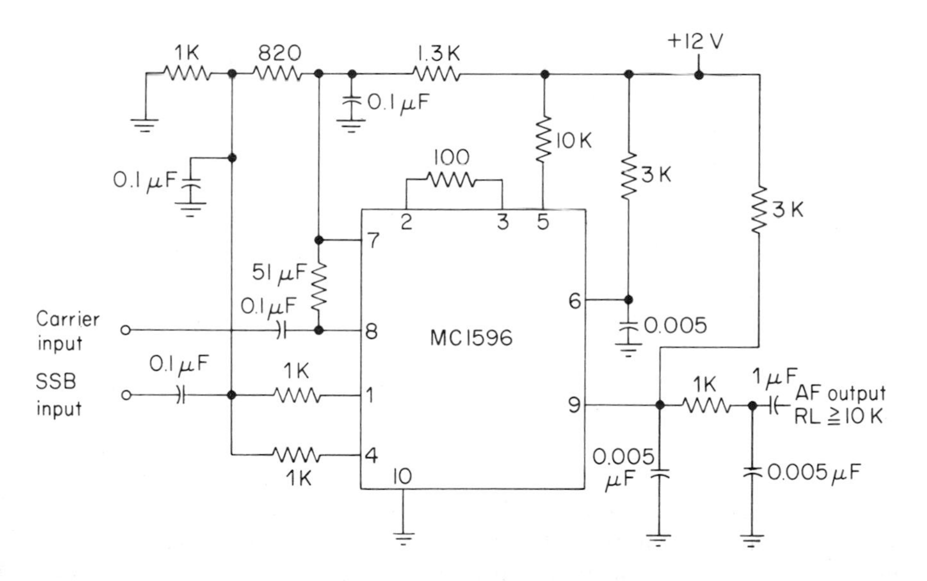

FIG. 5-89. Product detector using balanced modulator (Motorola).

Note that dual outputs are available from the product detector, one from pin 6 and another from pin 9. One output can drive the receiver audio amplifiers while a separate output is available for the AGC system.

5-8.5 AM detector

The product detector circuit of Fig. 5-89 can also be used as an AM detector. The modulated signal is applied to the upper differential amplifier, while the carrier signal is applied to the lower differential amplifier.

Ideally, a constant-amplitude carrier signal would be obtained by passing the modulated signal through a limiter ahead of the carrier input terminals. However, if the upper input signal is at a high enough level (typically greater than 50 mV), its amplitude variations do not appear in the output signal.

For this reason, it is possible to use the product detector circuit shown in Fig. 5-89 as an AM detector simply by applying the modulated signal *to both inputs* at a level of about 600 mV on modulation peaks, without using a limiter ahead of the carrier input. A small amount of distortion will be generated as the signal falls below 50 mV during modulation valleys, but it will probably not be significant in most applications. Advantages of the IC as an AM detector include linear operation and the ability to have a detector stage with gain.

5-8.6 Mixer

Since the IC generates an output signal consisting of the sum and difference frequencies of the two input signals only, the IC can be used as a double-balanced mixer. Figure 5-90 shows the IC used as a high frequency mixer with a broadband input, and a tuned output at 9 MHz. The 3-dB bandwidth of the 9-MHz output tank is 450 kHz.

The local oscillator signal is injected at the upper input with a level of 100 mV. The modulated signal is injected at the lower input with a maximum level of about 15 mV. Since the input is broadband, the mixer can be operated at any HF and VHF input frequencies. The same circuit can be used with a 200-MHz input signal, and a 209-MHz local oscillator signal, as an example. At the higher frequency, the circuit shows an approximate 9-dB conversion gain and a 14-μV sensitivity.

Greater conversion gains can be obtained by using tuned circuit with impedance matching on the signal input. Of course, the bandwidth is narrower when tuned circuits are used. The nulling circuit permits the local oscillator signal to be nulled from the output, and it can be eliminated when operating with a tuned output in many applications. Likewise, the tuned output tank can be replaced with a resistive load to form a broadband input and output

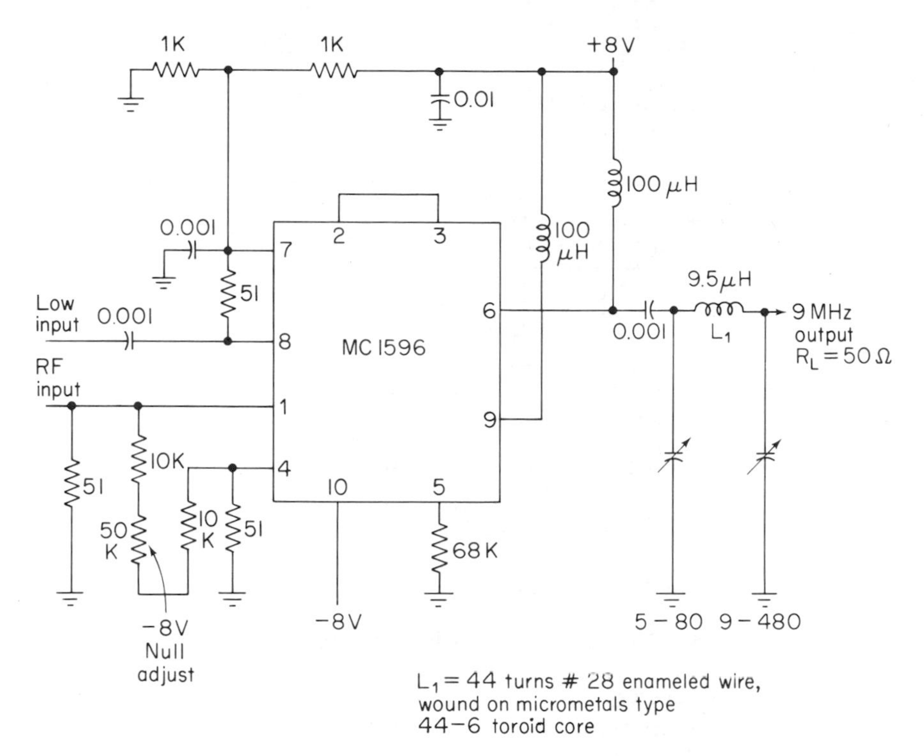

FIG. 5-90. Mixer using balanced modulator (Motorola).

doubly-balanced mixer. The magnitude of the output load resistance becomes a simple matter of tradeoff between conversion gain and output signal bandwidth.

5-8.7 Doubler

The IC balanced modulator can function as a frequency doubler when the same signal is injected in both inputs.

For operation as a broadband low frequency doubler, the balanced modulator circuit of Fig. 5-87 need be modified only by adding ac coupling between the two inputs, and reducing the lower differential amplifier emitter resistance between pins 2 and 3 to zero (tieing pin 2 to pin 3). This latter modification increases the circuit sensitivity and doubler gain.

A low frequency doubler with this modifications is shown in Fig. 5-91. This circuit will double in the range below 1 MHz. For best results, both upper and lower differential amplifiers should be operated within their linear ranges. Typically, this limits input signals to about 15 mV.

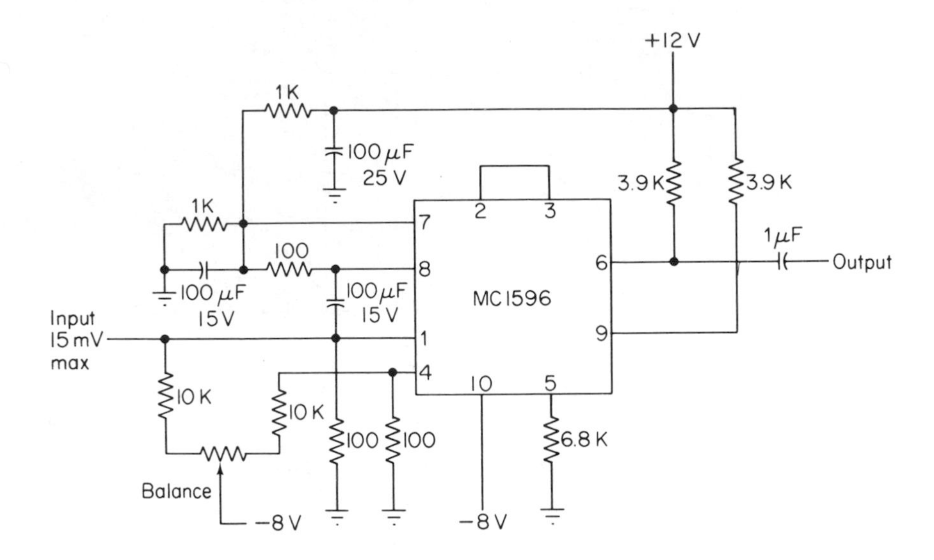

FIG. 5-91. Doubler using balanced modulator (Motorola).

5-8.8 FM detector and phase detector

A balanced modulator produces a dc output which is a function of the phase difference between two input signals of the same frequency. Thus, when one signal is fixed frequency, and the other signal is frequency-modulated, the dc output corresponds to the phase difference between the two signals, or to the modulation. The balanced modulator can therefore be used as a phase detector (or FM detector).

5-9. IC DIODE ARRAYS

An IC diode array provides several diodes on a single semiconductor chip. Because all of the diodes are fabricated simultaneously on the same chip, they have (nearly) identical characteristics. Particularly important in many applications, their parameters track each other with temperature variations as a result of their close proximity, and the good thermal conductivity of silicon. Consequently, such diode arrays are particularly useful in circuits which require either a balanced diode bridge, or identical diodes.

5-9.1 RCA CA3019 diode array

Figure 5-92 shows both the schematic and chip fabrication diagram of the unit. The IC provides four diodes internally connected in a

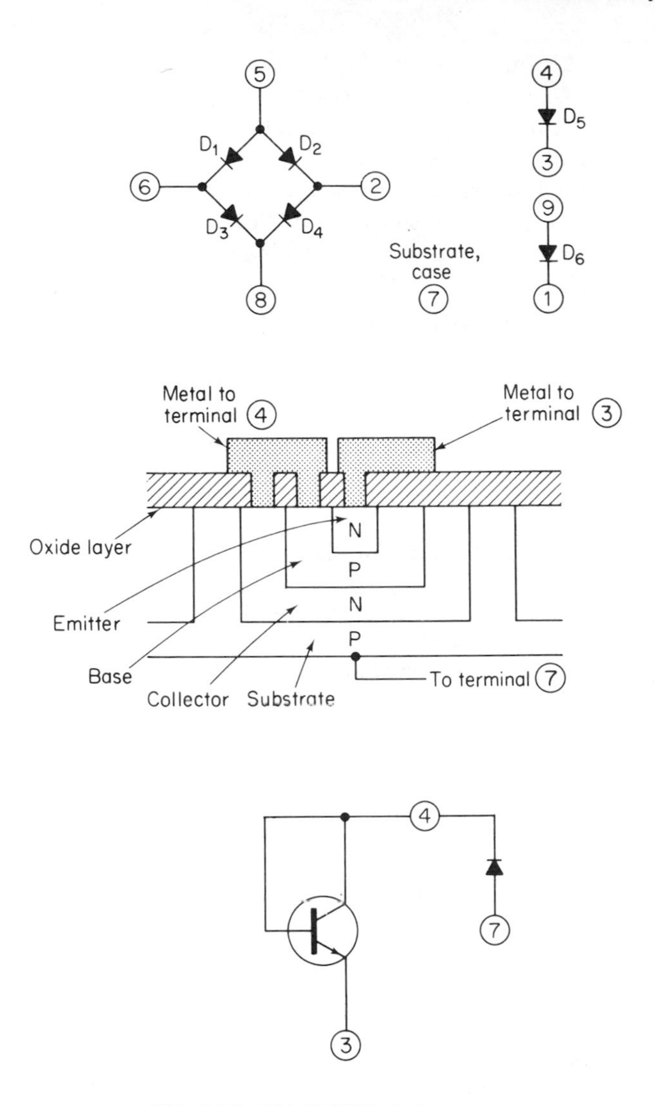

FIG. 5-92. RCA CA3019 diode array.

diode-quad arrangement, plus two individual diodes. Its applications include gating, mixing, modulating, and detecting circuits.

Each diode is formed from a transistor by connection of the collector and base to form the diode anode, and use of the emitter for the diode cathode. This diode configuration, in which the collector-base junction is shorted, is the most useful connection for a high-speed diode because it has the lowest storage time. The only charge stored is that in the base. The configuration also exhibits the lowest forward voltage drop, and is the only one which has no

PNP transistor action in the substrate. The diode has the emitter-to-base reverse breakdown voltage characteristic (typically 6 V).

Connected at each diode anode is the cathode of a substrate diode for which the anode is the substrate (terminal 7). In many applications, the substrate can be left floating because of a forward bias on any substrate diode creates a self reverse-bias on the other substrate diodes. However, the uncertainty of this bias and the capacitive feedthrough paths provided by the substrate must be considered.

The capacitive feedthrough paths can be eliminated if the substrate is either ac ground or returned to a dc voltage (or ground when possible). This dc voltage must be more negative than any anticipated ac or dc operating voltage present on any diode anode to assure that every substrate diode will always be reverse-biased. When a choice exists, the substrate diodes should be reverse-biased.

5-9.2 Application for diode arrays

There are many applications for diode arrays. Obviously, the IC can be used in any circuit where from one to six diodes are required, provided that the voltage and current are within the ICs capabilities. However, the IC is most useful when the circuit requires diodes of matched characteristics. The *high speed gate* of Fig. 5-93 is a classic example. Here, all six diodes are used, and the most is made of their matching characteristics.

In high-speed gates, the gating signal often appears at the output and causes the output signal to ride on a "pedestal". A diode-quad bridge circuit can be used to balance out the undesired gating signal at the output and reduce the pedestal to the extent that the bridge is balanced.

The diode-quad gate functions as a variable impedance between a source and a load, and can be connected either in series or in shunt with the load. The circuit configuration used depends on the input and output impedances of the circuits to be gated. A series gate is used if the source and load impedances are low compared to the diode back resistance, and a shunt gate is used if the source and load impedances are high compared to the diode forward resistance.

The circuit of Fig. 5-93 uses the six diodes as a series gate in which the diode bridge, in series with the load resistance, balances out the gating signal to provide a pedestal-free output. With a proper gating voltage (1 to 3 V, 1 to 500 kHz) diodes D_5 and D_6 conduct during one half of each gating cycle, and do not conduct during the other half of the cycle. When diodes D_5 and D_6 are conducting, the diode bridge (D_1–D_4) is not conducting, and the high diode back resistance prevents the input signal V_S from appearing across the load resistance R_L. When diodes D_5 and D_6 are not conducting, the diode bridge conducts and the low diode forward resistance allows the input signal to

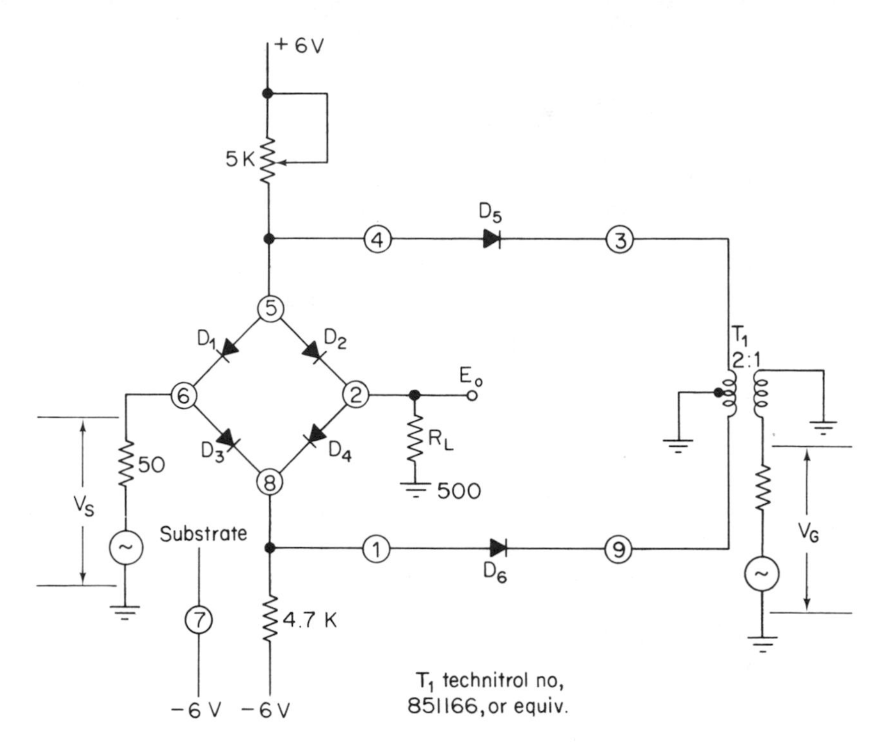

FIG. 5-93. Series gate using matched diode array (RCA).

appear across the load resistance. Resistor R_1 may be adjusted to minimize the gating voltage present at the output. The substrate (terminal 7) is connected to the −6-V supply.

5-10. IC TRANSISTOR ARRAYS

An IC transistor array provides several transistors on a single semiconductor chip. As in the case of diode arrays, transistor arrays are particularly suitable for applications in which closely matched device characteristics are required, or in which a number of active devices must be interconnected with external parts not possible to fabricate in IC form (tuned circuit, large-value resistors, variable resistors, large-value capacitors, etc.).

5-10.1 RCA CA3018 transistor array

Figure 5-94 shows the schematic of the array. The IC provides four silicon transistors on a single chip. The four active devices include two

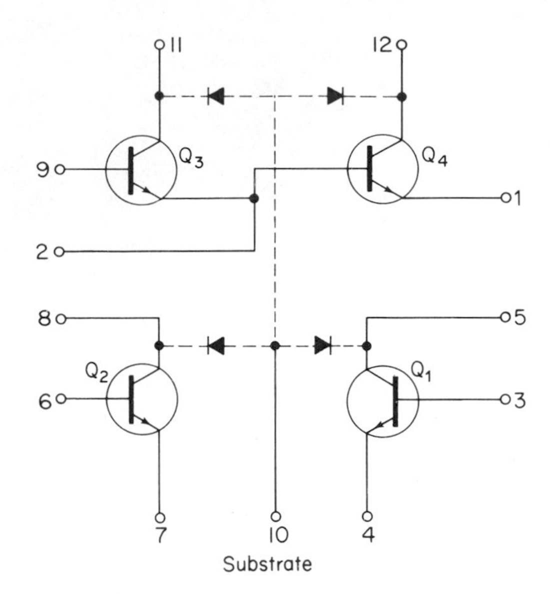

FIG. 5-94. RCA CA3018 transistor array.

isolated transistors plus two transistors with an emitter-base common connection. Because it is necessary to provide a terminal for connection to the substrate, two transistor terminals must be connected to a common lead. The particular configuration is useful in emitter-follower and Darlington circuit connections. Also, the four transistors can be used almost independently if terminal 2 is grounded or ac grounded so that Q_3 can be used as a common-emitter amplifier and Q_4 as a common-base amplifier.

In pulse video amplifiers and line-drivers, Q_4 can be used as a forward-biased diode in series with the emitter of Q_3. Likewise, transistor Q_3 can be used as a diode connected to the base of Q_4 or, in a reverse-biased connection, Q_3 can serve as a protective diode in RF circuits connected to operational antennas. The presence of Q_3 does not inhibit the use of Q_4 in a large number of circuits.

In transistors Q_1, Q_2 and Q_4, the emitter lead is interposed between the base and collector leads to minimize package and lead capacitances. In Q_3, the substrate lead serves as the shield between base and collector. This lead arrangement reduces feedback capacitance in common-emitter amplifiers, and thus extends video bandwidth and increases tuned-circuit amplifier gain stability.

5-10.2 Application for transistor arrays

The applications for transistors are many and varied. As in the case of diode arrays, the transistor array can be used in any application where four transistors are required. The following application has been selected to demonstrate the advantage of four matched devices available on a single chip.

Figure 5-95 shows a broadband video amplifier design using the four-transistor array. This amplifier can be considered as two dc coupled states, each consisting of a common-emitter, common-collector configuration. The common-collector transistor provides a low-impedance source to the input of the common-emitter transistor, and a high-impedance, low capacitance load at the common-emitter output.

Two feedback loops provide dc stability of the broadband video amplifier and exchange gain for bandwidth. The feedback loop from the emitter of Q_3

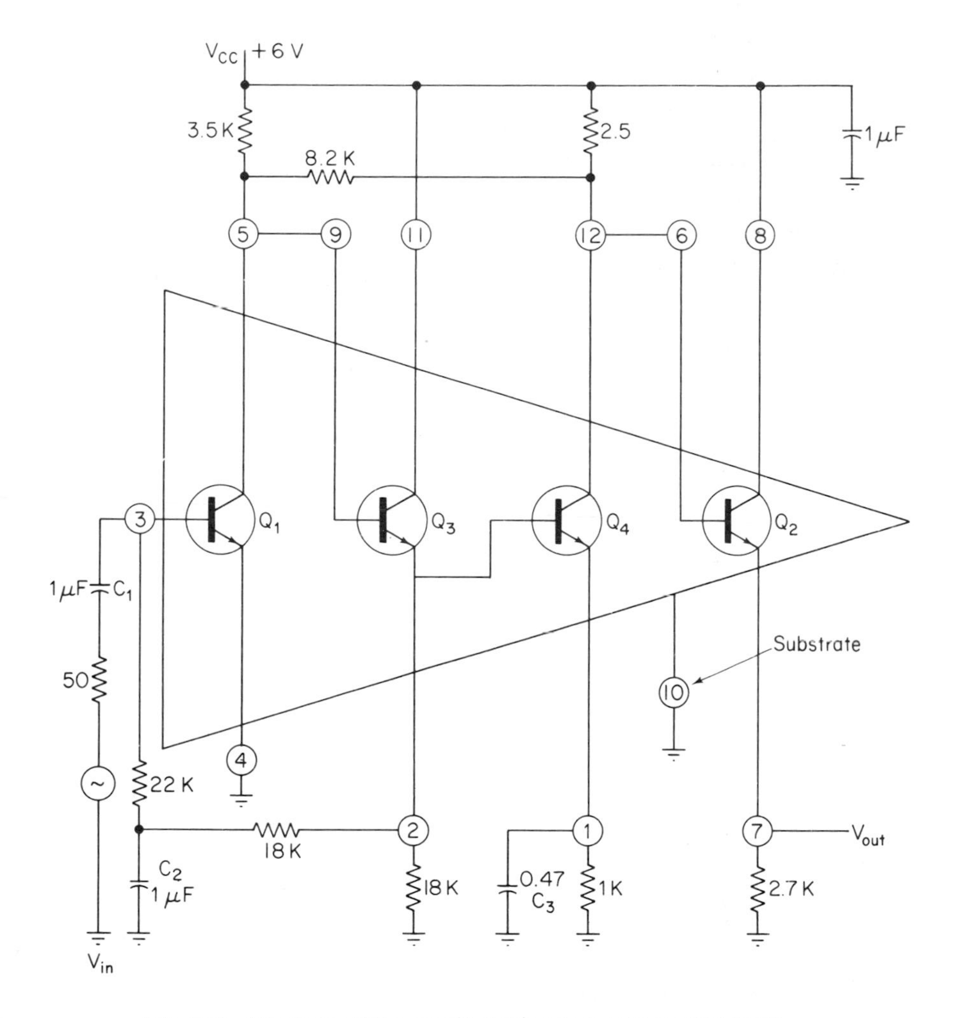

FIG. 5-95. Broadband video amplifier using IC transistor array (RCA).

to the base of Q_1 provides dc and low-frequency feedback. The loop from the collector of Q_4 to the collector of Q_1 provides both dc feedback and ac feedback at all frequencies.

5-11. IC DARLINGTON ARRAYS

As in the case of diode and transistor arrays, IC Darlington arrays provide several Darlington pairs on a single chip. Figure 5-96 shows both the schematic and typical application for an RCA CA3036 dual Darlington array. The IC can be used to provide two independent low-noise wideband amplifier channels, and is particularly useful for preamplifier and low-level amplifier applications in single-channel or stereo systems.

As shown in Fig. 5-96, the array consists of four transistors connected to form two independent Darlington pairs. The block diagram illustrates the use of the array in a typical stereo phonograph. The IC can be mounted directly on a stereo cartridge. Because of the low noise, high input impedance, and low output impedance of the array, only minimum shielding is required from the pickup to the amplifier. The buffering action of the IC also substantially reduces losses and decreases hum pickup.

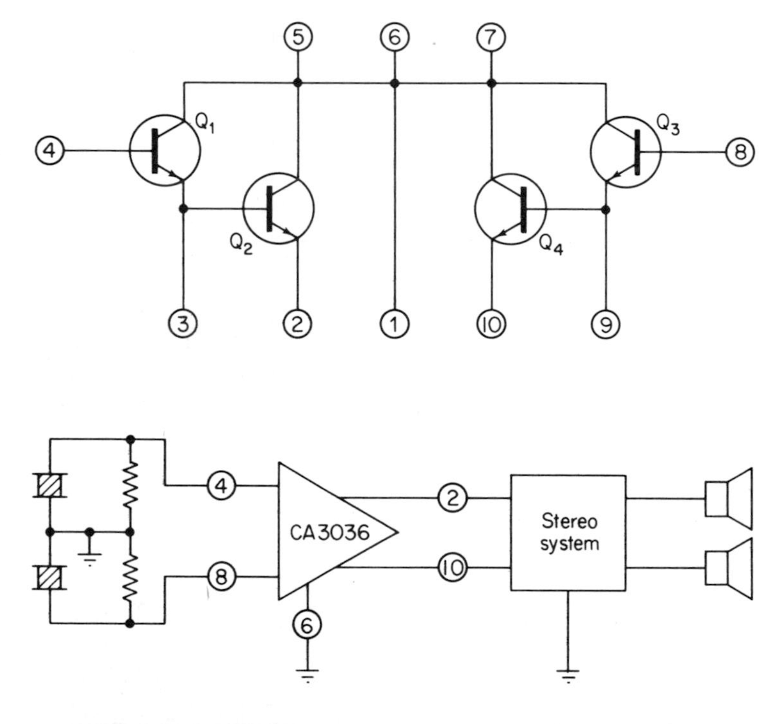

FIG. 5-96. RCA CA3036 Darlington array used as stereo preamp.

6. DIGITAL INTEGRATED CIRCUITS

Unlike linear ICs, which can be used in a great variety of applications, digital ICs are used primarily for logic circuits. To make full use of digital ICs, the user must be familiar with digital logic, including simplification and manipulation of logic equations, working with logic maps, and the implementation of basic logic circuits (counters, registers, etc.). All of these subjects are discussed in the author's *HANDBOOK OF LOGIC CIRCUITS*, Reston Publishing Company, Inc., P.O. Box 547, Reston, Virginia, 1972.

In this chapter, we shall concentrate on what type of digital ICs are available, their relative merits, and interpreting digital IC datasheets. Some typical applications for the IC elements, as well as notes on how one type of digital IC can be made to work with other types of ICs in a given system, are discussed for each of the five major types (RTL, DTL, TTL, HTL, and ECL). Refer to Chapter 1.

Also included in this chapter are discussions concerning special applications for digital ICs, such as forming multivibrators, pulse shaping circuits, etc. The chapter is rounded off by a discussion of special digital IC devices. Data on testing digital ICs are covered in Chapter 7.

6-1. SELECTING DIGITAL INTEGRATED CIRCUITS

One of the main problems for the digital IC user is to choose the right logic family for a given application. In some cases, price is the all-important factor. In other circumstances, a specific design problem (high speed operation, noise, etc.) exists that requires a certain logic family. Before selecting any logic family, the user should study the characteristics of all available types. At present, there are five basic logic families, plus special-purpose field effect logic ICs. Each of these is discussed in separate sections of this chapter.

The user will be in a much better position to choose the right digital IC after reading all of the data in this chapter. However, the user should keep the following points in mind when reading the data.

6-1.1 Availability and compatibility

TTL is the most available of all logic families. That is, at present, it is possible to obtain the greatest variety of off-shelf logic devices in the TTL family. Some designers consider TTL as the "universal" IC logic family, since they can obtain an infinite number of gates (with a variety of input/output combinations), buffers, inverters, majority logic, etc., from more than one manufacturer.

RTL and DTL are the next "most available" logic ICs. At one time, RTL was the most available, since its operation is nearest to that of discrete transistor logic, and thus familiar to most designers. Because of the special-purpose nature, both ECL and HTL are available in less variety than the other logic families. An exception to this is the field effect logic which, at present, is very limited. However, as the advantages of field effect logic becomes known, and as cost goes down, field effect logic could replace many other types.

TTL and DTL are compatible with each other. Thus, some designers use only these two families, taking the advantages of both as applicable.

RTL is the most compatible with linear and analog systems, or any discrete transistor application. This is because RTL is essentially an IC version of conventional solid-state circuits.

Because of their special nature, ECL and HTL are the least compatible with other families and with external devices. ECL requires a large supply voltage for a comparatively small output swing, while HTL produces a very large output swing which is generally too high for any other system.

Keep in mind that, barring some unusual circumstance, any logic family can be adapted for use with other logic families, or external equipment, by means of *interface circuits*. These interface circuits are discussed throughout this chapter.

6-1.2 Noise considerations

HTL has the highest noise immunity of any logic family. (Or, HTL has the least noise sensitivity, whichever term you prefer.) Typically, signal noise up to about 5 V will not affect HTL. Often, HTL can be used without shielding in noise environments where other families require extensive shielding.

RTL has the lowest noise immunity (or is the most noise sensitive). Typically, signal noise in the order of 0.5 V can affect RTL. Considering that RTL operates at logic levels of about 1 V, an RTL system is almost always operating near the noise threshold. As a result, RTL is not recommended for noisy environments.

DTL and TTL are about the same in regards to noise immunity or sensitivity. ECL can be operated so that the noise immunity is about equal to that of DTL and TTL. The input of ECL is essentially a differential amplifier. If one base is connected to a fixed bias voltage about half way between the logic 1 and 0 levels, this sets the noise immunity at the level of the bias voltage. (This is discussed further in Sec. 6-7.)

In addition to signal line noise, logic ICs are affected by noise on the power supply and ground lines. Most of this noise can be cured by adequate bypassing as described in Chapter 2. However, if there are heavy ground currents due to large power dissipation by the ICs, it may be necessary to use separate ground lines for power supply and logic circuitry.

In addition to noise immunity or sensitivity, the generation of noise by logic circuits must be considered. Whenever a transistor or diode switches from saturation to cutoff, and vice versa, large current spikes are generated. These spikes appear as noise on the signal lines, as well as the ground and power lines. Sincc ECL does not operate in the saturation mode, it produces the least amount of noise. Thus, ECL is recommended for use where external circuits are sensitive to noise. On the other hand, TTL produces considerable noise, and is not recommended in similar noise situations. DTL produces somewhat less noise than TTL. Since TTL and DTL are often used together, one solution is to use DTL in close proximity to the noise-sensitive external circuits, and TTL for all other circuits.

6-1.3 Propagation delay and speed

The speed of an IC logic system is inversely proportional to the propagation delay of the IC elements. That is, ICs with the shortest propagation delay can operate at the highest speed. Since ECL does not saturate, the delay is at a minimum (typically 2–4 ns), and speed is maximum. At the other end, HTL has the longest delay (80–90 ns), and operates at the slowest speed. TTL is the next-to-fastest logic family, with delays of about 10 ns, and can be used in any application except where the extreme high speed of ECL is required. RTL and DTL are about the same speed, with typical delays of 25 and 30 ns, respectively.

6-1.4 Power source and dissipation

HTL requires the largest power supply voltage, and dissipates the most power. A V_{CC} of 15 V and a power dissipation of 30 mW is typical.

Of the conventional logic ICs, DTL dissipates the least power, typically less than 10 mW. Average power dissipation factors for RTL, TTL and ECL are 12, 15 and 25 mW, respectively. From this, it will be seen that ECL is about on a level with HTL in regards to power consumption. In fact, some lines of ECL draw more power than other lines of HTL.

Except for HTL, the remaining logic families can operate with supply voltages in the range from about 3 to 5 V. However, the logic swing (or output voltage) varies considerably, and is dependent upon supply voltage. For example, a typical DTL unit can produce an output of about 4.5 V with a 5-V supply. TTL will produce an approximate 3.5 V output with the same supply. ECL and RTL will produce outputs of about 1.3 V with 5-V supplies.

Where absolute minimum power dissipation is wanted, field effect logic can provide the answer. Field effect circuits are available in MOS (metal-oxide semiconductor) ICs, and may be either single-channel (P-channel or PMOS), or complementary (CMOS). The CMOS is well suited for battery-operated systems since little standby power is required. In the future, when price is reduced and availability is increased, CMOS ICs may replace the now-popular TTL as the universal IC. Field effect ICs will operate over a wide range of power supply voltages, so they are adaptable to almost any system.

6-1.5 Fan-out

Fan-out, or the number of load circuits that can be driven by an output, is always of concern to logic designers. Some IC datasheets list fan-out as a simple number. For example, a fan-out of three means that the IC output will drive three outputs. While this system is simple, it may not be accurate. Usually the term fan-out implies that the output will be applied to inputs of the same logic family and the same manufacturer. Other datasheets describe fan-out (or load and drive) in terms or input and output current limits. (This is discussed in Sec. 6-2.)

Aside from these factors, the following typical fan-outs are available from the logic families: RTL 4–5, DTL 5–8, TTL 5–15, HTL 10, ECL 25.

6-2. INTERPRETING DIGITAL IC DATASHEETS

As in the case of linear ICs, digital IC datasheets are presented in various formats. However, there is a general pattern used by most manufacturers. For example, most digital IC datasheets are divided into four parts. The first part usually provides a logic diagram and/or circuit schematic, plus truth tables, logic equations, general characteristics, and a brief description of the IC. The second part is devoted to test, and shows diagrams of circuits for testing the ICs. These two parts are fairly straight-forward, and are usually easy to understand.

This is not necessarily true of the remaining two parts (or one large part in some cases), which have such titles as "maximum limits" or "maximum ratings", and "basic characteristics" or "electrical characteristics". These parts are generally in the form of charts, tables or graphs (or combinations of all three), and often contain the real data needed for design with logic ICs. The terms used by manufacturers are not consistent. Likewise, a manufacturer may use the same term to describe two slightly different characteristics, or use two different terms to describe the same characteristic, when different lines or different families are being discussed.

It is impractical to discuss all characteristics found on digital IC datasheets. However, the following notes should help the user interpret the most critical values.

6-2.1 Electrical characteristics

It is safe to assume that any electrical characteristic listed on the datasheet has been tested by the manufacturer. If the datasheet also specifies the test conditions under which the values are found, the characteristic can be of immediate value to the user. For example, if leakage current is measured under worst-case conditions (maximum supply voltage and maximum logic input signal), the leakage current shown on the datasheet can be used for design. However, if the same leakage current is measured with no signal input (inputs grounded or open), the leakage current is of little value. The same is true of such factors as output breakdown voltage and maximum power supply current.

To sum up, if an electrical characteristic is represented as being tested under "typical" (or preferably worst-case) operating conditions, it is safe to take that characteristic as a design value. If the characteristic is measured under no-signal conditions, it is probably included on the datasheet to show the relative merits of the IC. When in doubt, test the IC as described in Chapter 7.

6-2.2 Maximum ratings

Maximum ratings are values that must never be exceeded in any circumstance. They are not typical operating levels. For example, a maximum rating of 15 V for V_{CC} means that if the regulator of the system power supply fails, and the V_{CC} source moves up from the normal 8–10 V to 15 V, the IC will probably not be burned out. But never design the system for a normal V_{CC} of 15 V. Allow at least a 10 to 20 percent margin below the maximum, and preferably a 50 percent margin for power supply (voltage and current) limits. Of course, if typical operating levels are given, these can be used even though they are near the maximum ratings.

6-2.3 Drive and load characteristics

Generally, the most important characteristics of digital ICs (from the user's standpoint) are those that apply to the output drive capability, and the input load presented by an IC. This applies to both *combinational logic* and *sequential logic*.

Combinational logic involves circuits having outputs that are a direct function only of present inputs, and involve *no memory function*. Sequential logic involves circuits which contain at least one memory element. Thus, combinational logic includes such devices as adders, coders, decoders, selectors, and complementers, whereas sequential logic includes flip-flops, registers, and counters.

No matter which type of logic is involved, the user must know how many inputs can be driven from one IC output (without amplifiers, buffers, etc.). It is equally important to know what kind of load is presented by the input of an IC on the output of the previous stage (either IC or discrete component). As discussed in Sec. 6-1.3, fan-out is a simple, but not necessarily accurate term to describe drive and load capabilities of an IC.

A more accurate system is where actual input and output currents are given. There are four terms of particular importance. These are:

Output logic 1-state source current, I_{OH}

Output logic 0-state sink current, I_{OL}

Input forward current, I_F

Input reverse current, I_R

The main concern is that the datasheet value of I_{OL} must be equal to or greater than the combined I_F value of all gates (or other circuits) connected to an IC output. Likewise, I_{OH} must be equal to or greater than total I_R.

Unfortunately, the same condition exists for datasheet load and drive factors, as exists for other electrical characteristics; the values are not consistent.

6-2.4 Interfacing digital ICs

No matter what load and drive characteristics are given on the datasheet, it may be necessary to include some form of interfacing between digital ICs to provide the necessary drive current. This is especially true when the datasheet shows a very close tolerance. For example, assume that an IC with a rated fan-out of 3 is used to drive three gates. If the fan-out rating is "typical" or "average", and the three gates are operating in their "worst case" condition, the IC may not be able to supply (and dissipate) the necessary current.

Although it is not practical to have a "universal" interfacing circuit for all ICs, the circuits of Fig. 6-1 should provide enough information to design inter-

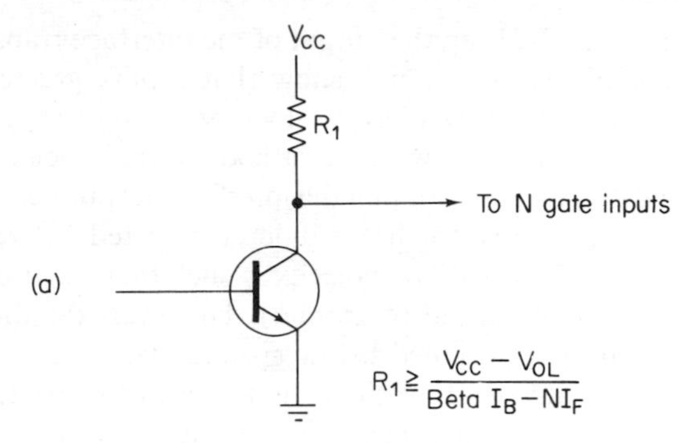

FIG. 6-1. Basic interfacing circuit for digital IC elements (Motorola).

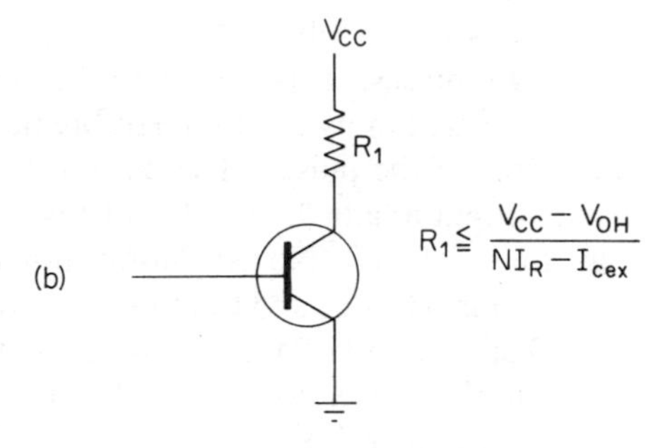

facing between IC logic elements. The equations shown on Fig. 6-1 are used to find the approximate or trial value of pull-up resistor R_1. The following is an example of how to use the equations.

Assume that the common-collector circuit is used, that V_{CC} is 5 V, that V_{OH} (high or logic-1 state voltage) is 3 V, there are 7 gate inputs and each has a forward current of 1 mA, and that I_{CEX} is 3 mA. Note that I_{CEX} is the input leakage current as tested with a voltage between base and emitter. I_{CEX} is a "worst-case" or active datasheet value, and is not to be confused with I_{CER} which is input leakage current with no signal (tested with a fixed resistance between emitter and base).

Using the equation of Fig. 6-1, the value of R_1 is:

$$R_1 = \left(\frac{5 - 3\text{ V}}{(7 \times 1\text{ mA}) + 3\text{ mA}}\right)\left(\frac{2\text{ V}}{10\text{ mA}}\right) = 200\ \Omega$$

The next lowest standard value is 180 Ω.

Two cautions must be observed when using the circuits of Fig. 6-1. Do not try to increase the output voltage of an IC (with an interface) to a level higher than the V_{CC} of either IC (input or output). For example, if both ICs have a

V_{CC} of 5 V, keep the output of the interface transistor below 5 V, even though the IC datasheet may show that inputs greater than 5 V are safe. This is because many ICs have a diode between resistors in the circuit and the power supply terminal (with the cathode of the diode connected to the power supply side). Also, do not pull the interface output below ground. In many ICs, each internal transistor has a diode connected between the collector and ground terminal. The diode polarity is such that the diode is normally reverse-biased (anode connected to ground). However, the diode will be forward-biased if the output is pulled below ground. Unfortunately, the datasheet schematics do not always show these diodes, even though they exist in the circuit. To be on the safe side, keep the interface inputs and outputs above ground and below V_{CC}.

Sometimes, there are signal problems with interface circuits, particularly when high speeds are involved. Interface circuits slow down rise and fall times of the logic pulses. This slowdown can cause saturated logic elements (which is about every family but ECL) to break into oscillation on the leading and trailing edges of the pulses. The basic rule to follow is that the rise and fall times of an input to any digital IC must be shorter than the typical propagation delay time of the IC. No oscillation should occur if this rule is followed. Note that the terms rise time and fall time refer to the time required to go from a 1-state to a 0-state, and vice versa. These rise times and fall times are somewhat longer than the conventional 10 and 90 percent values for pulse measurements. Refer to Chapter 7.

Also note that interfacing schemes and guidelines vary from logic family to family, as is discussed in the following sections.

6-3. RESISTOR-TRANSISTOR LOGIC (RTL)

Resistor-transistor logic, or RTL, was derived from Direct-Coupled-Transistor logic, or DCTL, and was the first integrated logic form introduced around 1960. The basic circuit is a direct translation from the discrete design into integrated form. This circuit was the most familiar to logic designers, easy to implement and, therefore, the first introduced by IC manufacturers.

6-3.1 Basic RTL gate

The RTL gate circuit shown in Fig. 6-2 is presented here to illustrate the basic building block type of logic. The most complex elements are constructed simply by the proper interconnection of this basic circuit. The circuit of Fig. 6-2 is actually a modified form of the DCTL circuit that does not have the resistor in the base lead. With DCTL, the problem of current

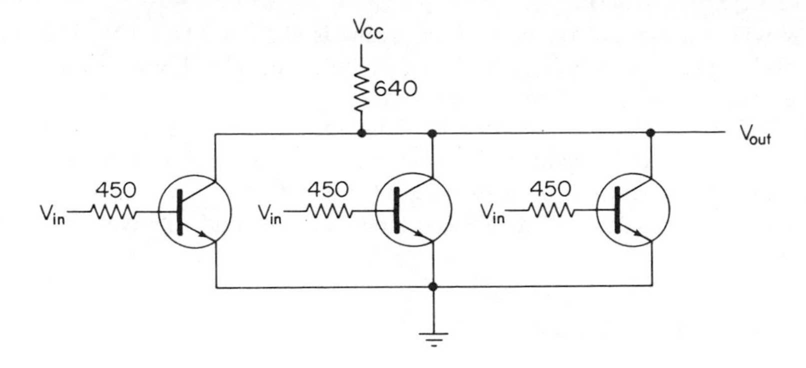

FIG. 6-2. Basic RTL gate.

"hogging" often arises. When several base-emitter junctions are driven from the same output, the input with the lowest base-emitter junction forward bias could severely limit the drive current to the other transistor bases. A small resistor added in the base circuit increases the input impedance assuring proper operation when driving more than one load.

6-3.2 Transfer characteristics

The transfer characteristic of a typical RTL gate circuit is shown in Fig. 6-3. For any logic circuit, the transfer characteristic is a plot of output voltage versus input voltage. These characteristics are useful in determining operating points, logic swing, and compatibility between inputs and outputs. They also may be used to determine the effect of noise signals at any of the inputs, and for worst case design in terms of temperature and power supply variation.

The typical RTL transfer characteristic in Fig. 6-3 indicates that operating

FIG. 6-3. Typical RTL transfer characteristics.

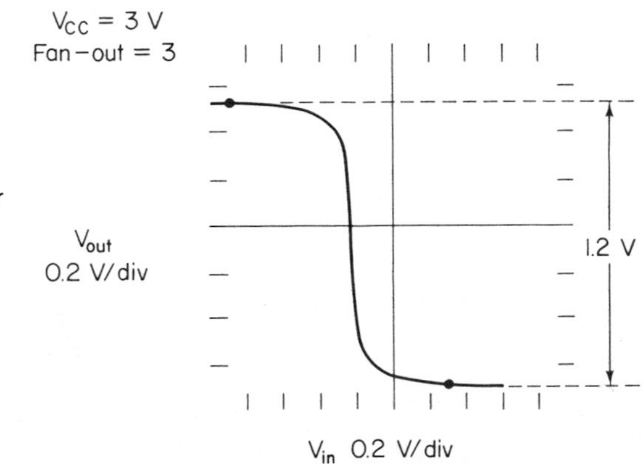

at 3 V with a fan-out of three, the logic swing is about 1.2 to 1.3 V. The voltage level required to switch an RTL circuit shifts considerably with variations in the operating temperature and fan-out.

A similar transfer characteristic for the DCTL circuit without the base resistor would show a reduced logic swing—on the order of 0.6 to 0.7 V. The high level operating point would be closer to the active region of the transfer curve, resulting in a reduced noise margin and relatively poor dc stability.

6-3.3 Fan-out

Fan-out is limited to five in the basic RTL gate. Larger fan-out drivers are available, capable of driving up to 25 gate circuits. But generally, fan-out is limited because logic swing and noise immunity are heavily dependent upon fan-out. The fact that the noise immunity of an RTL degrades rapidly with an increase in temperature is a limitation of this type of digital IC.

6-3.4 Noise thresholds

The relationship between the typical input dc noise threshold of an RTL and temperature is shown in Fig. 6-4. The top curve indicates the amount of noise immunity that a circuit has when the input is in the high logic state, or the "1" state, and a negative noise pulse is injected on the input signal line. The lower curve describes the noise immunity values during the input logic "0" condition when a positive signal pulse is injected on the input.

A plot of typical input ac noise threshold versus pulse width and temperature is shown in Fig. 6-5. Note that at an input pulse width of roughly 50 to 100 ns the noise immunity levels off and approaches the dc condition. At

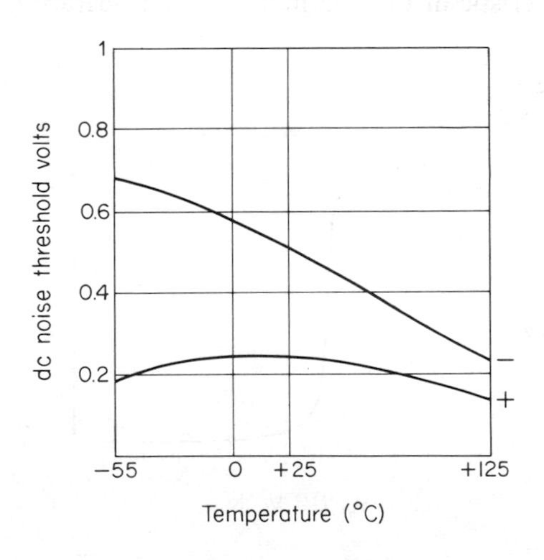

FIG. 6-4. Typical RTL input dc noise threshold versus temperature.

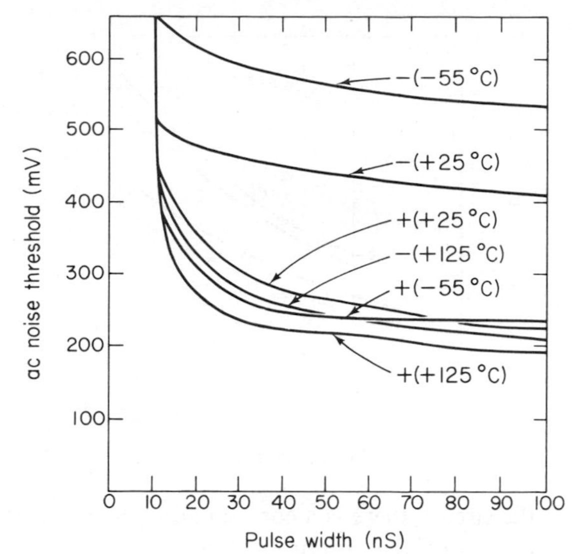

FIG. 6-5. Typical RTL input ac noise threshold versus pulse width and temperature.

smaller pulse widths, noise margins on the input signal line are appreciable. For example, with pulse widths on the order of 10 to 20 ns, the noise margin will be greater than 0.5 V.

6-3.5 Circuit speed

Temperature and fan-out variations also affect turn-on and turn-off time delay of an RTL. This is shown in Figs. 6-6 and 6-7. Note that at any given temperature, an increase in fan-out results in a slower turn-off time but faster turn-on time. As variations in fan-out put more or less load on

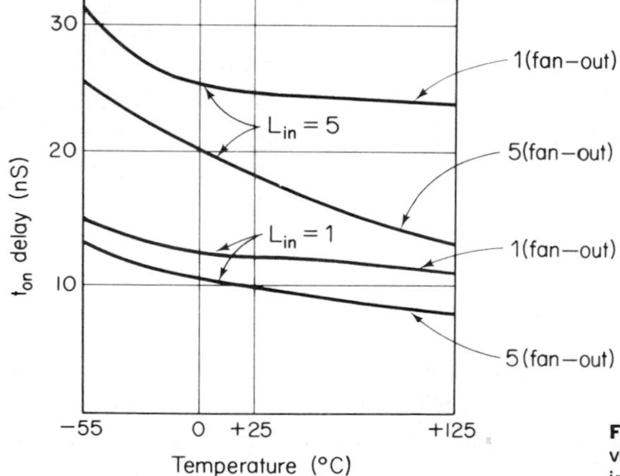

FIG. 6-6. Typical RTL t_{on} delay versus temperature and fan-out for input loading of 1 and 5.

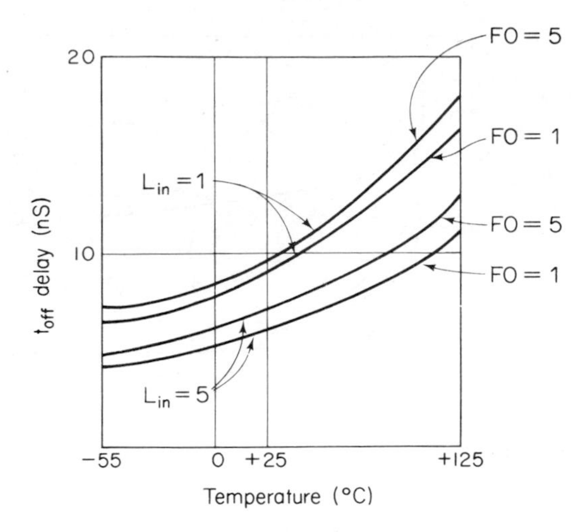

FIG. 6-7. Typical RTL t_{off} delay versus temperature and fan-out for input loading of 1 and 5.

the circuit, there is a corresponding increase or decrease in the time required to turn off the output inverter.

6-3.6 Primary advantages

The unusual flexibility of an RTL is one of the major advantages. Practically any type of integrated digital circuit that a designer might need is readily available in RTL. Also, RTL is backed by a long history of reliability and evaluation data. Another plus for RTL is the fact that most design engineers are familiar with this logic even before they encounter it in IC form, because RTL is a holdover from discrete design days.

At present, no other form of IC logic family can compete with RTL on the basis of cost. Of course, this picture will probably change in the future. The availability of multi-function and complex elements is another factor that results in reduced system costs with RTL. Devices such as decade counters, 4-bit shift registers, full adders, and dual J-K flip-flops are either presently available or will be in the very near future. These extremely complex devices are readily possible with RTL because of basic circuit simplicity and lack of stringent requirements on individual circuit components. Unlike most other logic designs, component requirements are at a minimum and resistor values are small, resulting in a small IC die size and a high yield. This means more logic function per package at less cost per function with RTL circuits.

6-3.7 Low power dissipation

For applications that require low power dissipation, Motorola has modified its basic MRTL (Motorola RTL) line, and has developed a low-power version of this IC logic form, called milliwatt (or mW) MRTL. In the modified version, the base resistors and the collector load resistor are increased significantly to reduce the power dissipated by a typical gate circuit to about 2 mW. As a result, if power dissipation is a main consideration, the mW MRTL offers the bid advantage in this area. The basic mW MRTL circuit is shown in Fig. 6-8.

6-3.8 Paralleling collector loads

The logic interconnection of a large system sometimes requires heavier than normal loading of gating circuits. The typical fan-out capability of an RTL is five unit loads. However, it is permissible to parallel collector loads to obtain increased fan-out, as is shown in Fig. 6-9.

The standard loading rules, which are found on the RTL datasheet must be adjusted when paralleling collector loads. With the gates in parallel each input now has a load factor of 2 when compared to the unit load on the basic gate input. This can be accounted for by the fact that each ON transistor can possibly conduct twice the normal collector load current. Since each output separately can fan out to five unit loads, an output at one-half the collector impedance should have the capability of driving twice normal loading, or 10 unit loads.

This can be shown through the relationships for the currents in Fig. 6-10. Referring to Fig. 6-10 for a fan-out of 5,

$$I_5 = \frac{V_1 - V_{BE}}{R_{1/5}}$$

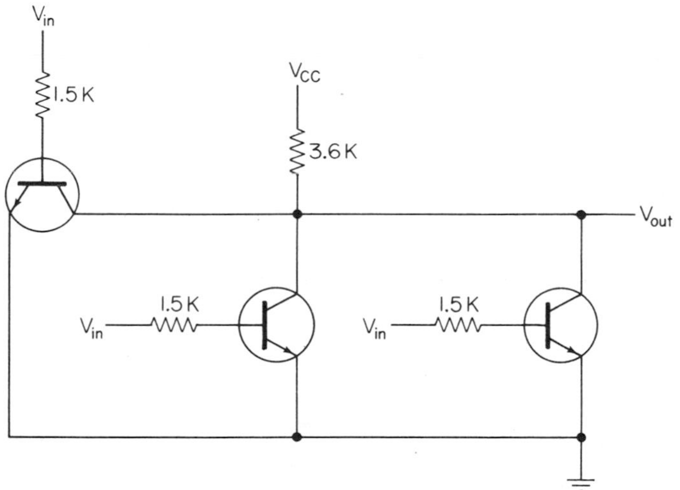

FIG. 6-8. Motorola milliwatt MRTL digital IC.

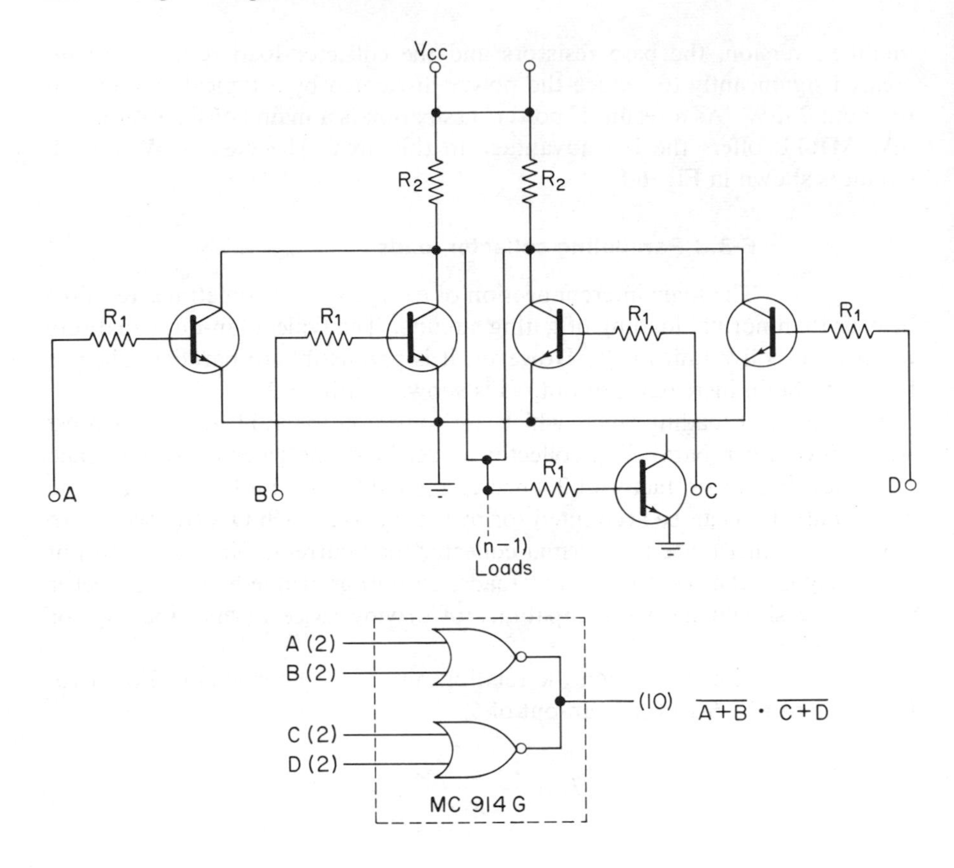

FIG. 6-9. Dual 2-input MRTL gate with parallel collector loads (Motorola).

where I_5 is the output gate current into five gate loads. V_1, the collector output voltage, is

$$V_1 = V_{CC} - R_2 I_{CC}$$

After expressing supply current, I_{CC}, in terms of leakage or output currents, we have:

$$I_5(R_{1/5}) = V_{CC} - V_{BE} - R_2(I_5 + I_{CEX}),$$

and finally

$$I_5 = \frac{V_{CC} - V_{BE} - R_2 I_{CEX}}{R_{1/10} + R_2}$$

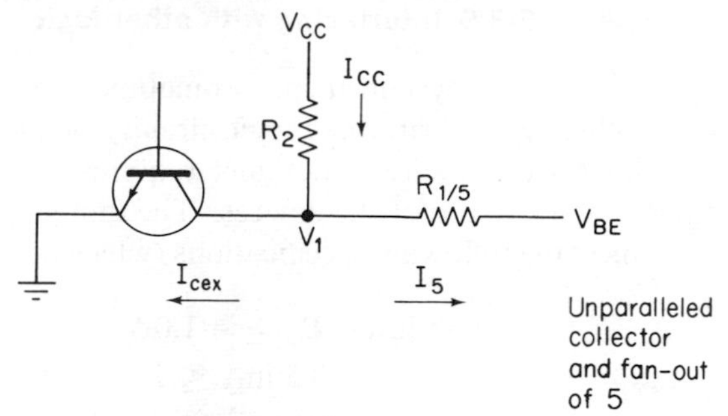

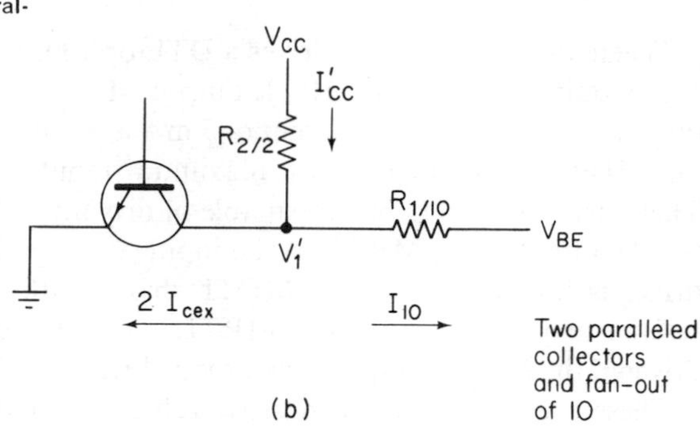

FIG. 6-10. Relationship of gate outputs for unparalleled and paralleled collector loads.

From Fig. 6-10b, since the collector load is at $R_{2/2}$ and the load resistance is $R_{1/10}$ for 10 loads, the output becomes:

$$I_{10} = \frac{V_{CC} - V_{BE} - R_{2/2}(2I_{CEX})}{R_{1/10} + R_{2/2}}$$

where I_{10} is the paralleled output gate current into 10 unit gate loads. Now the ratio of these two outputs will be the increased driving factor with parallel collector loads.

$$\frac{I_{10}}{I_5} = \frac{R_{1/5} + R_2}{R_{1/10} + R_{2/2}} = 2$$

This last equation says that if a single collector output can supply the I_{in} requirement to a 5-gate input, parallel collector loads can supply the I_{in} requirement to 10-gate inputs. We can arrive at the same conclusion by showing that $V_1 = V_1'$ under full load (or we can simply accept the manufacturer's word that paralleling two outputs doubles the drive capability).

6-3.9 Interfacing with other logic

In a system it may sometimes be desirable to interface RTL with another type of saturated logic circuit, such as DTL or TTL. Therefore, at the interface, the circuit designer requires a knowledge of the input and output characteristics of the devices. The input signals to an MRTL element should meet the following specifications (which are typical for most RTL ICs):

$$\text{High level: } E_{in} \geq +1.0 \text{ V}$$
$$0.5 \text{ mA} \leq I_{in} \leq 0.5 \text{ mA per load}$$
$$\text{Low level: } E_{in} \leq 0.3 \text{ V}$$

When an MRTL output drives a DTL or TTL type input, there is a current sink requirement on the MRTL output. To insure operation under all conditions, a maximum sink current of 5 mA is permissible. For example, Motorola MD930 series logic has a maximum input forward current of 1.6 mA. Therefore, an MRTL gate is capable of driving 3 MD930 loads.

When interfacing MRTL at the input, the high level output current of the driver is important. For an MDTL driver with an output of 6-kΩ pull-up resistor to a +5-V supply, one MRTL load is allowed. Five gate loads can be driven if the MDTL output resistor is reduced to 1.5 kΩ.

These specifications and resistor values are both typical and conservative requirements, and are given to insure operation over the normal temperature range.

6-4. DIODE TRANSISTOR LOGIC (DTL)

A logic circuit form that was very familiar to the discrete-component designer was one that used diodes and transistors as the main components (plus a minimum number of resistors). The diodes provided higher signal thresholds than could be obtained with resistor-transistor circuits (or RTL), the first popular discrete logic that was integrated. Therefore, it is not surprising that soon after the introduction of RTL, IC manufacturers began translating DTL into integrated devices also.

The basic DTL circuit is shown in Fig. 6-11. Note that it requires two power supplies—a positive supply and a negative supply which is used to improve the turn-off time of the output inverter. Although the circuit would operate with only one supply, an undesirable compromise in the design would be required to provide fairly symmetrical turn-on and turn-off times.

If a single power supply were used, the charge could be pulled out of the base of the inverter when the input goes to a low logic state, allowing the inverter to turn-off, by either putting a small resistor between base and emitter

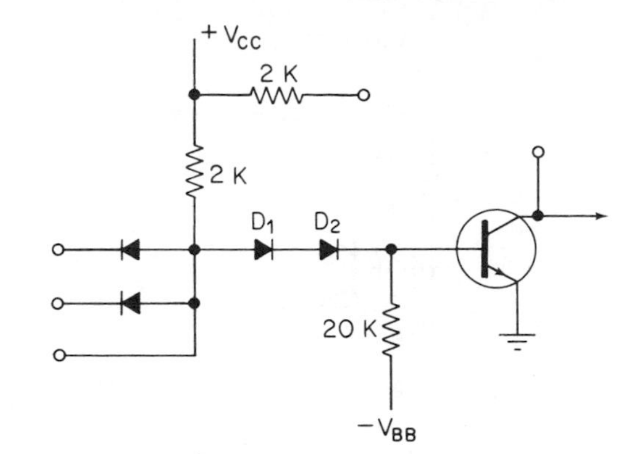

FIG. 6-11. Basic DTL gate.

or by making diodes D_1 and D_2 special slow-recovery diodes. In ICs, special-purpose diodes are not desirable because they result in a lower yield and, consequently, higher cost. On the other hand, a small resistor between the base-emitter would limit the amount of base current to the inverter, thus reducing the fan-out of the circuit. This fan-out could be improved by decreasing the size of the input pull-up resistance to supply the needed drive current. However, to do so, would increase the power dissipated by the circuit, as well as the current from each load circuit.

As a result, in basic DTL, a negative power supply is used to turn off the output inverter. The negative supply is used even though this second supply introduces additional complexity into the system.

6-4.1 Primary advantages

The main advantages of DTL include large logic swings and high noise immunities. Typical logic swings of 4.5 V (with a 5-V supply) are available. Input signal line noise margins run typically in excess of 1 V. DTL can be classified in the medium speed range with circuit delay times from 30 to 70 ns. A typical DTL binary element is capable of clocking rates in the 10-MHz range. Such elements include flip-flops, counters, and registers.

6-4.2 Modified DTL circuit

The need for a dual power supply, a feature adding considerable complexity to typical applications, has been eliminated in Motorola's version of the DTL, called MDTL. The basic gate circuit of the MDTL family is shown in Fig. 6-12. The circuit features an input transistor in place of one of the input offset diodes in conventional DTL. This is a stage of amplification that supplies turn-on drive to the output inverter.

This source of current in the base allows for possible increase in fan-out with the MDTL design. It also means that a much smaller base-emitter

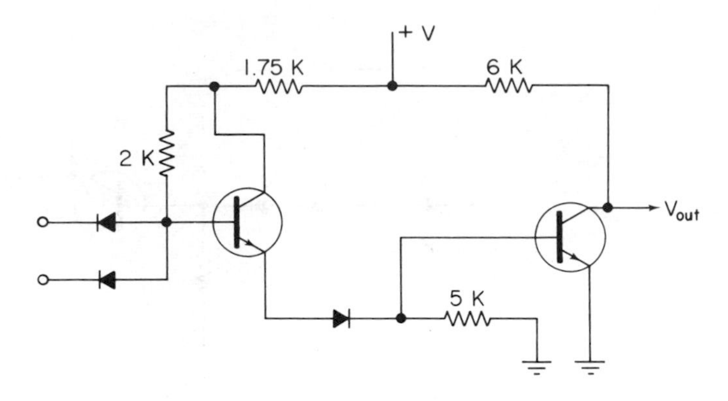

FIG. 6-12. Basic Motorola MDTL gate circuit.

resistor can be used and, consequently, a negative power supply is not required for rapid transistor turn-off. Further advantages of this approach are greater tolerances on resistor values, and relaxed gain specifications on the inverter stage (not as much gain is required).

6-4.3 Advantages of MDTL over DTL

From a consideration of the design requirements for the MDTL circuit, it is apparent that less demand is made on resistor tolerances, gain, fan-out, and power dissipation than in the conventional DTL circuit. Therefore, although the performance characteristics of DTL and MDTL circuits are similar, the latter offers many advantages over DTL such as:

Single power supply operation.

Lower power dissipation—typically 8 mW per single gate for a 5-V supply.

Larger drive capabilities—fan-out of 8 for a gate; 25 for the buffer.

Higher speed operation—20 to 50 ns typical propagation delay.

Better noise immunity—typically 1.2 V at 25 °C with worst case dc noise immunity specified at 350 mV over the full temperature range (−55 °C to 125 °C) with maximum fan-out.

6-4.4 Transfer characteristics

The transfer characteristics of a typical MDTL gate is given in Fig. 6-13. V_{IL} is the minimum low threshold on the input, while V_{IH} is the maximum high threshold of the input. The worst case specified noise margins with a low level input and a high level input are $V_{IL} - V_{OL}$ and $V_{OH} - V_{IH}$ respectively.

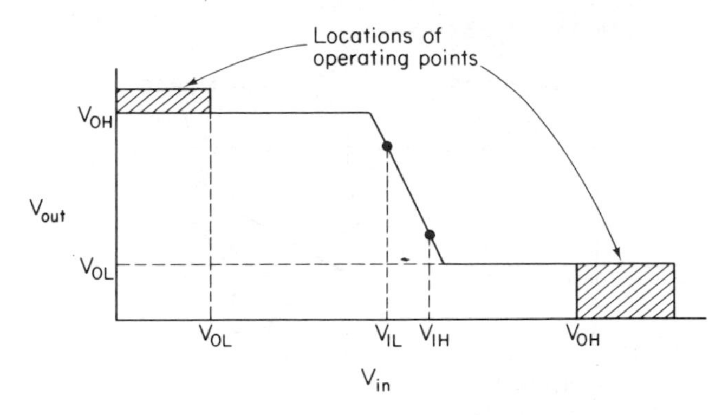

FIG. 6-13. Transfer characteristics of a typical MDTL gate circuit (Motorola).

The low level noise margins will always be in excess of 350 mV, while the high level noise margin will be greater than 400 mV over the full temperature range of − 55 °C to 125 °C.

Increased noise margins are obtained when circuits are not fully loaded. Worst case noise thresholds generally are better than 600 mV in systems that do not use the maximum drive capability in each circuit.

6-4.5 Clock rise and fall times for flip-flops

Although most DTL clocked flip-flops operate on the master-slave principle, it is advisable to limit clock rise and fall times, since noise margin is reduced during the transition periods. In general, the clock input to a DTL ripple counter or similar circuit should have a rise and fall time of less than 100 μs, while a clocked counter or shift register requires fall times of approximately 100 ns to effectively combat "race" problems.

Selection of the proper DTL flip-flop for the particular application can help reduce clock pulse input requirements. For example, Motorola MC948/MC848 flip-flop has clock input cricuitry which is better for higher-frequency operation than the MC945/MC845. The latter device normally represents a better choice for operation below 1 MHz, although the lower output pull-up resistances of the MC948/MC848 (2 kΩ versus 6 kΩ) sometimes makes an exception to this rule if heavy capacitive loads are to be driven. The same general rule applies to any flip-flop; choose the elements with 2-kΩ (or whatever is the lower value) pull-up resistors only if high frequency and/or high capacitive loads are required. For lower frequency and normally-resistive loads, use the 6-kΩ (or higher value) pull-up elements.

The circuit of Fig. 6-14 may be used to shape waveforms sufficiently to drive any DTL flip-flop. This circuit will reduce rise and fall times for clock waveforms to a safe duration. Do *not* use this circuit with RTL. Also if,

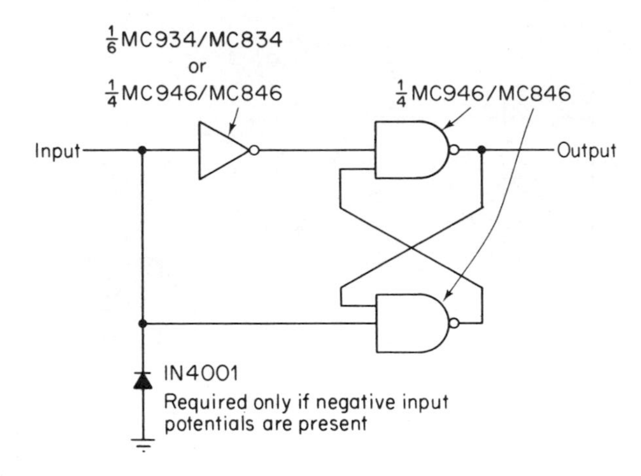

FIG. 6-14. MDTL pulse-shaping circuit (Motorola).

several DTL loads are to be driven, it may be necessary to use a buffer between the circuit of Fig. 6-14 and the clock inputs.

6-4.6 Input restrictions for DTL

As discussed in Sec. 6-2.4, there are restrictions on gate and flip-flop inputs which are not readily apparent from circuit diagrams. This is especially true in DTL where diodes are frequently used as protective devices. Figure 6-15 shows the circuit of a DTL gate as it normally appears on the datasheet, together with the same DTL circuit as it exists (with parasitic diodes and transistors used as diodes).

With the circuit of Fig. 6-15, negative input excursions should be less than 2 V (−1 V to +1 V, 0 V to +2 V, etc.) to limit current through the input diode path. Although input leakage current is tested at 4 V for this particular circuit, a low value is specified for leakage current to insure that an input voltage of 5.5 V will not damage the IC. Avalanche of the diodes shown in Fig. 6-15 is typically 7.5 V.

Note that these restrictions are not normally violated in systems designed exclusively with a single logic line (all DTL, all TTL, etc.), but precautions should be observed to prevent erratic operation and/or device failures when designing systems which require interfacing with other types of components.

6-5. TRANSISTOR-TRANSISTOR LOGIC (TTL)

TTL has become one of the most popular logic families available in IC form. Most IC manufacturers produce at least one line of TTL,

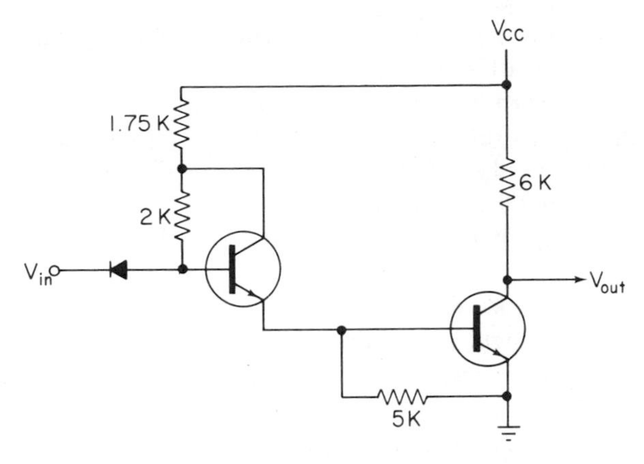

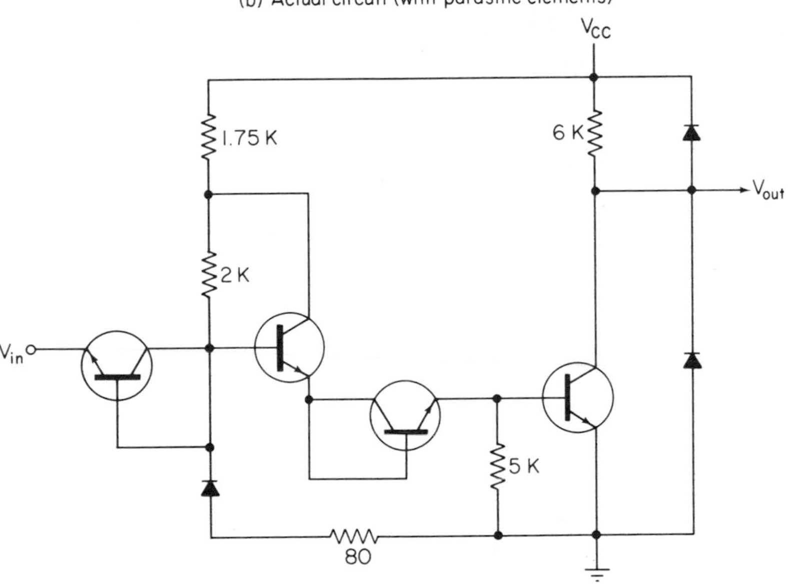

FIG. 6-15. Differences between datasheet schematic and actual circuit of DTL gate (Motorola).

and often several lines. This fact gives TTL the widest range of logic functions. It is impractical to cover all TTL logic lines here. Instead, we shall concentrate on Motorola's MTTL MC3000/MC3100 line. To point out the characteristics of TTL from a user's standpoint, we shall also draw comparisons between the selected TTL line, and a "typical" or "conventional" TTL line.

6-5.1 Typical TTL gate

Figure 6-16 shows the schematic of a typical TTL gate. By applying a "low" to either A or B inputs of the gate, the base-emitter junction of Q_1 becomes forward biased. No base drive is available for Q_2 which turns off, and causes Q_5 to turn off. The collector of Q_2 rises toward the supply voltage V_{CC}, and supplies base drive to turn on Q_3, causing Q_4 to turn on. Transistors Q_3 and Q_4 act as emitter-followers. This places the output V_O about two V_{BE} drops below V_{CC}, or about 3.5 V (for a V_{CC} of +5 V) in the "high" state.

Now consider the case where input A is "high" and input B begins to go from a "low" to a "high". Base drive is gradually applied to Q_2 which now starts to conduct with emitter current initially flowing through R_4. As Q_2 turns on, the collector current of Q_2 produces a voltage drop across R_2. The collector of Q_2 is connected to output V_O by means of the two emitter-followers Q_3 and Q_4. Thus, the output V_O tracks the voltage at the collector of Q_2. As the voltage at input B increases, the base of Q_2 tracks voltage by the difference of $V_{BE} - V_{BC}$ of transistor Q_1.

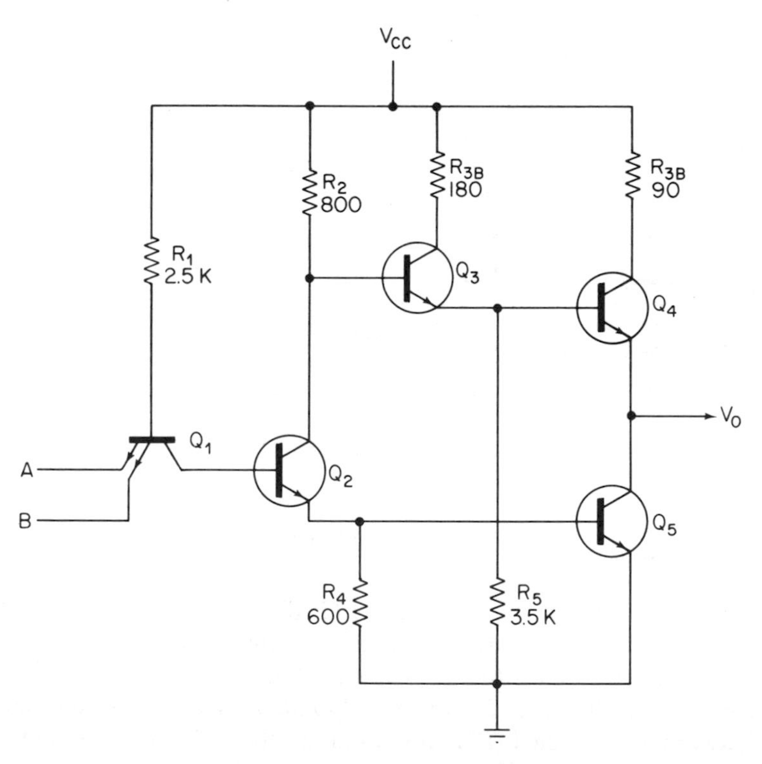

FIG. 6-16. Conventional high-speed TTL gate.

The collector current through Q_2 increases, causing a larger drop across resistor R_2. The output V_O now changes at a rate of:

$$\Delta V_B = \frac{R_2}{R_4}$$

These are shown on the conventional TTL curve of Fig. 6-17, as points A and B.

When the emitter current of Q_2 increases to the point where the drop across R_4 equals one V_{BE} drop, transistor Q_5 begins to conduct. Point B on the transfer characteristic has been reached. With a further increase in the voltage at input B, Q_5 saturates and point C on the transfer characteristic is reached. The collector of transistor Q_2 is now at the V_{BE} of Q_5, plus V_{CE} of Q_2. This voltage is not positive enough to sustain operation of Q_3 and Q_4. Thus, Q_3 and Q_4 turn off. The output is now at its "low" state; V_{CE} equal to about 0.2 V.

6-5.2 Motorola MC3000/MC3100 MTTL gate

One difference between the MTTL and conventional TTL is the replacement of resistor R_4 (Fig. 6-16) with resistors R_{4A} and R_{4B}, and transistor Q_6, as shown in Fig. 6-18.

With a "low" on either A or B, Q_1 is forward-biased and no base drive is available for Q_2. This keeps Q_2, as well as Q_5 and Q_6, in the off condition. The collector of Q_2 is approximately at V_{CC}, and base current is supplied to Q_3 and Q_4, keeping Q_3 and Q_4 on.

Now assume that input A is "high" and input B gradually goes from a "low" to a "high". The base of Q_2 tracks the voltage at input (the same as for the conventional TTL) by the difference of $V_{BE} - V_{BC}$ of Q_1. At the point where Q_2 turns on in conventional TTL, Q_2 does not turn on in MTTL, since the equivalent of an open circuit exists at the Q_2 emitter.

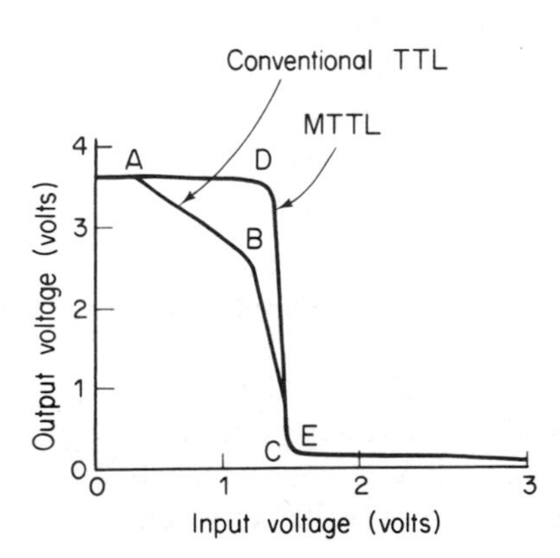

FIG. 6-17. Comparison of MTTL and conventional TTL characteristics (Motorola).

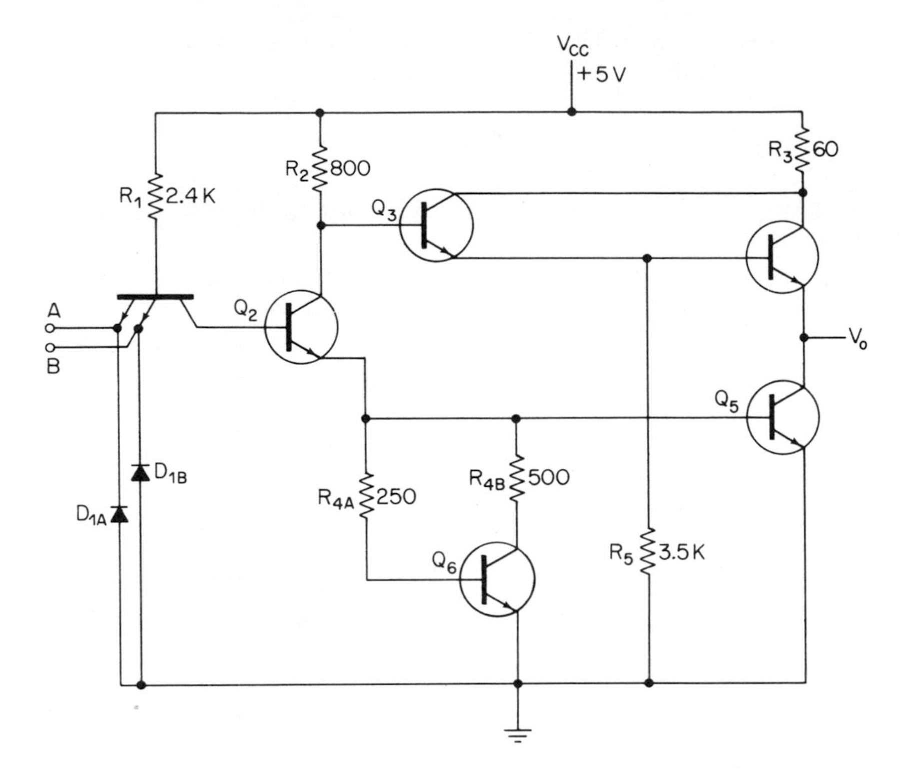

FIG. 6-18. Improved Motorola MTTL gate.

With no current flow, the collector of Q_2 remains near V_{CC}. Since Q_3 and Q_4 act as emitter-followers, the output remains at the "high" level, approximately two V_{BE} drops below V_{CC}. The bypass network turns on when the potential at the base of Q_2 is two V_{BE} drops above ground, causing the output transistor Q_5 to turn on, as well as bypass transistor Q_6. This is point D on the MTTL transfer characteristic (Fig. 6-17).

As the potential on input B increases further, output transistor Q_5 saturates, and point E is reached. The resistors in the bypass network are chosen so that the network conducts the same current as resistor R_4 in Fig. 6-16 when transistor Q_5 is saturated.

6-5.3 Function of input diodes

Due to high speeds of operation, TTL generates large values of current and voltage rates-of-change. A 1 V in approximately 1.3 ns rise time, and a 1 V in approximately 1 ns fall time provides dV/dt rates on the order of 10^9 V/s (volts per second).

With these ranges-of-change, undershoot exceeding 2 V can develop in the system. Such undershoot can cause two serious problems. First, false triggering of the following stage is possible since a positive overshoot follows the large undershoot. This positive overshoot may act as a "high" signal, and turn on the following stage for a short period of time. The diodes D_{1A} and D_{1B} on inputs A and B, respectively, in Fig. 6-18, limit the negative value of the undershoot, dissipating the ringing energy, and thereby reducing the positive overshoot.

The second problem created by high TTL speed results if the unused inputs of a gate are returned to the supply voltage, and a negative undershoot in excess of 2 V occurs. The reverse-biased emitters of the inputs may break down, and draw excessive current, generating noise in the system.

6-5.4 MTTL AND gate

The basic gate in most TTL systems is a NAND. In some logic designs, an AND gate is more convenient. Such an AND gate is available in the MTTL line, as shown in Fig. 6-19. The technique used to

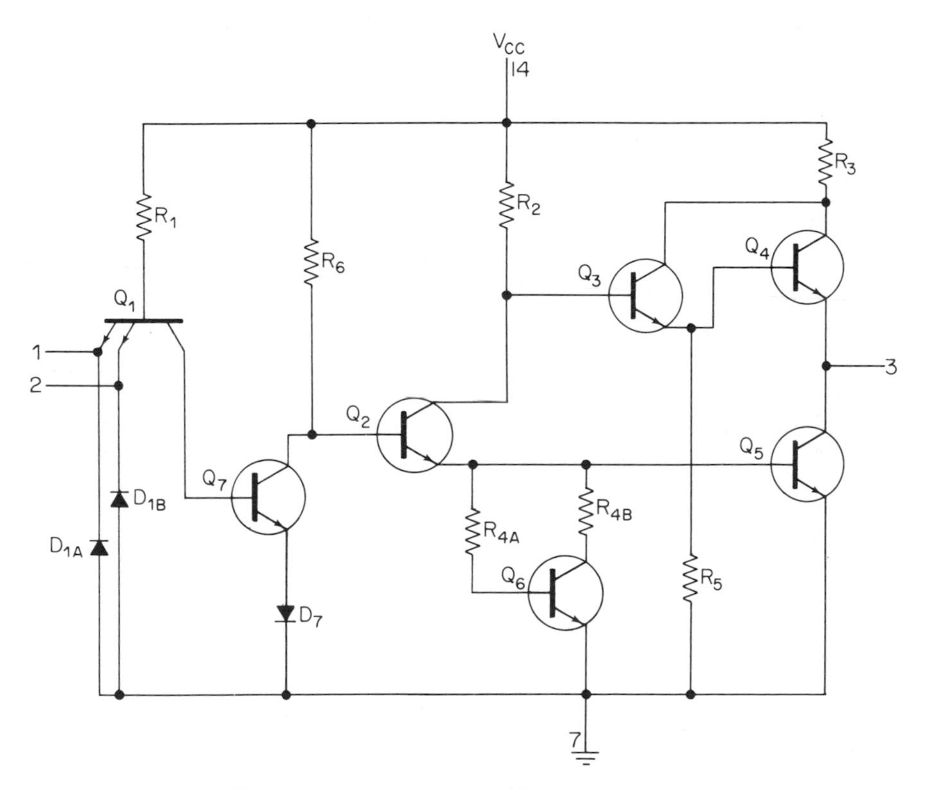

FIG. 6-19. Motorola MTTL positive logic AND gate.

generate the AND function consists of adding the network composed of Q_7, D_2 and R_6 to the basic NAND gate (Fig. 6-18).

With this network inserted between transistors Q_1 and Q_2, the voltage levels presented to Q_2 are the same as in the case of the NAND function, only inverted in phase. The inversion of a NAND function is an AND function, and vice versa. This additional inversion stage adds about 3 ns propagation delay, and around 6-mW power dissipation to the basic gate characteristics.

6-5.5 AND-OR-INVERT gate

The wired-OR function is normally not available in most TTL lines. This is due to the active pull-up circuit on the output. Any gates with active pull-up elements should not be used as wired-OR, for when one has a 0 output and the other a 1 output, the resulting logic output is unpredictable.

The MTTL series offers an AND-OR-INVERT gate, shown in Fig. 6-20, to overcome this problem. This gate uses two 2-input AND functions (Q_{1A}

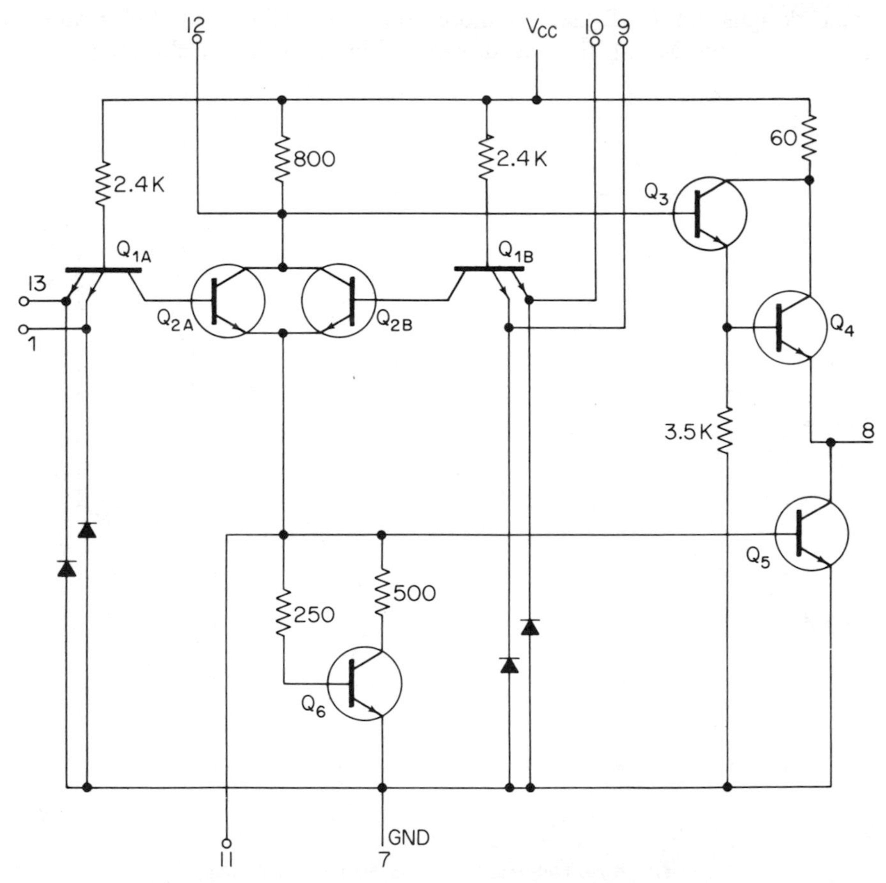

FIG. 6-20. Motorola MTTL AND-OR-INVERT gate.

and Q_{1B}) and two inverters (Q_{2A} and Q_{2B}). The inverter transistors Q_{2A} and Q_{2B} operate in parallel and perform the invert functions.

With "highs" on both of the inputs to either Q_{1A} and Q_{1B}, either Q_{2A} or Q_{2B} is on, thus causing Q_5 to be on with the output being low. With "highs" on all inputs, both transistors Q_{2A} and Q_{2B} are on again, causing transistor Q_5 to be on, and the output is "low". With "lows" on both emitter inputs to Q_{1A} and Q_{1B}, both transistors will be off, thus causing Q_3 and Q_4 to be on, and the output is "high".

6-5.6 NOR gate

The NOR function is available in MTTL by a slight modification to the AND-OR-INVERT gate. If one emitter is removed from each of the input transistors Q_{1A} and Q_{1B} of Fig. 6-20, we have a gate which performs the NOR function, as shown in Fig. 6-21.

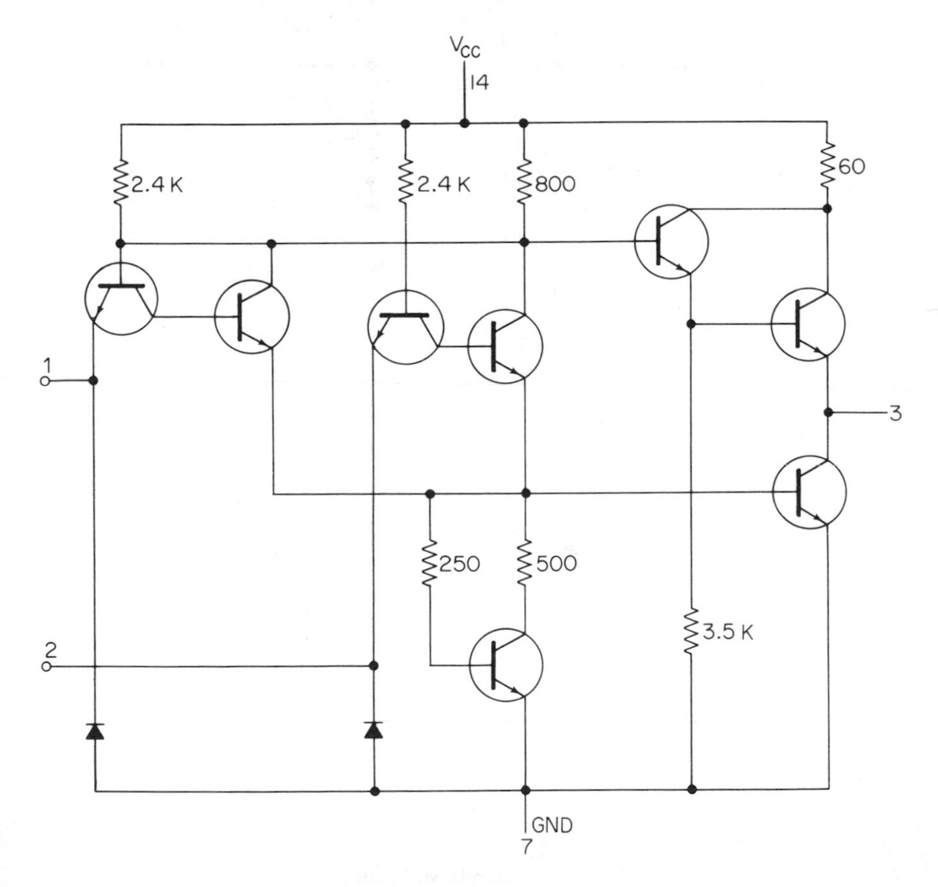

FIG. 6-21. Motorola MTTL NOR gate.

6-5.7 OR gates

By the addition of a transistor and a resistor to the NOR gate, the OR function can be performed. Fig. 6-22 is the same as the NOR gate of Fig. 6-21, with the addition of transistor Q_7 and resistor R_7 to perform the necessary phase inversion for the OR function.

6-5.8 Line drivers

Using an unterminated line driver, a line appears essentially as an open circuit at each end. (IC line drivers are generally used as amplifiers to increase the fan-out capability of gates, without the use of power gates.) Any pulse traveling down the line will see a reflection almost equal

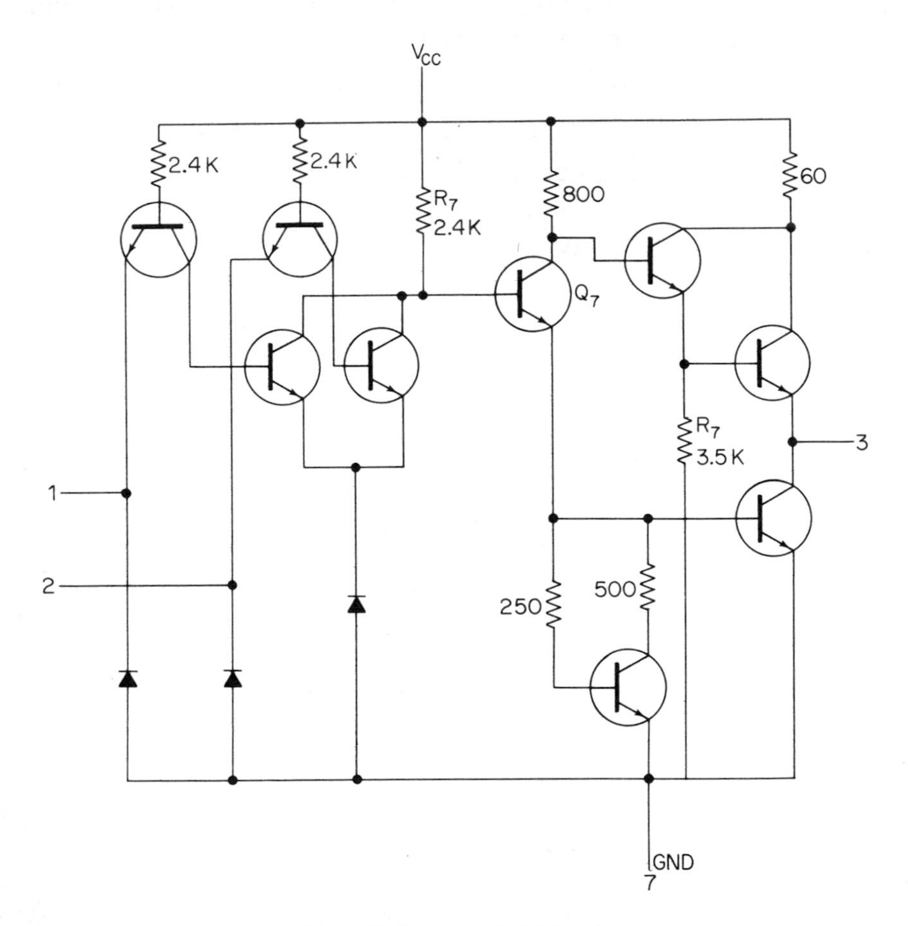

FIG. 6-22. Motorola MTTL OR gate.

in magnitude to the original pulse. By terminating at the loaded end, reflections are minimized.

To minimize switching transients on long lines, the MTTL family includes a pair of series-terminated gates. Each of these gates has three outputs. Two of the outputs have 75-Ω resistors in series with the standard output, and one connected directly to the output. For driving 93-Ω coaxial cable, or 120-Ω twisted pair, a good match can be made at the output of each resistor. For loads of 50 to 93 Ω, the two resistive outputs are shorted together for better impedance matching. The non-resistive output can be used to drive gates in a normal fashion.

Figure 6-23 shows the circuit of the NAND line driver, while Fig. 6-24 shows a typical application of the circuit.

6-5.9 Power gates

In some logic systems, there are fan-out requirements that exceed the capability of a standard gate. Power gates are designed to meet these requirements with a minimum of additional circuitry. The MTTL AND power gate is shown in Fig. 6-25. The IC power gate is also available in the NAND configuration. With either AND or NAND, the output circuitry is designed to provide twice the fan-out of conventional gates (in this case, 20 standard gate loads instead of 10).

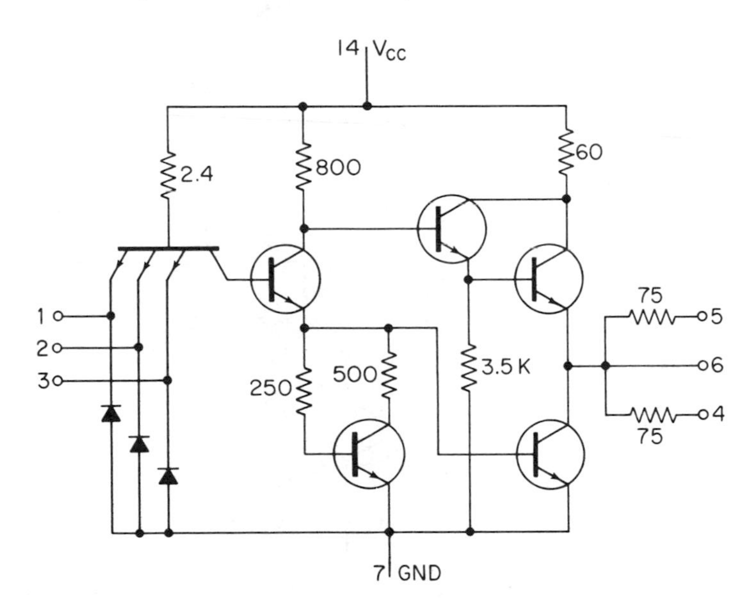

FIG. 6-23. Motorola MTTL terminated line driver.

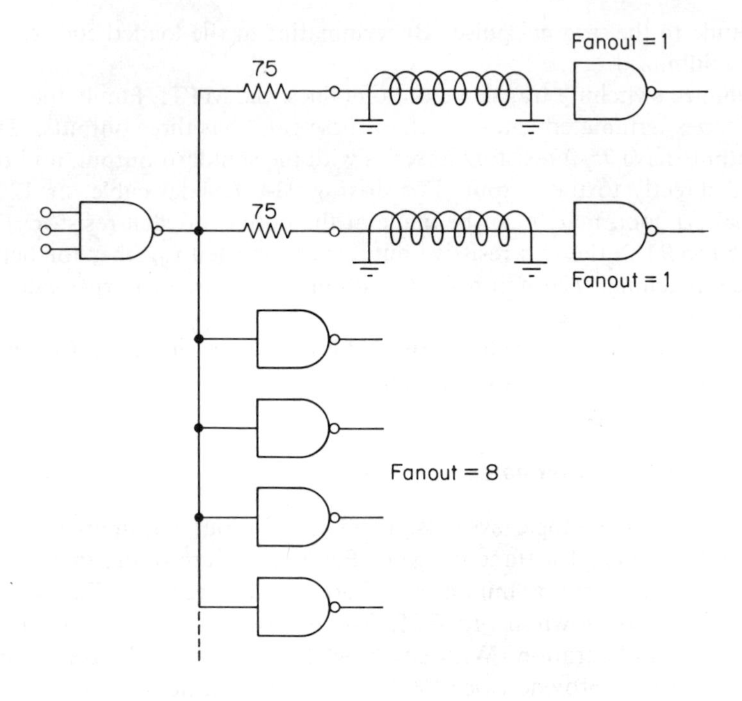

FIG. 6-24. Typical application of line driver (Motorola).

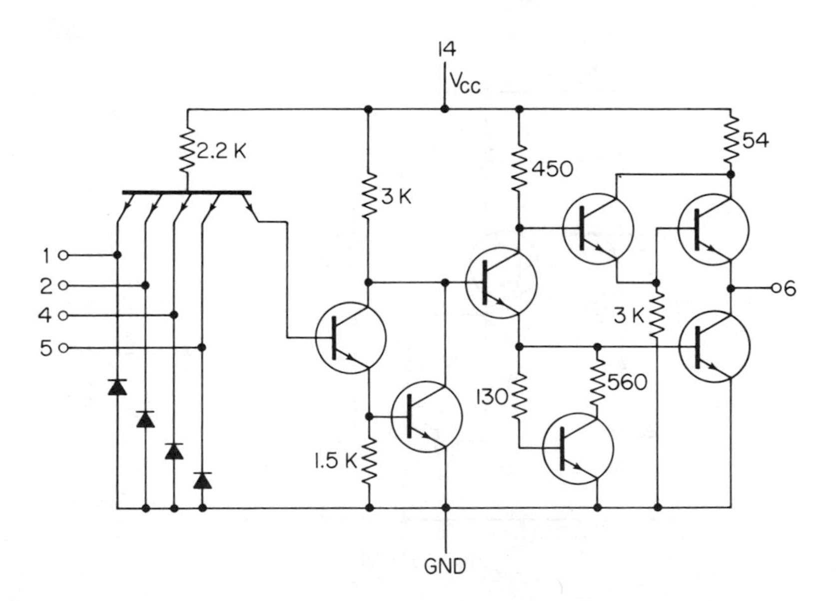

FIG. 6-25. Motorola MTTL power gate.

6-5.10 Open collector gates

To overcome the TTL limitation of not permitting wired-OR operation, a special MTTL gate is available without an active pull-up resistor. The circuit is shown in Fig. 6-26. The output of this circuit can be used for wired-OR, or to drive discrete components.

6-5.11 DLT compatible

The MC3000/MC3100 series of TTL is pin compatible with the MDTL (Motorola DTL) family. Supply voltage is applied to pin 14, and ground is applied to pin 7, in both families. More important, because of its "square" transfer characteristic, the MTTL family is the only TTL line that has a DTL-type transfer function. It should be noted that most TTL lines of a given IC manufacturer are compatible with the DTL lines of the same manufacturer, but may or may not be compatible with DTL lines of other manufacturers.

6-5.12 Output impedance

The Darlington output configuration provides extremely low output impedance in the "high" state. The low impedance results in excellent noise immunity, and allows high-speed operation while driving large capacitive loads. Typically, the "high" state output impedance varies from about 10 Ω at outputs of 3.5 V, to about 64 Ω at lower output voltages.

The "low" state output impedance is typically 6 Ω at an output of 0.2 V, and goes on up to about 500 Ω if the output increases to 0.5 V.

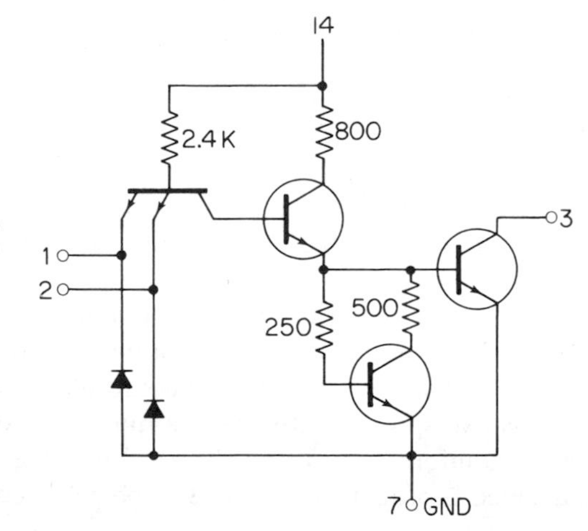

FIG. 6-26. Open collector Motorola MTTL gate for implementing the wired-OR function.

6-5.13 Totem-pole output

One disadvantage of conventional TTL is the so-called "totem-pole" output. As shown in Fig. 6-16, both output transistors are on during a portion of the switching time. Since the turn-off time of a transistor is normally greater than the turn-on time, the following occurs.

In going from a "high" state to a "low" state on the output, transistor Q_4 is initially on, and is in the process of turning off. Transistor Q_5, at the same instant in time, is off and is attempting to turn on. Transistor Q_5 turns on before transistor Q_4 can turn off. The result is a current spike through both transistors and the load resistor. The same effect takes place when the conditions are reversed, and transistor Q_4 turns on before transistor Q_5 can turn off. The active bypass network in the MTTL line, shown in Fig. 6-18, helps to limit this problem.

6-6. HIGH-THRESHOLD LOGIC (HTL)

HTL is designed specifically for logic systems where electrical noise is a problem, but where operating speed is of little importance (since HTL is the slowest of all IC logic families).

6-6.1 Noise sources for ICs

Electrical noise has always been a source of trouble for electronic systems whether they are composed of discrete components or integrated circuits. Electrical noise can come from many sources, both external to the electronic system under consideration, and noise self-induced by the circuitry itself. Some typical examples of external noise sources are switching of inductive circuits, rotating machinery, and various electrical control circuits.

Internal noise may be caused by the switching of one circuit affecting the state of another circuit. The amount of noise induced into the passive circuit is a function of the voltage swing, current change, the switching speed of the active circuit, and the inductive-capacitive coupling between the two circuits. Coupling may also take place by the use of a common path for the active and passive devices such as a power supply or ground lead. Noise from external sources is induced into the system under similar conditions.

Generally, noise is a random combination of many sources and, as such, is extremely hard to analyze. The net result, however, is that induced positive and negative spikes relative to the quiescent condition of a line may cause erroneous information to be absorbed into the system. This condition must be avoided if proper operation is to be achieved.

6-6.2 Noise reduction for ICs

Several schemes have commonly been used to reduce the effect of electrical noise on a system composed of ICs. Physical shielding of the ICs and the associated wiring prevents external electromagnetic radiation from inducing noise into the circuitry. Special buffering circuits can be used between the electronic circuits and signal leads. In many cases, signal leads require special routing considerations and special shielding. Extra filtering of the power supply leads may be required to reduce the noise introduced by this routing. Internal noise generation may require special spacing and routing considerations, as well as maintaining short lead lengths. In some cases, the power supply may need to be by-passed at several points on a board.

The amount of additional components and equipment necessary to protect ICs from electrical noise can increase to a point where it is economically desirable to seek other methods of operation to obtain the desired results. It is advantageous to have an IC family with a high degree of inherent noise immunity for the economical construction of an electronic system. This will minimize the amount of special care needed for proper circuit operation in areas with a high electrical noise environment.

6-6.3 Motorola MHTL

As discussed in Chapter 1, the basic MHTL gate is very similar to the MDTL. This is shown in Fig. 6-27, along with the transfer characteristics of the MHTL.

The basic difference between MHTL and MDTL is in diode D_1, resistor values, and the collector supply voltage V_{CC}. In MDTL, D_1 is a base-emitter diode operated in its forward direction and having a drop of approximately 0.75 V. The input threshold level of MDTL is a net of two forward diode drops (the input diode offsets the diode drop in the other direction) or about 1.5 V.

In MHTL, D_1 is a base-emitter junction that is operated in its reverse direction (or Zener condition). Conduction occurs when the junction has approximately 6.7 V across it. Thus, the threshold voltage for MHTL is one forward diode drop, plus one reverse diode drop, or about 7.5 V. The normal supply voltage for MHTL is 15 V. In order to keep power dissipation down, the gates have higher resistance values than comparable resistors in MDTL devices.

The MHTL gate of Fig. 6-27a provides the same positive-logic NAND function as the MDTL gate of Fig. 6-27b. If either of the A or B inputs is below the threshold level, possible base current to transistor Q_1 is routed to the low input. If both inputs are above the threshold level, Q_1, D_1 and output transistor Q_2 all turn on, and the output goes low. Thus, the output is true or high if A or B is not true. That is, $F = \overline{A \cdot B} = \overline{A} + \overline{B}$.

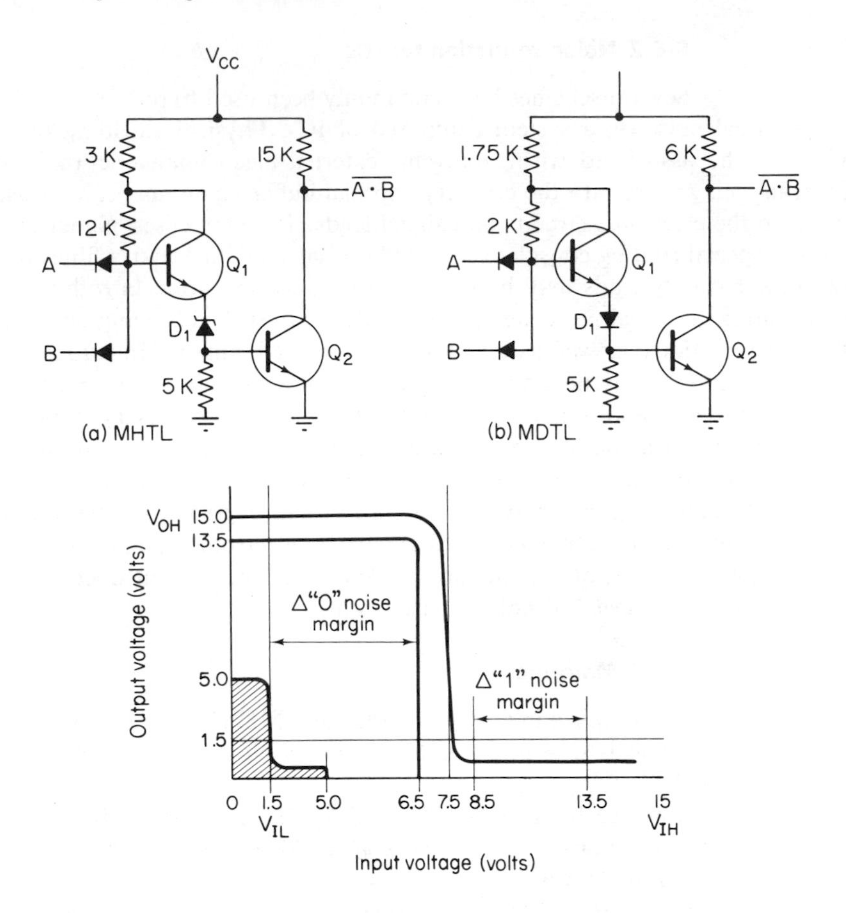

FIG. 6-27. Motorola MHTL gate and transfer characteristics.

6-6.4 MHTL transfer characteristics

The typical MHTL transfer curve is shown in Fig. 6-27c. For normal input low voltages, less than 1.5 V, it can be noted that the output exceeds $V_{CC} - 1.5$ V, and will continue to do so for any input up to 6.5 V. A transition width is specified from 6.5 V to 8.5 V, and once the input exceeds 8.5 V, the output is guaranteed to be below 1.5 V. This will remain true for any further increase in the input voltage.

With a 15-V supply, worst-case noise margin in either the high or low state is 5 V. Normally, the low input voltage is 1 V, the transition region is between 7 and 8 V, and the high output voltage is better than 14 V, thus typical noise margins of 6 V are obtained in either state. As a comparison, the transfer region for other forms of IC logic generally lies within the shaded area shown in the lower left-hand portion of Fig. 6-27c.

6-6.5 Active versus passive pullup configurations

Although the basic MHTL gate shown in Fig. 6-27 is available with the 15-kΩ pullup resistor, the MHTL IC line is normally supplied with the active pullup configuration shown in Fig. 6-28. In this cricuit, when Q_2 is off, base current is supplied to Q_3 from the 15-kΩ resistor, and load current is effectively supplied through the 1.5-kΩ resistor. When Q_2 is on, load current flows through D_2 and Q_2. Base current is also shunted from Q_3, cutting Q_3 off in this state. The diode drop across D_2 accounts for the somewhat higher low-state voltage of MHTL as compared to other forms of logic families.

Each form of output has its advantages and disadvantages, and the particular application will determine which device to use.

The active pullup configuration has a lower output impedance in the high state, and consequently will provide a higher degree of noise immunity from an energy point of view. This lower impedance can also better drive a load when in the high state. Thus, it is a superior interface for discrete components such as NPN transistors. The outputs of the functions with active pullup should not be connected together unless all inputs are also paralleled to insure simultaneous operation of all devices. If one device is turned on, while another device is off, the device in the low state will be required to sink current from the active pullup configuration of the high state unit. This will not damage the devices, but it leaves very little margin for providing load capacity to other devices.

The main advantage of the passive pullup configuration is its ability to have outputs of separate devices connected together. For each additional gate

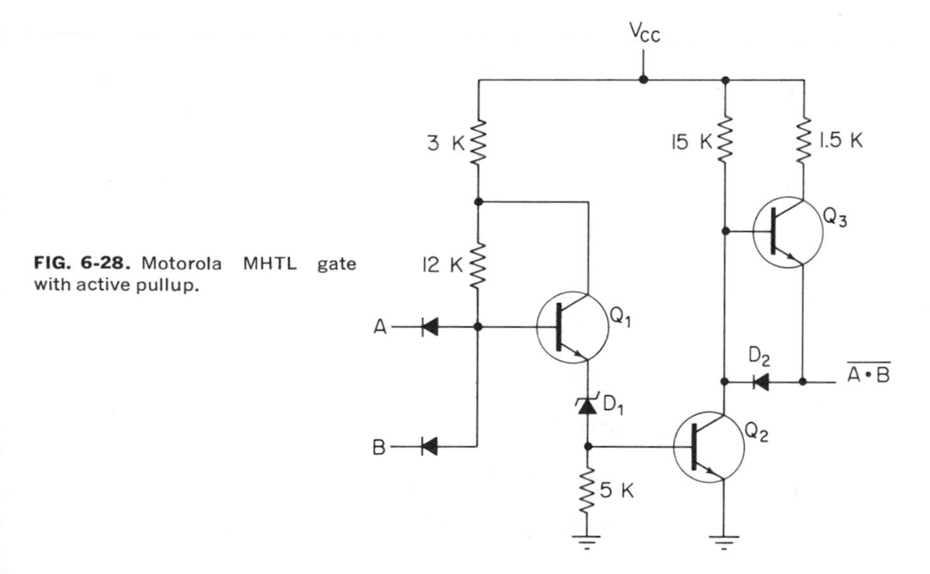

FIG. 6-28. Motorola MHTL gate with active pullup.

connected to the output of a gate, the original output loading factor of that gate must be reduced 1.25 because of the additional current that will be handled when a device is in the low state. When passive outputs are connected together, the impedance in the high state is reduced and, correspondingly, the noise immunity from an energy standpoint is increased as compared to that for a single gate. The passive gate also normally has a lower V_{OL} than the active pullup configuration since only a $V_{CE(\text{sat})}$ is involved although V_{OL} is still tested at 1.5 V with an I_{OL} of 12 mA.

6-6.6 MHTL translators

The MHTL logic line includes the usual devices: standard gates, multiple-input gates, expandable gates, flip-flops, and monostable multivibrators. In addition, the MHTL line includes two translators. One translator is used for interfacing from HTL to RTL, DTL, or TTL, whereas the other translator provides interface from the three logic families to HTL.

The translator for interface from MHTL to the other three logic families is shown in Fig. 6-29. For conversion to DTL and TTL, a 5-V supply is connected to a 2-kΩ pullup resistor through pin 13. (Note that only $\frac{1}{3}$ of the

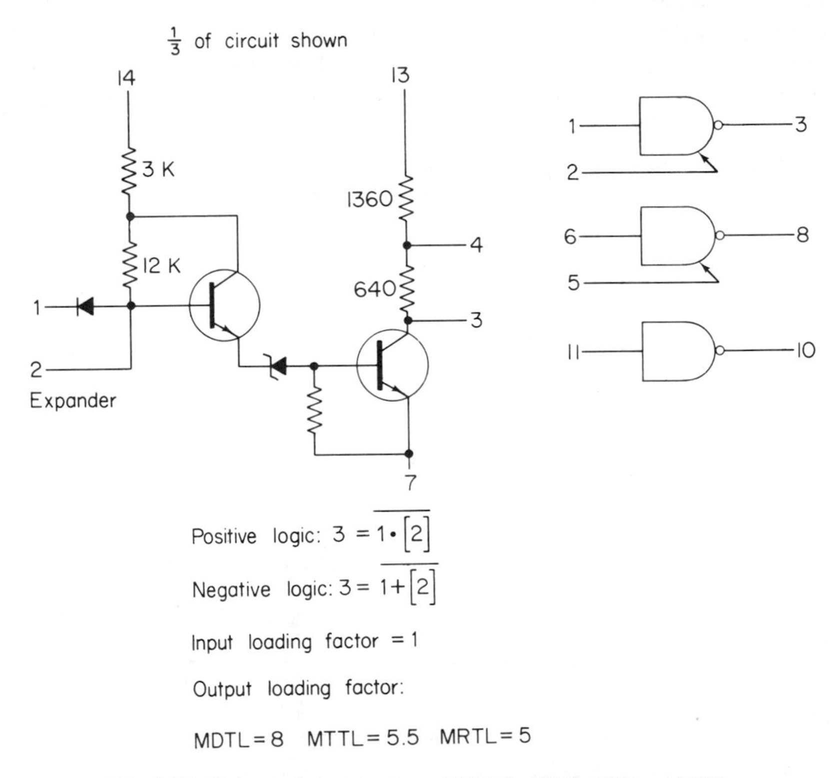

FIG. 6-29. Motorola translator from MHTL to MRTL, MDTL or MTTL.

schematic is shown. However, all three logic elements are contained in one IC package.) For interface with RTL, pins 4, 9 and 12 are connected to the RTL supply voltage (nominally 3.6 V). Expander points (pins 2 and 5) without diodes are present at the inputs of two of the units, but not on the third unit. This is because of the need for additional leads for the RTL power supply, and pin limitation of the 14-lead package.

The translator for interface from RTL, DTL and TTL to MHTL is shown in Fig. 6-30. This translator is also a triple unit. Signals from DTL/TTL

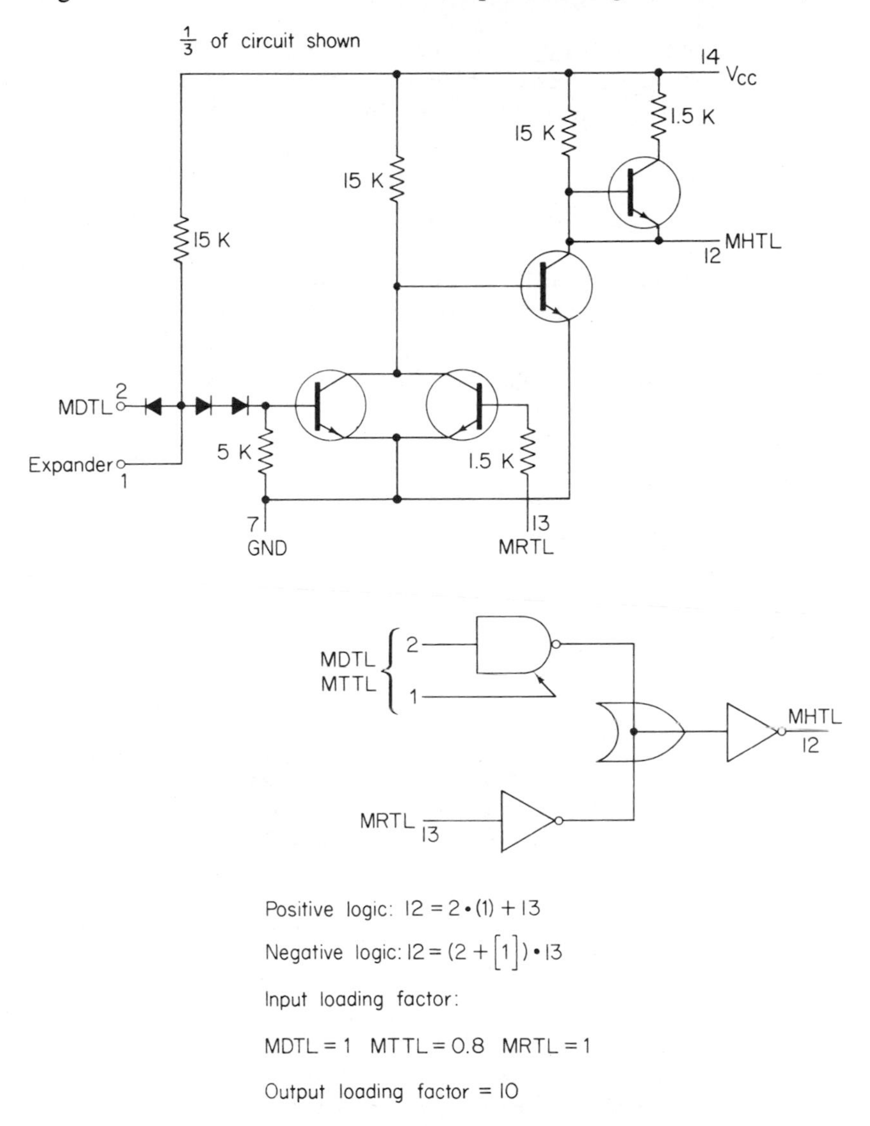

FIG. 6-30. Motorola translator from MRTL, MDTL, or MTTL to MHTL.

sources are applied to one set of input terminals, while signals from RTL sources are applied to another terminal. The different inputs provide threshold levels and characteristics compatible with MDTL/MTTL and MRTL families. Each DTL/TTL section is also an input expander terminal (pins 1, 6 and 9) without diodes. These terminals may be used to expand input logic capability or to use high-voltage diodes to readily interface high-voltage relay or switch circuits to HTL levels. Both types of inputs may be applied simultaneously, with one output going high if the logic function of either input goes high.

If the RTL input is used by itself, the DTL/TTL input must be grounded for proper operation. This is not necessary if the DTL/TTL input is being used by itself, but it is advisable under this condition to ground the RTL input to reduce any possible noise pickup.

6-6.7 Line driving problems

Applications exist where it is desirable to transmit data over an appreciable distance. The large logic swing of HTL provides the means of transmitting data, while minimizing the effects of noise. Unfortunately, most transmission lines have impedances below 150 Ω. The output impedance of any HTL line is much higher than this value, and is generally greater than 500 Ω. This mismatch can cause reflections to appear on the transmission line.

The circuit of Fig. 6-31 can be used to minimize mismatch (and thus minimize reflections or distortion). The value of R_1 is selected (or adjusted) to match the impedance of the IC output, while the value of R_2 is set to match the line impedance. In practice, R_1 and R_2 are set to approximate values, a pulse input is applied to the gate (or other digital IC element), and the pulse waveform is monitored with an oscilloscope at both ends (originating and

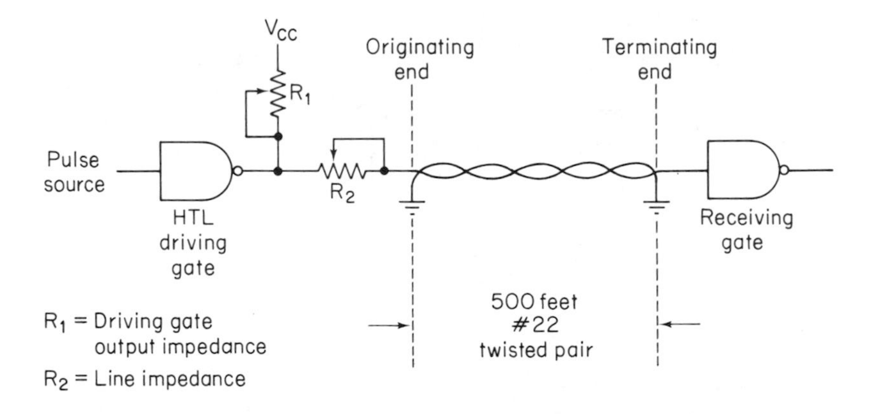

FIG. 6-31. Line driving circuit for digital ICs.

terminating) of the line. The waveform should be substantially the same at both ends, if R_1 and R_2 are properly adjusted to provide impedance match.

One precaution must be observed. The IC output must be capable of sinking the current passed through R_1. For example, if V_{CC} is 5 V, and R_1 is 500 Ω, a 10-mA current is applied to the IC output.

6-6.8 Driving discrete components from HTL

In some cases, HTL must be used to drive discrete transistors. This creates a problem because of the high output voltage from HTL. The output in the high state is not as much of a problem as the output of an HTL in the low state. Assuming positive logic, the output of an HTL in the low state (or "0" condition) is about 1 V. (This is generally listed as V_{OL}.) If 1 V is applied to an NPN in the low state, the NPN will probably never turn off. However, an HTL with an active pullup output can be used for driving NPN transistors if the circuit of Fig. 6-32 is used.

The higher V_{OL} voltage of the HTL is partially due to the extra diode D_1 on the output. However, if the gate is not required to sink current, then the voltage on the base of the NPN transistor will be very close to ground. The base voltage of the NPN is equal to the leakage current of the collector-base junction times the value of R_1. Thus, resistance R_1 should be chosen to provide a base voltage of about 0.2 V (or less), using I_{BC} leakage current as the factor. With such a circuit, the HTL should not be required to sink any current when the output is in the low state.

6-7. EMITTER-COUPLED LOGIC (ECL)

Since ECL uses transistors in the non-saturating mode, it is inherently the fastest type of logic available. Delay times range from about

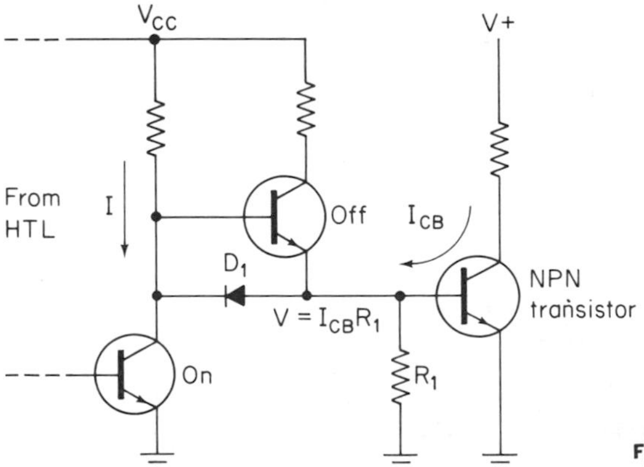

FIG. 6-32. Driving discrete transistors from HTL digital IC outputs.

2 to 10 ns. ECL generates the least amount of noise, and has considerable noise immunity. However, as a tradeoff for the non-saturating mode, ECL is the least efficient. That is, ECL dissipates the most power for the least output voltage. Power dissipation of the ECL is exceeded only by HTL (which must operate at higher power levels to produce the high input/output voltage levels).

A typical ECL gate is shown in Fig. 6-33. Note that a single, negative supply V_{EE} is used (V_{CC} is grounded), but logic is positive. Also note that a bias V_{BB} is required.

The gate of Fig. 6-33 uses a differential amplifier input, resulting in high input impedance, and good rejection of power supply variations. The very low output impedance of the emitter followers results in high fan-out and fast rise time for capacitive loads. Resistors and logic levels are chosen to prevent saturation of the input transistors, thus eliminating storage time. Resistor R_1 stabilizes circuit operation for wide variations in transistor gain.

A logical 1 for the circuits is −0.75 V, which corresponds to one base-emitter voltage drop below ground. Logical 0 is −1.55 V, which yields a nominal voltage swing of 0.8 V. (However, ECL lines are available with logic swings up to about 2 V.) Some typical transfer characteristics are shown in Fig. 6-34.

6-7.1 Interfacing ECL logic

It can be seen from the transfer characteristics that the ECL circuit of Fig. 6-33 is not directly compatible with typical levels of the other logic families. As in the case of HTL (Sec. 6-6.6), ECL level translators are

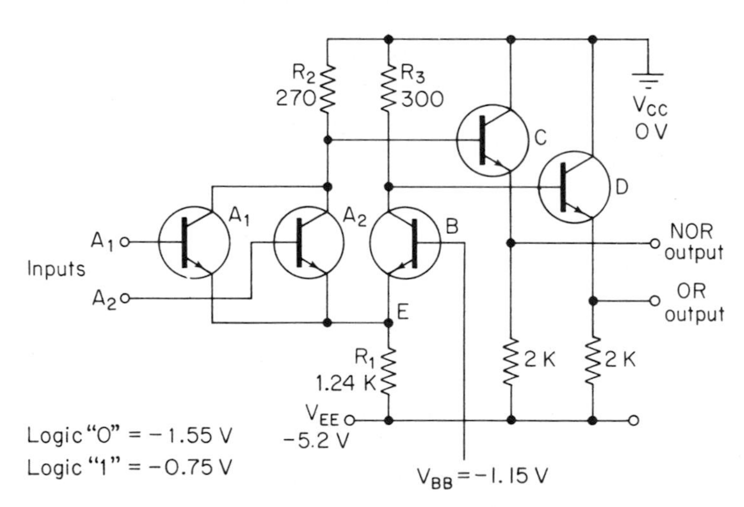

FIG. 6-33. Motorola MECL logic gate.

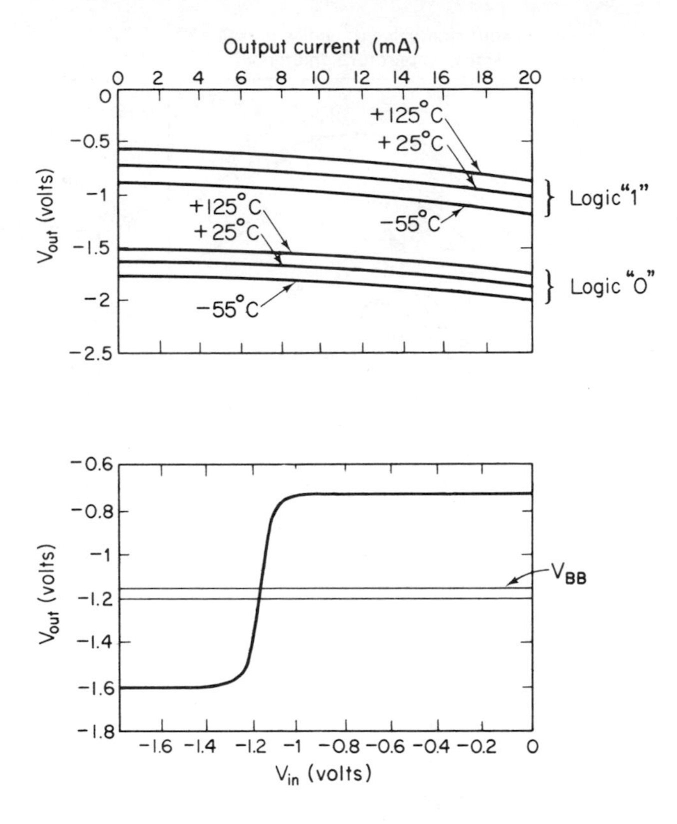

FIG. 6-34. Typical ECL transfer characteristics.

available for interfacing with other logic lines. However, if the only problem is one of interfacing from an input to ECL, it is possible that the circuit of Fig. 6-35 will solve the problem. The input (either discrete component or IC) must be on the order of 1 V. As shown by the equations, the values of R_1 and R_2 are selected to give a logic 0 (-1.5 V) at the output (to the ECL input). The value of C can be determined by the fact that the RC time constant should be several times greater than the duration of the input pulse.

6-7.2 Bias problems with ECL

Referring to the circuit of Fig. 6-33, as the power supply voltage is increased, the 0-level will move more negative, while the 1-level remains essentially constant. It is essential that the bias voltage V_{BB} track any variations in the power supply voltage. For this reason, Motorola supplies a bias driver with their MECL line. The bias driver circuit is shown in Fig. 6-36.

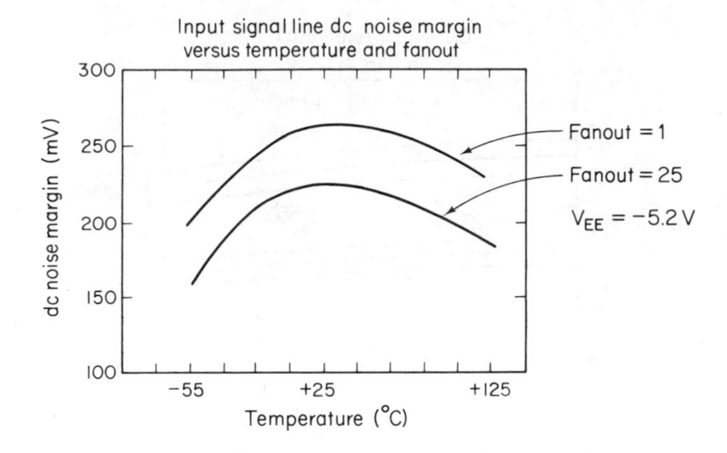

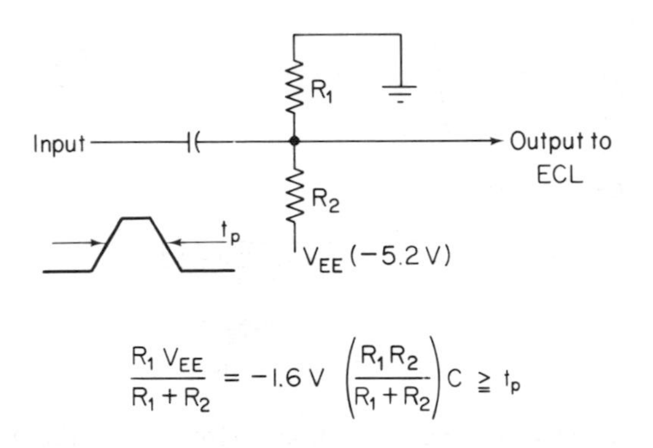

$$\frac{R_1 V_{EE}}{R_1 + R_2} = -1.6\text{ V} \qquad \left(\frac{R_1 R_2}{R_1 + R_2}\right) C \geqq t_p$$

FIG. 6-35. Interfacing from non-ECL input to ECL.

The bias driver provides a temperature and voltage compensated reference for MECL logic. Any of the three MECL voltages may be grounded, but the common voltage of the bias driver must correspond to that of the logic system.

The bias driver has a unique application when coupling a signal through a capacitor to pin 4. First, the gate acts as a level translator from any voltage to MECL levels. If the input is 0.8 V (peak to peak), the standard MECL levels will appear at pin 1. For low-level ac and RF inputs, the output will be centered in the active region of the MECL gate, which may then be used as an amplifier. A low-quality, high-bandwidth "differential amplifier" may be obtained by using two bias drivers, one connected to a normal gate input, and the other connected to the V_{BB} input of a MECL gate. The OR and NOR outputs of the MECL gate then provide a very low impedance differential output.

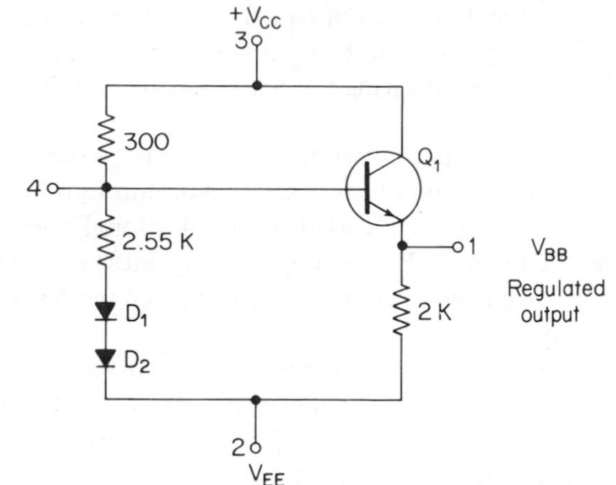

FIG. 6-36. Motorola MECL bias driver.

The ac input impedance of the bias driver is 250 Ω in parallel with about 5-pF input capacitance. This will terminate a low impedance line fairly well. If V_{CC} contains excessive noise, the output on pin 1 can be filtered by connecting a capacitor between pins 4 and 2. Typical power dissipation is 12 mW. The bias driver will fan-out to 25 unit input loads.

Referring back to the circuit of Fig. 6-33, if V_{BB} is obtained from the bias driver connected to the same supply, the bias or reference voltage will track with supply voltage changes or temperature variations, thus keeping V_{BB} in the center of the logic levels.

6-7.3 Using ECL units

A fairly complete line of ECL devices is available in IC form. A typical ECL line includes: OR, NOR gates with multiple inputs, flip-flops of all types, adders, input expanders, gates with or without *pulldown* resistors (emitter resistors in the output), line drivers, and lamp drivers. The datasheets (and complete brochures in some cases) supplied with the devices provide adequate information for both application and design rules pertaining to ECL. Such data will not be repeated here. Instead, we shall list the highlights or general rules for using ECL.

1. The maximum recommended ac fan-out for typical ECL is about 15 input loads. Direct-current fan-out is about 25 loads. The ac fan-out is lower than dc fan-out because of the increase in rise time and fall time with high fan-out. Also, if high fan-outs and long leads are used, overshoot due to lead inductance becomes a problem.

2. A circuit such as the bias driver (Fig. 6-36) will fan-out to about 25 loads. Note that a dual gate or half adder is equivalent to two gate input loads for a circuit such as the bias driver.

3. Each $\bar{J}$ or $\bar{K}$ input to a flip-flop is equivalent to one and one-half loads. For example, a $\bar{J}$ and $\bar{K}$ input tied together as a flip-flop clock input would be a load of three, allowing a gate to drive five flip-flops. All other inputs are a load of unity (or one).

4. The output of two ECL gates may be tied together to perform the wired-OR function, in which case a maximum fan-out of 5 is allowed. If only one pulldown resistor is used, each additional common output is equivalent to one gate load. For example, if 6 gates are wired together with only one pulldown resistor connected, the fan-out would be (15 − 5), or a fan-out of 10 remaining.

5. All unused inputs must be tied to V_{EE} for reliable operation (assuming that the power connections are as shown in Fig. 6-33). As seen from the gate input characteristics, the input impedance of a gate is very high when at a low level voltage. Any leakage to the input and/or wiring capacity of the gate will gradually build up a voltage on the input. This may affect noise immunity of the gate or hinder switching characteristics at low repetition rates. Returning the unused inputs to V_{EE} insures no buildup of voltage on the input, and a noise immunity dependent only upon the inputs used.

6. A recommended maximum of three input expanders should be used (assuming that each input expander provides 5 inputs). Thus, the recommended maximum input to any ECL gate is 15. If this is exceeded, the NOR output rise and fall times suffer noticeably because of the increased capacitance at the collector of the input transistors. For low frequencies, higher fan-ins may be used, if rise and fall times are of no significance.

7. Each gate in the IC package must have external bias supplied (except for certain ECL gates which have an internal bias scheme). ECL flip-flops do not normally require an external bias.

6-7.4 ECL crystal-controlled oscillators

The ECL circuit (typically that shown in Fig. 6-33), makes an ideal basic element for a crystal-controlled oscillator. If a circuit is to produce sustained oscillations, it must contain an amplifier with a feedback network capable of 360° phase shift from output to input. The use of an ECL gate as an amplifier may be explained with the aid of a basic ECL circuit, rearranged as shown in Fig. 6-37. The circuit consists of a differential amplifier, with emitter-follower stages on either side of the amplifier. As the input makes small excursions around V_{REF}, the OR output will amplify this signal, while the NOR output will amplify and invert the signal. Thus, the ECL gate may be treated as an amplifier, or an inverting amplifier, depending on which output is under consideration.

For the feedback network, a quartz crystal is used in the series resonant mode of oscillation, providing a low impedance path from output to input

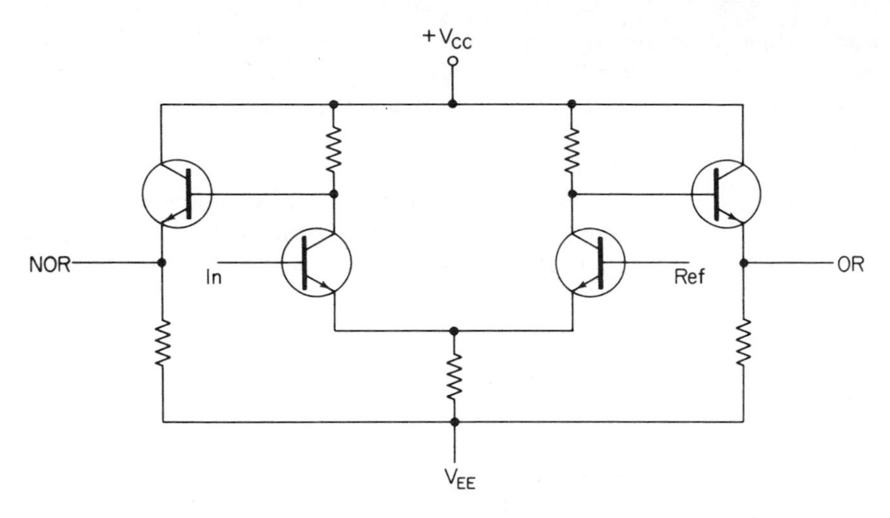

FIG. 6-37. Basic MECL gate configuration (Motorola).

when operating at or near resonance. An electrical equivalent circuit of a quartz crystal is shown in Fig. 6-38. L, R, and C_S represent the piezoelectric characteristics of the quartz crystal, while C_O represents the static capacitance between leads. Two modes of resonance are possible: the series resonant frequency of L and C_S, and the parallel resonant, or anti-resonant frequency of L and C_O. The series resonant mode is used. With the crystal operating in this manner, the reactances of L and C_S effectively cancel each other, and the resulting feedback path then consists of the resistance R, shunted by C_O.

Due to processing limitations during manufacture, the maximum resonant of most quartz crystals is about 20 MHz. For frequencies above 20 MHz, crystals are used which will oscillate at a harmonic, or "overtone" of their fundamental frequency. When the desired oscillator frequency requires that an overtone crystal be used, additional precautions must be taken in the design of the circuit to insure that it will oscillate only at the desired frequency. This is because an overtone crystal will operate at harmonics other than the one intended, as well as the fundamental frequency.

The use of ECL ICs as crystal oscillators has several advantages. When a second gate is used as a buffer for the oscillator, frequency does not change

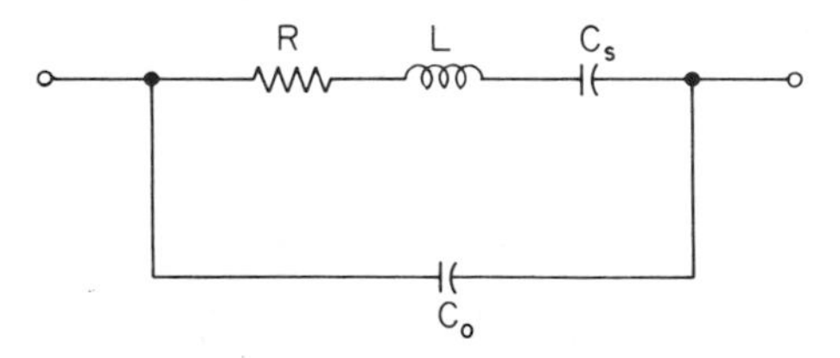

FIG. 6-38. Equivalent circuit of a quartz crystal.

with loading. Also, the high input impedance and approximate 0.9-V logic swing prevent the possibility of overdriving the crystal. In addition, both circuits described in the following paragraphs are frequency insensitive to ±20 percent power supply variations, and the frequency versus temperature characteristics closely follow the crystal characteristics.

In making physical layout of the following circuits, follow the general practices of Chapter 2. Also, connect all unused inputs of the IC package to V_{EE}. Since high frequencies are involved, keep *all leads* as short as possible.

6-7.4.1 ECL overtone crystal oscillator

Figure 6-39 shows a circuit using an adjustable tank circuit which insures operation at the desired crystal overtone. C_1 and L_1 form the resonant tank circuit, which with the values specified, has a resonant frequency adjustable from approximately 50 MHz to 100 MHz. Overtone or harmonic operation is accomplished by adjusting the tank circuit frequency at or near the desired frequency. The tank circuit exhibits a low impedance shunt to off-frequency oscillations, and a high impedance to the desired fre-

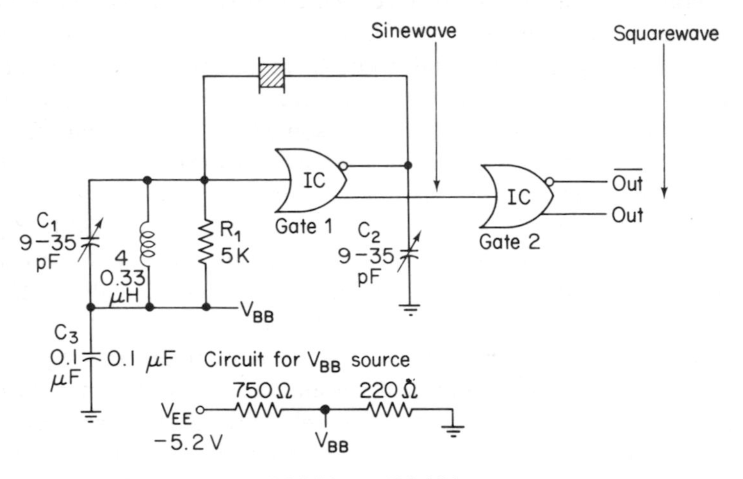

Frequency range: 50 MHz to 100 MHz

$$F \approx \frac{1}{6.28\sqrt{L_1 C_1}} \qquad L(\mu H) \approx \frac{2.54 \times 10^4}{F(MHz)^2 \times C(pF)}$$

$$F_{max} \approx \frac{2.86}{t_r} \qquad C(pF) \approx \frac{2.54 \times 10^4}{F(MHz)^2 \times L(\mu H)}$$

t_r = risetime

Gate 1 = Gate 2 = $\frac{1}{2}$ MC 1023

FIG. 6-39. MECL crystal controlled oscillator (overtone crystal).

quency, allowing feedback from the output. Operation in this manner guarantees that the oscillation will always start at the correct overtone.

The reference voltage for the differential amplifier is supplied internally in the Motorola gate shown (MC1023) and has a nominal value of -1.2 V when V_{CC} is at ground, and V_{EE} is -5.2 V. Other ECL ICs can be used. However, the reference voltage may or may not be supplied internally.

The exact reference voltage configuration (internal or external) is not critical. However, for the oscillator to function properly, the input, which is the other side of the differential amplifier, must be biased near the reference voltage (-1.2 V in this case). The bias is produced by R_1 which is connected to an external source V_{BB} of -1.2 V. The high resistance of R_1 (about 5 kΩ typical) is necessary to prevent the feedback signal from being shunted to V_{BB}.

The V_{BB} source can be obtained by a simple resistive voltage divider as shown in Fig. 6-39. The capacitor C_3 is a filter for the V_{BB} supply.

Capacitor C_2 is used to load the output of the gate in order to increase the rise time, fall time, and the propagation delay. This is necessary to provide the proper phase relationship in the feedback loop.

The second gate serves the dual purpose of a buffer and waveshaper. The output of the first gate, which is approximately a sinewave, is fed into the second gate which shapes the output into a square wave with rise and fall times of approximately 2 ns. This shaping is due to overdriving of the second gate. It should be noted that both the square wave and its complement are simultaneously available at the second gate, due to the OR/NOR configuration of an ECL gate.

Although the circuit of Fig. 6-39 is usable in the 50- to 100-MHz range, it is restricted to these frequencies only by the specified values of L_1 and C_1. The operating frequency range can be moved in either direction by changing the value of L_1 and/or C_1. For example, if the value of L_1 is changed to 1 μH, and C_1 remains unchanged, the usable frequency range will now cover approximately 27 to 50 MHz. The equations shown in Fig. 6-39 can be used to find the approximate frequency range for other values of L_1 and C_1.

The high frequency limit of the circuit is set by the rise and fall times (or delay) of the ECL IC involved. The maximum frequency equation shown on Fig. 6-39 can be used to find the approximate maximum operating frequency. For example, assuming rise and fall times of 1.8 ns (which is about minimum for an ECL), the maximum operating frequency is equal to about 150 MHz (2.86/1.8). In practice, this can be verified by experimental results with an appropriate crystal, and correct values of L_1 and C_1.

6-7.4.2 ECL fundamental frequency crystal oscillator

For frequencies below about 20 MHz, where a fundamental frequency crystal can be used, the resonant tank is no longer required. Also, at this lower frequency range, the typical ECL propagation delay of 2 ns becomes small compared to the period of oscillation, and it becomes necessary to use the non-inverting output for feedback.

The fundamental crystal oscillator circuit is shown in Fig. 6-40. Since the non-inverting output is used for feedback, the ECL now functions as an amplifier. R_1, C_1 and the reference voltage V_{BB} are the same value, and serve the same purpose as for the circuit of Fig. 6-39. Also, the second gate again serves as a buffer and waveshaper. However, if the frequency is quite low, the rise and fall times of the square waveform from the second gate may be increased slightly. The trimmer capacitor C_2, in series with the crystal, is used for minor frequency adjustment.

The circuit of Fig. 6-40 has a maximum operating frequency of approximately 20 MHz, and a minimum frequency of about 1 MHz.

6-7.5 ECL Schmitt triggers

The ECL circuit (typically that shown in Fig. 6-33), makes an ideal basic element for a Schmitt trigger. Such a trigger is a regenerative circuit which changes state abruptly when the input signal crosses specified dc triggering levels.

The circuit of Fig. 6-41 is a basic Schmitt trigger. The following initial state is assumed: Q_1 is cut off, and Q_2 is in saturation. V_{out} is in the low-level state. V_{in} is starting to increase.

As long as Q_2 is saturated, V_{out} will remain in the low-level state. V_{in} is increased until Q_1 conducts some minimum current. As this point, the voltage at the collector of Q_1 has lowered until Q_2 is just at the active-saturation point. At this point, V_{in} equals V_U, the "upper triggering level".

As V_{in} is further increased, the regenerative action occurs as follows. The positive increase on the base of Q_1 is felt as a negative increase on the base

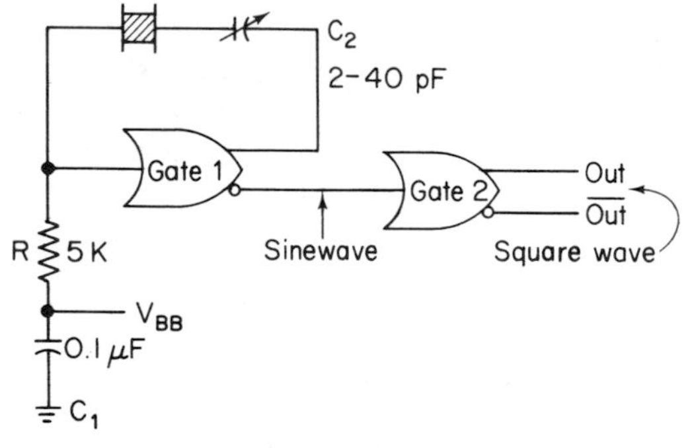

FIG. 6-40. MECL crystal controlled oscillator (fundamental crystal).

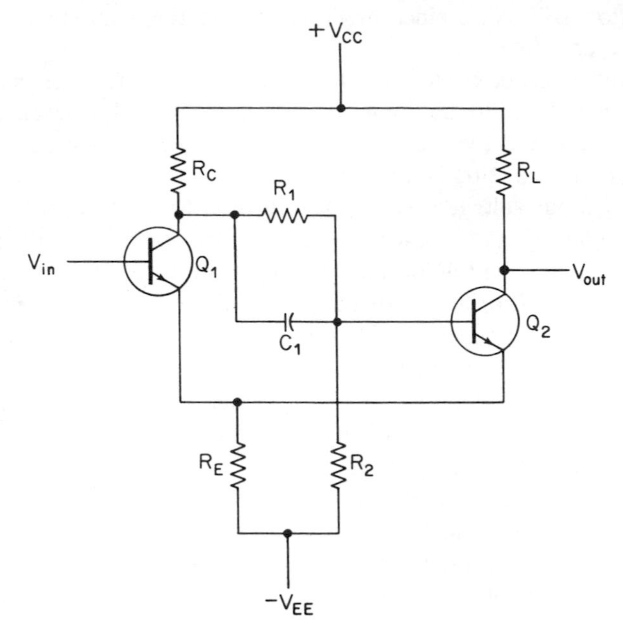

FIG. 6-41. Basic Schmitt trigger circuit.

of Q_2. Q_2 is no longer saturated, and the negative voltage at the base tends to cut Q_2 off, allowing a negative voltage increase to be felt at the common emitters of Q_1 and Q_2. This negative at the emitter of Q_1 tends to turn Q_1 on more, and the regenerative cycle will continue until Q_1 is saturated and Q_2 is cut off, which occurs without a further increase in $V_{\text{in}} \cdot V_{\text{out}}$ now goes to the high state, $V_{\text{out}} = V_{CC}$.

The state of the circuit has been reversed. Q_1 is saturated, Q_2 is cut off, and $V_{\text{out}} = V_{CC}$. To return the circuit to the original state, V_{in} is now decreased until Q_1 is no longer in saturation, and Q_2 is at the cut-off active threshold point. At this point, V_{in} equals V_L, the "lower triggering level". The addition of a negative voltage to V_{in} causes the regeneration action to start. Without further decrease in V_{in}, the regenerative action will continue until the circuit is once again in the initial state (Q_1 cut off, Q_2 saturated, and V_{out} at the low level).

V_U and V_L are, in general, not the same voltage, and the differential between them is called the hysteresis voltage, $V_H = V_U - V_L$. By proper selection of resistor values, one can obtain desired values of V_U, V_L and V_H.

An ECL Schmitt trigger is shown in Fig. 6-42. The circuit is accomplished by tying the NOR output of an ECL gate to the biased reference input through a feedback resistor R_F. The circuit shown is that of a Motorola MECL gate. Even though the gate is being used as a Schmitt trigger, the logic function still exists since any one of the three inputs will cause the Schmitt trigger action.

The operation of the circuit in Fig. 6-42 is basically the same as that for the circuit of Fig. 6-41. With all the input voltages at some low level, current will flow through Q_1, and the input transistors (Q_6, Q_7, Q_8) will be cut off since each base-emitter junction is reversed biased.

When any input voltage (say $V_{in\ 3}$) is allowed to increase, the current will switch from Q_1 to Q_8 as $V_{in\ 3}$ exceeds the reference potential on the base of Q_1. In a typical circuit, the OR output (V_{out}) will go to the high level when the potential on pin 6, 7, or 8 exceeds the potential on pin 1. Similarly, when the OR output is in the high state, and all input potentials go below the reference potential, current will switch back to Q_1, and the output will go to its low value. In typical Motorola MECL gate operation, the reference potential is V_{BB} taken directly from a bias driver. Thus, the reference voltage is the same for increasing and decreasing input voltages.

In using an ECL gate as a Schmitt trigger, two external resistors R_F and R_I are added to the gate and bias driver combination, as shown in Fig. 6-43 (the block diagram equivalent of Fig. 6-42).

One characteristic of a Schmitt trigger is that the upper triggering level V_U, and the lower triggering level V_L, on the input are not at the same level. This is shown in Fig. 6-44.

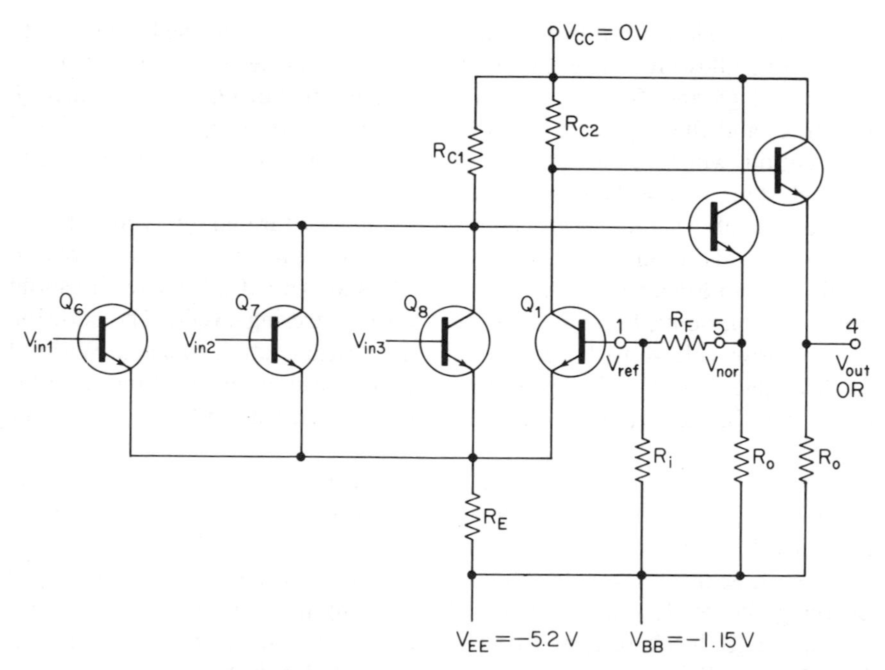

FIG. 6-42. MECL Schmitt trigger (using MC306).

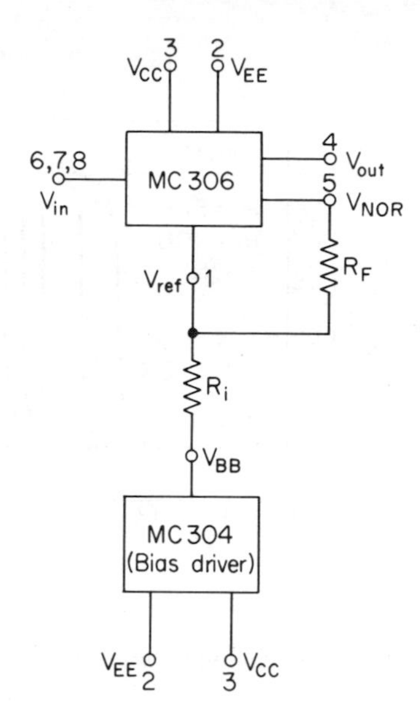

FIG. 6-43. Block diagram of MECL Schmitt trigger.

The V_L and V_U points (and thus the hysteresis or V_H) are set by the reference voltage V_{BB}, and the ratio of R_I and R_F. Figure 6-45 shows some typical values of V_L and V_U for typical R_I/R_F ratios. As shown, the V_U and V_L points are not symmetrical about the V_{BB} point. Also, minimum hysteresis occurs when R_I is low and R_F is high. An opposite ratio, R_F low and R_I high, produces the maximum hysteresis.

6-8. MOSFET DIGITAL INTEGRATED CIRCUITS

As discussed in Sec. 1-8.11, the most common form of field-effect transistor (FET) used in digital ICs is the metal-oxide FET, or MOSFET. At present, the most common type of digital IC MOSFET is the complementary or CMOS type. Such complementary CMOS ICs are a combination of P-channel and N-channel devices on a common chip.

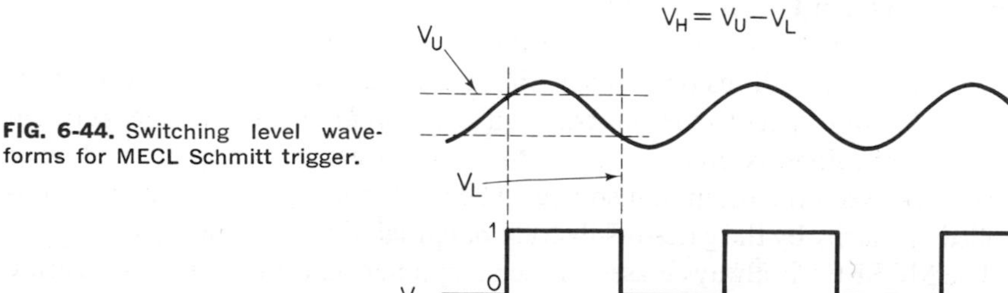

FIG. 6-44. Switching level waveforms for MECL Schmitt trigger.

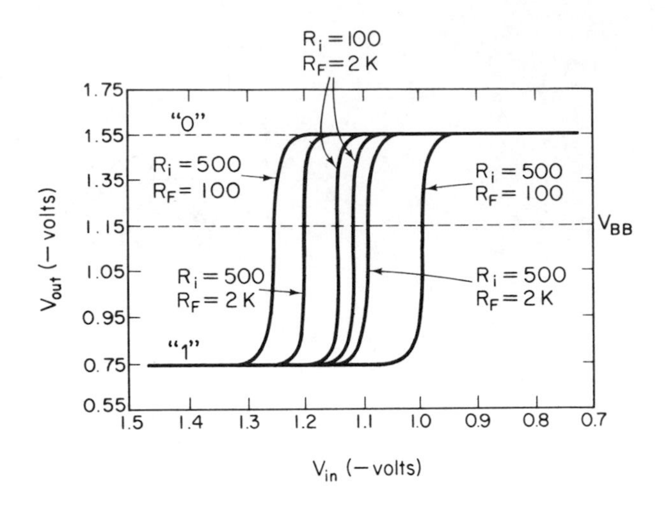

FIG. 6-45. Typical hysteresis voltage curves for MECL Schmitt trigger.

The basic physical construction, and corresponding diagrams, for both P- and N-channels of a CMOS IC are shown in Fig. 6-46. It is assumed that the reader is already familiar with the operating principles of discrete FETs. Such data is found in the author's *PRACTICAL SEMICONDUCTOR DATABOOK FOR ELECTRONIC ENGINEERS AND TECHNICIANS*, Prentice-Hall, Inc., Englewood Cliffs, New Jersey, 1970. The following is a summary of P- and N-channel operation, and MOSFET bias relationships, as they relate to digital ICs.

The P-channel device consists of a lightly N-doped silicon substrate with heavily-doped P-type diffusions into this substrate. Between the drain and source is the gate-oxide region which serves as the insulation between the metal gate and the substrate. The basic operation of the P-channel device involves placing the metal gate at a negative potential with respect to the substrate. The induced electric field causes an inversion of the N-type substrate into a P-type region. This inversion occurs only between the drain and the source diffusions. The inverted area of the substrate is called the *channel.* The carriers in a P-channel are *holes.*

The N-channel device consists of a P-doped silicon substrate with N-dopes drain and source diffusions. When the metal gate is placed at a positive potential with respect to the substrate, an *electron-dominated* channel between the two diffusions is created in the P-type substrate, resulting in the flow of current between the drain and source. The magnitude of current flow is controlled primarily by the gate-to-substrate potential difference, or bias.

The MOSFET is always biased in such a manner that the drain-to-substrate junction, and the source-to-substrate junction are *reversed-biased.* Thus, the

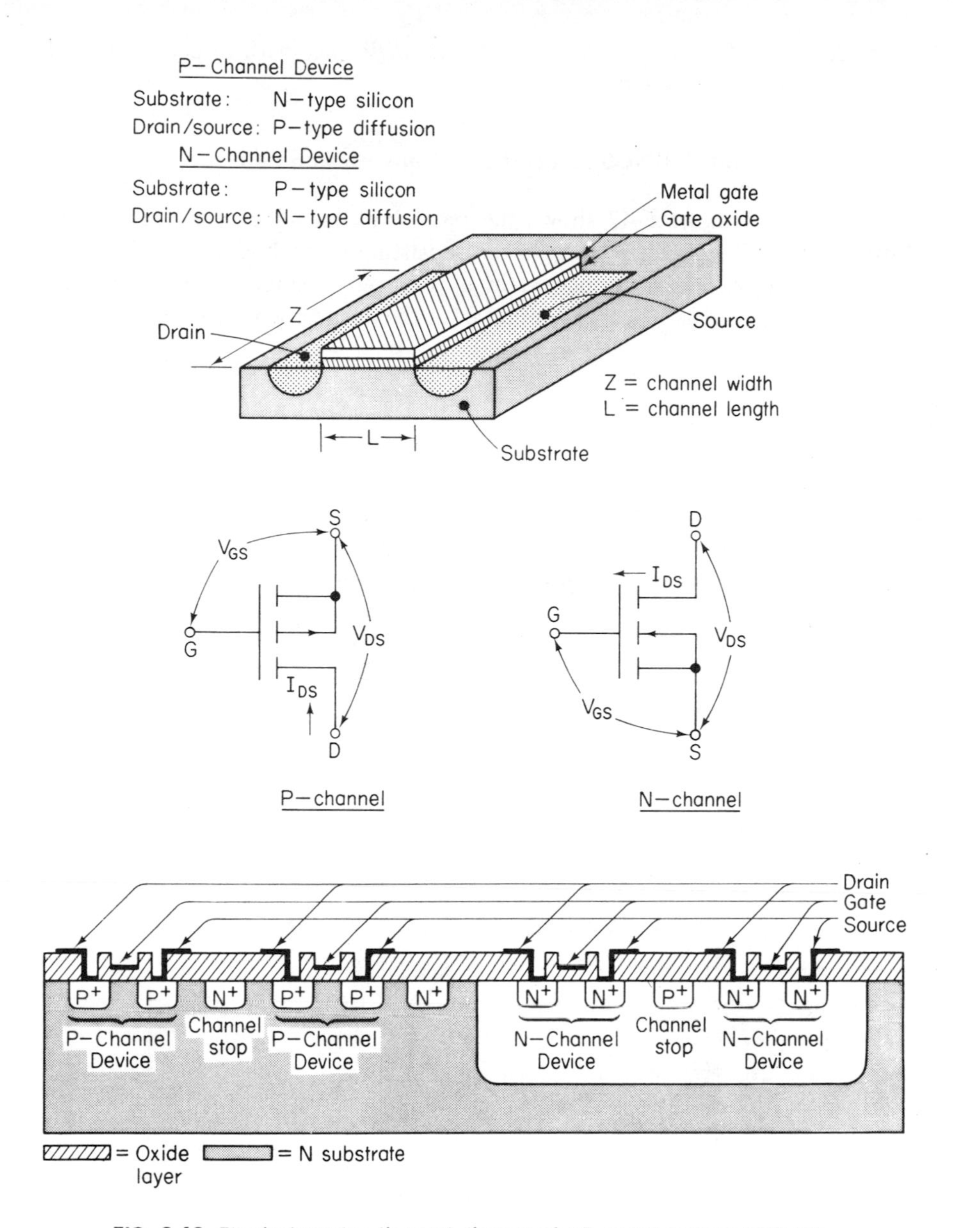

FIG. 6-46. Physical construction and diagrams for P- and N-channel CMOS IC.

substrate is always the most positive voltage on the P-channel device, and is always the most negative voltage on the N-channel device.

Physically, the drain diffusion is identical to the source diffusion. The two diffusions, however, are usually distinguished when the device is used in a circuit. The diffusion at the *least potential difference* with respect to its substrate is called the *source*.

6-8.1 CMOS inverter and transmission gate

Figure 6-47 shows the basic building block for most CMOS digital ICs. The circuit is an inverter consisting of an N-channel (Q_2) and a P-channel (Q_1) device connected in a push-pull fashion. Figure 6-47 also shows the V_{in} versus V_{out} transfer curve for the inverter when the power supply is +10 V.

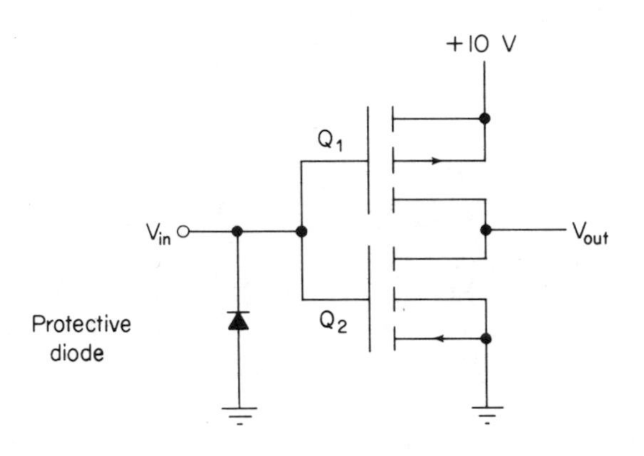

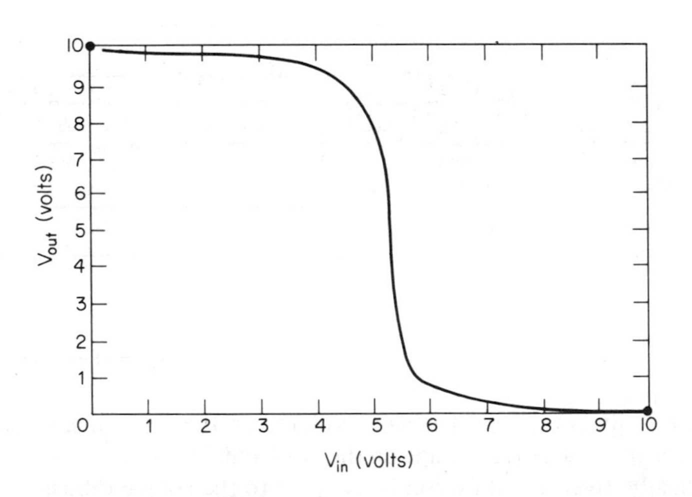

FIG. 6-47. Typical CMOS inverter and transfer characteristics.

When V_{in} is between +8 V and +10 V, the potential difference between the gate and substrate of Q_1 is less than the threshold voltage, and Q_1 is in an off condition. The potential difference between the gate of Q_2 and its substrate is 8 V to 10 V; thus Q_2 is on. At the output, the inverter appears as an approximate 1- or 2-kΩ resistance to ground, and an extremely high (greater than 1000 MΩ) to +10 V. The current drain from the +10-V supply is less than 10 nA, resulting in very low static power dissipation. This low power consumption is one of the primary advantages of MOSFETs in digital ICs.

When V_{in} is between 0 V and +2 V, Q_2 is off and Q_1 is on. In this case, the output appears as a low resistance of +10 V. As V_{in} makes the transition from +2 V to +8 V, both Q_1 and Q_2 are in an on condition, resulting in current flow from the +10 V supply.

Figure 6-48 illustrates this current flow as a function of V_{in}. This graph is generated by applying various dc voltages at V_{in} and then measuring I_D.

In a situation involving the application of a switching waveform to V_{in}, the current I_D does not flow through both Q_1 and Q_2. The reason for this may be explained as follows. Assume that Q_1 is off, and Q_2 is on. As V_{in} switches quickly from V_{CC} to ground, there is a period of time when both Q_1 and Q_2 are on. However, because of stray and load capacitance on the output, the output voltage is still nearly 0 V when Q_1 and Q_2 are both on. Thus, very little current flows through Q_2 to ground.

In CMOS switching circuits, the increase in power dissipation with increasing switching speed is due almost entirely to capacitance loading. The effect of a capacitive load on the power dissipation is shown in Fig. 6-49. The relationship between the switching frequency and the power dissipation is also

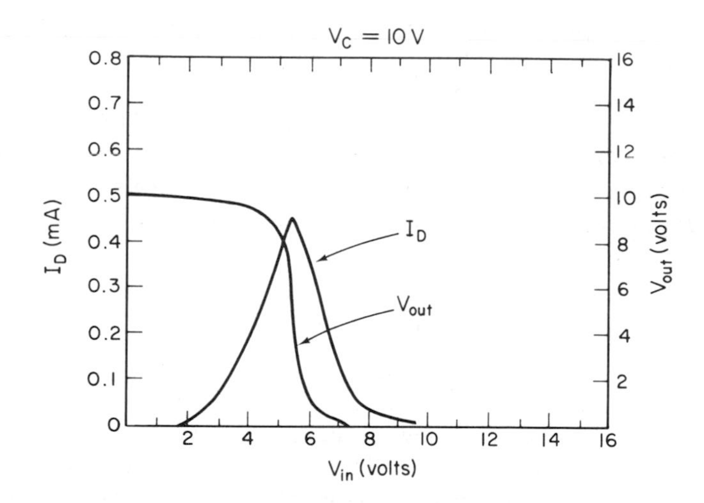

FIG. 6-48. Power supply current versus input voltage curve for CMOS inverter.

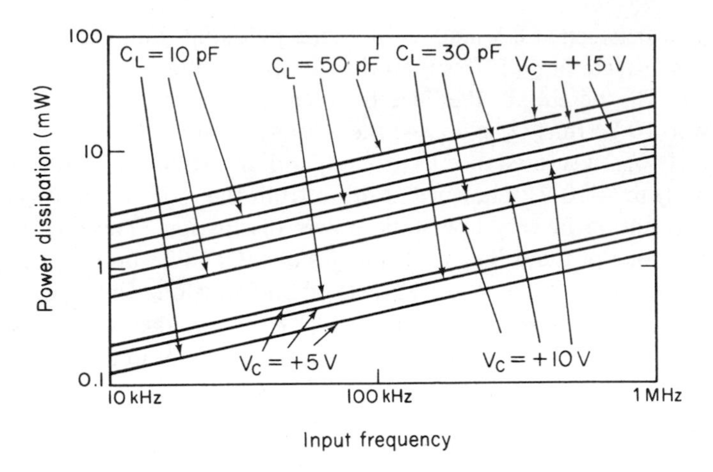

FIG. 6-49. Effect of capacitive load on power dissipation of CMOS inverter.

illustrated. This relationship holds true for more complicated CMOS logic systems since the inverter is the basic building block.

The effects of capacitive loading and power supply variations on the inverter switching times are given in Fig. 6-50. The switching times decrease as the power supply voltage increases. At higher operating voltages, the channel resistance of the devices is reduced, resulting in greater switching speed.

The input capacitance of the inverter is greatly affected by the input voltage, as shown in Fig. 6-51. The specification of the gate input capacitance on a datasheet is generally made with $V_{in} = 0$ V. However, the input capacitance increases in the transition region to a value which is three to four times greater than the specified value on the datasheet. Because of feedback capacitances from the output, it is difficult to say what average capacitive load will be presented by a gate input to a switching signal.

Figures 1-24 and 1-25 show how the basic inverter is used to form a NOR gate and a NAND gate respectively. Since either the NOR function or the NAND function can be used to generate any Boolean algebra function (as well as storage elements), the simple inverter circuit is truly a basic building block.

6-8.2 CMOS transmission gate

The other important building block for the construction of CMOS digital ICs is the transmission gate shown in Fig. 6-52. When the transmission gate is turned on, a low resistance exists between the input and the output. This allows current flow in either direction through the gate. The voltage on the input line must always be positive with respect to the substrate on the N-channel device, and negative with respect to the substrate

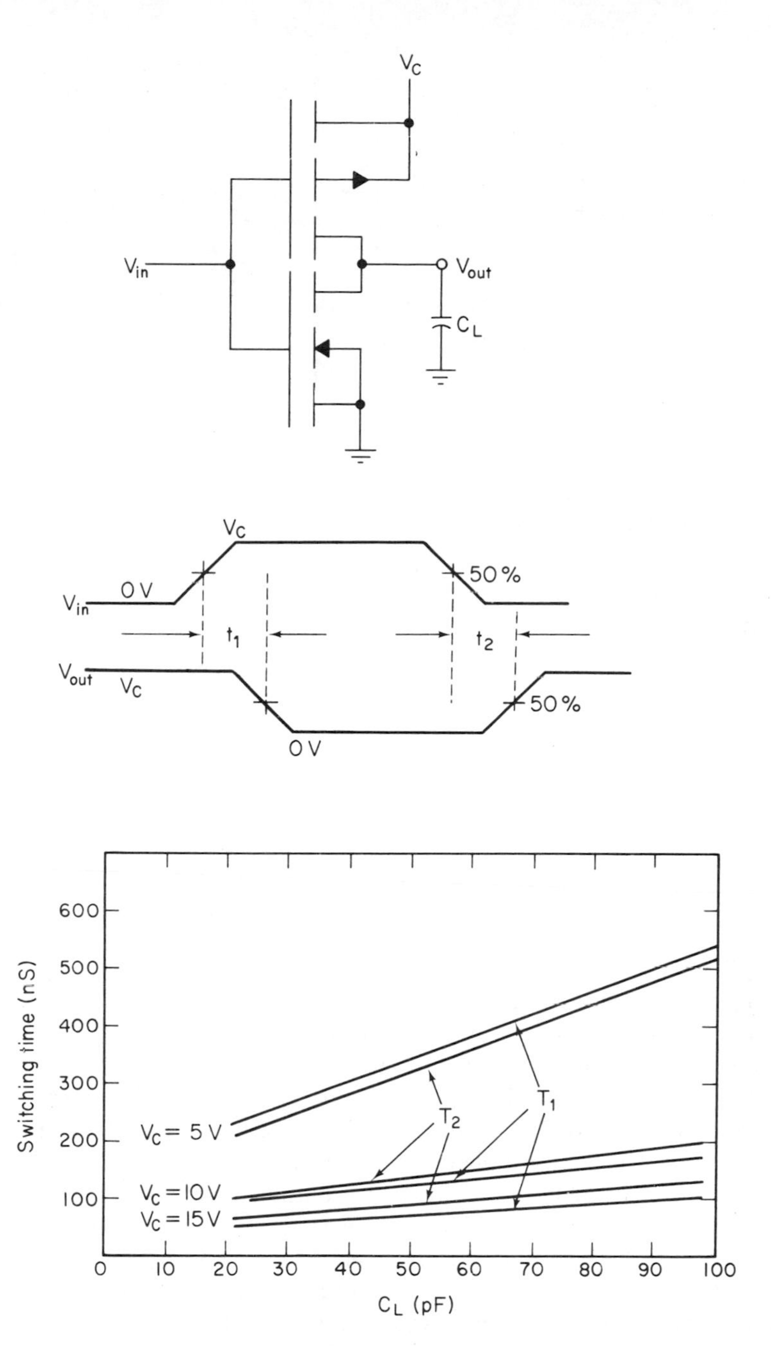

FIG. 6-50. Effects of capacitive loading and power supply variations on switching times of CMOS inverter.

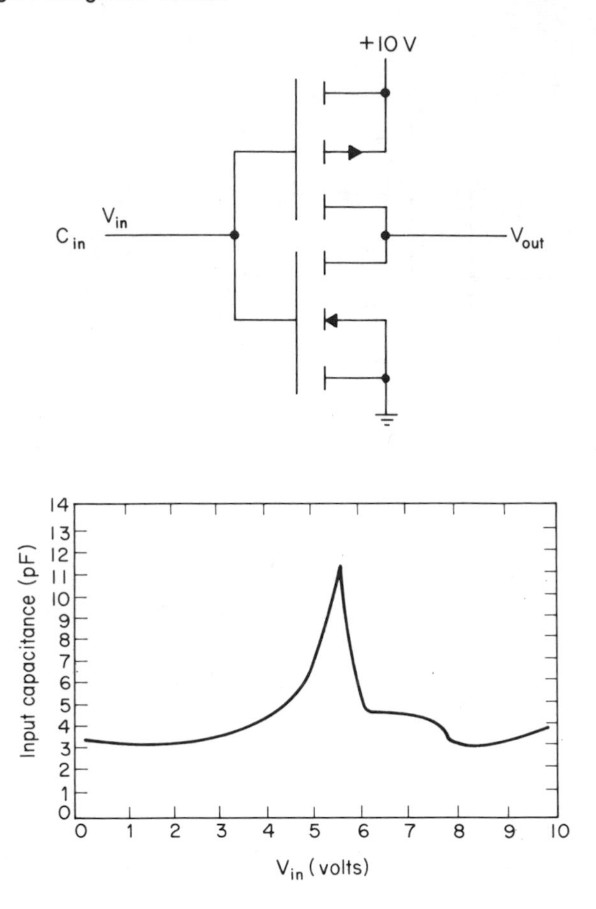

FIG. 6-51. Effects of input voltage on input capacitance of CMOS inverter.

of the P-channel device. The gate is turned on when G_1 is at ground and G_2 is at V_{CC} (or V_C as it is often called in CMOS ICs). When G_2 is at ground and G_1 is at V_{CC}, the transmission gate is off, and an extremely high resistance exists between the input and the output.

The resistance between the input and output of a transmission gate in the on condition (R_{ON}) is dependent upon the voltage applied at the input, the potential difference between the two substrates, and the current from input to output. R_{ON} is defined to be the input-to-output resistance when current of 100 μA is flowing. The two graphs of Fig. 6-52 illustrate the influence of the power supply voltage, the input voltage, and the temperature.

A peaking effect occurs in the R_{ON} versus V_{in} curves. When V_{in} is at or near V_{CC}, the P-channel device is providing the low resistance, and the N-channel device is off, since the potential difference between G_2 and the drain or source of the N-channel device is less than the threshold voltage. When V_{in}

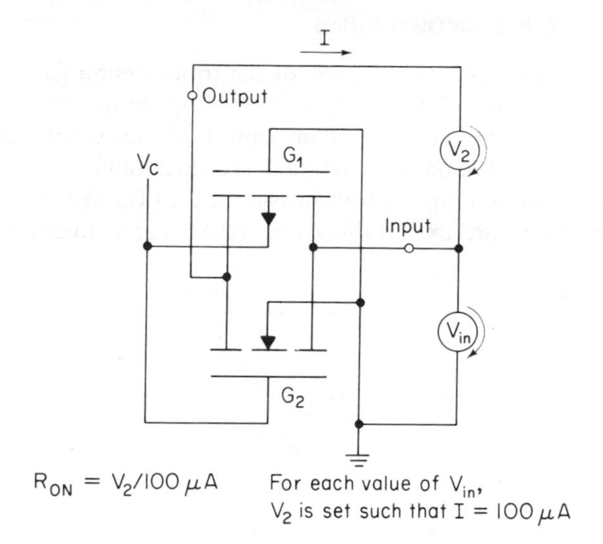

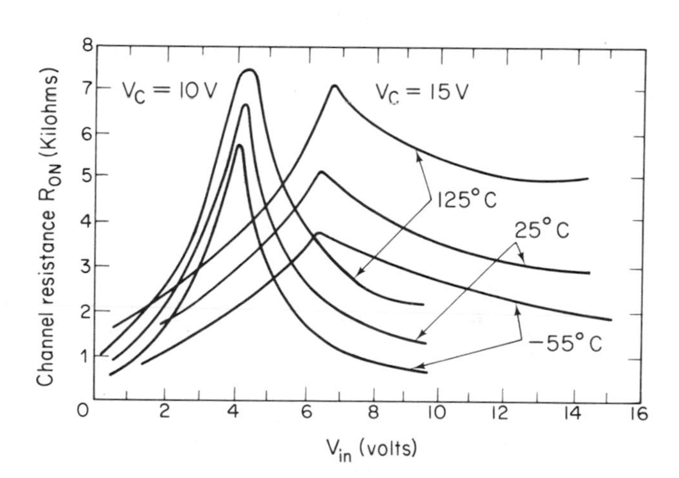

FIG. 6-52. Typical CMOS transmission gate and transfer characteristics.

is at or near ground, the N-channel device is conducting and the P-channel device is off. At voltages between the two extremes, both devices are partially on, and the value of R_{ON} is due to the net resistance of the N- and P-channels connected in parallel. The different slope of the curve on either side of the peak is due to the greater sensitivity of the N-channel resistance to substrate degeneration (or substrate bias) than the P-channel. Thus, the rate of increase in R_{ON} with respect to V_{in} is greater for input voltages between 0 V and the "R_{ON} peaking voltage" than for input voltages greater than the "R_{ON} peaking voltage".

6-8.3 CMOS flip-flop

An illustration for use of the transmission gate is given by the MCMOS (Motorola CMOS) Type-D flip-flop shown in Fig. 6-53. This flip-flop works on the Master-Slave principle and consists of four transmission gates, four NOR gates, one inverter, and one clock buffer.

When the clock is a logic 0, transmission gates (TG) #2 and #3 are off, and TG #1 and TG #4 are on. In this case, the Master is logically disconnected

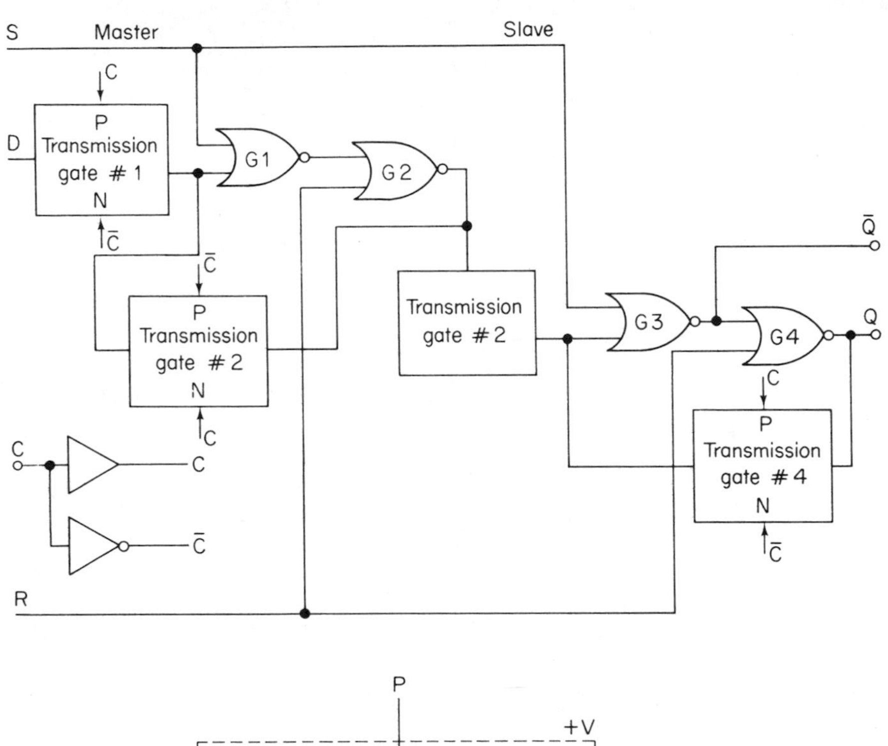

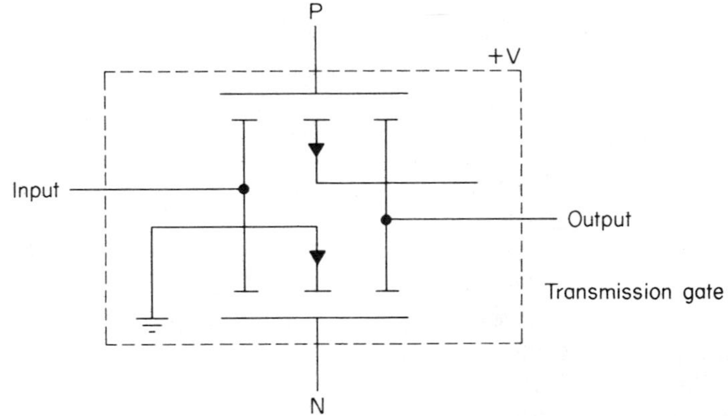

FIG. 6-53. Motorola MCMOS Type D flip-flop.

from the Slave. Since TG #4 is on, G_3 and G_4 are cross-coupled, and latched into a stable state. Assuming the S and R inputs, which provide for asynchronous setting and resetting, are at a logic 0, the states of G_1 and G_2 are completely determined by the logic signal applied to G_1 from input D.

When the clock changes to a 1, TG #2 and TG #3 turn on, and TG #1 and TG #4 turn off. G_1 and G_2 are cross-coupled through TG #2, and the two gates latch into the state in which they existed before the clock changed from a 0 to a 1. Since TG #3 is on and TG #4 is off, the output of the Master stage is fed through G_3 and G_4 to $\overline{Q}$ and Q.

When the clock returns to a 0, TG #3 turns off, and TG #4 turns on. This disconnects the Slave stage from the Master stage, and latches the Slave stage into the state existing before the clock changed from a 1 to a 0. Thus, data is entered into the Master when the clock changes from a 0 to a 1.

When the clock is in the 1 state, the output of the Master is transmitted directly through the Slave, which then provides the Q and $\overline{Q}$ outputs. While the clock is a 0, the Master and Slave are disconnected and the Master is connected to the D input. This Master-Slave Type-D flip-flop has been built by Motorola using MCMOS technology, and has been observed to toggle at frequencies approaching 13 MHz when using a 15-V power supply.

6-8.4 Proper handling of CMOS ICs

As in the case of any insulated-gate FET, the CMOS ICs must be handled with care. The following are some general precautions.

Store unused devices in conductive foam, or short the leads with aluminum foil. Use grounded tip soldering irons. If possible, use grounded assembly tables. Always touch ground with one hand when picking up a device. Do not exceed maximum ratings (voltage, current, temperature, etc.).

The oxide insulator between the metal gate and the substrate of a CMOS device breaks down at a gate-to-substrate potential of about 100 V. If this oxide insulation breaks down once, the IC is destroyed. Although IC inputs that are connected to MOSFET gates are usually protected with *clamp diodes*, care should be taken to ensure that excessive voltages due to static charge, or power supplies, are not applied to gate inputs.

The protection scheme used in Motorola's MCMOS is illustrated in Fig. 6-54. Since the P tub and the N substrate are lightly doped, the breakdown of this junction is high—about 100 V. Therefore, the N^+ region is diffused into a P tub. The junction between the heavily doped N^+ region, and the P tub, breaks down at about 30 V. This 30-V junction breakdown is used to protect the gates of CMOS devices that interface with the outside world. Under normal operation, the protective diode is back-biased by a voltage which is never greater than the power supply potential difference.

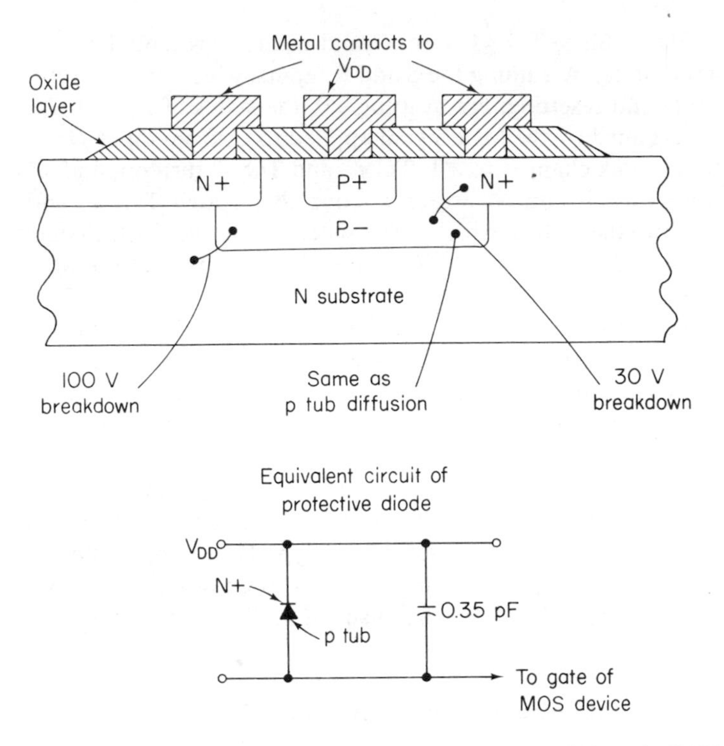

FIG. 6-54. Protection diode for input going to metal gate of Motorola CMOS devices.

6-8.5 Interfacing with MOS ICs

As in the case of ECL and HTL, level translators are available for interfacing MOS ICs with other logic families. Always follow the design notes found on the datasheets supplied with the translators. Although no special rules can be applied to all MOS IC translators, it is generally best to place the substrate at a potential equal to the *most positive voltage* applied to the logic family with which the IC is to be interfaced.

In addition to the packaged IC translators, it is possible to interface MOS ICs with other logic families using discrete components. The following is a summary of such circuits.

6-8.5.1 Interfacing MOS IC with ECL

The simplest ECL to MOS interface circuit is shown in Fig. 6-55. This circuit can be used to interface with MOS when the substrate is at ground (0 V). The value of R must satisfy the inequality $R \geq [V(\text{volts})/1.7]$ kΩ since the collector of the PNP is effectively a current source. Switching times for several load capacitances are included in Fig. 6-55.

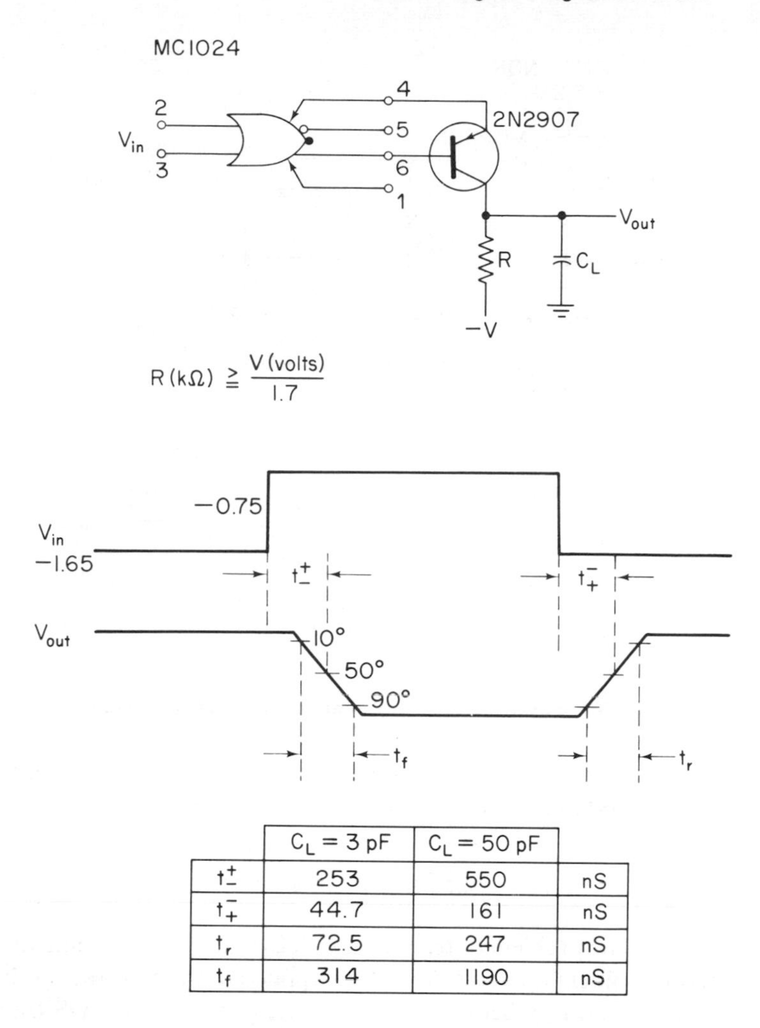

	$C_L = 3$ pF	$C_L = 50$ pF	
t_-^+	253	550	nS
t_+^-	44.7	161	nS
t_r	72.5	247	nS
t_f	314	1190	nS

FIG. 6-55. MECL to MOS interface circuit (Motorola).

Driving ECL gates with MOS circuitry is less expensive than driving MOS with ECL. Figure 6-56 shows a MCMOS NOR gate driving an MECL NOR gate. In one case, the MCMOS NOR gate is operating between +5.2 and −5.2 V. At this point, a clamp diode to V_{BB} is required to prevent the MCMOS gate from pulling the MECL gate input above 0.6 V.

In the other case, the MCMOS gate is operating between 0 V and −10 V. Here, a clamp diode is required to the −5.2-V supply to prevent the voltage on the ECL gate from falling more than a diode drop below −5.2 V. If the supply voltages for the MCMOS gate were 0 V and −5.2 V, the MCMOS gate could directly drive the ECL gate. However, the MCMOS speed is consider-

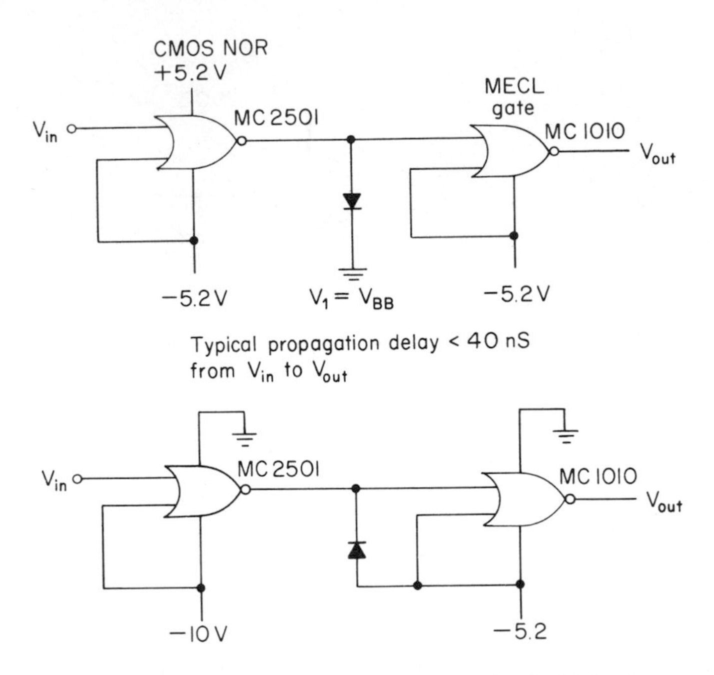

FIG. 6-56. CMOS to MECL interface circuits (Motorola).

ably reduced at the 5-V potential difference. The basic requirement on the ECL gate input is to maintain its voltage potential between 0 V and −5.2 V.

6-8.5.2 Interfacing MOS IC with TTL/DTL

The problem of translating TTL/DTL to MOS, and vice versa, is strongly dependent upon the dc voltage applied to the substrate of the MOS device. In general, it is better to use one of the packaged MOS translators when converting from MOS to TTL/DTL. The same holds true when going from TTL/DTL to MOS, when the MOS substrate is not at ground. However, if the MOS substrate is at ground, it is possible to go from TTL/DTL to MOS with fairly simple discrete circuits.

Figure 6-57 shows two discrete circuits that are useful in converting TTL/DTL logic levels to MOS logic (when the MOS substrate is at ground). The circuit of Fig. 6-57a uses a passive pull-up resistor R_1. The value of R_1 is selected on the basis of power dissipation, and the rise and fall times desired on the MOS capacitive load. Q_1 acts as a level shifter, and provides drive to Q_2. If $-V_2$ is less than -1 V, Q_1 is never in saturation, and therefore has little effect on the speed of the circuit. The circuit speed is dependent primarily on Q_2. The operation of the circuit in Fig. 6-57b is essentially the same as in

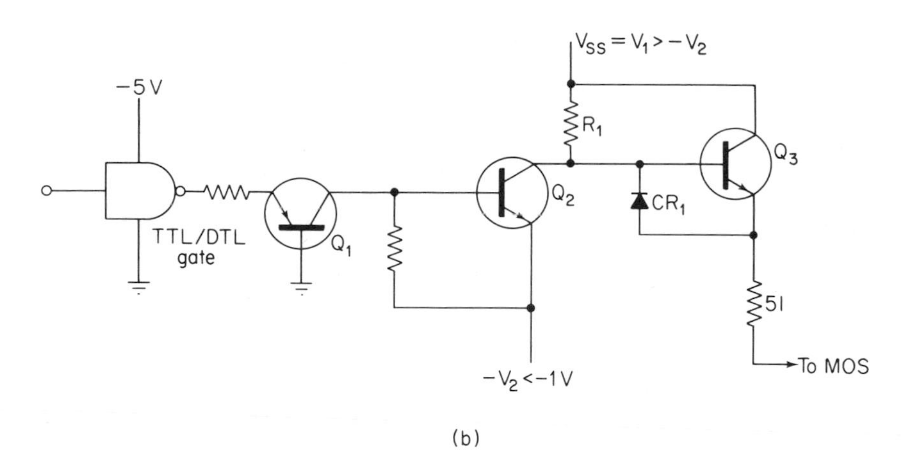

FIG. 6-57. TTL/DTL to MOS interface circuits (Motorola).

Fig. 6-57a, except for the active pull-up configuration consisting of Q_3, CR_1 and R_2. The circuit of Fig. 6-57b is useful when driving large capacitive loads.

6-8.5.3 Interfacing MOS IC with RTL

Since IC RTL is most similar to discrete transistor logic, discrete interface circuits are generally required when translating between RTL and MOS.

Figure 6-58 gives three discrete circuits that can be used in converting RTL signals to MOS logic levels. Note that the circuits of Fig. 6-58a and 6-58b are essentially the same as the circuits of Fig. 6-57, except for value changes. These two circuits are particularly useful when the MOS substrate is at the RTL power supply voltage, or at ground.

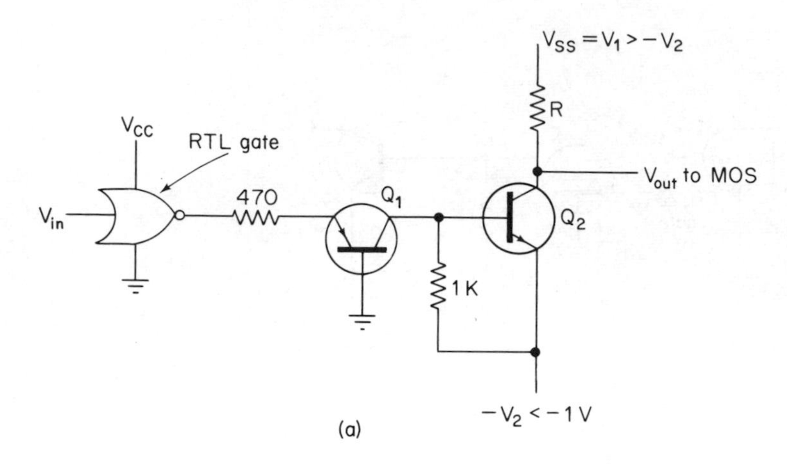

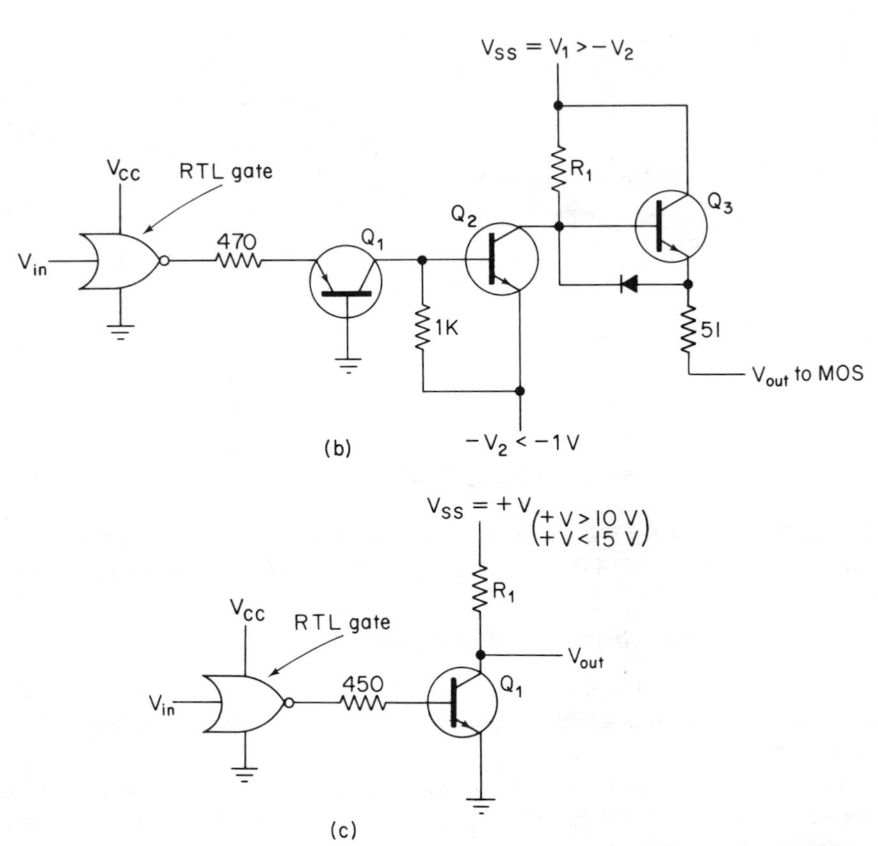

FIG. 6-58. RTL to MOS interface circuits (Motorola).

The circuit of Fig. 6-58b has an active pull-up, and should be used when driving large capacitive loads. Resistor R_1 should be selected on the basis of power dissipation and the desired rise and fall time at the output. If the MOS substrate is at a voltage in the range of +10 V to +15 V, the circuit of Fig. 6-58c provides an efficient interface. Again, the value of R_1 is selected on the basis of available drive from the RTL gate, desired rise and fall times, and power dissipation.

The circuit of Fig. 6-59 illustrates conversion from MOS to RTL when the MOS substrate voltage is at +13 V. The diode clamps the gate input to the RTL power supply voltage, which prevents the MOS output driver from pulling the gate input above the maximum specified rating at +4 V. When V_{CC} is 3.6 V, the clamp diode should be germanium.

In the circuit of Fig. 6-60, the MOS substrate is at the RTL power supply voltage. In this mode of operation, the clamp diode is used to prevent the voltage on the RTL gate input from falling below the −4-V rating.

6-9. SUMMARY OF INTERFACING PROBLEMS

As discussed in preceding sections of this chapter, each digital IC line or family has its own set of design or usage problems. Most of these problems can be resolved using the datasheets and brochures supplied with the device. One area where the datasheets leave a gap is the interfacing between IC logic forms. For that reason, we shall summarize the interfacing problems here.

If either ECL or HTL are involved, the best bet is to use the various level translators supplied with most ECL and HTL lines, or use the specific translator circuits shown in the datasheets and brochures.

DTL and TTL are compatible with each other, thus eliminating the need for special translators or interface circuits. However, the load and drive characteristics may be altered somewhat. These changes are generally noted on the datasheets.

If field effect digital ICs (CMOS, MOSFET) are involved, follow the recommendations of Sec. 6-8.

If RTL must be interfaced, follow the recommendations of Sec. 6-3.9 and 6-8.5.3, and the following special notes:

The specifications of most RTL and DTL/TTL are not compatible, so worst-case interfacing can not be guaranteed between the two families without performing special tests and/or adding discrete components. However, by using conservative design principles and making one relatively simple test, interfacing can be accomplished.

The recommended circuitry for driving RTL gates from DTL outputs is shown in Fig. 6-61a. In order to insure that this technique is valid for opera-

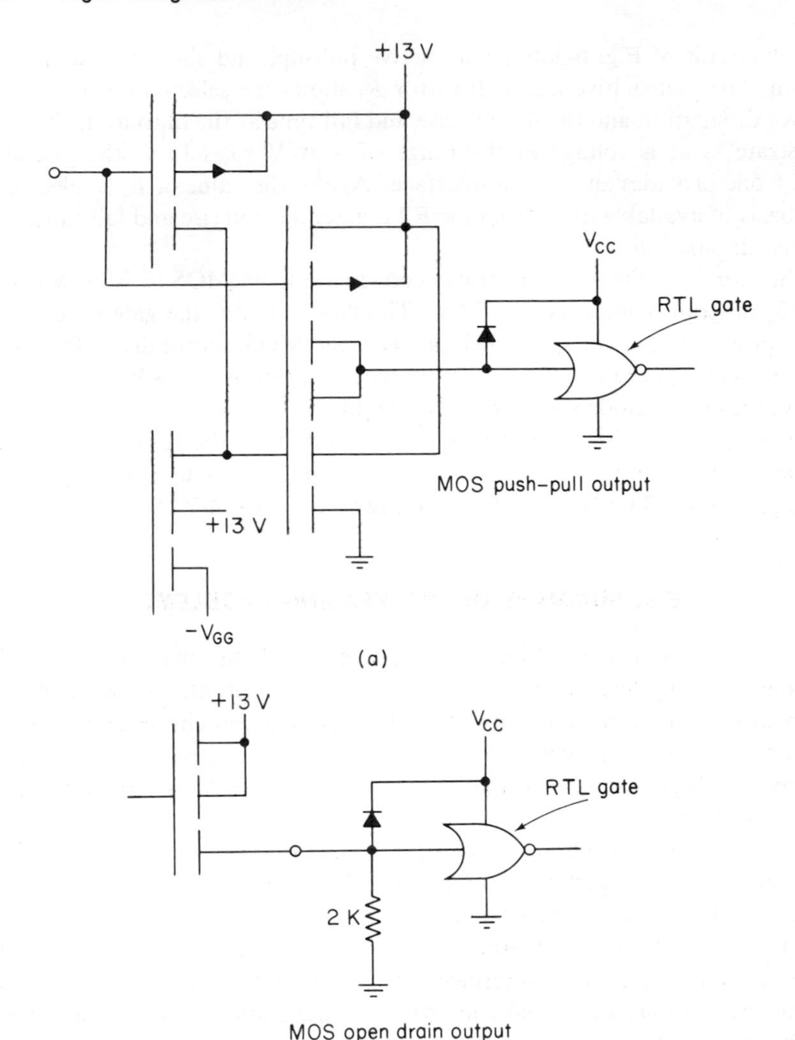

FIG. 6-59. MOS to RTL interface (with MOS substrate at +13 V) (Motorola).

tion under worst-case conditions, it is necessary to test V_{OL} of the DTL circuit at an I_{OL} of 6.5 mA. Use only those DTL components (gates) with a V_{OL} less than 0.4 V for this application.

The circuit of Fig. 6-61a should not be used to drive both RTL and DTL from a common output. If both must be driven from a common point, use

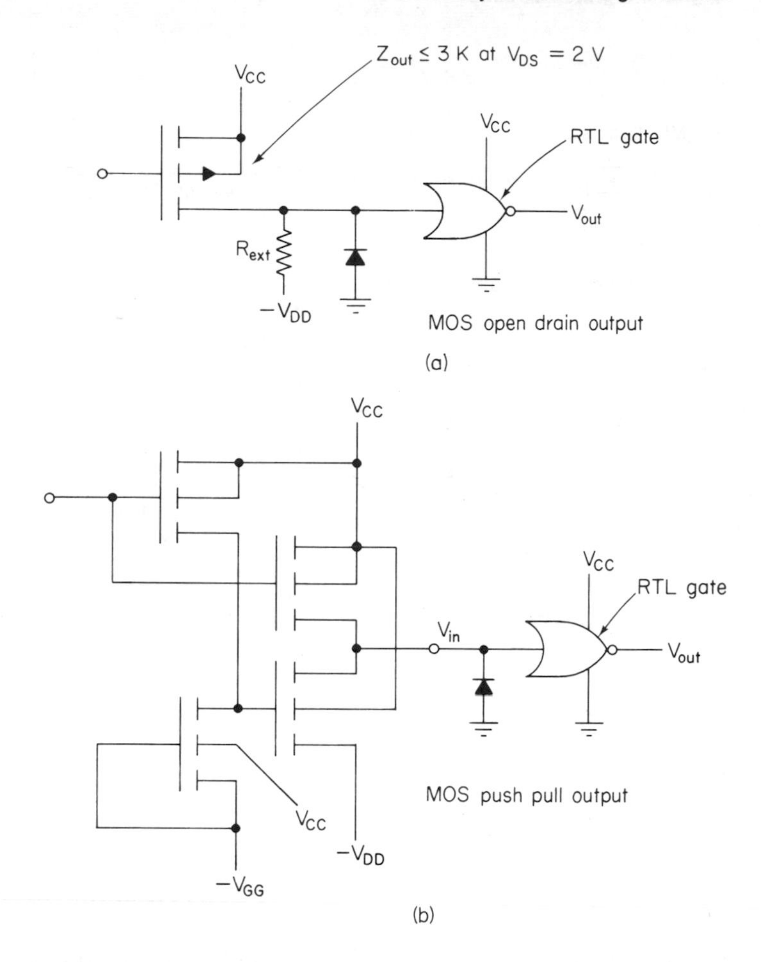

FIG. 6-60. MOS to RTL interface (with MOS substrate at V_{CC}) (Motorola).

the circuit of Fig. 6-61b. The load current seen by the output transistor of the DTL gate in Fig. 6-61b is greater than that of Fig. 6-61a. The leakage current of the RTL gate may increase slightly also. Neither of these factors should cause significant problems.

Driving DTL from RTL outputs is not a situation where conservative design can result in no-test circuit limits. The primary reason for this is that the high level output of RTL is tested only to the level of V_{on} (about 0.67 to 1 V, depending on type and temperature). Although a typical RTL gate exhibits a high level output well above the DTL threshold, a test is always recommended.

The test required is relatively simple—with the datasheet value of V_{off} applied to *all inputs* of the driving RTL gate, the unloaded output voltage

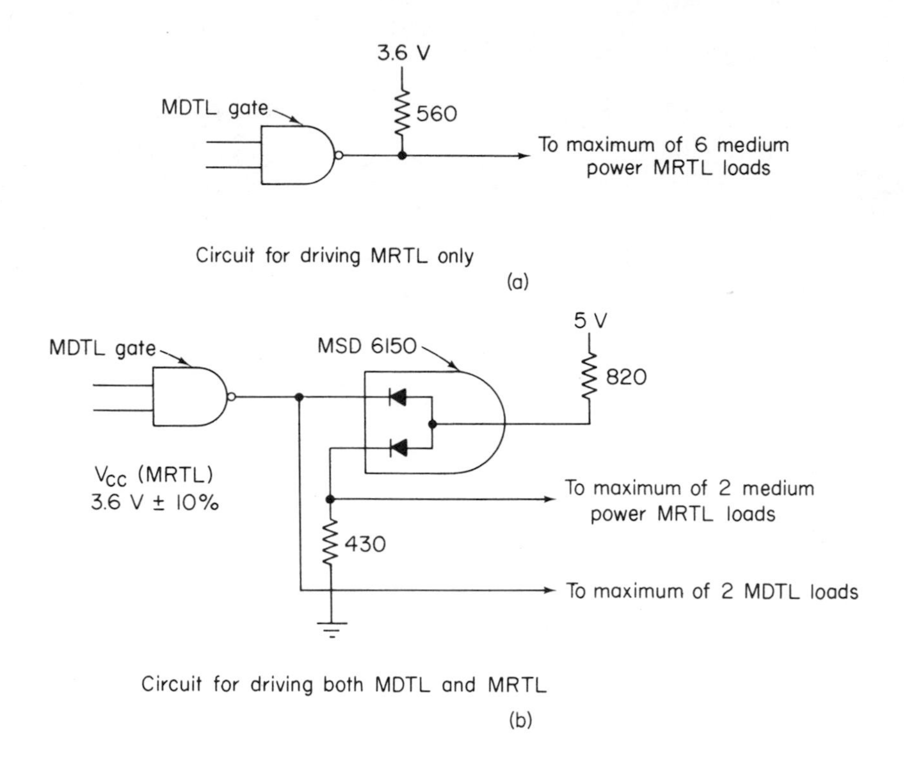

FIG. 6-61. DTL to MRTL interface (Motorola).

should be a minimum of 2.5 V at the maximum operating temperature. If so, RTL can be used to drive DTL directly without special interface.

In cases where driving both RTL and DTL from an RTL output is required (including the case of an unbuffered flip-flop output driving a DTL component), use the circuit of Fig. 6-62a. This is also the recommended alternative when the RTL fails the 2.5-V output test.

There are several techniques for insuring that an RTL gate can sink enough current to drive a number of DTL inputs. The most common method is to double the input loading factors of the driving (RTL) gate (which is equivalent to allowing up to a typical 5 mA of additional sink current (enough for 3 DTL gates). This is not a guaranteed condition, of course, but the incidence of devices which will not perform in this manner is very small. Even so, the conservative designer will drive at least two inputs of the RTL gate as shown in Fig. 6-62b, thus insuring low-level drive capability under worst-case conditions.

Another worst-case design technique is shown in Fig. 6-62c. This circuit requires an RTL expander, rather than an RTL gate, and an external resistor.

5 V ± 10%
430
2N4264
(a)
Using discrete components

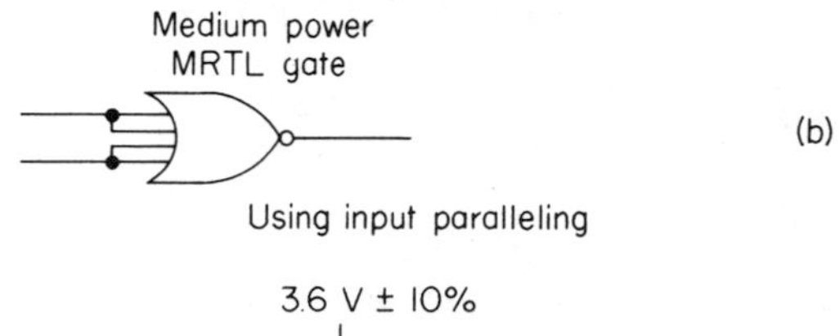

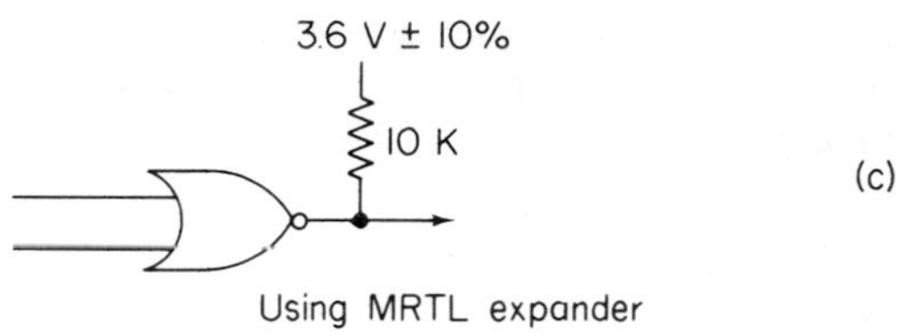

FIG. 6-62. MRTL to DTL interface (Motorola).

7. TESTING INTEGRATED CIRCUITS

This chapter is devoted to test procedures for integrated circuits. These tests provide the user with a means of determining the actual or true characteristics of an IC. The determination of true characteristics is more important than many users realize. There are several reasons for this.

First, a datasheet may not be available. More likely, the datasheet may not list all of the characteristics needed by the user. Even if all required characteristics are given, the procedures for finding the values may not be given or are not clear.

When test information and characteristics are available in datasheet form, keep in mind that the values given are "typical". They can vary from IC to IC, and with different operating conditions. There is no substitute for testing each IC under actual operating conditions (temperature, power supply, noise, etc.) of the intended use.

Just as important, as described in previous chapters, an IC can often be adapted to many uses other than the application intended by the manufacturer. Often, the IC manufacturers are surprised at the uses to which their units are adapted. Therefore, they would have no reason for supplying test data (values or procedures) for such applications. The user must devise his own tests, and find his own values.

Finally, some IC manufacturers assume that all users will automatically know how to test for all characteristics. As a result, they simply omit test data from their literature.

It is assumed that the reader is thoroughly familiar with basic electronic test procedures. It is particularly important that the reader be familiar with the oscilloscope, voltmeter, signal generator, and pulse generator.

The most important tests for linear ICs are for frequency and phase response. These tests involve the use of signal generators and oscilloscopes (and/or voltmeters). The most important tests for digital ICs involve the use of pulses. Thus, the digital IC user must be able to monitor and measure pulses using pulse generators and oscilloscopes.

Also included at the end of this chapter are some selected notes on IC troubleshooting problems.

7-1. TESTING IC OPERATIONAL AMPLIFIERS

The following sections describe test procedures for IC operational amplifiers. As a minimum, the following tests should be made on the basic IC, with power sources connected as described in Chapter 2, operating in an open-loop circuit. This will confirm (or deny) the IC characteristics found on the datasheet. The procedures can also be used to establish a set of characteristics for an IC where the datasheet is missing or inadequate.

7-1.1 Frequency response

The frequency response of an IC op-amp can be measured with a signal generator and a meter or oscilloscope. When a meter is used, the signal generator is tuned to various frequencies, and the resultant IC output response is measured at each frequency. The results are then plotted in the form of a graph or *response curve*, such as shown in Fig. 7-1. The procedure is essentially the same when an oscilloscope is used to measure IC frequency response. However, an oscilloscope gives the added benefit of visual distortion analysis (as discussed in Sec. 7-1.6.1).

The basic frequency response measurement procedure (with either meter or oscilloscope) is to apply a *constant amplitude* signal while monitoring the IC output. The input signal is varied in frequency (but not amplitude) across the entire operating range of the IC. The voltage output at various frequencies across the range is plotted on a graph similar to that shown in Fig. 7-1. Generally, an oscilloscope is a better instrument for response measurement of an IC at higher frequencies. However, an electronic digital voltmeter can be used. Both open- and closed-loop frequency response should be measured with the same load.

1. Connect the equipment as shown in Fig. 7-1. Set the generator, meter, and oscilloscope controls as necessary.

2. Initially, set the generator output frequency to the low end of the frequency range. Then set the generator output amplitude to the desired input level. For example, a typical IC op-amp may require 10 mV to produce full output voltage (5 V).

3. In the absence of a realistic test input voltage, set the generator output to an arbitrary value. A simple method of finding a satisfactory input level

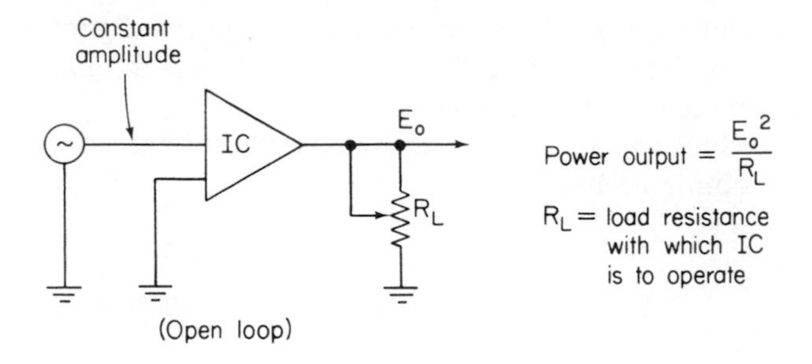

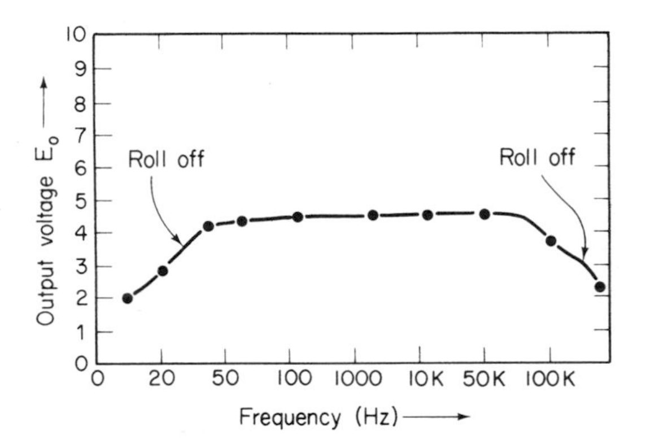

FIG. 7-1. Linear IC frequency response test connections and typical response curve.

is to monitor the circuit output (with the meter or oscilloscope) and increase the generator output at the IC center frequency (or at 1 kHz) until the IC is overdriven. This point is indicated when further increases in generator output do not cause further increases in meter reading (or the output waveform peaks begin to flatten on the oscilloscope display). Set the generator output *just below* this point. Then return the meter or oscilloscope to monitor the generator voltage (at IC input) and measure the voltage. Keep the generator at this voltage throughout the test.

4. Record the circuit output voltage on the graph. Without changing the generator output amplitude, increase the generator frequency by some fixed amount, and record the new circuit output voltage. The amount of frequency increase between each measurement is an arbitrary matter. Use an increase of 10 Hz at the low end and high end (where rolloff occurs), and an increase of 100 Hz at the middle frequencies. If the IC frequency range is very broad, use 1-kHz increments at the middle frequencies.

5. Repeat the process, checking and recording the IC output voltage at each of the check points in order to obtain a frequency response curve. Depending upon IC design, the curve will resemble that of Fig. 7-1, with a flat portion across the middle frequencies, and rolloff at each end. Theoretically, there should be no rolloff at the low end, since an IC op-amp is direct coupled. However, there is usually a capacitor at the output of the generator which combines with the IC input impedance or resistance to form a high-pass (low-cut) filter. This produces rolloff at the low-frequency end.

6. Note that generator output may vary with changes in frequency, a fact often overlooked in making a frequency response test of any circuit. Even precision laboratory generators can vary in output with changes in frequency, thus resulting in considerable error. Therefore, it is recommended that the generator output be monitored after each change in frequency (some generators have a built-in output meter). Then, if necessary, the generator output amplitude can be reset to the correct value. Within extremes, it is more important that the generator *output amplitude remain constant* rather than at some specific value when making a frequency response check.

7-1.2 Voltage gain

Voltage gain measurement in an IC is made in the same way as frequency response. The ratio of output voltage to input voltage (at any given frequency, or across the entire frequency range) is the voltage gain. Since the input voltage (generator output) must be held constant for a frequency response test, a voltage gain curve should be identical to a frequency response curve.

Keep in mind that the basic IC has a maximum input and output voltage limit, neither of which can be exceeded without possible damage to the IC and/or clipping of the waveform. In general, the maximum rated input should be applied, and the actual output measured. (See Fig. 3-17). Check the output for clipping at the maximum level. If clipping occurs, decrease the input until clipping just stops, and note the input voltage. Record these values as a basis for design.

Note the frequency at which the open-loop voltage gain drops 3 dB from the low-frequency value. This is the open-loop bandwidth. (See Fig. 3-18.)

Keep in mind that the open-loop voltage gain and bandwidth are characteristics of the basic IC. Closed-loop gain is (or should be) dependent upon the ratio of *feedback and input* resistances, while closed-loop bandwidth is essentially dependent upon *phase compensation* values.

Closed-loop characteristics are generally lower than open-loop characteristics (voltage gain is lower, frequency response narrower, etc.). However, closed-loop characteristics are modifications of open-loop characteristics.

7-1.3 Power output, gain and bandwidth

Most IC op-amps are not designed as power amplifiers. However, their power output, gain, and bandwidth can be measured. Keep in mind that an IC has a power dissipation of its own which must be subtracted from the *total device dissipation* to find the available power output.

The power output of an IC op-amp is found by noting the output voltage E_o across the load resistance R_L (Fig. 7-1), at any frequency, or across the entire frequency range. Power output is E_o^2/R_L.

To find power gain of any amplifier, it is necessary to find both the input and output power. Input power is found in the same way as output power, except that the impedance at the input must be known (or calculated). This is not always practical in some ICs, especially where impedance is dependent upon frequency or gain. With input power known (or estimated), the power gain is the ratio of output power to input power.

Generally, a power gain is not required by IC design specifications. Instead, an *input sensitivity* specification is used. Input sensitivity specifications require a minimum power output with a given voltage input (such as 100-mW output with 10-mV RMS input).

Some power IC op-amp design specifications include a power bandwidth factor. Such specifications (generally limited to audio range IC op-amps) require that the IC deliver a given power output across a given frequency range. For example, an IC might produce full power output up to 20 kHz, even though the frequency response is flat up to 100 kHz. That is, voltage (without a load) will remain constant up to 100 kHz, while power output (across a normal load) will remain constant up to 20 kHz.

7-1.4 Load sensitivity

Since an IC op-amp is generally not used as a power amplifier, load sensitivity is not critical. However, if it becomes necessary to measure load sensitivity, use the following procedure.

Any amplifier will produce maximum power when the output impedance of the amplifier is the same as the load impedance. For example, if the load is twice the IC output impedance, the output power is reduced to approximately 50 percent. If the load is 40 percent of the IC output impedance, the output power is reduced to approximately 25 percent.

The circuit for load sensitivity measurement is the same as for frequency response (Fig. 7-1), except that load resistance R_L is variable. (Never use a wirewound load resistance. The reactance can result in considerable error.) If a non-wirewound variable resistance of sufficient wattage is not available, use several fixed resistances (carbon or composition) arranged to produce the desired resistance values).

Measure the power output at various load impedance/output impedance ratios. To make a comprehensive test of the IC, repeat the load sensitivity test across the entire frequency range.

7-1.5 Input and output impedance

Dynamic input and output impedance of an IC op-amp (Figs. 3-15 and 3-16) can be found using the following procedures. Keep in mind that the closed-loop impedances will differ from open-loop impedances.

7-1.5.1 Dynamic output impedance measurement

The load sensitivity measurement test just described can be reversed to find the dynamic output impedance of an IC. The connections and procedure (Fig. 7-1) are the same, except that the load resistance R_L is varied until maximum output power is found. Power is removed and R_L is disconnected from the circuit. The dc resistance of R_L (measured with an ohmmeter) is equal to the dynamic output impedance. Of course, the value applies only at the frequency of measurement. The test should be repeated across the frequency range.

7-1.5.2 Dynamic input impedance measurement

To find the dynamic input impedance of an IC op-amp, use the circuit of Fig. 7-2. The test conditions should be identical to those for frequency response, power output, etc. That is, the same generator, operating load, meter, or oscilloscope, and frequencies should be used.

The signal source is adjusted to the frequency (or frequencies) at which the circuit will be operated. Switch S is moved back and forth between position A and B, while resistance R is adjusted until the voltage reading is the same in both positions of the switch. Resistor R is then disconnected from the

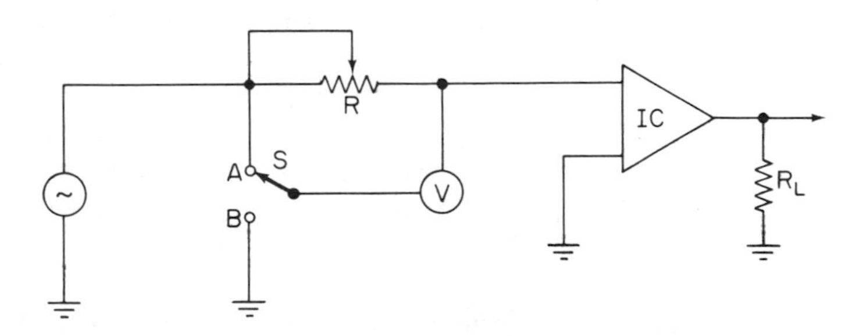

FIG. 7-2. Linear IC dynamic input impedance test connections.

circuit, and the dc resistance of R is measured with an ohmmeter. The dc resistance of R is then equal to the dynamic impedance at the IC output.

Accuracy of the impedance measurement is dependent upon accuracy with which the dc resistance is measured. A non-inductive resistance must be used. The impedance found by this method applies only to the frequency used during the test.

7-1.6 Distortion

Distortion requirements for IC op-amps are usually not critical. Therefore, no detailed discussion of distortion test procedures is given here. Distortion in IC op-amps can be measured by means of sinewaves or square waves.

7-1.6.1 Checking distortion by sinewave analysis

The procedure for checking IC distortion by means of sinewaves is to connect the equipment as shown in Fig. 7-1, and then monitor both the input and output waveforms with an oscilloscope. The primary concern is deviation of the IC output waveform from the input waveform. If there is no change (except in amplitude), there is no distortion. If there is a change in the waveform, the nature of the change will often reveal the cause of distortion.

In practice, analyzing sinewaves to pinpoint distortion is a difficult job, requiring considerable experience. Unless the distortion is severe, it may pass unnoticed. Thus, if an oscilloscope is to be used alone (without intermodulation or harmonic distortion analyzers), square waves provide the best basis for distortion analysis.

7-1.6.2 Checking distortion by square wave analysis

The procedure for checking distortion by means of square waves is essentially the same as for sinewaves. Distortion analysis is more effective with square waves because of their high odd-harmonic content, and because it is easier to see a deviation from a straight line with sharp corners, than from a curving line.

As in the case of sinewave distortion testing, square waves are introduced into the IC input, while the output is monitored on an oscilloscope. (See Fig. 7-3.) The primary concern is deviation of the IC output waveform from the input waveform (which is also monitored on the oscilloscope). If the oscilloscope has a dual-trace feature, the input and output can be monitored simultaneously.

If there is a change in waveform, the nature of the change will often reveal the cause of distortion. For example, a comparison of the square wave response at the output of an IC op-amp against the "typical" patterns of

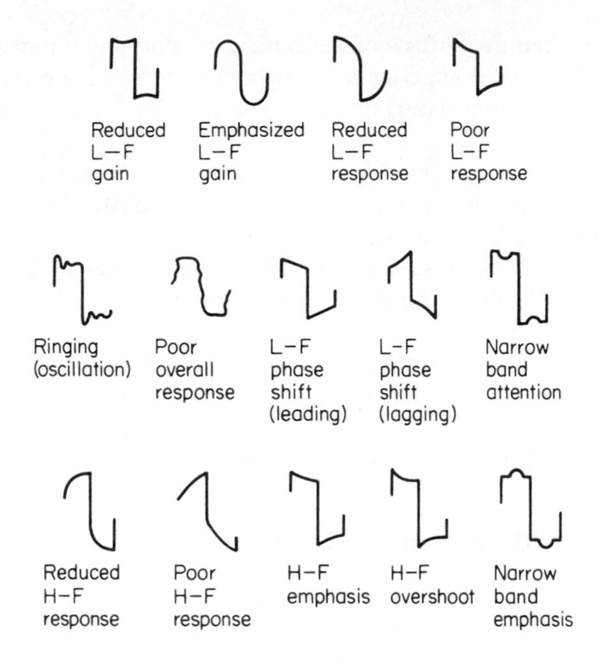

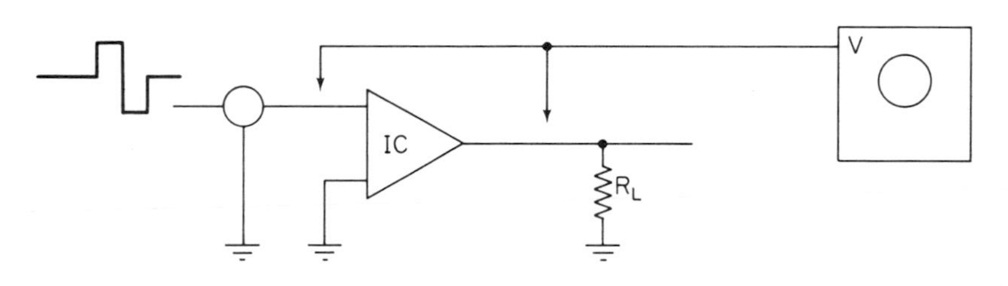

FIG. 7-3. Linear IC square wave distortion analysis.

Fig. 7-3 can show such faults as poor frequency response, overshoot, ringing, phase shift, and emphasis or attenuated gain at certain frequencies.

7-1.7 Background noise

If the vertical channel of an oscilloscope is sufficiently sensitive, an oscilloscope can be used to check and measure the background noise level of an IC op-amp, as well as to check for the presence of hum, oscillation, etc. The oscilloscope vertical channel should be capable of a measurable deflection with about 1 mV (or less) since this is the background noise level of many IC op-amps.

The basic procedure consists of measuring IC output without an input signal. The oscilloscope is superior to a voltmeter for noise level measurement since the frequency and nature of the noise (or other signal) are displayed visually.

The basic connections for background noise level measurement are shown in Fig. 7-4. The oscilloscope gain or sensitivity control is increased until there is a noise or "hash" indication.

It is possible that a noise indication could be caused by pick up in the lead wires. If in doubt, disconnect the leads from the IC but not from the oscilloscope.

Generally, background noise should be measured under open-loop conditions. Some datasheets specify that both input and output voltages be measured. When input voltage is to be measured, a fixed resistance (typically 50 Ω) is connected between the input terminals.

7-1.8 Feedback measurement

Since IC op-amp characteristics are based on the use of feedback signals, it is often convenient to measure feedback voltage at a given frequency with given operating conditions.

The basic feedback measurement connections are shown in Fig. 7-5. While it is possible to measure the feedback voltage as shown in Fig. 7-5a, a more

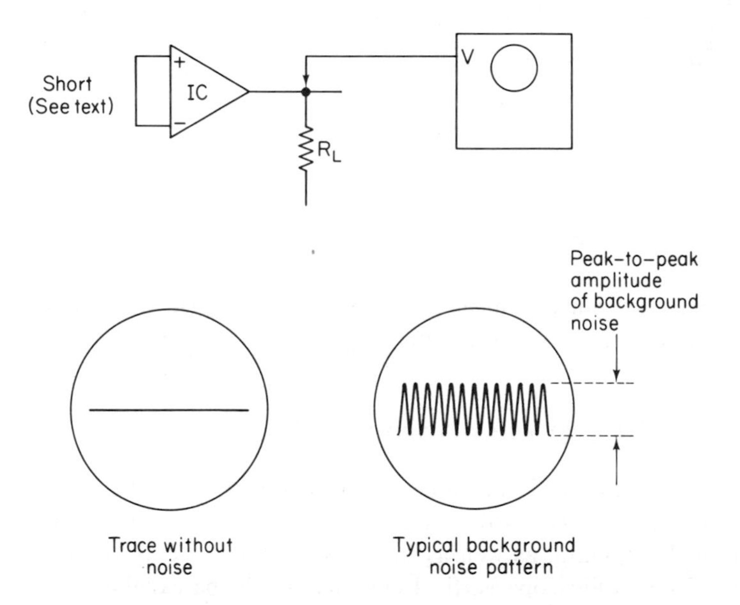

FIG. 7-4. Linear IC background noise test connections.

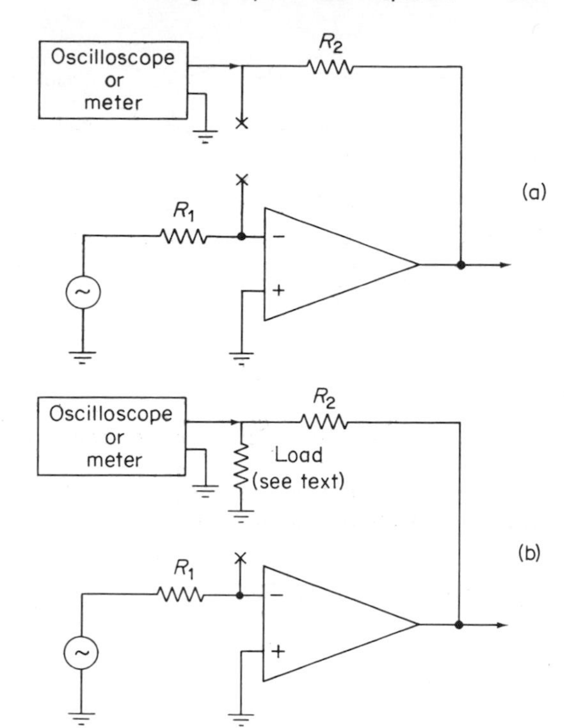

FIG. 7-5. Linear IC feedback measurement.

accurate measurement is made when the feedback lead is terminated in the normal operating impedance.

If an input resistance is used in the normal circuit, and this resistance is considerably lower than the IC input impedance, use the resistance value.

If in doubt, measure the input impedance of the circuit (Sec. 7-1.5), then terminate the feedback lead in that value to measure feedback voltage.

7-1.9 Input bias current

Input bias current (Fig. 3-21) can be measured using the circuit of Fig. 7-6. Any resistance values for R_1 and R_2 can be used, provided the value produces a measurable voltage drop. A value of 1000 Ω is realistic for both R_1 and R_2.

Once the voltage drop is found, the input bias current can be calculated. For example, if the voltage drop is 3 mV across 1 kΩ, the input bias current is 3 μA.

In theory, the input bias current should be the same for both inputs. In practice, the bias currents should be *almost equal* for a well-designed IC op-amp. Any great difference in input bias is the result of unbalance in the input differential amplifier of the IC, and can seriously affect design.

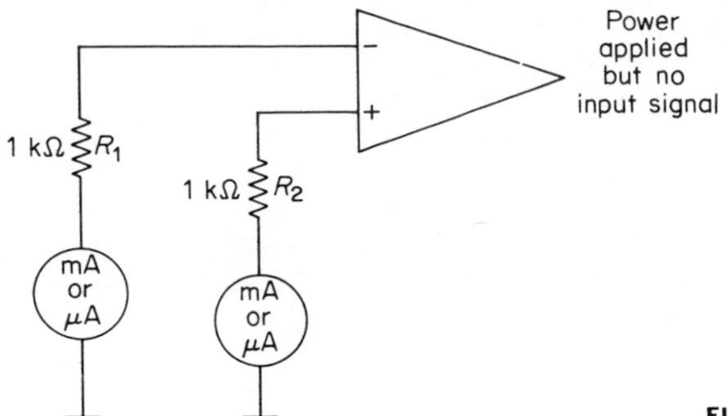

FIG. 7-6. Linear IC input bias current measurement.

7-1.10 Input-offset voltage and current

Input-offset voltage and current (Figs. 3-22 and 3-24) can be measured using the circuit of Fig. 7-7.

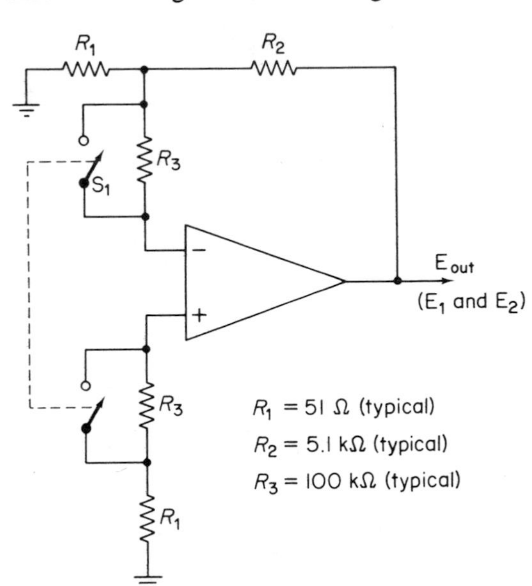

FIG. 7-7. Linear IC input-offset voltage and current measurement.

$E_1 = V_{out}$ with S_1 closed (R_3 shorted)

$E_2 = V_{out}$ with S_1 open (R_3 in circuit)

$$\text{Input offset voltage} = \frac{E_1}{\left(\frac{R_2}{R_1}\right)}$$

$$\text{Input offset current} = \frac{(E_2 - E_1)}{R_3\left(1 + \frac{R_2}{R_1}\right)}$$

As shown, the output is alternately measured with R_3 shorted and with R_3 in the circuit. The two output voltages are recorded as E_1 (S_1 closed, R_3 shorted), and E_2 (S_1 open, R_3 in the circuit).

With the two output voltages recorded, the input-offset voltage and input-offset current can be calculated using the equations of Fig. 7-7. For example, assume that $R_1 = 51\ \Omega$, $R_2 = 5.1\ \text{k}\Omega$, $R_3 = 100\ \text{k}\Omega$, $E_1 = 83\ \text{mV}$, $E_2 = 363\ \text{mV}$ (all typical values).

$$\text{Input-offset voltage} = 83\ \text{mV}/100 = 0.83\ \text{mV}$$

$$\text{Input-offset current} = \frac{280\ \text{mV}}{100\ \text{k}\Omega(1+100)} \approx 0.0277\ \mu\text{A}$$

7-1.11 Common mode rejection

Common mode rejection (Fig. 3-20) can be measured using the circuit of Fig. 7-8.

First, find the open-loop gain under specific conditions of frequency, input, etc., as described in Sec. 7-1.1 and 7-1.2.

Then connect the IC in the common-mode circuit of Fig. 7-8. Increase the common-mode voltage V_{in} until a measurable V_{out} is obtained. Be careful not to exceed the maximum specified input common-mode voltage swing. If no such value is specified, do not exceed the normal input voltage of the IC.

To simplify calculation, increase the input voltage until the output is 1 mV. With an open-loop gain of 100, this will provide a differential input signal of 0.00001 V. Then measure the input voltage. Move the input voltage decimal point over 5 places to find the CMR.

7-1.12 Slew rate

An easy way to observe and measure the slew rate of an IC op-amp is to measure the slope of the output waveform of a square-wave input signal, as shown in Fig. 7-9. The input square wave must have a *rise*

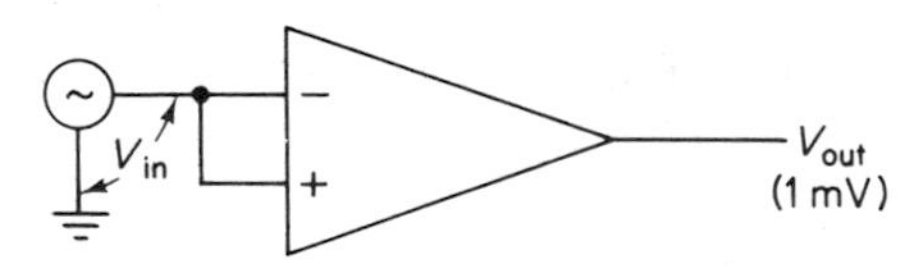

$$\frac{V_{out}\ (1\ \text{mV})}{\text{open loop gain}} = \text{equivalent differential input signal}$$

$$\text{Common mode rejection} = \frac{V_{in}}{\text{equivalent differential input signal}}$$

FIG. 7-8. Linear IC common-mode rejection measurement.

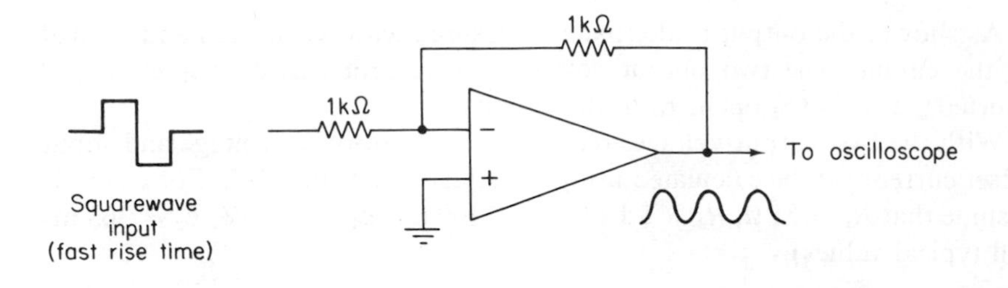

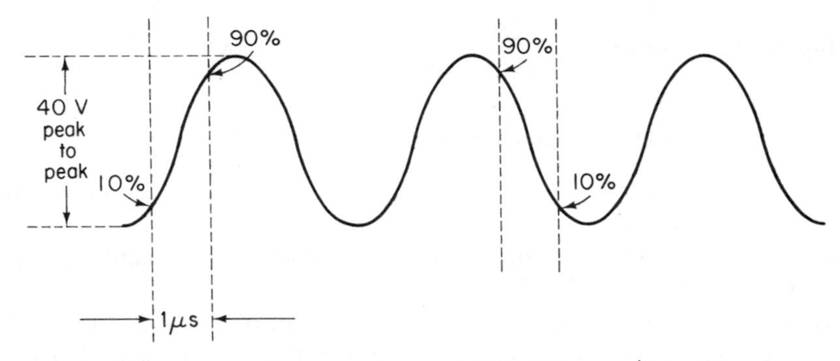

Example shows a slew rate of approximately 40 (40 Volts / μs) at unity gain.

FIG. 7-9. Linear IC slew rate measurement.

time that exceeds the slew rate capability of the IC. Thus, the output will not appear as a square wave, but as an integrated wave.

In the example shown, the output voltage rises (and falls) about 40 V in 1 μs.

Note that the slew rate is usually measured in the closed-loop condition. Also, note that the slew rate increases with higher gain.

7-1.13 Power-supply sensitivity

Power-supply sensitivity (Fig. 3-25) can be measured using the circuit of Fig. 7-7 (the same test circuit as for input-offset voltage, Sec. 7-1.10).

The procedure is the same as for measurement of input-offset voltage, except that one supply voltage is changed (in 1-V steps) while the other supply voltage is held constant. The amount of change in input-offset voltage for a 1-V change in one power supply is the power-supply sensitivity (or input-offset voltage sensitivity, as it is sometimes called).

For example, assume that the normal positive and negative supplies are 10 V, and the input offset voltage is 7 mV. With the positive supply held constant, the negative supply is reduced to 9 V. Under these conditions, assume that the input offset voltage is 5 mV. This means that the negative power-supply sensitivity is 2 mV/V. With the negative power supply held constant at 10 V, the positive supply is reduced to 9 V. Now assume that the input-offset voltage drops to 4 mV. This means that the positive power-supply sensitivity is 3 mV/V.

The test should be repeated over a wide range of power supply voltages (in 1-V steps), if the IC is to be operated under conditions where the power supply may vary by a large amount.

7-1.14 Phase shift

Because an IC op-amp uses the principle of feeding back output signals to the input, the phase shift between input and output is quite critical. All of the IC op-amp phase compensation schemes are based on the feedback principle. Under ideal open-loop conditions, the output should be 180° out-of-phase with the negative input, and in-phase with the positive input.

The following sections describe two procedures for the measurement of phase shift between input and output of an IC op-amp. The same procedure can be used for any amplifier, provided the signals are of a frequency that can be measured on an oscilloscope.

The oscilloscope is the ideal tool for phase measurement. The most convenient method requires a dual-trace oscilloscope, or an electronic switching unit, to produce a dual trace. If neither of these is available, it is still possible to provide accurate phase measurements up to about 100 kHz using the single-trace or X-Y method.

7-1.14.1 Dual-trace phase measurement

The dual-trace method of phase measurement provide a high degree of accuracy at all frequencies, but is especially useful at frequencies above 100 kHz where X-Y phase measurements may provide inaccurate results (owing to inherent internal phase shift of the oscilloscope).

The dual-trace method also has the advantage of measuring phase differences between signals of different amplitudes and waveshapes, as is usually the case with input and output signals of an IC op-amp. The dual-trace method can be applied directly to those oscilloscopes having a built-in dual-trace feature or to a conventional single-trace oscilloscope using an electronic switch or "chopper" unit. Either way, the procedure is essentially one of displaying both input and output signals on the oscilloscope screen simultaneously, measuring the distance (in screen scale divisions) between related points on the two traces, then converting this distance into phase.

The test connections for dual-trace phase measurement are shown in Fig. 7-10. For the most accurate results, the cables connecting input and output

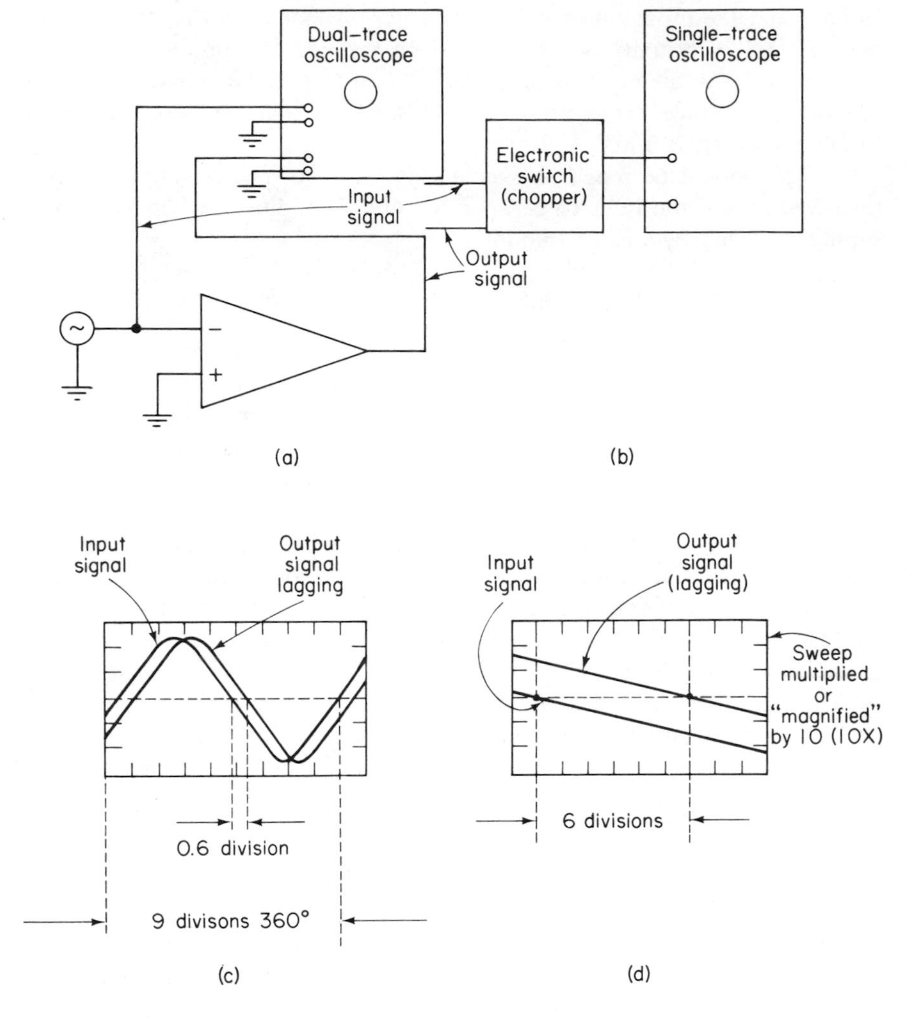

FIG. 7-10. Linear IC dual trace phase shift measurement.

signals should be of the same length and characteristics. At higher frequencies, a difference in cable length or characteristics could introduce a phase shift.

The oscilloscope controls are adjusted until one cycle of the input signal occupies exactly nine divisions (9 cm horizontally) of the screen. Then the phase factor of the input signal is found. For example, if 9 cm represents one complete cycle or 360°, 1 cm represents 40° (360/9 = 40).

With the phase factor established, the horizontal distance between corresponding points on the two waveforms (input and output signals) is measured. The measured distance is then multiplied by the phase factor of 40 °/cm to find the exact amount of phase difference. For example, assume a horizontal difference of 0.6 cm with a phase factor of 40° as shown in Fig. 7-10. Multiply the horizontal difference ($0.6 \times 40 = 24°$ phase shift between input and output signals).

If the oscilloscope is provided with a sweep magnification control where the sweep rate is increased by some fixed amount (5×, 10×, etc.) and only a portion of one cycle can be displayed, more accurate phase measurements can be made. In this case, the phase factor and the approximate phase difference is found as described. Without changing any other controls, the sweep rate is increased (by the sweep magnification control or the sweep rate control), and a new horizontal distance measurement, as shown in Fig. 7-10d.

For example, if the sweep rate is increased 10 times, the adjusted phase factor is $40\,°/10 = 4\,°/\text{cm}$. Figure 7-10d shows the same signal as used in Fig. 7-10d, but with the sweep rate set to 10×. With a horizontal difference of 6 cm, the phase difference would be $6 \times 4° = 24°$.

7-1.14.2 Single-trace (X-Y) phase measurement

The single-trace (or X-Y) phase measurement method can be used to measure the phase difference between input and output of an IC op-amp, at frequencies up to about 100 kHz. Above this frequency, the inherent phase shift (or difference between the horizontal and vertical systems of the oscilloscope) makes accurate phase measurements difficult.

In the X-Y method, one of the signals (usually the input) provides horizontal deflection (X), and the other signal provides the vertical deflection (Y). The phase angle between the two signals can be determined from the resulting pattern.

The test connections for single-trace phase measurement are shown in Fig. 7-11.

Figure 7-11a shows the test connection necessary to find the inherent phase shift (if any) between the horizontal and vertical deflection systems of the oscilloscope. Inherent phase shift (if any) should be checked and recorded. If there is excessive phase shift (in relation to the signals to be measured), the oscilloscope should not be used. A possible exception exists when the signals to be measured are of sufficient amplitude to be applied directly to the oscilloscope deflection plates, bypassing the horizontal and vertical amplifiers.

The oscilloscope controls are adjusted until the pattern is centered on the screen as shown in Fig. 7-11c. With the IC output connected to the vertical input, it is usually necessary to reduce vertical channel gain (to compensate for the increased gain through the amplifier). With the display centered in

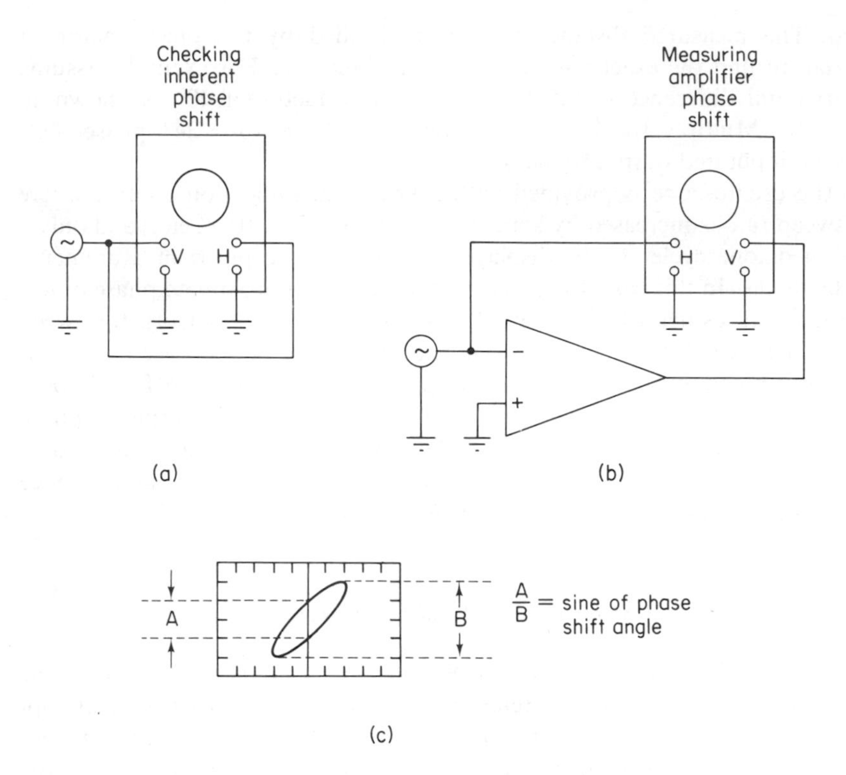

FIG. 7-11. Linear IC single-trace (X-Y) phase shift measurement.

relation to the vertical line, distances A and B are measured, as shown in Fig. 7-11c.

Distance A is the vertical measurement between two points where the traces cross the vertical centerline. Distance B is the maximum vertical height of the display. Divide A by B to find the *sine of the phase angle* between the two signals. This same procedure can be used to find inherent phase shift (Fig. 7-11a) or phase angle (Fig. 7-11b).

If the display appears as a diagonal straight line, the two signals are either in phase (tilted from the upper right to the lower left, for positive slope), or 180° out-of-phase (tilted from the upper left to lower right, for negative slope). If the display is a circle, the signals are 90° out-of-phase.

Figure 7-12 shows the displays produced between 0° and 360°. Notice that above a phase shift of 180°, the resultant display will be the same as at some lower frequency. Therefore, it may be difficult to tell whether the signal is leading or lagging. One way to find correct polarity (leading or lagging) is to introduce a small, known phase shift into one of the signals. The proper angle may then be found by noting the direction in which the pattern changes.

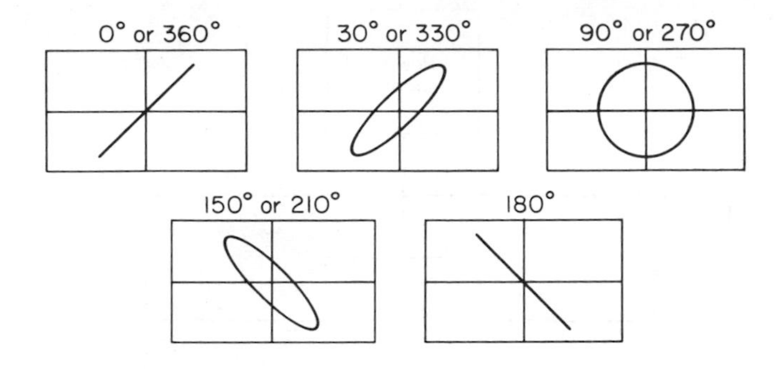

FIG. 7-12. Approximate phase of typical X-Y displays.

Once the oscilloscope's inherent phase shift has been established (Fig. 7-11a), and the IC phase shift measured (Fig. 7-11b), subtract the inherent phase difference from the phase angle to find the true phase difference.

For example, assume an inherent phase difference of 2°, a display as shown in Fig. 7-11c, where A is 2 cm and 4 is 4 cm, Sine of phase angle = A/B, or 2/4, or 0.5. From a table of sines, 0.5 = 30° phase angle. To adjust for the phase difference between X and Y oscilloscope channels, subtract the inherent phase factor (30° − 2°) = 28° true phase difference.

7-2. IC OPERATIONAL AMPLIFIER TESTER

From a user's standpoint, there are four important IC op-amp tests: open-loop voltage gain (A_{VOL}), input offset voltage (V_{io}), output voltage limits (V_o max, plus and minus), and the transfer function. Motorola has developed a basic circuit that will test these four primary functions. The Motorola tester is simple and inexpensive, as is discussed in the following paragraphs. Also included in the following discussion is an elementary discussion of the parameters measured, and their relationship to closed-loop performance of the IC op-amp.

7-2.1 The test fixture

As with almost any tester, there must be a drive circuit, a power source, a device under test (DUT), and a display unit. These are shown in Fig. 7-13.

Since a display unit will duplicate the function of an X-Y oscilloscope, which is usually available in most laboratories, this function is not included in the tester.

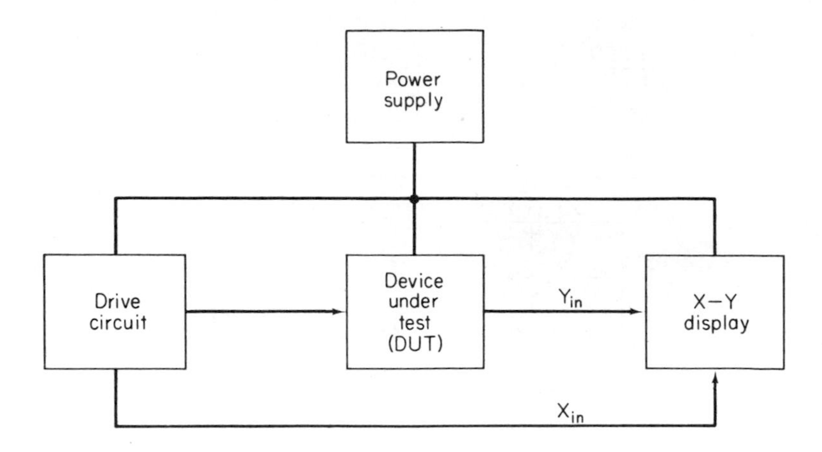

FIG. 7-13. IC op-amp tester block diagram (Motorola).

7-2.2 Power supply

The power supply shown in Fig. 7-14 is a straight-forward shunt Zener regulated power supply. The pilot lamp is bridged across the filtering capacitors as an on-off indicator, and as a bleeder to remove the capacitor charge after power is turned off. The resistive divider in the transformer secondary serves as the input to the drive circuit.

7-2.3 Drive circuit

The drive circuit shown in Fig. 7-15 provides synchronized horizontal and vertical signals. In any test of this sort, it is essential that the vertical drive signal be synchronized with the horizontal sweep signal. The most straight-forward way of accomplishing this is to drive the DUT input and the horizontal sweep with the *same input.* Since a typical IC op-amp will have a gain of at least 1000, a 1000-to-1 precision resistive divider is placed at the input of the DUT. This brings into line the relative amplitudes of the X and Y inputs presented to the oscilloscope. For ICs which show much greater gains, say greater than 100,000, another divider can be added. At this level, it is recommended that the divider be placed at the individual test socket to lessen the possibility of interference on the line.

It is desirable, although not necessary, that the retrace not appear on (or be blanked from) the visual presentation. This is accomplished by the drive circuit as a form of "intensity modulation". Note that this is not true intensity modulation, where a signal is applied to the oscilloscope Z-axis, but produces the same effect. That is, the relative intensity of the oscilloscope trace is determined by the sweep rate, given a fixed set of other trace variables.

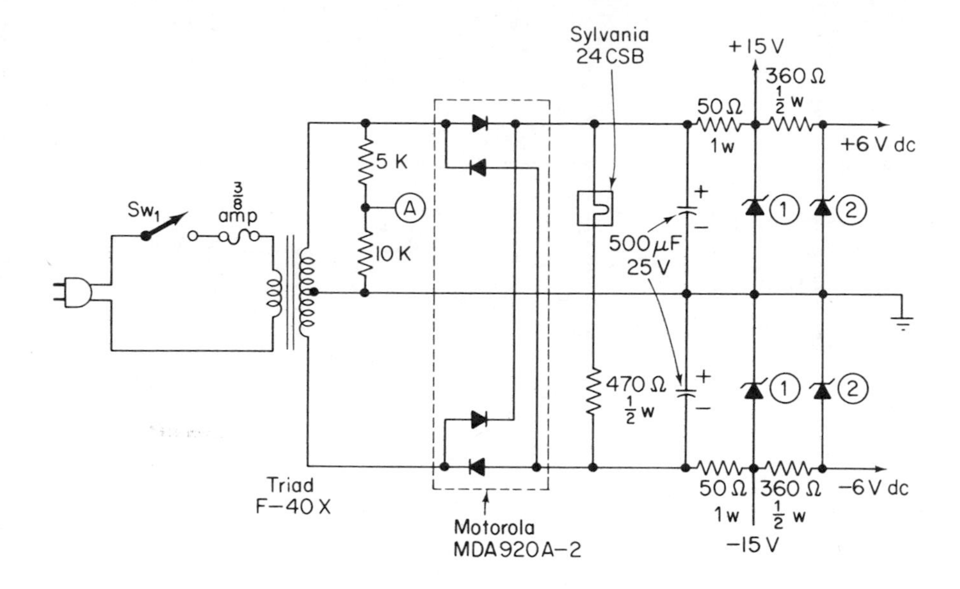

FIG. 7-14. Power supply for IC op-amp tester (Motorola).

Operation of the Fig. 7-15 drive circuit is simple. While the transistor is off, the *RC* combination charges toward the +15-V supply. The time constant is chosen such that an ample amplitude results, while deleting the more exponential portion of the charge curve. The time constant must be sufficiently longer than the period of the reset rate (which in this circuit is 60 Hz).

A portion of the power transformer secondary voltage is used as the drive circuit input (Point A on Figs. 7-14 and 7-15). As the input sinewave exceeds a predetermined threshold voltage (approximately $2V_{BE} + V_Z$) the transistor is turned on, thus discharging the capacitor. This results in the waveform shown in Fig. 7-15.

The waveform is coupled to the IC and oscilloscope via the amplitude adjust potentiometer and the divider resistors. As shown by the Fig. 7-15 waveforms, the *relative intensities* of the trace and retrace are determined by the ratio of time T_1 to T_2. Thus, with the oscilloscope properly adjusted, the retrace "disappears".

The difference between the horizontal drive voltage and the IC input is a precise 1000-to-1. This permits the horizontal sweep voltage to be of sufficient amplitude to drive the oscilloscope, while retaining a low input at the IC to match realistic values of input offset voltage.

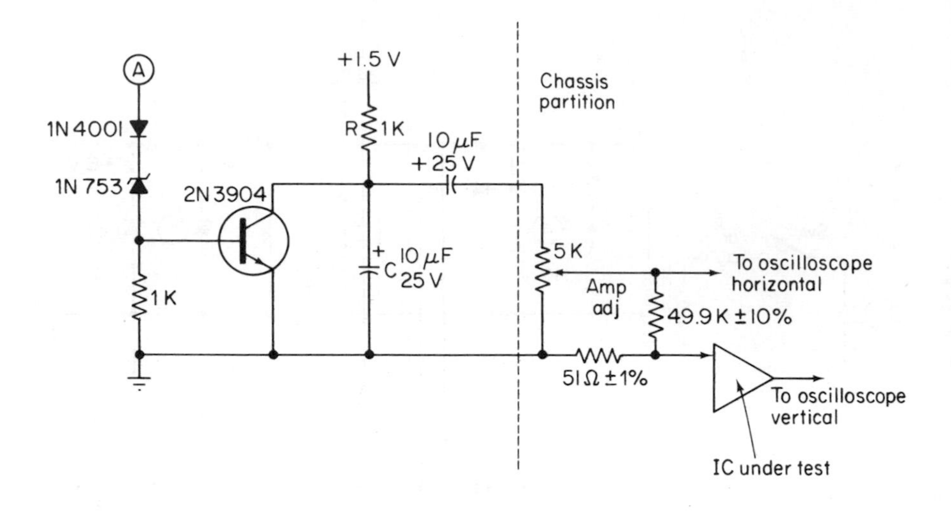

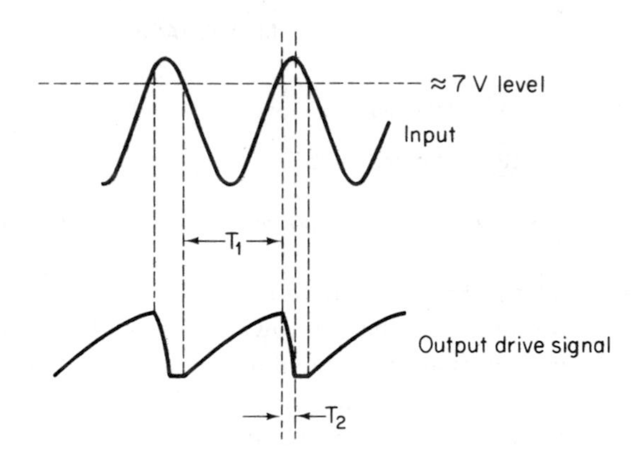

FIG. 7-15. Drive circuit for IC op-amp tester (Motorola).

7-2.4 Tester wiring

The wiring of all tester sockets is shown in Fig. 7-16. Note that these socket pin arrangements are for Motorola ICs. However, the socket connections can be rearranged to match other ICs. In any event, the power supply terminals should be bypassed at each individual socket, and each socket should be provided with a proper frequency compensation network (capacitor, or capacitor and resistor, as necessary). Note that a single 2-kΩ resistor is used as a common load for all ICs. The exact test load can be tailored individually, if desired. Also note that a switch is provided for three of the sockets. This is to test the hi-lo gain option of certain Motorola ICs (the MC1433/MC1533).

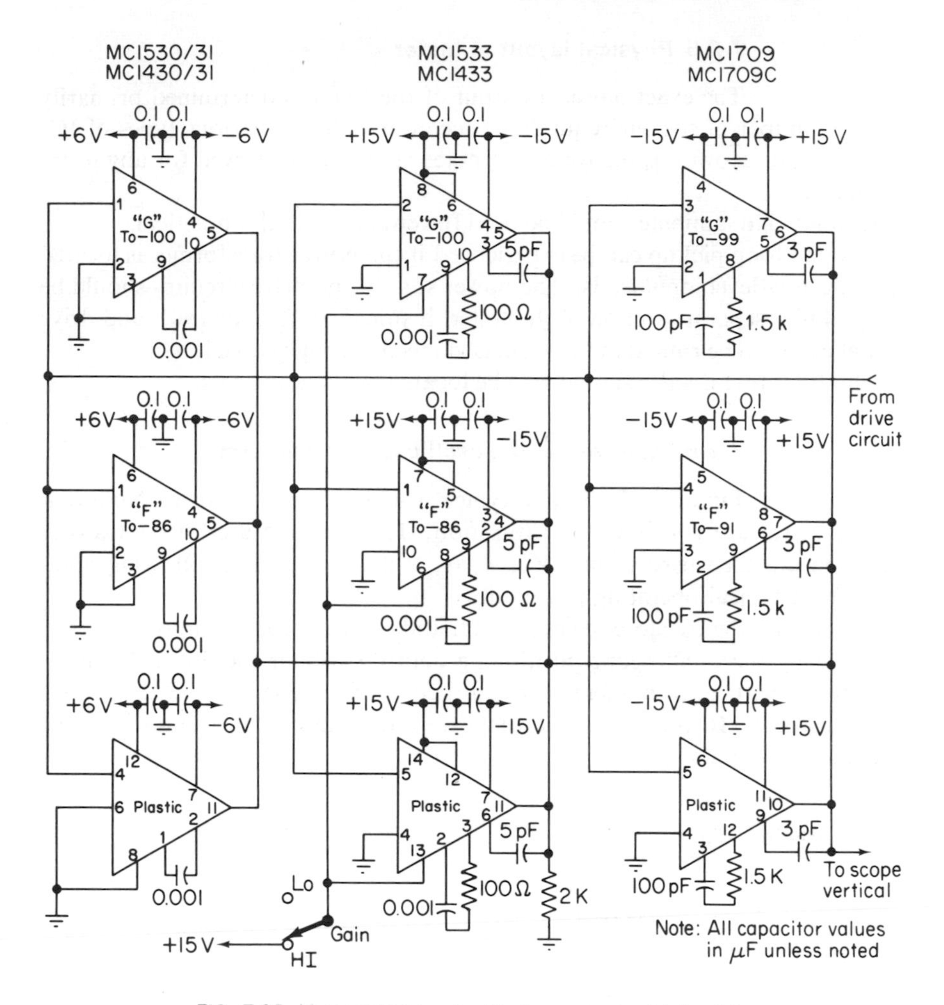

FIG. 7-16. IC socket wiring for IC op-amp tester (Motorola).

Additional sockets for other ICs can be incorporated as the demand warrants. Likewise, only one socket is needed if only one type of IC is to be tested. The maximum drive signal available is 8 V peak-to-peak, open-circuit from the 5-kΩ amplitude adjust potentiometer. This should be sufficient to test most IC op-amps.

In order to add other sockets, it is necessary to install the power supply bypass and frequency compensation components, route the IC output to the oscilloscope vertical input, route the IC input to the drive circuit, and adjust the drive voltage as needed. Keep in mind that the amplitude of the drive signal (to the IC) must be sufficient to overcome any input offset voltage, as well as driving the IC into saturation.

7-2.5 Physical layout of tester

The exact physical output of the tester is determined primarily by the number of sockets required to accommodate the various types of ICs being tested. However, the following notes should be observed for any tester layout.

Copper-clad laminate should be used for circuit wiring, if practical.

Ground loop pickup can be experienced if the power transformer is located near the low-level input leads. The power supply and drive circuitry should be located in a separate section of the tester, if practical. The dc leads and drive signal can then be routed to the sockets via feed-through capacitors.

The 1000-to-1 divider should also be located near the sockets.

7-2.6 Interpreting the oscilloscope waveform

Figure 7-17 shows a typical oscilloscope waveform. Note that the IC under test is driven into saturation. The main features of the transfer function are then used to find V_{io}, A_{VOL}, and V_o max. The following notes describe a typical operating procedure.

1. Ground the oscilloscope horizontal and vertical inputs temporarily.

2. Using the oscilloscope positioning controls, center the dot at the horizontal and vertical zero reference point (center of screen).

3. Insert the IC into the appropriate socket. Remove the ground from the oscilloscope inputs.

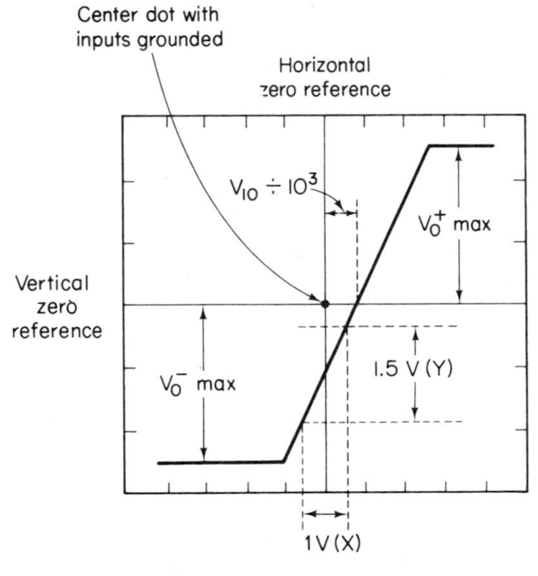

FIG. 7-17. Interpreting oscilloscope waveforms of IC op-amp tester (Motorola).

4. Increase the amplitude adjust potentiometer setting until the IC is in deep saturation (top and bottom of the output waveform flattened, as shown).

5. Read V_o max (both plus and minus) directly from the transfer function trace. For example, assume that each screen division equals one volt. The minus V_o max is then about 2.5 V, whereas the plus V_o max is about 2.3 V. Total output voltage swing is then about 4.8 V.

6. A_{VOL} is equal to the calculated slope of the transfer function, multiplied by 1000. Any part of the transfer function can be used. However, the voltage required for deflection across both horizontal and vertical screen divisions must be known. For example, assume that both horizontal and vertical screen divisions equal one volt. Under these conditions, the X dimension of the transfer function sampled in Fig. 7-17 is equal to 1 V, whereas the Y dimension is equal to approximately 1.5 V. Thus, Y/X = 1.5, and A_{VOL} = 1500 (1.5 × 1000). If the oscilloscope horizontal and vertical screen division dimensions are identical (say both divisions equal 1 centimeter), the exact voltage need not be considered. Instead, only the ratio of Y/X for any given portion of the transfer function is used.

7. V_{io} is equal to the horizontal displacement from the horizontal zero reference, to the point where the function crosses the vertical zero reference, divided by 1000. For example, assume that each screen division equals one volt. Under these conditions, the function crosses the vertical zero reference about 0.7 V from the horizontal zero reference in Fig. 7-17. Thus, V_{io} = 0.7 mV (0.7 V/1000).

8. A measure of the ICs linearity can also be obtained by noting the linearity of the transfer function between saturation points. For example, if the line is perfectly straight, the IC is perfectly linear.

9. In reading A_{VOL} and V_{io} from the oscilloscope presentation, it is advisable to increase the horizontal sensitivity so that resolution is increased for better accuracy. Keep in mind that if the oscilloscope voltage ranges are changed, it may be necessary to check the zero references (both horizontal and vertical). However, it is usually not necessary to check the zero reference when testing one IC after another.

7-2.7 Analyzing measured parameters

Open-loop voltage gain (A_{VOL}) can be defined as the ratio of a change in output voltage to a change in input voltage. Ideally, for an IC op-amp, A_{VOL} should be infinitely high, since the primary function is to amplify In general, the higher the gain the better the accuracy. The significance of open-loop gain is many times misapplied. From a user's standpoint, open-loop gain determines closed-loop *accuracy limits*, rather than the ultimate accuracy.

Referring to Fig. 7-18 (an ideal op-amp), the closed-loop gain of the circuit is:

$$\frac{e_o}{e_{in}} = -\frac{A_{VOL}\dfrac{R_2}{R_1 + R_2}}{1 + A_{VOL}\dfrac{R_1}{R_1 + R_2}} \tag{1}$$

If the op-amp of Fig. 7-18 shows infinite open-loop gain, equation 1 reduces to:

$$\frac{e_o}{e_{in}} = -\frac{R_2}{R_1} \tag{2}$$

the ratio of the two passive feedback elements, which is the ideal closed-loop gain of an op-amp connected as shown in Fig. 7-18.

The error in the closed-loop gain (error_{CL}) of an op-amp may be represented as:

$$\%(\text{error}_{\text{CL}}) = \frac{\dfrac{e_o}{e_{\text{in IDEAL}}} - \dfrac{e_o}{e_{\text{in ACTUAL}}}}{\dfrac{e_o}{e_{\text{in IDEAL}}}} \times 100 \tag{3}$$

which, after insertion of equations 1 and 2, reduces to:

$$\%(\text{error}_{\text{CL}}) = \frac{100}{1 + A_{VOL}\dfrac{R_1}{R_1 + R_2}} \tag{4}$$

The closed-loop gain error is a direct function of the loop gain $[A_{VOL}(R_1/(R_1 + R_2))]$ rather than solely open-loop gain. Open-loop gain is the limiting factor in closed-loop gain error; loop gain establishes the accuracy.

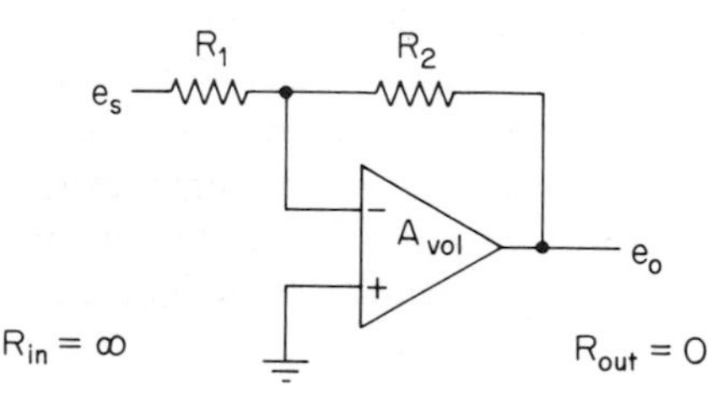

FIG. 7-18. Idealized operational amplifier (Motorola).

Output voltage swing (V_o max) can be defined as the peak output voltage swing, referred to zero, that can be obtained without clipping (due to saturation). A symmetrical voltage swing is generally implied. However, if V_o^+ max and V_o^- max differ, the maximum symmetrical voltage swing is limited by the *lesser absolute value*. Output impedance, load current, and frequency directly affect V_o max.

In addition to the limiting factor of V_o max on the output swing, the transfer linearity affect the maximum output swing, within distortion limits. The deviation of the transfer function from a perfect straight line, within the saturation limits, indicates the degree of distortion that can be expected in the output signal, as well as the peak voltage at which this distortion occurs.

Input offset voltage (V_{io}) can be defined as that voltage which must be applied at the input terminals to obtain zero output voltage. V_{io} also indicates matching tolerances in the differential amplifier stages of the IC op-amp. V_{io} is primarily determined by the V_{BE} difference in the input stage, and unbalance in the second stage, attenuated by the gain of the first stage.

In general, V_{io} will be the major source of offset voltage error in low source impedance circuits. IC op-amps with minimum V_{io} are better matched, and will generally track well with temperature variations. Thus, such ICs will show minimum output drift with temperature variations.

A factor not always considered when determining the contribution of V_{io} in closed-loop operation is that the error is not simply increased by the ratio of feedback resistance to input summing resistance but by unity (or 1) plus this ratio. At high closed-loop gain levels, the difference is of little concern. However, at unity gain operation, the difference is a factor of 2.

7-3. TESTING IC DIGITAL LOGIC ELEMENTS

As is the case with any logic element, the only practical test of an IC logic circuit is its response to various combinations of inputs. For example, an AND gate should produce a true output when both of the inputs are true simultaneously. A sequential network (such as an IC flip-flop or counter) requires the additional element of proper timing to perform its function.

From the user's standpoint, it is possible to predict with some accuracy the response of logic circuits to various input and output combinations by means of truth tables, timing diagrams, and flow charts found on the datasheets. Ultimately, however, the user must test the circuit under simulated (or actual) operating conditions. Signals and/or pulses must be applied to inputs (many input simultaneously in the case of some circuits), and the resultant outputs noted. The basic IC logic test tools are the oscilloscope (to monitor inputs and outputs), and the pulse generator (to simulate clock or other inputs).

Unlike most linear circuits, IC logic elements should also be tested for response to *improper inputs*. Due to the high operating speeds of today's IC logic elements (particularly ECL), and the unknown delays produced by various elements, it is quite possible for two inputs (each one "normal" in its own right) to be applied simultaneously to a circuit or element, and produce an undesired result.

A good example of this is a JK flip-flop (or FF) which should change states (or toggle) with each clock pulse. The same FF should set when a J input is applied, and reset or clear when a K input is applied. But what happens if a J and K input are applied simultaneously? Will the FF toggle? Will the FF set without a toggle? If so, to what state (set or reset) will the FF go? Or will the FF assume some undesired intermediate state? The answers to these questions can be predicted with carefully prepared truth tables (as appear on some IC manufacturer's data sheets), *but can be proved only by actual test.*

Another common problem is that IC logic elements may respond properly to fixed input signals, but fail to produce correct outputs with pulsed inputs (particularly at high speeds). Generally, this condition is due to delay introduced by the various logic elements. Thus, it is essential in the test of any logic circuit that the delay of individual elements, as well as groups of elements, be measured accurately.

The following paragraphs of this section describe practical IC logic element tests. It is assumed that the user is familiar with operation procedures of oscilloscopes (particularly sampling oscilloscopes) and pulse generators.

It is also assumed that the user has made a sufficient study of basic logic circuits to apply *improper inputs*, as well as the proper inputs, throughout the following tests. For example, if a datasheet truth table for a particular circuit indicates a "don't care" or "not allowed" condition, apply the corresponding inputs, and note the results, good or bad.

7-3.1 Basic pulse measurement techniques

No matter what IC logic element is involved, proper test requires that the user be able to work with pulses. That is, the user must be able to monitor pulses on an oscilloscope, measure their amplitude, duration or width, and frequency or repetition rate. The user must also be able to measure delay between pulses and to check operation of basic logic circuits or building blocks, such as NAND gates, FFs, and the like.

The following measurement techniques apply to all types of pulses, including square waves, used in logic circuits.

7-3.1.1 Pulse definitions

The following terms are commonly used in describing pulses found in IC logic elements. The terms are illustrated in Fig. 7-19. The input

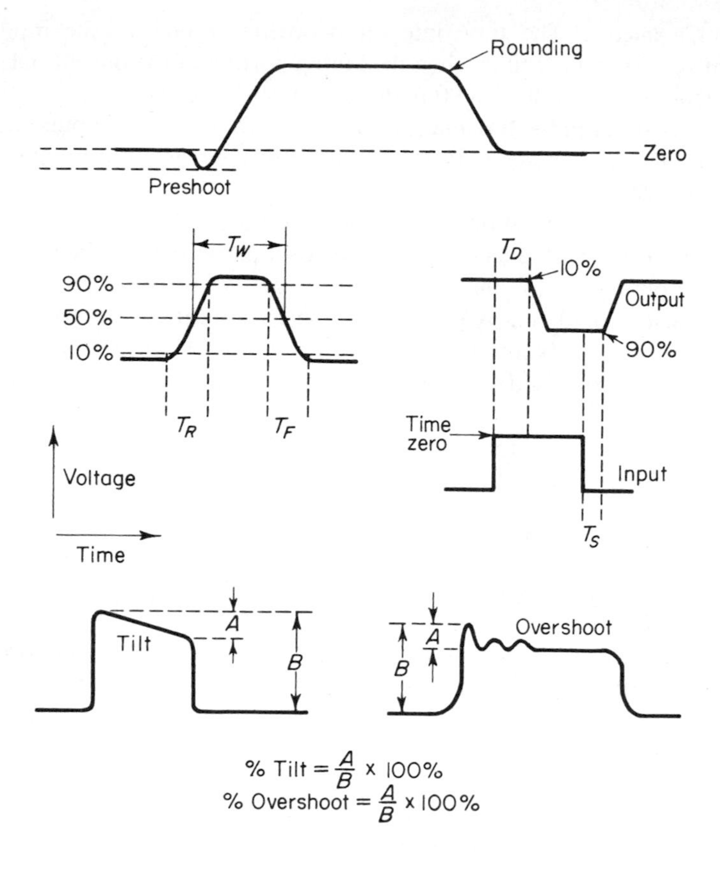

FIG. 7-19. Basic pulse and square wave definitions.

pulse represents an ideal input waveform (almost never found in any IC logic element) for comparison purposes. The other waveform in Fig. 7-19 represents the shape of pulses that may appear in IC logic elements, particularly after they have passed through many gates, delays, and so on. The terms are defined as follows:

Rise time, t_r. The time interval during which the amplitude of the output voltage changes from 10 percent to 90 percent of the rising portion of the pulse.

Full time, t_f. The time interval during which the amplitude of the output voltage changes from 90 percent to 10 percent of the falling portion of the waveform.

Time delay, t_d. The time interval between the beginning of the input pulse (time zero) and the time when the rising portion of the output pulse attains an arbitrary amplitude of 10 percent above the baseline.

Storage time, t_s. The time interval between the end of the input pulse (trailing edge) and the time when the falling portion of the output pulse drops to an arbitrary amplitude of 90 percent from the baseline.

Pulse width (or pulse duration), t_w. The time duration of the pulse measured between the two 50 percent amplitude levels of the rising and falling portions of the waveform.

Tilt. A measure of the tilt of the full-amplitude, flat-top portion of a pulse. The tilt measurement is usually expressed as a percentage of the amplitude of the rising portion of the pulse.

Overshoot or preshoot. A measure of the overshoot or preshoot occurring generally above or below the 100 percent amplitude level. These measurements are also expressed as a percentage of the amplitude of the rising portion of the pulse.

These definitions are for guide purposes only. When pulses are very irregular (such as excessive tilt, overshoot, etc.), the definitions may become ambiguous.

7-3.1.2 Rule of thumb for rise-time measurements

Since rise-time measurements are of special importance in testing any pulse circuit, and especially in testing sequential logic IC elements, the relationship between the oscilloscope rise time, and the rise times of the IC logic elements under test must be taken into account. Obviously, the accuracy of rise time measurements can be no greater than the rise time of the oscilloscope. Also, if the IC is tested by means of an external pulse from a pulse generator, the rise time of the pulse generator must also be taken into account.

For example, if an oscilloscope with a 20-ns rise time is used to measure the rise time of a 15-ns IC logic element, the measurement will be hopelessly inaccurate. If a 20-ns pulse generator and a 15-ns oscilloscope are used to measure the rise time of a component, the fastest rise time for accurate measurement will be something greater than 20 ns. Two basic rules of thumb can be applied to rise-time measurements.

The first method is known as the *root of the sum of the squares.* It involves finding the squares of all the rise times associated with the test, adding these squares together, and then finding the square root of the sum. For example, using the 20-ns pulse generator and the 15-ns oscilloscope, the calculation is:

$$20 \times 20 = 400,\ 15 \times 15 = 225,\ 400 + 225 = 625,\ \sqrt{625} = 25 \text{ ns}$$

One major drawback to this rule is that the coaxial or shielded cables required to interconnect the test equipment are subject to *skin effect.* As frequency increases, signals tend to travel on the outside or skin of the conductor.

This decreases conductor area and increases resistance. In turn, this increases cable loss. (Keep in mind that IC logic elements often operate at clock speeds in excess of 100 MHz). The losses of cables do not add properly to apply the root-sum-squares method, except as an approximation.

The second rule or method states that if the logic element or signal has a rise time ten times slower than the test equipment, the error is 1 percent. If this amount is small it can be considered as negligible. If the IC being measured has a rise time three times slower than the test equipment, the error is slightly less than 6 percent. By keeping these relationships in mind, the results can be interpreted intelligently.

7-3.1.3 Measuring pulse amplitude

Pulse amplitudes are measured on an oscilloscope, preferably a laboratory-type oscilloscope where the vertical scale is calibrated directly in a specific deflection factor (such as volts per centimeter). Such oscilloscopes usually have a step attenuator (for the vertical amplifier) where each step is related to a specific deflection factor. Typically, the pulses used in IC logic elements are 5 V or less. Often the pulse amplitude is critical to the operation of the circuit.

For example, an AND gate may require two pulses of 3 V to produce an output. If the two pulses are slightly less than 3 V, the AND gate may act as if there is no input, or that both inputs are false. Thus, when testing any IC logic element, always check the actual pulse amplitude against those shown in the datasheet (such as required input levels for gates, FFs, etc.) Generally, IC logic elements are designed with considerable margin in pulse amplitude. However, actual values should always be checked.

The basic test connections and corresponding oscilloscope display for pulse amplitude measurement are shown in Fig. 7-20.

As an example, assume that a vertical deflection of 4.7 cm is measured, using a probe with no attenuation, and a vertical deflection factor (step attenuator setting) of 1 V/cm. Then the peak-to-peak pulse amplitude is 4.7 V.

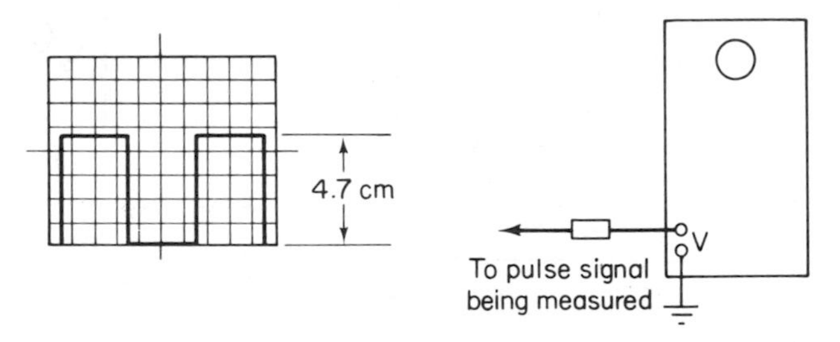

FIG. 7-20. Measuring pulse amplitude.

7-3.1.4 Measuring pulse width or duration

Pulse width or pulse duration is measured on an oscilloscope, preferably a laboratory type, where the horizontal scalc is calibrated directly in relation to time. The horizontal sweep circuit of a laboratory oscilloscope is usually provided with a selector control that is direct reading in relation to time. That is, each horizontal division on the screen has a definite relation to time at a given position of the horizontal sweep rate switch (such as 1 μs/cm). Laboratory oscilloscopes are also provided with sweep magnification controls. Such magnification is generally required since the pulses used in IC logic elements are on the order of nanoseconds or microseconds (but can be as long as a few milliseconds).

The basic test connections and corresponding oscilloscope display for pulse width measurement are shown in Fig. 7-21.

As an example, assume that the horizontal sweep rate is 0.1 μs/cm, the sweep magnification is 100, and a horizontal distance of 5 cm is measured (between the 50 percent points). Then the pulse width is $5 \times 0.1/100 = 0.005$ μs (or 5 ns).

7-3.1.5 Measuring pulse frequency or repetition rate

Pulse frequency or repetition rate is measured on an oscilloscope, preferably a laboratory type, where the horizontal scale is calibrated in relation to time. Pulse frequency is found by measuring the time duration of *one complete pulse cycle* (not pulse width) and then dividing the time into 1 (since frequency is the reciprocal of time).

The basic test connections and corresponding oscilloscope display for pulse frequency measurement is shown in Fig. 7-22.

As an example, assume that the horizontal sweep rate is 0.1 μs/cm, the sweep magnification is 100, and a horizontal distance of 20 cm is measured

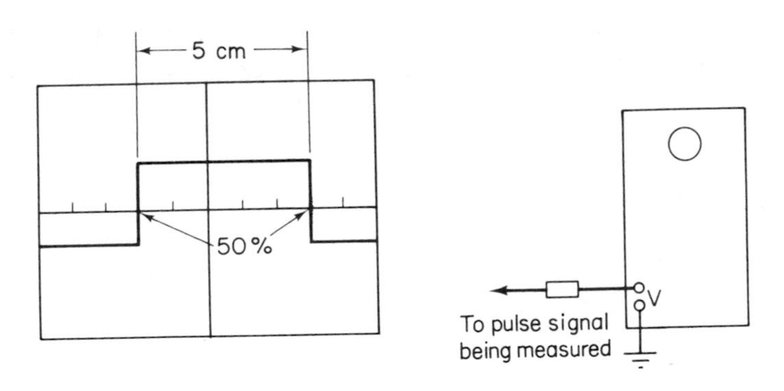

FIG. 7-21. Measuring pulse width.

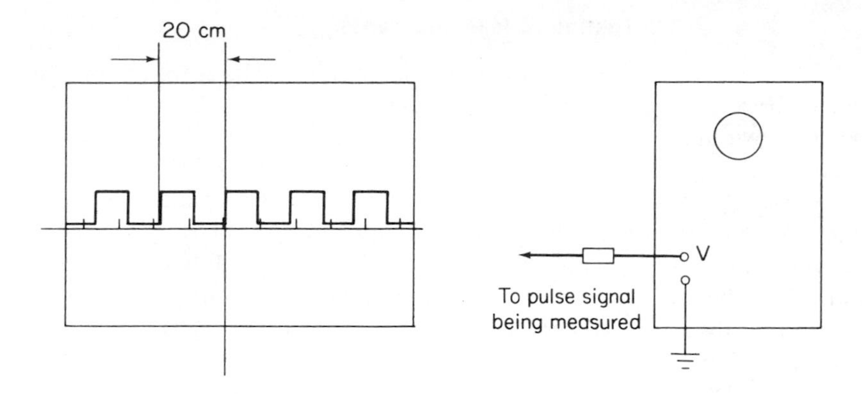

FIG. 7-22. Measuring pulse frequency or repetition rate.

between the beginning and end of a complete cycle. Then the frequency is $20 \times 0.1/100 = 0.020\ \mu s$ (or 20 ns); 1/20 ns = 50 MHz.

7-3.1.6 Measuring pulse delay

The time interval or delay between two pulses (say an input pulse and an output pulse) introduced by a gate, FF, or complete logic circuit can be measured most conveniently on an oscilloscope with a *dual trace*, or *multiple-trace*. It is possible to measure delay on a single-trace oscilloscope, but it is quite difficult.

The basic test connections and corresponding oscilloscope display for pulse-delay measurements are shown in Fig. 7-23.

As an example, assume that the horizontal sweep rate is 100 ns/cm, there are seven screen divisions between the input and output pulses, and each horizontal screen division is 1 cm. Then the delay is $7 \times 100 = 700$ ns.

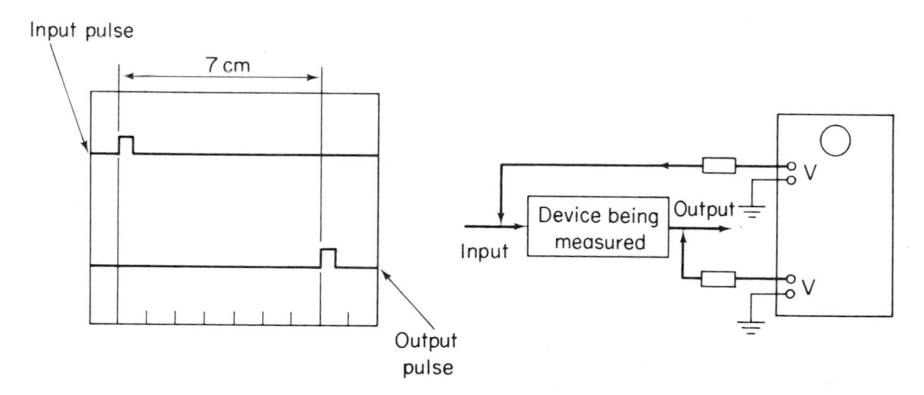

FIG. 7-23. Measuring pulse delay.

7-3.2 Testing IC logic elements

The following paragraphs describe procedures for test or check-out of basic IC logic elements (gates, delays, FFs, etc.). Such elements can be tested with fixed voltages, or with pulse trains.

An oscilloscope is the best tool for pulse-train testing. In its simplest form, pulse-train testing consists of measuring the pulses at the input and output of a logic element or circuit. The pulses are supplied by a pulse generator, and should simulate those with which the IC logic element will be used (in amplitude, duration, and frequency). If the input pulses are normal, but the output pulses are absent or abnormal, the problem is localized to the IC being tested.

IC logic elements can also be tested by applying fixed voltages at the inputs, and noting the change in output voltage levels. For example, assume that an IC AND gate operates with a +5-V true level. A fixed +5 V can be applied to both inputs, and the output monitored with a voltmeter for +5 V. While this method is simple, it is generally not realistic (unless the IC is always to operate with fixed voltages).

Pulse-train testing is most practical for the majority of IC logic elements. However, the pulse train can be slowed down to a frequency lower than that with which the IC is to operate. If the IC performs properly at the lower frequency, but not at the normal frequency, the IC is probably not suitable for the particular design requirements.

Before going into detailed test procedures, the *marginal test technique* should be considered. Most IC logic elements are designed to operate in systems with power supplies that are adjustable (typically over a ±10 percent range). Thus, the IC should be tested with various operating voltage levels. In many cases, this will show marginal failures in the IC. The technique is not recommended by all IC manufacturers, since the results are not certain. However, it is usually safe to apply the technique to any IC on a temporary basis. Of course, the logic element datasheet should be consulted to make sure.

7-3.2.1 Testing AND gates

Figure 7-24 shows the basic connections for testing an AND gate. Ideally, both inputs and the output should be monitored simultaneously, since an AND gate produces an output only when two inputs are present. If pulse trains are monitored on an oscilloscope, check that an output pulse is produced each time there are two input pulses of correct amplitude and polarity. If this is not the case, check carefully that both input pulses arrive at the same time (not delayed from each other).

Note that in Fig. 7-24 the input pulse trains are not identical. Input 2 has fewer pulses for a given time interval than input 1. Also, input 2 has one

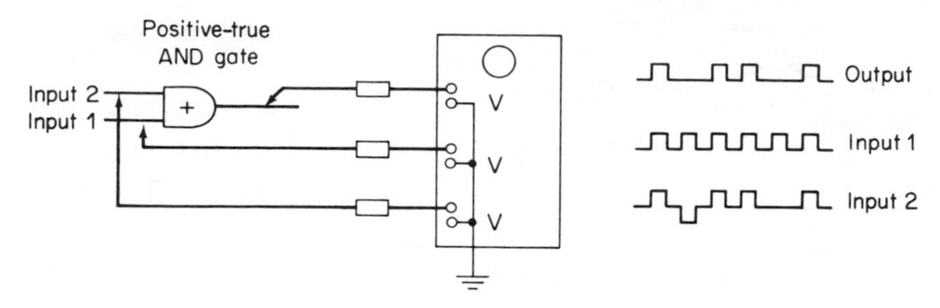

FIG. 7-24. Testing AND gates.

negative pulse simultaneously with one positive pulse at input 1. This *should not* produce an output pulse. Often, one input will be a long pulse (long in relation to the pulses at the other input). In other cases, one input will be a fixed dc voltage.

The complete truth table of the circuit can be checked using the basic connections of Fig. 7-24. Always check the IC element for response to improper inputs, as well as proper inputs. For example, an output pulse train should not appear if only one input is present. Likewise, the output should remain at the false (or 0) level, with either input false.

7-3.2.2 Testing OR gates

Figure 7-25 shows the basic connections for testing an OR gate. Ideally, all inputs and the output should be monitored simultaneously. An OR gate produces an output when any inputs are present. If pulse trains are monitored on an oscilloscope, check that an output pulse is produced for each input pulse (of appropriate amplitude and polarity).

Note that in Fig. 7-25, the input pulse trains are not identical. That is, the input pulses do not necessarily coincide. However, there is an output pulse when either input has a pulse, and when both inputs have a pulse. This last condition marks the difference between an OR gate and an EXCLUSIVE OR gate, discussed in Sec. 7-3.2.3.

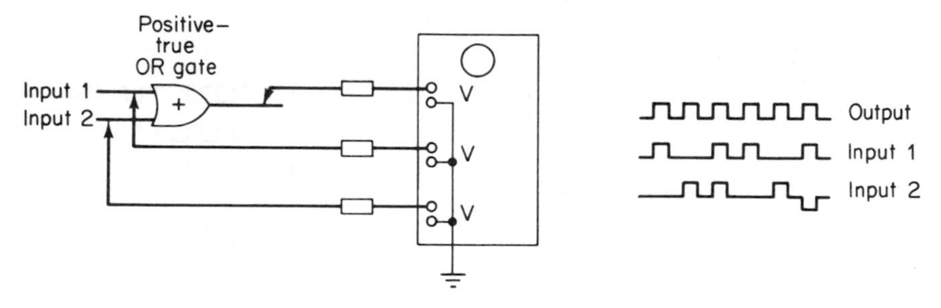

FIG. 7-25. Testing OR gates.

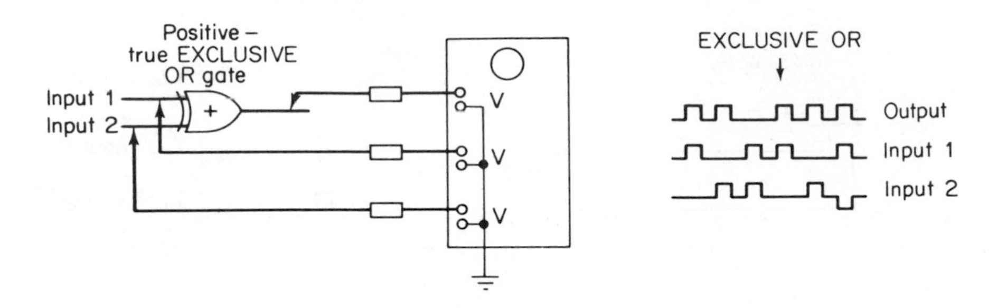

FIG. 7-26. Testing EXCLUSIVE OR gates.

The complete truth table of the IC can be checked using the basic connections of Fig. 7-25. To check the OR gate for response to improper inputs, remove all inputs, and check that there are no output pulses, and that the output remains at the false (or 0) level.

7-3.2.3 Testing EXCLUSIVE OR gate circuits

Figure 7-26 shows the basic connections for testing an EXCLUSIVE OR gate. Ideally, all inputs and the output should be monitored simultaneously. An EXCLUSIVE OR gate produces an output when any one input is present, but not when both inputs are present. If pulse trains are monitored on an oscilloscope, check that an output pulse is produced for each input pulse that does not coincide with another input pulse.

The complete truth table of the IC can be checked using the basic connections of Fig. 7-26. To check the EXCLUSIVE OR gate for response to improper inputs, remove all inputs, and check that there are no output pulses, and that the output remains at the false (or 0) level. Then apply both input pulses simultaneously (connect both inputs to the same pulse source), and check that the output remains false or 0.

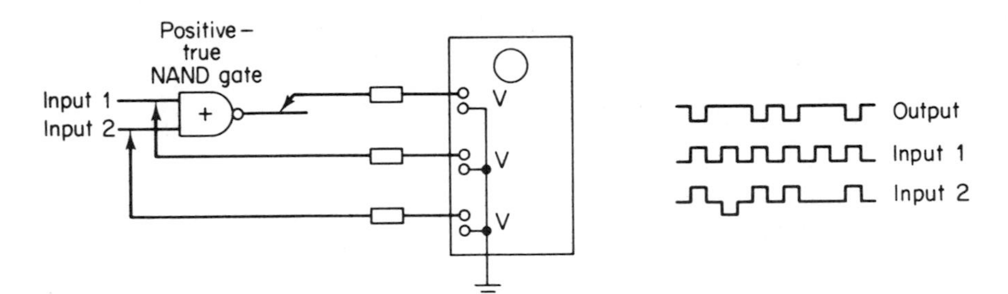

FIG. 7-27. Testing NAND gates.

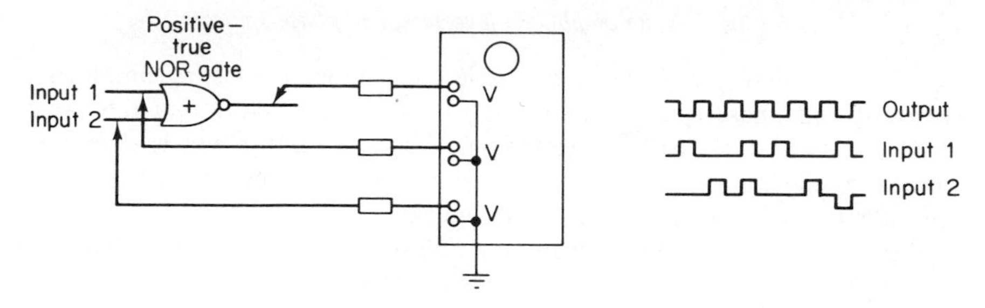

FIG. 7-28. Testing NOR gates.

7-3.2.4 Testing NAND, NOR, and EXCLUSIVE NOR gates

Figures 7-27, 7-28 and 7-29 show the basic connections for testing NAND, NOR and EXCLUSIVE NOR gates, respectively.

Note that the connections are the same for corresponding, AND, OR, and EXCLUSIVE OR gates, as are the test procedures. However, the output pulse is inverted. That is, a NAND gate produces an output under the same logic conditions as an AND gate circuit, but the output is inverted. For example, if the logic is positive true, and a positive pulse is present at both inputs of a NAND gate simultaneously, there is an output pulse, but the pulse is negative (or false).

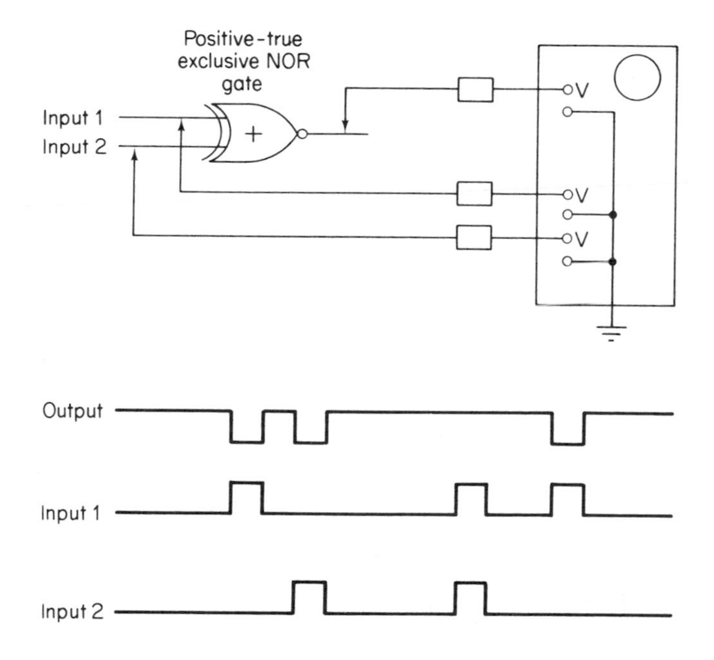

FIG. 7-29. Testing EXCLUSIVE NOR gates.

7-3.2.5 *Testing amplifiers, inverters, and phase splitters*

Figure 7-30 shows the basic connections for testing amplifiers, inverters, and phase splitters used in IC logic circuits. Note that the same connections are used for all three circuits. That is, each circuit is tested by monitoring the input and output with an oscilloscope. However, the relationship of output to input is different for each of the three circuits.

With the amplifier, the output pulse amplitude should be greater than the input pulse amplitude; with the inverter, the output pulse may or may not be greater in amplitude than the input pulse, but the polarity is reversed. In the case of the phase splitter, there are two output pulses (of opposite polarity to each other) for each input pulse.

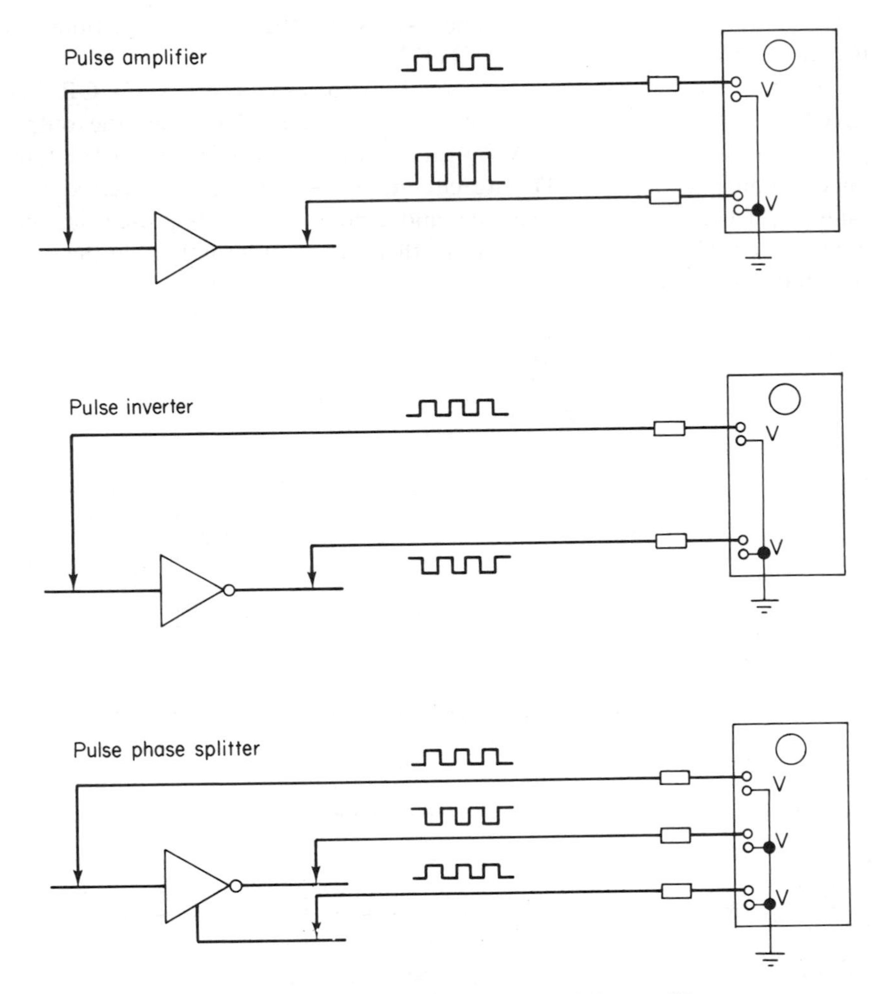

FIG. 7-30. Testing amplifiers, inverters, and phase splitters.

7-3.2.6 Testing FFs

Figure 7-31 shows the basic connections for testing FFs. There are many types of FFs (RS, JK, toggle, latching, delay, etc.) available in IC form. Each FF responds differently to a given set of pulse conditions. Thus, each type of FF must be tested in a slightly different way. However, all FFs have two states (even though there may be only one output, and it may not be possible to monitor both states directly). Some FFs require only one input to change states. Other FFs require two simultaneous pulses to change states (such as a clock pulse and a reset–set pulse).

The main concern in testing any FF is that the states change when the appropriate input pulse (or pulses) are applied. Thus, at least one input and

Basic FF test connections and waveforms

Input
Output
+FF
V
V
Input
Output

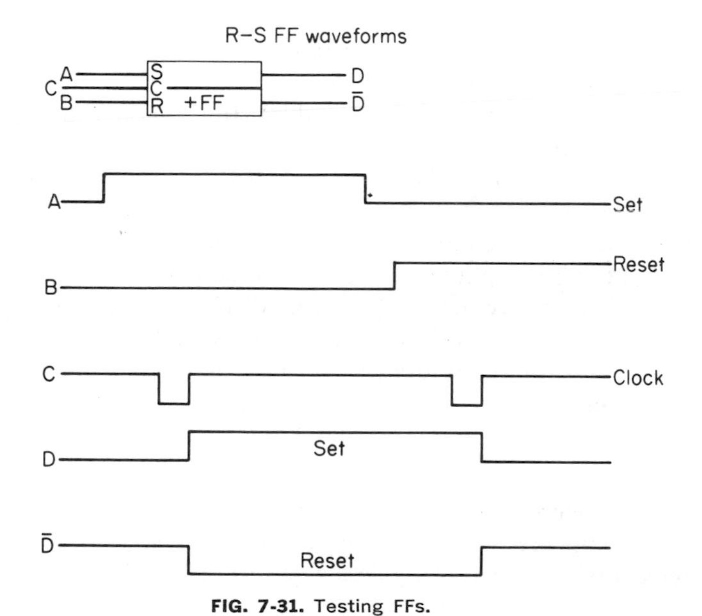

FIG. 7-31. Testing FFs.

one state (output) should be monitored simultaneously. For example, an RS FF (without clock) can be tested by connecting the pulse train at the set input and set output. If there is a pulse train at the set input and set output, it is reasonable to assume that the FF is operating. If the set output remains in one state, either the FF is defective and not being reset, or there are no reset pulses. The next step is to monitor the reset input and reset output. If both the set and reset pulse trains are present, but the FF does not change states, the FF is defective.

There are many exceptions, of course. For example, if the reset and set pulses arrive simultaneously, due to an unwanted delay in other circuits, the FF may remain in one state.

If the same RS FF requires a clock pulse, first check that both the clock pulse and a set or reset pulse arrive simultaneously. Then check that the FF changes states each time there is such a pulse coincidence. For example, assume that pulse trains are measured on a dual-trace oscilloscope, and that a set pulse occurs for every other clock pulse. Next, monitor the clock pulse and the set output, and note that the set output occurs for every two clock pulses. Or, monitor the set input and output, noting that the FF changes states on a pulse-for-pulse basis.

In testing any FF, it is always advisable to check for response to improper inputs. For example, a classic FF test is to apply simultaneous set and reset inputs, and check the result. Typically, an FF will remain in some unknown state, and then follow the last input to be removed.

7-3.2.7 Testing MVs

Figure 7-32 shows the basic connections for testing an MV. There are three basic types of IC MVs: free-running (astable), one-shot (monostable), and Schmitt trigger (bistable). In the case of the free-running MV, only the output need be monitored since the circuit is self-generating. Both the one-shot (OS) and Schmitt trigger (ST) require that the input and output be monitored.

If MV pulse trains are monitored on an oscilloscope, the frequency, pulse duration, and pulse amplitude can be measured as described in Sec. 7-3.1. While these factors may not be critical in all IC logic elements, quite often one or more of the factors is important. For example, the logic symbol for an OS MV (when properly drawn) includes the output pulse duration or width, thus indicating that width is a critical factor.

7-3.2.8 Testing delays

Figure 7-33 shows the basic connections for testing delay lines or delay elements used with IC logic. The test procedure is the same as for measurement of delay between pulses described in Sec. 7-3.1.6.

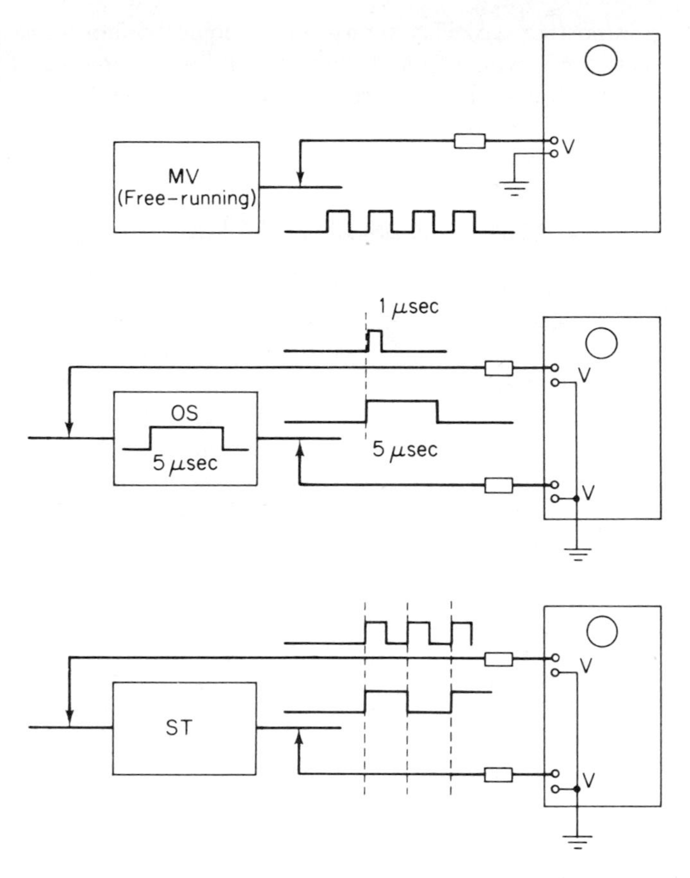

FIG. 7-32. Testing MVs.

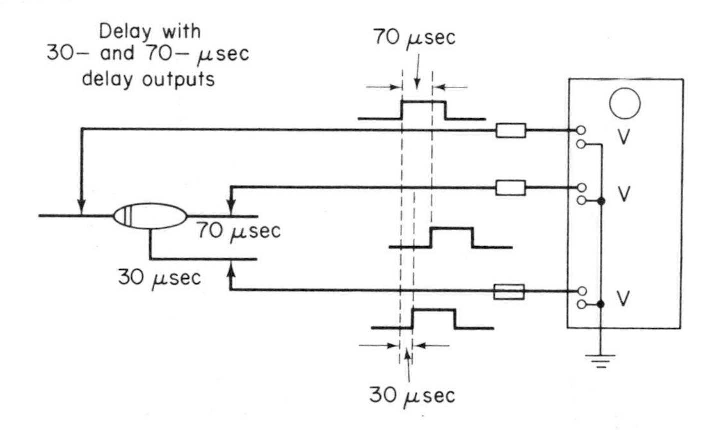

FIG. 7-33. Testing delays.

Keep in mind that the delay shown on the logic symbol *usually* refers to the delay between leading edges of the input and output pulses. This can be assumed if the symbol is not further identified. Some IC manufacturers identify their delay symbols as to leading or trailing edge.

7-4. TESTING HIGH-SPEED IC LOGIC ELEMENTS

In addition to the basic tests described in Sec. 7-3, high-speed logic elements must be tested for performance at higher operating speeds (higher clock pulse frequencies). The basic characteristics of IC logic elements can be altered at higher speeds. That is, an IC may perform properly at low speeds, but not at high speeds. Likewise, the high-speed characteristics of an IC can be affected by circuit conditions, such as fan-out, temperature, load capacitance, voltages, etc. It is essential that the user know the *actual performance* of an IC logic element, under both normal and abnormal conditions, to make effective use of the IC.

Although an IC can be subjected to an infinite variety of tests, the major high-speed characteristics are propagation delay, switching time, and noise margin. The following paragraphs describe a series of procedures to test these characteristics. Although the tests are specifically designed for use with ECL ICs, the same tests (or similar tests) can be adapted to other logic families. Of course, the power supply voltages, logic levels, etc., must be adapted as necessary to match the characteristics of the particular IC logic family.

7-4.1 Noise margin tests

A primary concern of the logic system designer is the noise rejection capability of the IC logic elements used in the system. A knowledge of the false-triggering levels of the circuits allows the user to provide the necessary protection against spurious ground line and signal line voltages.

Since logical operations in ECL (and most other IC logic families) are performed at the collector, disturbances that appear on the ground line are critical in establishing logic levels and system stability. The test configuration shown in Fig. 7-34 is used to determine ECL system susceptibility to ground line noise. As shown, the ECL gate under test is cross-coupled with another gate to form a bistable FF. The noise threshold level is detected by applying a voltage to the ground input connection, and observing the voltage at which the FF changes state.

The circuit of Fig. 7-35 is used to find ECL sensitivity to voltage disturbances on the input signal lines. The same method of detection used for ground-line noise is also used for signal-line noise.

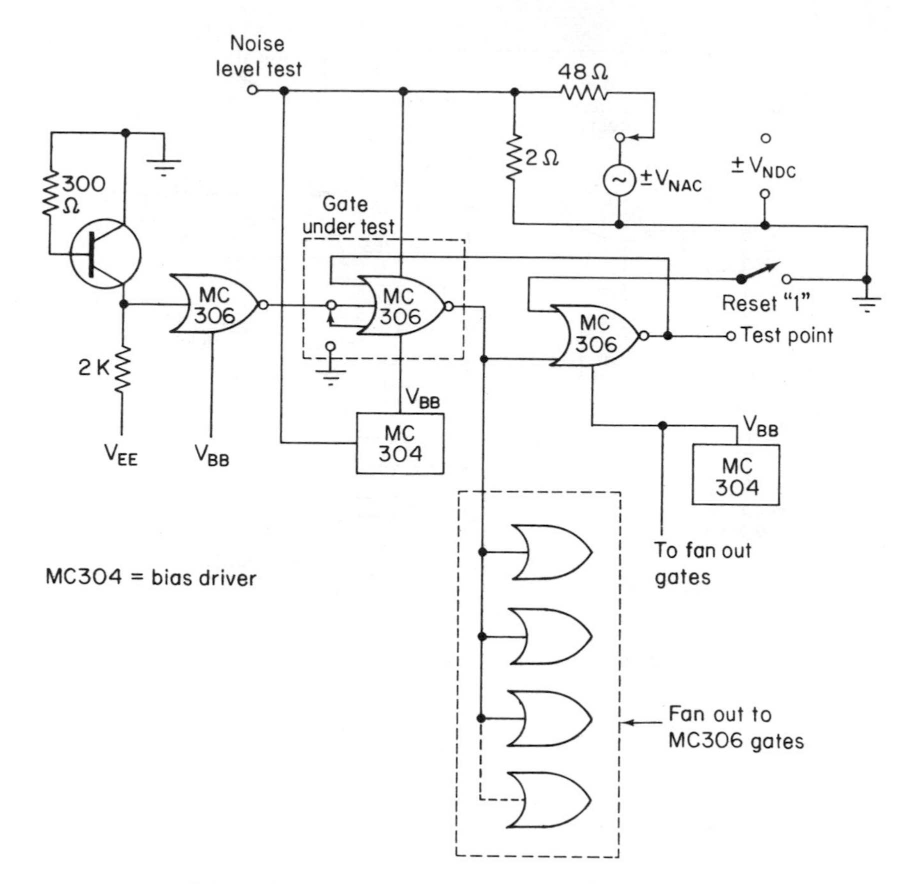

FIG. 7-34. Ground noise test circuit for MECL (Motorola).

The circuit of Fig. 7-36 is typical of an ECL IC logic element. All of the following tests are made on the assumption that the device under test is similar to the circuit of Fig. 7-36.

7-4.1.1 Noise test circuit operation

In both Figs. 7-34 and 7-35, means are provided for applying either ac (V_{nac}) or dc noise (V_{ndc}) voltages to the ground line and signal line inputs. By means of switches, the FF can be placed into either stable state, depending upon which polarity of noise input is being tested.

A separate bias regulator supply is used for the gate under test, while another regulator supplies the bias for the remaining gate circuits. The reason for the separate bias supply is to isolate the threshold of the IC under test from the remainder of the system.

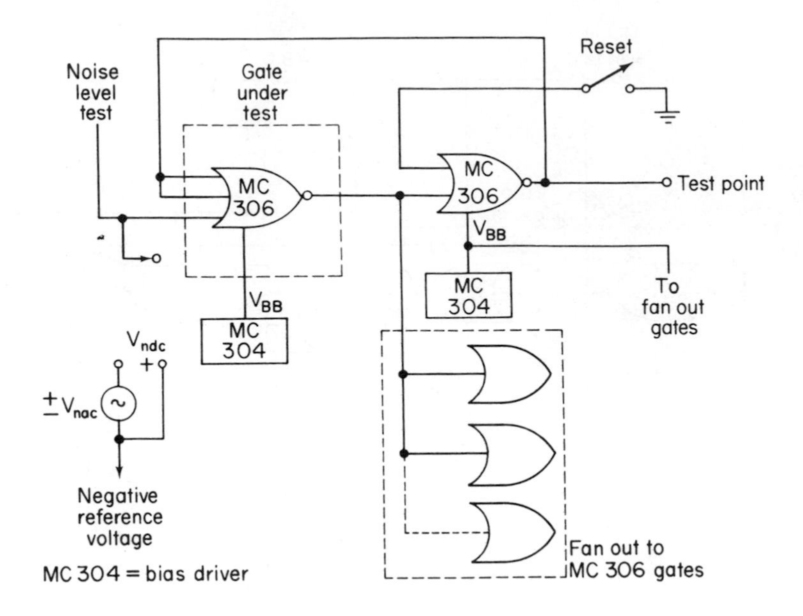

FIG. 7-35. Input noise test circuit for MECL (Motorola).

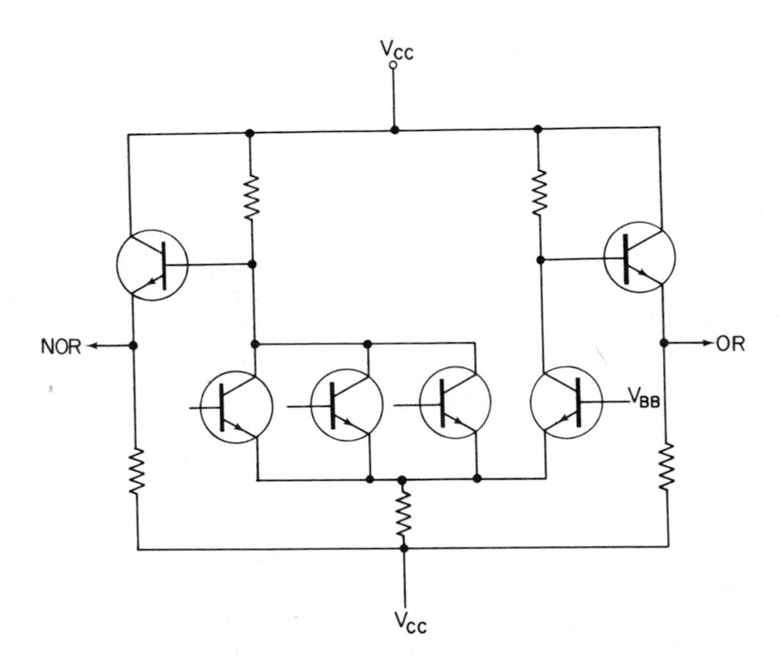

FIG. 7-36. MC306 MECL gate circuit (Motorola).

It is desirable for the operating levels of the IC test gate to vary without changing the levels of the other gate circuits. This is a worst-case situation that, under most operating conditions, would not exist. For example, ground potential variations in a localized part of a system will affect bias (V_{BB}) circuits and logic circuits similarly. Thus, as the operating points vary, the threshold levels tend to track, and system remains stable. So, with similar noise conditions existing in a system, the same disturbing results may not occur.

Figure 7-37 shows a reduced schematic of an ECL gate with the bias circuit. This circuit is used to show what goes on within the test gate and bias circuit when noise voltages are applied. The circuit also shows that ground line voltage variations are transferred through the divider circuit to the V_{BB} output. If V_{in} is low, I_1 current flows in the current switch, and the amplifier variations have a cancelling effect with the ground disturbances at the OR collector. The negative variations are observed directly on the NOR collector (NOR output). These voltages also reduce the common emitter potential which drives the input device toward the "on" condition.

Positive variations on the ground input do not have the same serious effects on circuit stability when I_1 current flows. Since the NOR output is already in the high state this level simply increases, and the system is not upset. It takes a greater positive voltage change (greater than the magnitude of the negative voltage change) to cause the low OR level to reach the threshold of the following stage.

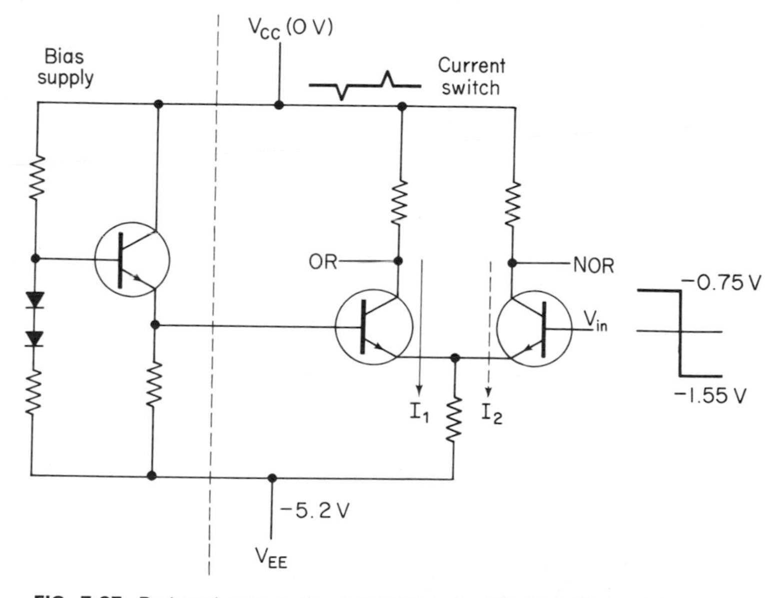

FIG. 7-37. Reduced schematic of MC306 gate with bias (Motorola).

With V_{in} in the high state and I_2 current flowing, negative ground level changes will produce the most disturbing effects at the OR collector. Negative ground level disturbances will not affect the NOR since this output is already in the low state. Also, with a high V_{in}, positive ground voltages attempt to turn on the OR side of the current switch, as well as raise the low NOR level.

7-4.1.2 Noise test procedures and conclusions

The basic procedure for ground noise test is as follows:

1. Set the NOR output of the IC gate under test to the high logic level.
2. Inject negative noise on the ground line.
3. Observe the input level that causes the FF to change state.

Figure 7-38 shows probability distribution curves for dc noise margins over the temperature range of −55 °C to +125 °C. Similar distribution curves will result for the situation where a high OR logic level exists, and positive noise is applied on the ground line. Greater ground noise margins exist on the NOR output for positive ground voltages, and also on the OR output for negative ground voltages.

Figure 7-39 shows the allowable ground noise amplitude as a function of the noise pulse width. The worst case conditions apply for pulse widths greater than approximately 35 ns. That is, the thresholds are much greater for short pulse widths.

Figures 7-40, 7-41 and 7-42 are plotted, worst-case, noise margins for ground-line disturbances as they vary with changes in supply voltage and fan-out at −55 °C, +25 °C, and +125 °C. These curves are for logic 1 only, since fan-out does not affect the noise level of the low logic output. (A logic 0 to an input transistor biases that transistor off.) The conditions for fan-out are shown in Figs. 7-34 and 7-35. For convenience, the curves of Fig. 7-43 are

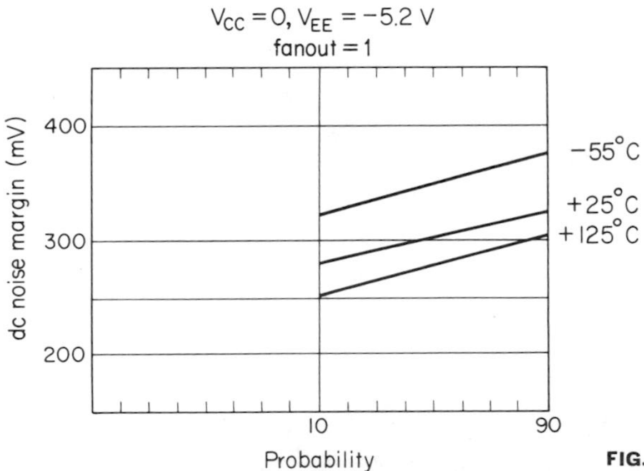

FIG. 7-38. Probability data for ground line dc noise margin versus temperature (Motorola).

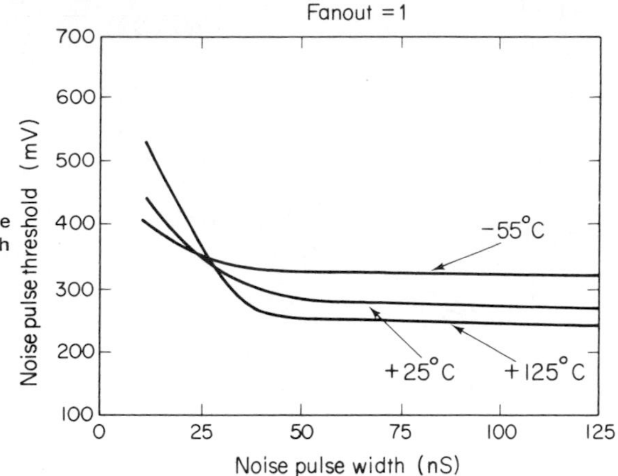

FIG. 7-39. Ground line noise pulse threshold versus noise pulse width and temperature (Motorola).

given, showing typical worst-case noise margins as functions of temperature and fan-out.

The tests for input noise immunity use the same general procedure as followed for ground noise testing. The test circuit of Fig. 7-35 shows that only positive input voltages are applied since the same noise margins will result for negative noise inputs to the other side of the FF.

Figures 7-44 through 7-46 show typical variations in signal line noise margin with noise pulse width, temperature, fan-out, and supply voltage. These noise margins are with respect to worst-case input logic levels. Once the threshold of the IC is established, it is assumed that the driving circuit output is at the highest specification value for the logic 0 level (in this case, −1.465 V at −5.2 V and +25 °C).

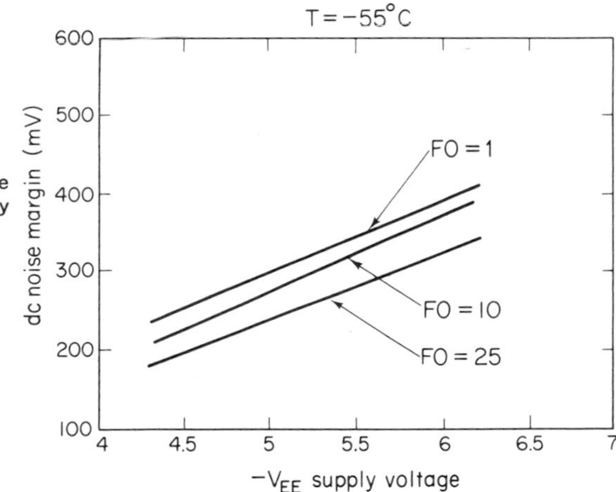

FIG. 7-40. Ground line dc noise margin versus fan-out and supply voltage variations (Motorola).

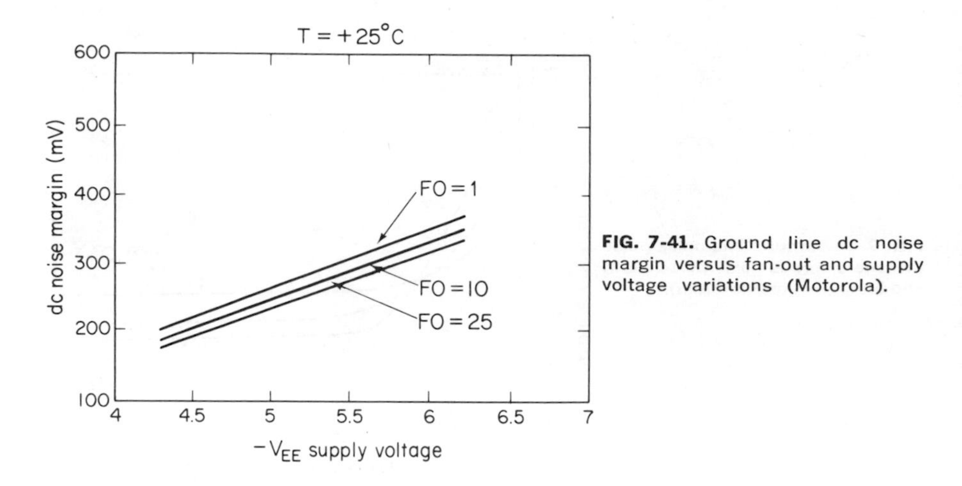

FIG. 7-41. Ground line dc noise margin versus fan-out and supply voltage variations (Motorola).

7-4.2 Propagation delay, switching time, and power supply sensitivity tests

The following paragraphs describe procedures for test of high-speed IC logic elements, under various operating conditions. To obtain satisfactory test information or operation in a specific application, certain layout and operational rules must be followed.

Power supply distribution, both V_{EE} and V_{CC}, should be done in one of two ways. The most desirable method is to provide two voltage planes of which one is V_{EE} (-5.2 V) and the other is V_{CC} (ground). Both V_{CC1} and V_{CC2} must be connected directly to the ground plane. If a ground plane is not possible, a ground buss must be used. The buss should be as wide as possible, with both ends connected to a common ground.

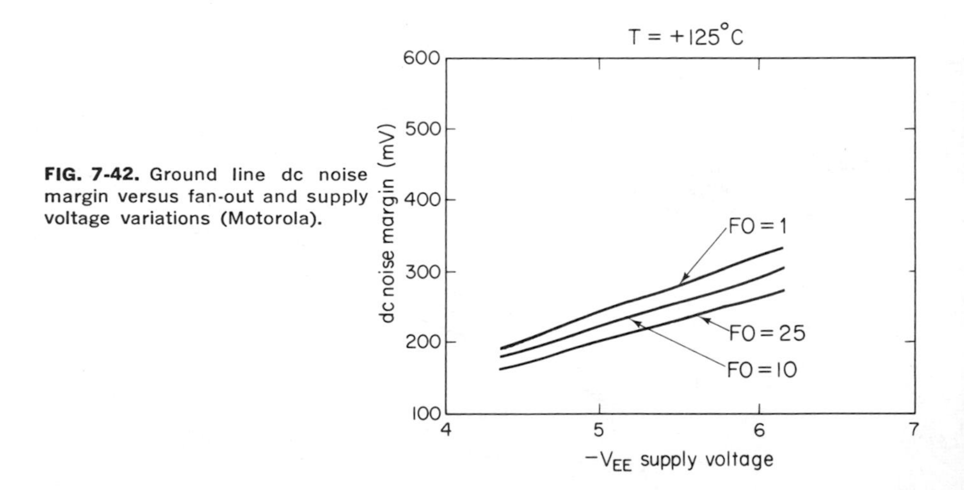

FIG. 7-42. Ground line dc noise margin versus fan-out and supply voltage variations (Motorola).

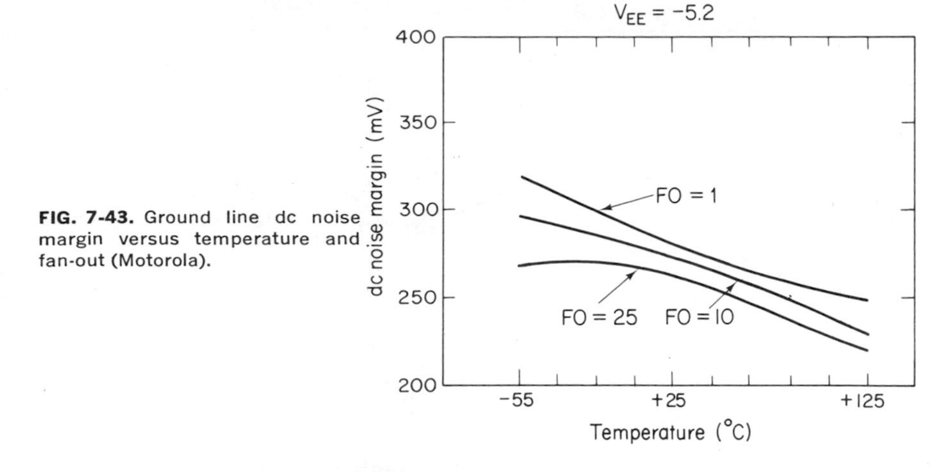

FIG. 7-43. Ground line dc noise margin versus temperature and fan-out (Motorola).

For breadboard testing, the ICs should be mounted over a ground plane, with signal interconnects held to a minimum length. The packages may be mounted on their back next to the ground plane with the signal leads held close to the ground plane. A V_{EE} supply buss can be distributed by means of a wide strap of copper ribbon to each mounting stud, after all signal and ground connections are completed.

As with any high-frequency circuit, good power supply decoupling is required with power distribution. A good quality RF bypass capacitor in the range of 0.01 to 0.1 μF should be placed between the V_{EE} buss and the V_{CC} ground plane every 4 to 6 IC packages.

For signal interconnects longer than $1\frac{1}{2}$ inches on breadboards, a modified twisted pair may be used. This involves twisting a ground wire with the

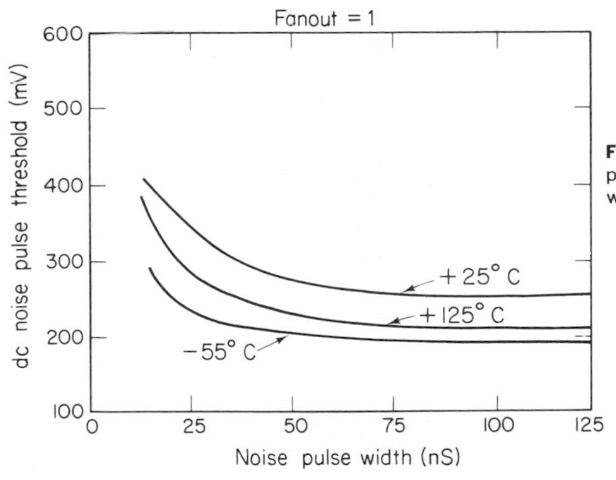

FIG. 7-44. Input signal line noise pulse thresholds versus noise pulse width and temperature (Motorola).

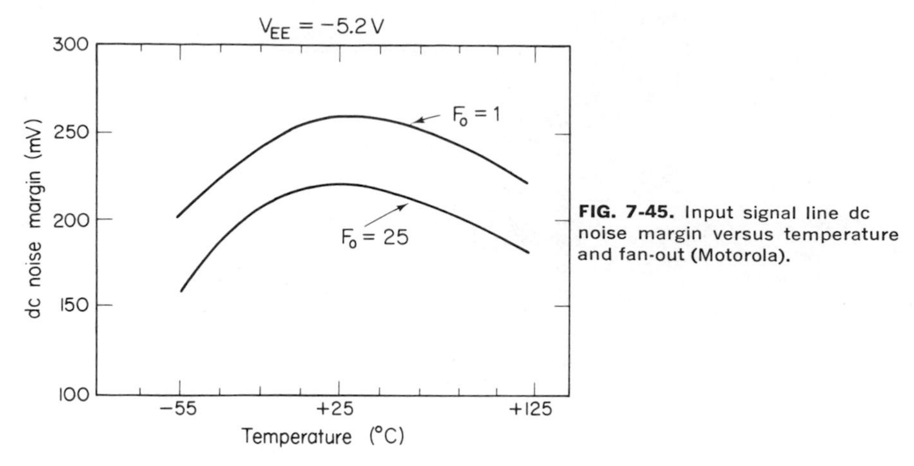

FIG. 7-45. Input signal line dc noise margin versus temperature and fan-out (Motorola).

signal lead, grounding the ground lead at both ends, and terminating the line at the receiving end.

The most ideal test fixture is one where microstrip transmission lines are used on two-sided printed circuit board or strip line used with multi-layer board. Included in Figs. 7-47 and 7-48 are design aids for microstrip and strip line applications. Figure 7-49 illustrates a typical multi-layer test fixture used for testihg Motorola ECL gates. Such test fixtures provide an IC junction-to-ambient temperature (T_{JA}) of about 50 to 55 °C/W in still air.

When a test socket is used for high-speed IC logic circuits, the input/output propagation delays must be measured physically where the *device contacts the socket*, and not at the end of the socket terminals. Figure 7-49 is a test circuit for Motorola logic gates. This particular test circuit measures ac parameters with 50-Ω terminations on all outputs. When using terminations less than 200 Ω, use a terminating voltage V_{TT} of −2.0 V.

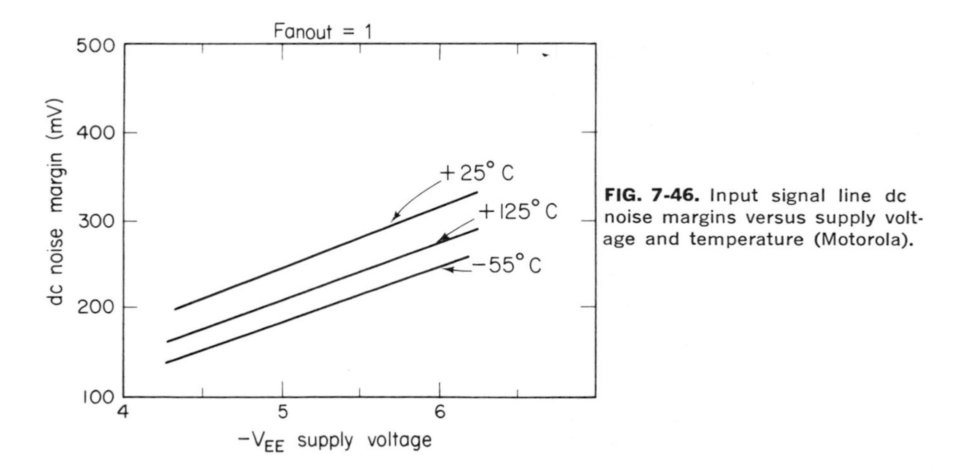

FIG. 7-46. Input signal line dc noise margins versus supply voltage and temperature (Motorola).

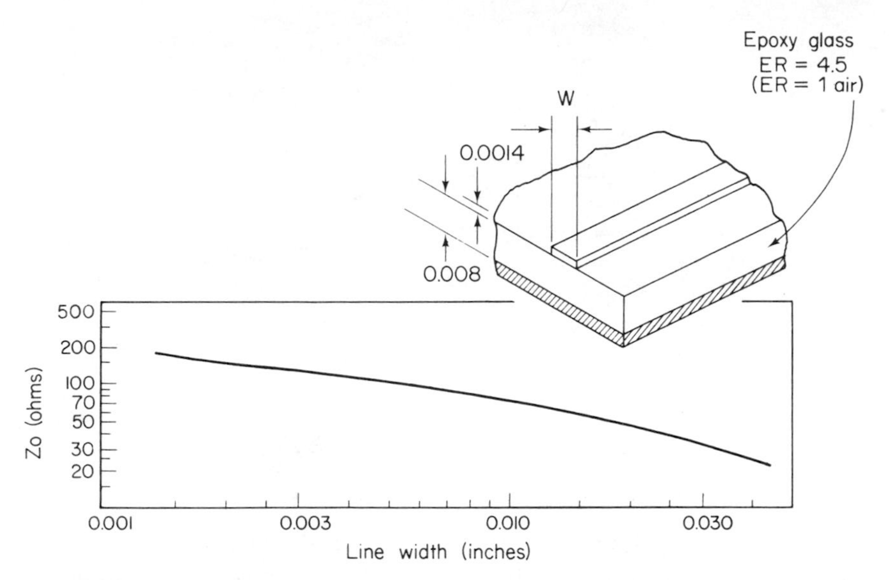

FIG. 7-47. Characteristic impedance of surface conductors (Motorola).

The limiting value for V_{TT} is due to a maximum output current capability from the emitter followers of about 25 mA (in the case of the Motorola ECLs). Rather than obtaining V_{TT} from a separate -2-V supply, an alternative is to use two resistors, one to ground and the other to V_{EE}. The parallel equivalent of the two resistors should equal the terminating impedance, and their equivalent voltage should equal -2 V.

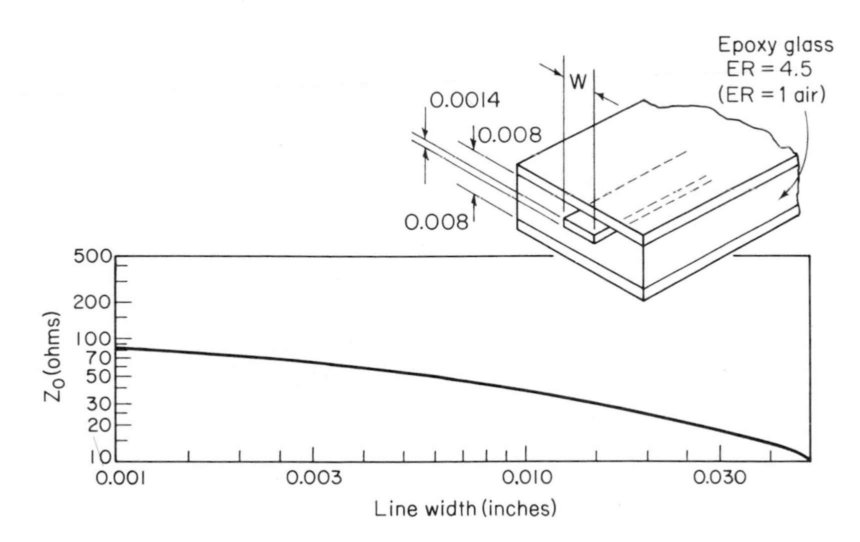

FIG. 7-48. Characteristic impedance of buried conductors (Motorola).

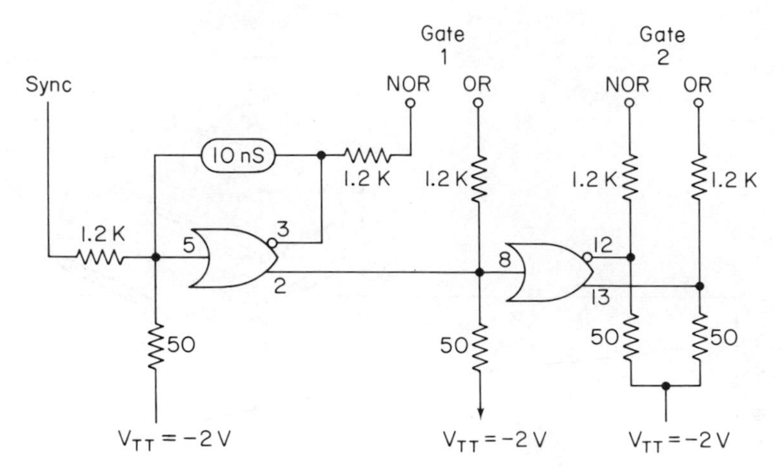

FIG. 7-49. Test circuit for MC1660 gate (Motorola).

When using IC gates with dual outputs, both the OR and NOR emitter followers should be symmetrically loaded to V_{TT}, even though only one output is used. Low impedance gates may be used in applications such as those illustrated in Fig. 7-50. Here, the 2-kΩ input resistor of the receiving gate is

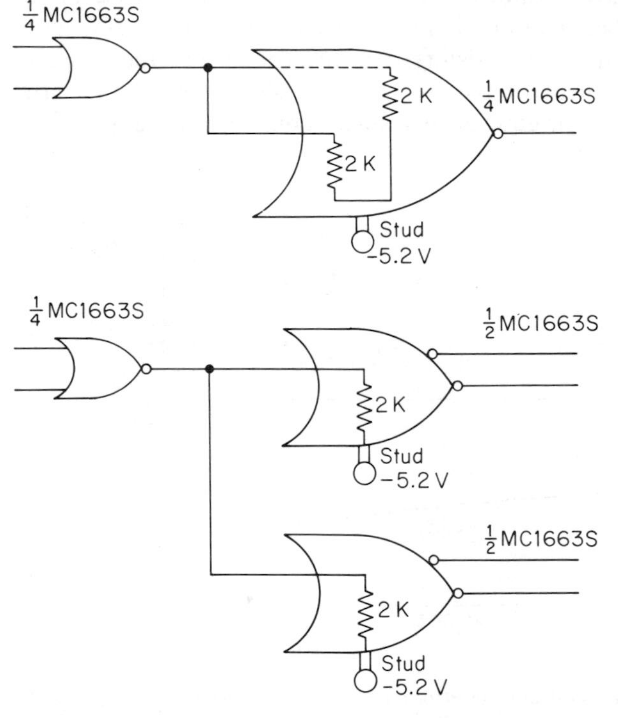

FIG. 7-50. Two methods of using low Z_{in} (2Ω) gates (Motorola).

parallel with two or more other 2-kΩ input resistors, thus becoming the pull-down or terminating resistor for the drive gate. This principle may be used in applications where lead lengths are less than $1\frac{1}{2}$ inches, and there are more than two (but less than seven) fan-outs. If the user can meet these conditions, no external pull-down resistors are required.

It should be noted that *fan-out characteristics are not always consistent*, due to the fact that there are both high and low input impedance IC elements. Three factors to be concerned with are the input current requirements of an IC gate, the current consumed in the emitter-follower pull-down resistor, and the output current capabilities of the driving gate. The following currents are typical of Motorola ECL gates.

Both high and low impedance gates have a maximum output source current I_O of -25 mA. Each input of the high impedance gates requires 350 μA of current, maximum. Each input of the low impedance gates requires a maximum current of 3.1 mA.

The following illustrates the fan-out loading rules to be used with Motorola MECL III gates. These rules are given as a typical example of how fan-out characteristics can be matched. The drive gate may be either a low or high input impedance gate. Theoretically, these gates will drive up to 75 high impedance inputs, and up to 7 low impedance inputs, depending upon external pull-down resistors. Both high and low input impedance gates show about 3 pF of capacitance per input. When interfacing with other MECL devices, such as MECL II logic functions, or MECL III flip-flops, the only requirement is that the user add up all input current requirements along with the pull-down resistor current drain, and make sure that I_O maximum of -25 mA is not exceeded. Also, when driving MECL II from MECL III logic, the output of the MECL III device should be lightly loaded (510 Ω). This method will maintain a more uniform noise immunity at the interface.

7-4.2.1 High speed test circuits and typical responses

Figures 7-51 through 7-61 show test circuits and typical responses or plots for high speed IC logic elements. The circuits provide for test of propagation delay, switching time, power supply characteristics and output voltage, all under various conditions. The circuits are straightforward and should be self-explanatory as to use. Note that the response curves are for Motorola ECL gates. The response curves of other logic families, and other IC manufacturer, will be different, but the test connections can be used with little or no modification.

7-5. TROUBLESHOOTING INTEGRATED CIRCUITS

There are four steps in the basic troubleshooting sequence for any electronic equipment: (1) analyze the symptoms of failure, (2) localize

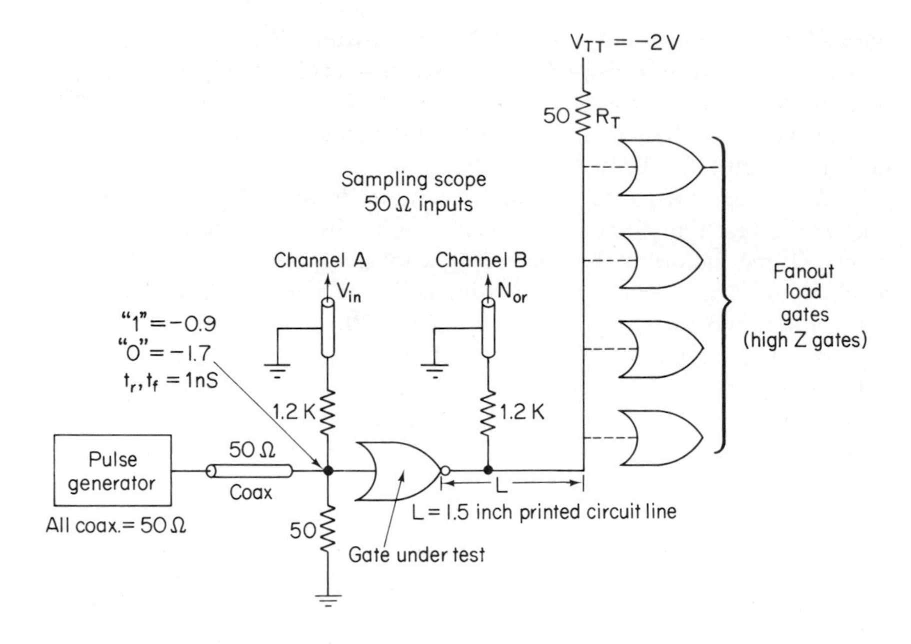

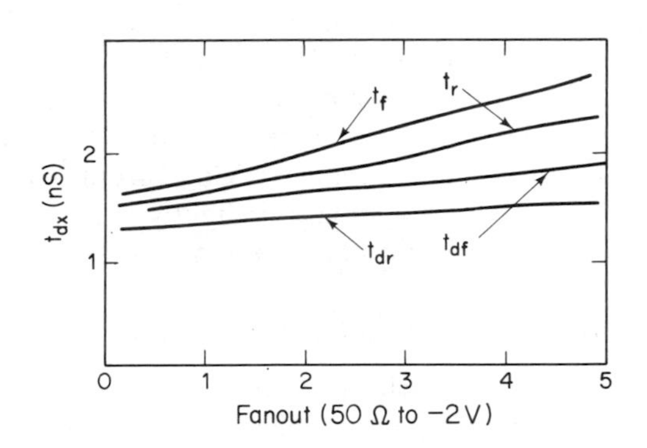

FIG. 7-51. Test circuit–t_{dx} versus fan-out.

the trouble to a complete functional unit or module, (3) isolate the trouble to a circuit within the module, and (4) locate the defect to a part.

The troubleshooting sequence must be modified when ICs are used. The last step (locate defect to a part) is unnecessary (if not impossible) to accomplish. ICs are made up of many parts forming one or several circuits. The ICs are sealed and are replaced as a package. Therefore, once trouble is located in the IC, the troubleshooting sequence is complete, and the repair/checkout phase starts. For example, if an IC is used to form the complete

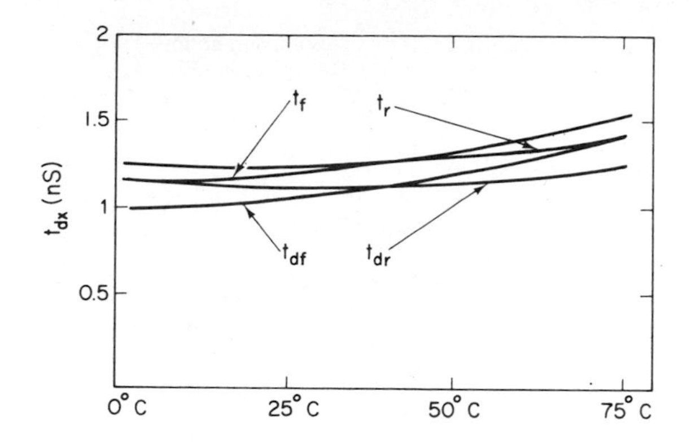

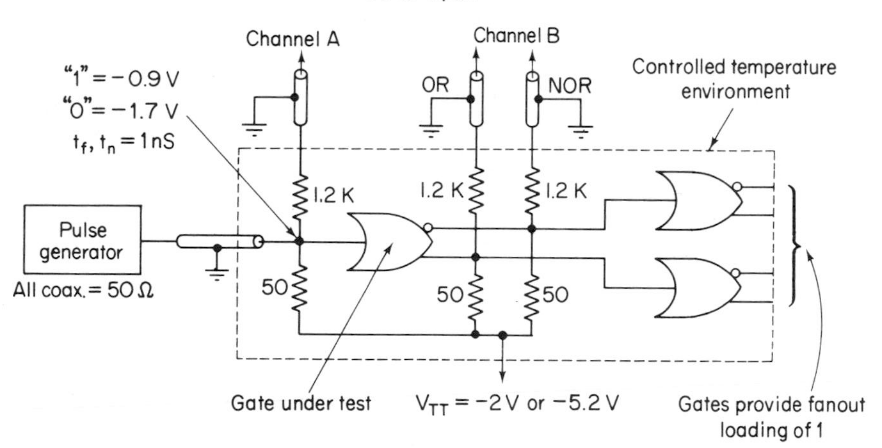

FIG. 7-52. Test circuit–t_{dx} versus temperature.

IF section of a TV set, and trouble has definitely been isolated to the IF section (by means of signal tracing, voltage checks, or whatever), the next logical step is to replace the IC and check the set.

As discussed in Chapter 2, it is often difficult to remove an IC from a board or other mounting. From a practical standpoint, it is far more convenient to test the IC "in circuit".

If the IC is a general-purpose op-amp (such as described in Chapters 3 and 4), it can be tested using the basic op-amp test of Sec. 7-1, and possibly with the aid of a tester as described in Sec. 7-2.

If the IC is a special-purpose linear device (such as described in Chapter 5), it must be tested in the same way as a corresponding discrete component

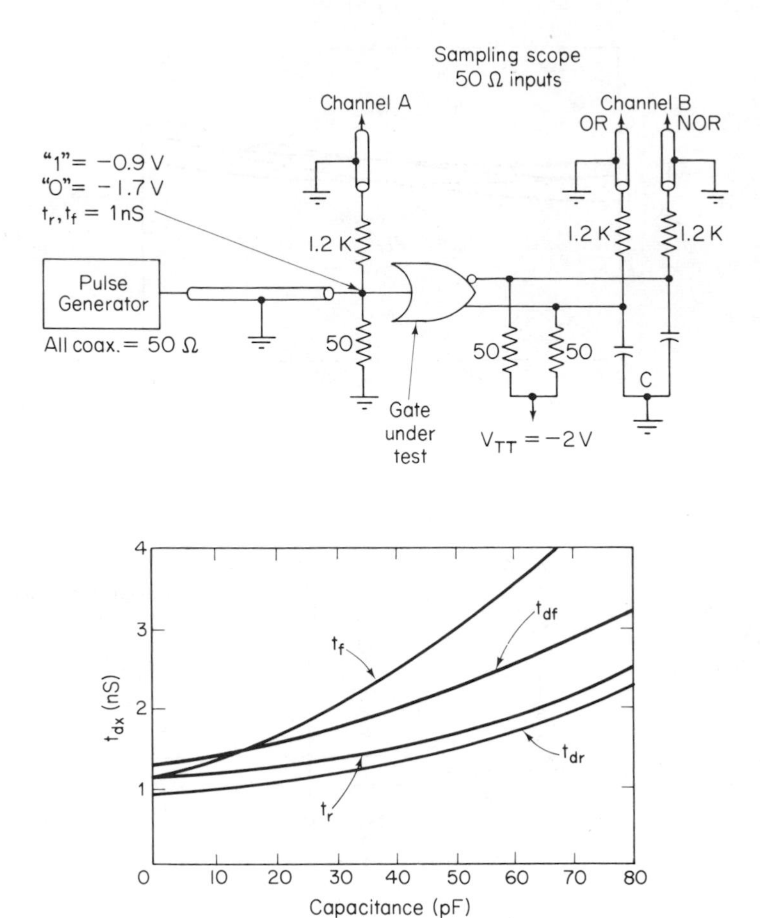

FIG. 7-53. Test circuit–t_{dx} versus load capacitance.

circuit. For example, if the IC forms the complete IF section of a TV set, test the IC as you would a discrete-component IF section.

If a digital IC is involved, it can be tested using the basic tests of Sec. 7-3 and 7-4. In addition to these basic tests, it is possible to use some of the troubleshooting aids designed specifically for digital ICs. The following paragraphs describe three such devices.

7-5.1 Logic probe

Hewlett-Packard has developed a logic probe for digital troubleshooting. This probe, shown in Fig. 7-62, can detect and indicate logic levels or states (0 or 1) in any digital circuit, and is particularly useful for use with digital ICs. The probe can also detect the presence (or absence)

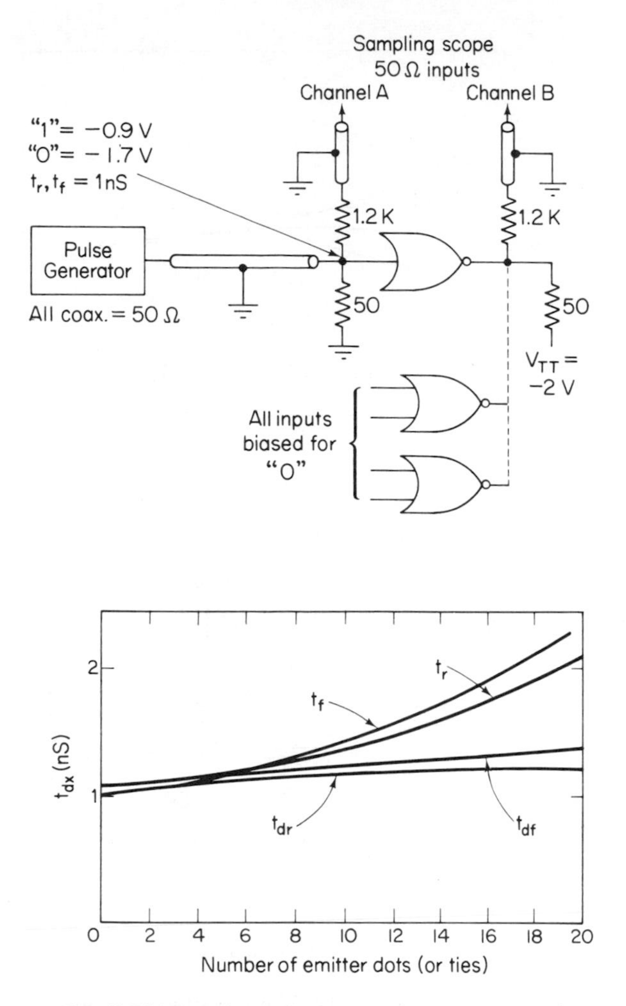

FIG. 7-54. Test circuit–t_{dx} versus emitter dots (or ties).

and the polarity of any single pulse 30 ns or greater in duration. The probe receives power (5 V) from an external source (or the equipment under test) through the cable. The input impedance is 10-kΩ nominal, which makes the probe compatible with most TTL and DTL ICs now in common use. There are no operating or adjustment controls. The only indicator is a lamp which appears as a band of light near the probe tip. The probe has a preset threshold of 1.4 V.

When the probe is touched to a high level or is open-circuited, a band of light appears around the probe tip. When the probe is touched to a low level, the light goes out. Single pulses of about 30 ns or wider are stretched to give a light indication time of 0.1 s. The light flashes on, or blinks off, depending

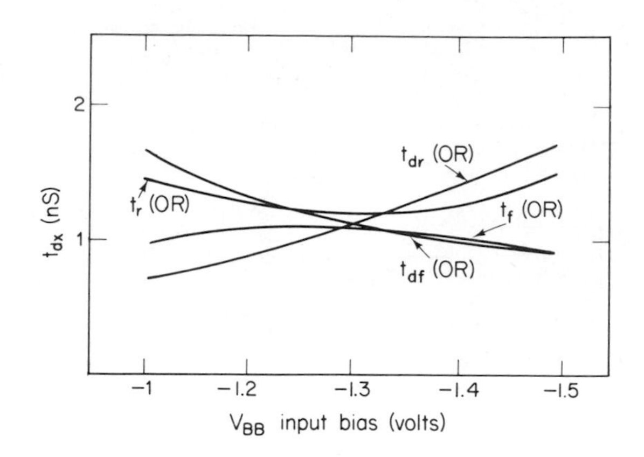

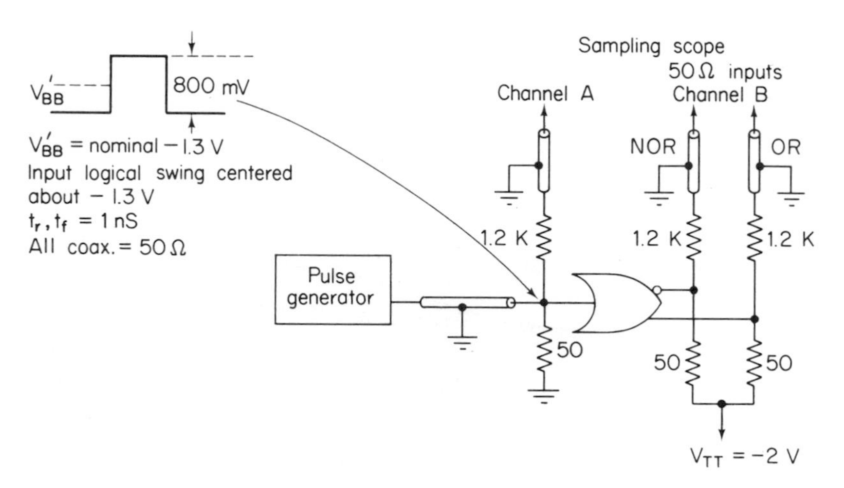

FIG. 7-55. Test circuit–t_{dx} versus input bias voltage (V_{BB}).

on the pulse polarity. When the probe is connected to a pulse train, partial illumination (partial brilliance) is displayed by the probe lamp. Pulse trains up to about 1 MHz produce partial brilliance. Pulse trains from 1 to 20 MHz produce either partial brilliance or momentary extinction, depending upon the duty cycle of the pulse. The probe response to different inputs is shown in Fig. 7-63.

When the logic probe is used to detect logic levels, the indicator lamp is on when the input is high, and is off when the input is low, giving an indication of a logical 1 or 0, respectively. With power applied and no connection to a circuit, the probe lamp will normally be on.

As shown in Fig. 7-63, the logic probe is ideal for detecting pulses of short duration and low repetition rates that would normally be very difficult to observe on an oscilloscope. Positive pulses cause the probe to flash on, while

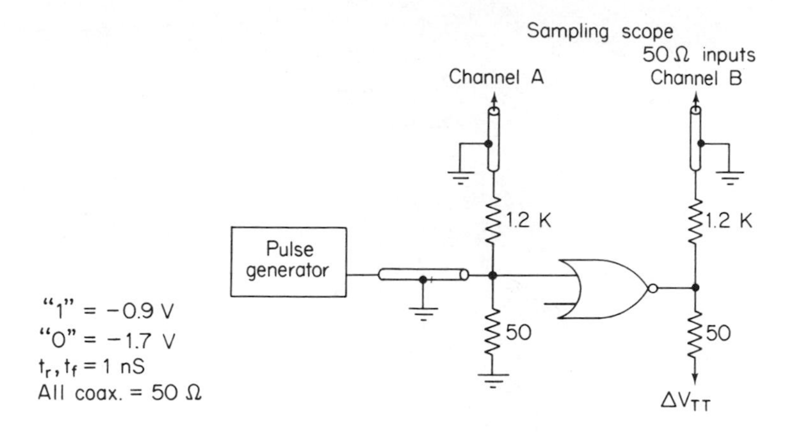

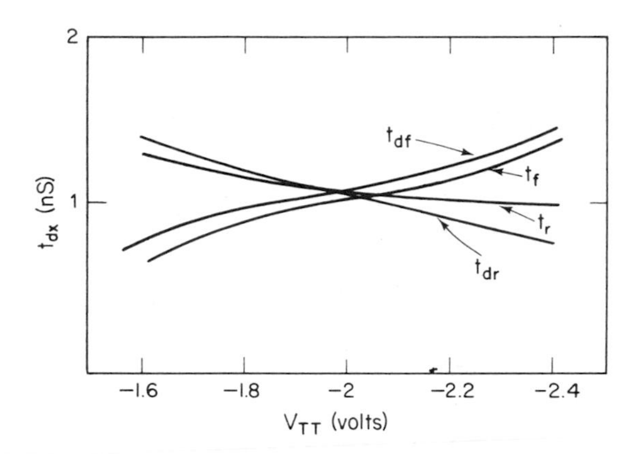

FIG. 7-56. Test circuit–t_{dx} versus termination voltage V_{TT} (load 50 Ω to V_{TT}).

negative pulses cause the probe to extinguish momentarily. High-frequency pulse trains, too fast for the eye to follow, are indicated by partial illumination. The minimum on-time of the indicator lamp for positive pulse (or off-time for negative pulses) is 0.1 s. The maximum time depends on input pulse width.

The logic probe can be used with any IC logic analysis technique, but is particularly effective for *pulse-train analysis* and *real-time analysis*.

The basis of pulse-train analysis using the logic probe is to run the IC under test at its normal clock rate, while checking for key logic pulses such as reset, start, transfer, and clock. Questions such as "Is a particular IC decade counting?" are easily resolved by noting if the probe indicator lamp is partially lit (which occurs only when fast repetition pulse trains are monitored).

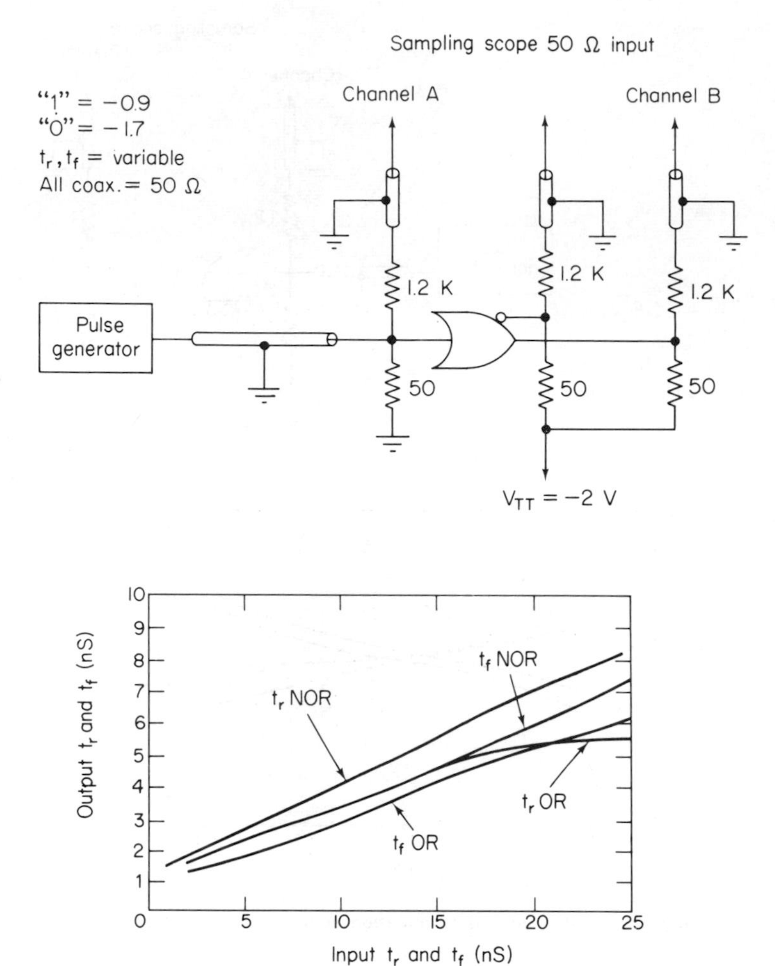

FIG. 7-57. Test circuit–t_{dx} versus input rise and fall times (load 50 Ω to −2 V).

With real-time analysis, the normal fast clock signal is replaced by a very slow clock signal from a pulse generator or by manual pulse triggering. The logic changes in the circuit under test will occur at a rate sufficiently slow to that individual changes and proper pulse occurrences can be observed on a real-time basis.

The logic probe can be tested at any time by touching the tip to a variable dc source. When the source is above the threshold level, the indicator lamp should be on. When the source is dropped below the threshold level, the indicator lamp should go out.

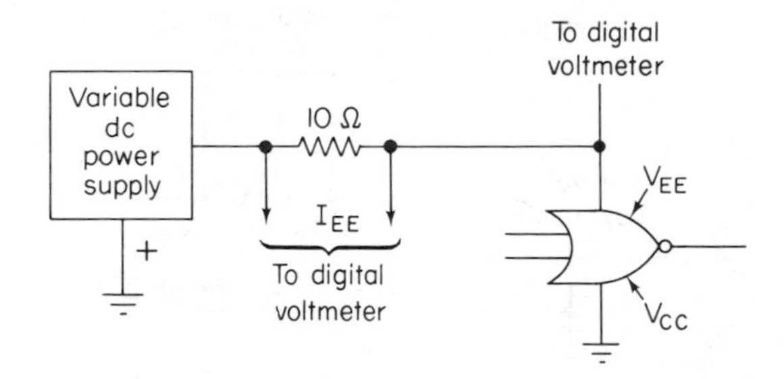

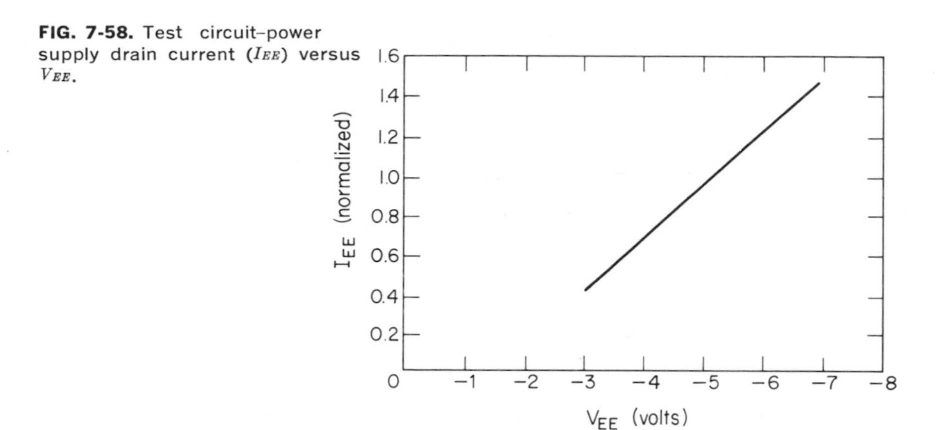

FIG. 7-58. Test circuit–power supply drain current (I_{EE}) versus V_{EE}.

7-5.2 Logic clip

Hewlett-Packard has also developed a logic clip for digital IC troubleshooting. Although the logic clip will perform all of its functions without additional equipment, it is particularly effective when used with the logic probe.

The logic clip, shown in Fig. 7-64, clips onto TTL and DTL ICs (dual in-line packages) and instantly displays the logic states of all 14 or 16 pins. Each of the clip's 16 light-emitting diodes (LED) independently follows level changes at its associated input. A lighted diode corresponds to a high, or 1, logic state.

The logic clip's real value to the IC user is in its ease of operation. The clip has no controls to be set, needs no power connections, and requires practically no explanation as to how it is used. Since the clip has its own gating logic for locating the ground (V_{EE}) and 5-V (V_{CC}) pins, it works equally well upside down or right-side up. Buffered inputs ensure that the IC under test is not loaded down. Simply clipping the unit onto a TTL or DTL dual

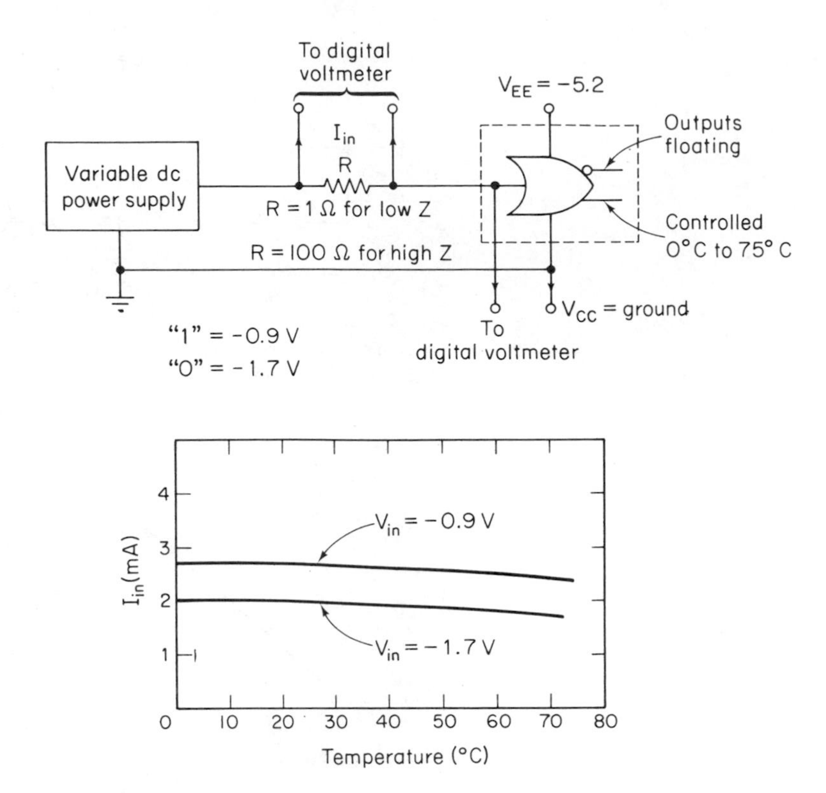

FIG. 7-59. Test circuit–input current (I_{in}) versus temperature.

in-line package of any type or manufacturer makes all logic states visible at a glance.

The logic clip is much easier to use than either an oscilloscope or a voltmeter when the troubleshooter is interested in whether a lead is in the *high or low state* (1 or 0), rather than the lead's *actual voltage*. The clip, in effect, is 16 binary voltmeters, and the user does not have to shift his eyes away from his circuit to make the readings. The fact that lighted diodes correspond to high logic states greatly simplifies the IC troubleshooting procedure. The user is free to concentrate his attention on his circuits rather than on measurement techniques.

When the clip can be used on a real-time basis (when the clock is slowed to about 1 Hz or is manually triggered) timing relationships become especially apparent. The malfunction of gates, FFs, counters, and adders then becomes readily visible as all of the inputs and outputs of an IC are seen in perspective.

When pulses are involved, the logic clip is best used with the logic probe. Timing pulses can be observed on the probe, while the associated logic-state changes can be observed on the clip.

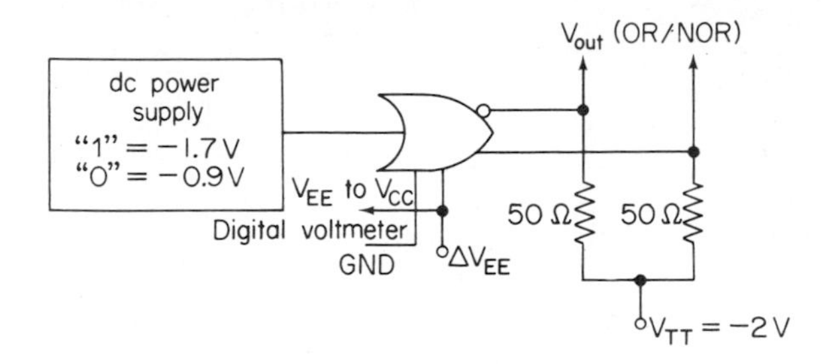

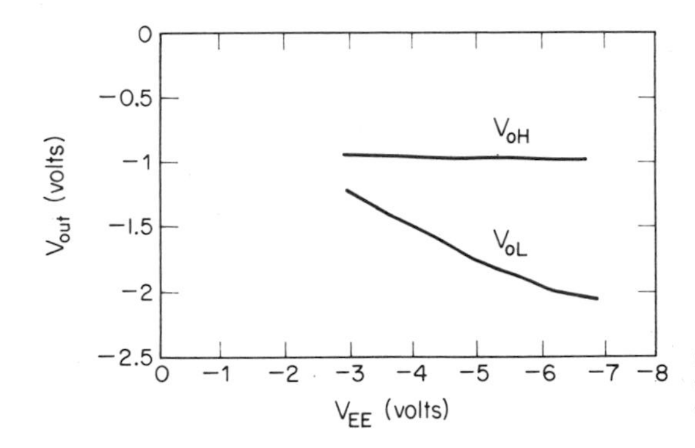

FIG. 7-60. Test circuit–output voltage versus V_{EE} power supply voltage.

7-5.3 Logic comparator

Another Hewlett-Packard IC troubleshooting device is the logic comparator. This instrument, shown in Fig. 7-65, clips onto TTL or DTL ICs and, through a comparison scheme, instantly displays any logic state difference between the test IC and a reference IC. Logic differences are identified to the specific pin(s) on 14- or 16-pin dual in-line packages with the comparator's display of 16 line-emitting diodes (LED). A lighted diode corresponds to *logic difference*.

As in the case of the logic clip, the real value of the logic comparator is the time it can save in locating a faulty IC. There are no controls to set, and no power connections are needed.

In use, an IC to be tested in a powered, but malfunctioning, module, instrument, or system is first identified. A reference board with a *good IC of the same type* is then inserted in the comparator. The comparator is clipped onto the IC in question, and an immediate indication is given if the test IC operates differently from the reference IC. Even very brief dynamic errors are detected, stretched, and displayed.

The logic comparator operates by connecting the test and reference IC inputs in parallel. Thus, the reference IC sees the same signals that are inputs to the test IC. The outputs of the two ICs are compared, and any difference in outputs greater than 200 ns in duration indicates a failure. A failure on an

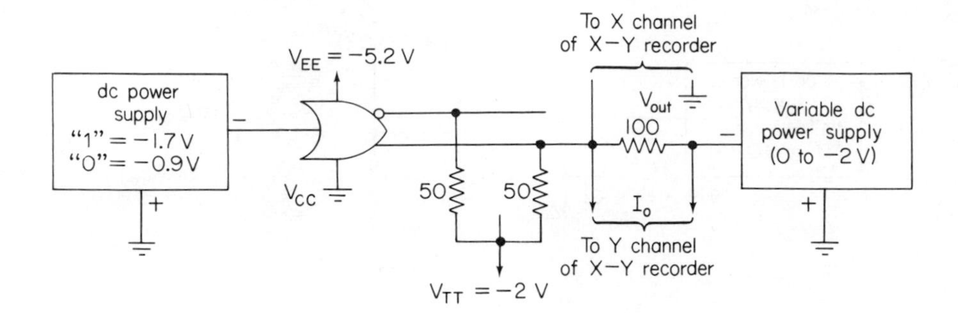

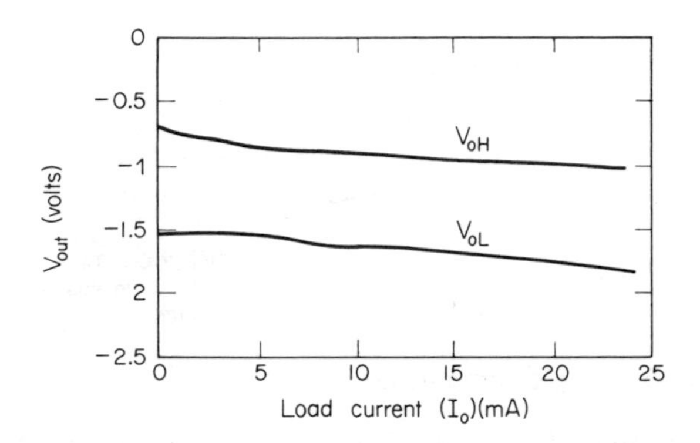

FIG. 7-61. Test circuit–output voltage versus load current.

input pin, such as internal short, will appear as a failure on the IC driving the failed IC. Thus, a failure indication actually pinpoints the malfunctioning pin.

All operating power is obtained from the test IC's power supply pins. Thus, the logic comparator is a great aid in determining whether an IC has failed within a given circuit. The user need not correlate readings from an

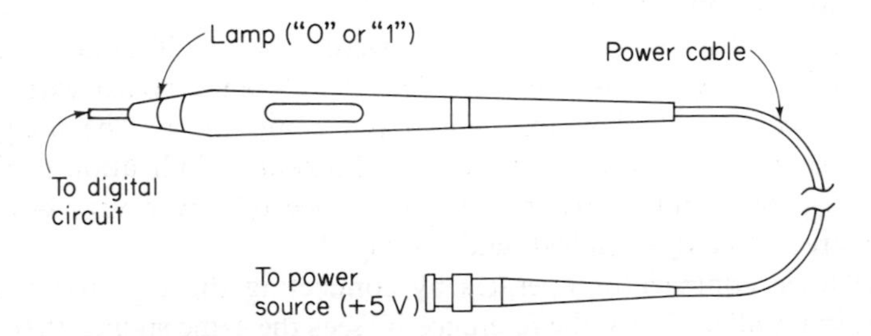

FIG. 7-62. Hewlett-Packard logic probe.

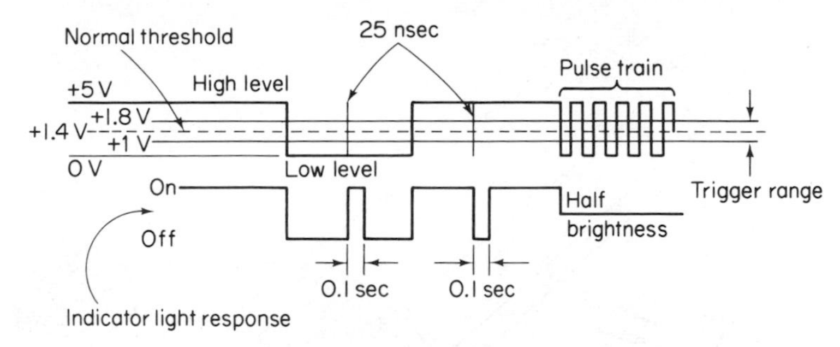

FIG. 7-63. Logic probe response.

oscilloscope or voltmeter to data on a schematic or logic diagram. At a glance, the comparator indicates a failed IC, as well as the faulty pins.

When troubleshooting any logic, it is reassuring to know that the logic tester is operating properly. A test board is supplied with the comparator for this purpose. When used in place of a reference board, the test board sees all of the comparator's circuitry, test leads, and display elements to verify proper operation.

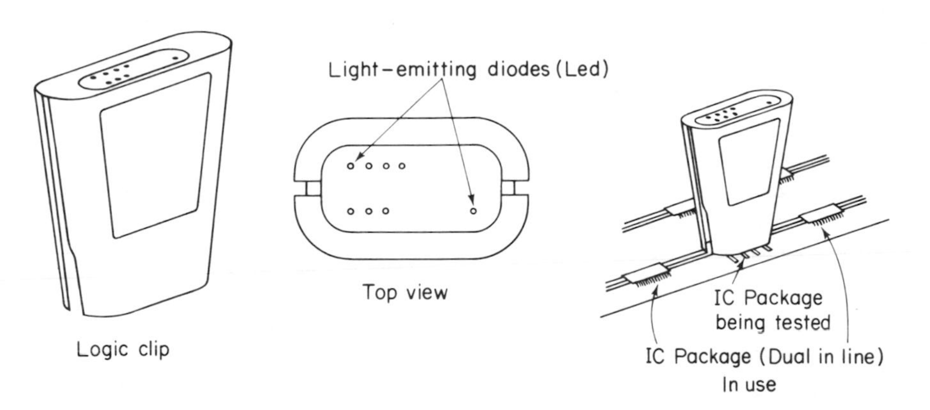

FIG. 7-64. Hewlett-Packard logic clip.

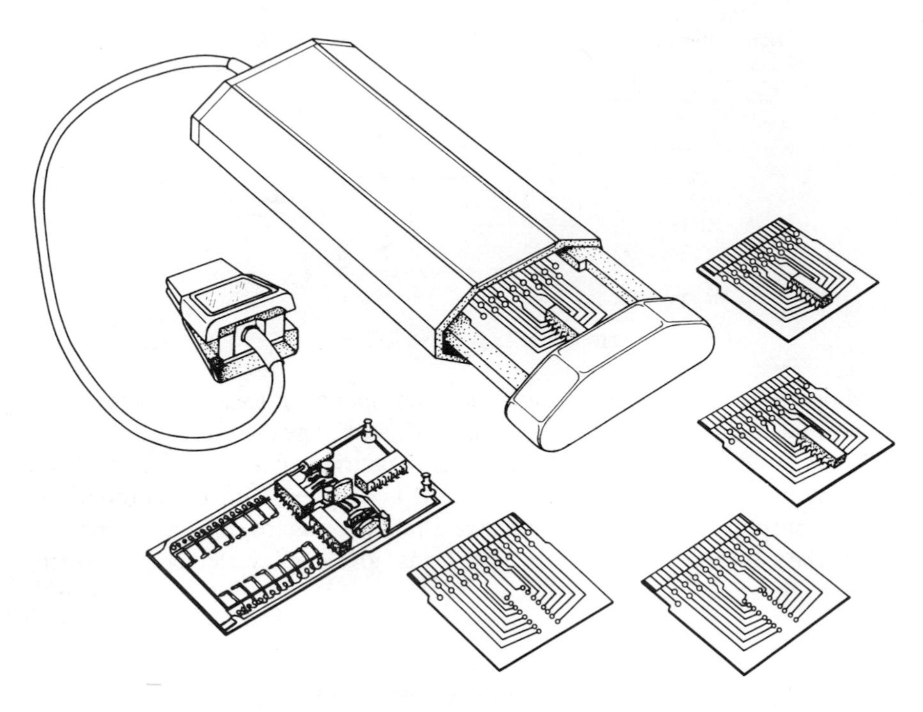

FIG. 7-65. Hewlett-Packard logic comparator.

INDEX